Humic Substances

Structures, Models and Functions

"...there is no need or experimental evidence for specific molecular structure control in the formation of humic substances."

Patrick MacCarthy, p 23

Humic Substances
Structures, Models and Functions

Edited by

Elham A. Ghabbour
Soil, Water and Environmental Research Institute, Alexandria, Egypt and Northeastern University, Boston, USA

Geoffrey Davies
Northeastern University, Boston, USA

RS•C
ROYAL SOCIETY OF CHEMISTRY

Based on the proceedings of the fifth Humic Substances Seminar held on 21–23 March 2001 at Northeastern University, Boston, Massachusetts.

The front cover illustration is taken from the contribution by M.S. Diallo, J.-L. Faulon, W.A. Goddard III and J.H. Johnson, p 232.

Special Publication No. 273

ISBN 0-85404-811-1

A catalogue record for this book is available from the British Library

Published by The Royal Society of Chemistry,
Thomas Graham House, Science Park, Milton Road,
Cambridge CB4 0WF, UK
Registered Charity No. 207890

For further information see our web site at www.rsc.org

Printed and bound by Athenaeum Press Ltd, Gateshead, Tyne & Wear, UK

Preface

This book is a companion to the volumes *Humic Substances: Structures, Properties and Uses, Humic Substances: Advanced Methods, Properties and Applications* and *Humic Substances: Versatile Components of Plants, Soils and Water*, published by the Royal Society of Chemistry, Cambridge in 1998, 1999 and 2000, respectively. These volumes report the best and most recent research on humic substances (HSs), the remarkable brown biomaterials in animals, coals, plants, sediments, soils and waters. HSs buffer pH, retain water, are photo-active, are redox catalysts, sorb solutes and bind metals. Nature can switch these properties on or off, depending on the circumstances. In other words, HSs can respond when nature demands. As such they are crucial components of the carbon cycle and other life processes. Although HSs have been acknowledged to be 'present and accounted for' for more than two centuries, they are nowadays acknowledged to be vital for life on earth and eminently worthy of study and use.

The structures responsible for HSs' remarkable versatility are out of sight but by no means out of mind. Sophisticated analytical, chemical and physical studies of reproducible HSs samples are paying dividends. The work is important because HSs are major functional components of soils and there is no life without soil.

This book and its companions are based on the Humic Substances Seminars, which have been held at Northeastern University, Boston in the Spring of each year since 1997. The Seminars have the character of Gordon Conferences (but with publication of the proceedings, as here). They are the major forum for discussion of HSs as biomaterials of high importance to the productivity, health and safety of the world's ecosystems, humans, land and water. The Seminars are helping to move HSs research from being a 'descriptive' science towards being a 'hard' science.

Better understanding of HSs is driving their study to new levels of excellence. Reliable analytical data are prompting models of HSs molecules and systems that give us something to look at and think about. Modeling is described in several of the contributions in this book. We also can share the fruits of C1s NEXAFS data together with properly conducted NMR measurements, mass spectrometry, capillary electrophoresis, flow field flow fractionation and other thoughtful work as we concentrate on HSs as targets to be marveled at and understood. Chemists are conferring with agronomists and other professionals who are solving ever-harder problems of productivity and safety in the field. Increasing participation of young men and women from many disciplines is making us all appreciate how important HSs are.

History, Philosophy and Spectroscopy

We start with a fascinating personal history of 45 years on the 'Organic Matter Trail' by C. Edward Clapp. This account recognizes much that has been done and sets present efforts in perspective. A strong desire to understand HSs is shored up philosophically by 'principles' based on similarities and differences between HSs samples from Patrick MacCarthy, a co-founder of the International Humic Substances Society (IHSS). HSs are hard to fractionate, have never been crystallized and are described by Patrick as 'supermixtures'. We can know HSs precursor structures, as demonstrated by beautiful sharp spectral peaks found with a new NMR probe in fresh and senescing leaves by Robert Wershaw and Igor Goljer.

Spectroscopic- and mass-based approaches (especially NMR and newly-made-to-work-for-HSs mass spectrometry) are driving progress. Scientists are demanding quantitative information and want to understand the relation between data from different measurements on the same sample. As an example, a study by Wershaw *et al.* published last year in *Humic Substances: Versatile Components of Plants, Soils and Water* correlated functional group contents from liquid and solid state NMR measurements. In this book we see correlations between quantitative near edge X-ray absorption fine structure (NEXAFS) and NMR data for different HSs demonstrated by Scheinost *et al.* NEXAFS is just one of many useful measurements that can be made at synchrotron facilities (others include metal speciation studies). We can look forward to deeper insights into HSs structures as more synchrotrons come on line.

Data, Mobility and Stability

The 'workhorse instruments' section of the book begins with an impressive paper by Wolf *et al.* that tackles discrepancies due to the choice of standards between information from high performance size exclusion chromatography, organic–aqueous size exclusion chromatography, flow field flow fractionation and time-of-flight secondary ion mass spectrometry. Then, quantitative one- and two-dimensional NMR data are used by colleagues at the University of Massachusetts, Amherst and Northeastern to find the two domains in solid HAs necessary to support the 'dual mode' model of sorption and to demonstrate different HA conformations and mobility in aqueous NaOH and DMSO. Humic acids have long been known to be stable free radicals. Now, through selective methylation of HA functional groups, Shinozuka *et al.* seem to have found that the amount of free radicals in a HA sample increases with its COOH content, which is one measure of the state of oxidation of an HSs sample.

Hatcher and co-workers are contributing strongly to the use of mass spectrometry and NMR to at least compare spectral features and hopefully identify the components of fulvic and humic acid samples. Here, ESI-QqTOF mass spectrometry is used to identify long chain saturated and unsaturated fatty acids from mass defects in what must be the most informative mass spectra obtained to date. We reported in *Humic Substances: Versatile Components of Plants, Soils and Water* that Xing and co-workers used quantitative ^{13}C NMR peaks at 31.0 and 32.9 ppm to measure the relative amounts of amorphous and crystalline long alkyl chains, respectively, in solid HAs. In the paper by Lenka Pokorná and colleagues later in the present volume we see MALDI-TOF mass spectral and capillary electrophoresis evidence that long term exposure to aqueous NaOH gives electropherograms with distinct peaks (instead of the usual 'hump') and MALDI-TOF peaks that differ by 14 mass units. It is reasonable to expect that long alkyl chains are the 'last to go' when HA samples are attacked by nasty reagents like strong base. So it seems that long alkyl chains are 'recalcitrant' components of HSs and especially of HAs and humins.

De Nobili and co-workers demonstrate that the mobilities of HSs fractions in entangled capillary zone electrophoresis gels increase with decreasing molecular size. This paper sets the scene for two papers from Havel's group at Brno, The Czech Republic that both use capillary electrophoresis together with MALDI-TOF mass spectrometry as investigative tools. The first paper asks whether there are HAs in

Antarctica (there are but they have very low C contents and high ash contents, as found by Chin, McKnight and co-workers who call them 'NOM'). As mentioned above, the paper by Pokorná *et al.* adds to the evidence that HAs prefer not to be left for a long time in strong aqueous bases and that what survives is long chain alkyl components. This might be turned to advantage in 'cleaning up' HA samples, though the left-overs would be of little practical value.

Pokorná *et al.* report that the products of reaction of the HAs with aqueous hydrazine could not be ionized by MALDI-TOF, leaving their identities as a mystery (perhaps they are 'capped' derivatives with less of a tendency to ionize?).

Masses, Similarities and Properties

Three important observations run through the papers in this volume. Firstly, there is hardly a mention of HSs as 'polymers'. That is, the reported molecular masses are in the hundred and low thousand Daltons, not the tens and hundreds of thousands. Secondly, except for Antarctic OM the compositions and spectra of the HSs samples from different sources often are similar, which is an encouragement to work harder. Thirdly, the words supermixture and supramolecular occur several times and people are beginning to see that HSs are nimble species with a strong tendency to congregate. One procedure affected by this question is the use of ultrafiltration to fractionate HSs into nominal molecular weight ranges. These ranges could refer to the masses of HSs congregations rather than to those of large covalently linked molecules.

In any event, the HA fractions obtained by hollow fiber ultrafiltration and assigned to particular molecular mass ranges by Kretschmar and Christl behave very much alike when it comes to proton, Cu(II) and Pb(II) binding and there is hardly a shade of difference in their FTIR and NMR spectra. Again, this is encouraging. The spirit of progress spills over into the paper by Amarasiriwardena and co-workers, whose flow field flow fractionation data indicate molecular masses M_p of 5 kDa (similar to their earlier FFFF results with different HA samples), similar hydrodynamic diameters and almost identical diffusion coefficients. This is in spite of different DRIFT spectra and polydispersities. Fractionation is also a feature of the paper by Takacs, Alberts and co-workers that examines the effects of climate and season on the dissolved organic carbon fed to six rivers of the southeastern United States by quite different watersheds.

Models and Theories

The next section in this volume has been placed after the latest experimental work on HSs structures and properties has been described because theory should follow experiment. The three theoretical papers are among the most fascinating treatments of HSs molecular modeling to be found in the current literature.

Bruccoleri and co-workers improve on the previous treatment of the so-called Steelink and Temple-Northeastern-Birmingham HA building blocks by turning to semi-empirical quantum mechanical and Monte-Carlo search methods that correct overemphasis of structures stabilized by aromatic ring stacking. The result is very flexible, open structures, several of which have essentially the same minimum energy. Bruccoleri *et al.* then show that these structures are affected by solvation and that they naturally self-assemble to give not helices but full and half tennis balls. Speaking as individuals with a strong liking for molecules and visual images, we think this paper will have lasting impact in HSs science.

Of equal interest is an excellent presentation of density function theory-based molecular modeling for two HSs related purposes by Haberhauer and co-workers: (1) to understand binding of the important soil element aluminum by carboxylic acids like those exuded by plants to stop an aluminum attack and like those that exist in HSs. Water plays a major role in determining what kind (monodentate or bidentate) of complex is formed and the associated thermodynamics; and (2) to see how a polar pesticide interacts with the functional groups found in HSs as a function of solvent dielectric constant (the extension of thought here might be the polar and hydrophobic surfaces and interiors of HSs).

Rounding out the modeling part of the book we have a courageous paper by Diallo *et al.* that takes a feather from MacCarthy's cap by letting a hypothetical HS have its likely constituents linked together in a random (but in every case chemically reasonable) way. The resulting molecules are isomers with the same molecular mass (2801 Da) but different calculated densities, strain energies and solubility parameters because of their different structures. These calculated parameters can be inserted into the Flory-Huggins equations to predict the equilibrium constants for binding of PAHs and other hydrophobic organic compounds by Chelsea HA. The comparison between theory and experiment for interaction with 18 such compounds in Figure 5 of the Diallo *et al.* paper is striking. Diallo *et al.* have put MacCarthy's First Principle to the test and the results are fascinating. Even better, the calculated average solubility parameter of Chelsea humic acid isomers compares favorably with experimental estimates for different HAs.

Taken together, the theory in these three papers is predicting relatively low HSs molecular weight, random/chaotic structures, both polar and hydrophobic character, a tendency to aggregate, selective behavior as sorbents and metal binders and it is not too far from the ideas of MacCarthy, Piccolo and Conte. We'll have to see.

Images, Oxidation and Humification

We do see as Tugulea and co-workers show images of composites formed by HSs on mica and Liu and Huang visualize the results of catalyzed catechol humification on metal oxide surfaces. Both studies make great use of the surface mapping resolution of an atomic force microscope (AFM). Care is taken in the work of Tugulea *et al.* to focus on HSs adsorption rather than sedimentation/precipitation by having the clay samples vertical. This work will help the efforts of biogeochemists to understand how HSs coatings affect ion exchange and other reactions of clays and minerals. The fine work of Liu and Huang shows that one effect of a humified catechol coating on metal oxides is a decreased surface area because the coated products tend to aggregate. The coating also changes the point of zero charge of the oxide surface, which is expected because protons are exchanged when the coatings are formed.

The next three papers are motivated by interest in oxidative humification. Based on previous work by Hatcher, MacCarthy, Rice and others, many HSs researchers see humins as humic-clay composites like the models studied by Tugulea *et al.* and Liu and Huang. All the papers in this group make good use of spectroscopy to see the effects of cover crops (or no cover) on HSs in the underlying horizons.

Ding and co-workers show that fertilization puts more polysaccharide into humin, which makes sense because HSs come from plants and fertilized plants grow faster. The humins are sensibly different with different cover systems, as revealed by NMR and DRIFTS data, but the differences are linked to compositional differences in the plants themselves. Ussiri and Johnson report a major study of the effects of tree removal ('clear cutting') on the character of HSs in the underlying soil. The work benefits from access to the Hubbard Brook Experimental Forest in New Hampshire, the data are long term and good use is made of chemical analysis and NMR. The message from this excellent study is that clear cutting depletes the soil of HSs (there are fewer inputs) and that the depletion is long-term. Biomass oxidation (as in humification) can be slow or fast, as in grass fires. The products are charred plant fragments, which are common in volcanic soils. Shindo and Honma use spectroscopy and analysis to show that oxidation of the charred plant fragments with nitric acid gives products like the very dark colored HSs in volcanic soils. Nitric acid converts the fragments to HSs a bit faster than air.

Organic Ores and Analysis of Commercial HSs

The next four papers are important to anyone involved in the commercial applications of HSs derived from coals. Ozdoba *et al.* map different 'organic ore' sources in western North America. The ores are lignite-like with moderate to high ash contents. Schnitzer *et al.* use chemical analysis, FTIR and NMR spectroscopy and Curie-Point Pyrolysis-GC-MS to learn about the similarities and differences between ores from different locations. The high ash contents make NMR problematic, but the assignment of aromatic and aliphatic contents is quantitated with Bloch decay data rather than conventional CPMAS NMR. Long chain alkanes and alkenes are prevalant and the ores are like humins in the early stage of coalification. Dinel *et al.* use two-step principal component analysis to analyze and interpret the pyrolysis-field ionization mass spectra and to sub-classify the organic ores of the two preceding papers. The most discriminatory mass regions are shown to be between m/z 250 and 310 and between 600 and 619. Peaks in these regions are reported by Hatcher and co-workers in the earlier paper on ESI-QqTOF spectra. Dinel *et al.* attribute the peaks in the m/z 250 to 350 mass region to saturated and unsaturated fatty acids, alkanes, alkenes, lignin dimers, alkyl aromatics and n-alkyl diesters with similar qualitative distributions in all the ores. The relative ages of the ores are deduced from m/z 250–310 data. The paper by Fataftah *et al.* demonstrates that current HSs industry analysis methods give different HSs contents of the same commercial samples.

Plant Growth Stimulation and Antimutagenesis

The last three papers of this collection are clear descriptions of potential large scale HSs applications. Paré *et al.* report that foliar spray application on alfalfa of the calcium derivative of one of the Luscar organic ores described in the earlier papers matches the fertilizing benefits of $CaCl_2$ but without depletion of N uptake and with the potential of stimulating absorption of other plant nutrients. Seyedbagheri and

Torell demonstrate the optimum levels of commercial HSs application in potato production and confirm that the major effect is stimulation of root growth. Adoption of a uniform code of commercial HSs product analysis and labeling will accelerate the use of such products in agriculture and for other purposes. The paper by Ferrara *et al.* is the first report at a HS Seminar of the protection of plants against mutagens and toxins by HSs. The Feulgen staining method is used to detect genetic damage in broad beans, onions, peas and durum wheat by maleic hydrazide (MH) and other mutagens in the absence and presence of IHSS reference and standard HSs. The best protection against the strongest broad bean and onion mutagen MH is from HAs and FAs from aquatic, peat and leonardite sources.

Taken together, the three papers that round out this outstanding collection of HSs research demonstrate the versatility and effectiveness of HSs in many applications. They amply justify continued study and use of these remarkable materials.

Participants and Authors

This book is derived from Humic Substances Seminar V, which was held at Northeastern University, Boston, Massachusetts, USA on March 21–23, 2001. We were honored by the presence of Drs Fritz Frimmel (President), James Alberts, Russell Christman, Michael Hayes and Nicola Senesi (past Presidents) and Yona Chen (President-Elect) of IHSS together with other IHSS officers. Immediate past President Donald Sparks represented the Soil Science Society of America. We welcomed nearly 100 eminent participants from fourteen countries. One reviewer says "…future meetings at Northeastern University will profoundly contribute to our knowledge of the subject…stimulating the shift from a more descriptive 'soft' science to a more mathematically formulated and experimentally-tested discipline".* We heartily thank the authors who bring their best work to Northeastern for discussion.

Acknowledgements

Editing this book was a pleasure. It gave us a bird's eye view of the best work on HSs from around the world. The authors and reviewers did everything we asked at short notice and the co-operation of everyone involved in the production of this collection of papers is appreciated. Financial support from Arctech, Inc., Luscar, Ltd., the Barnett Institute of Chemical and Biological Analysis, the Seminar V advertisers and our other sponsors is gratefully acknowledged. Northeastern University provides excellent facilities for the Humic Substances Seminars and Michael Feeney ably manages the Seminar presentations. The staff of the Barnett Institute are invaluable and the members of the Humic Acid Group are well-known hosts. We thank Janet Freshwater and her staff at the Royal Society of Chemistry for timely publication of the latest and best in humic substances research. We hope all who read this work will be encouraged to make their own contribution to this important subject.

Elham A. Ghabbour, Geoffrey Davies
Boston, Massachusetts
July, 2001

*J. Poerschmann, *Chromatographia*, 2001, **53**, 582.

Contents

History, Philosophy and Spectroscopy

An Organic Matter Trail: Polysaccharides to Waste Management to Nitrogen/Carbon to Humic Substances 3
C. Edward Clapp

The Principles of Humic Substances: An Introduction to the First Principle 19
Patrick MacCarthy

NMR Characterisation of the Mobile Components in Intact Green and Senescent Leaves as a Means of Studying the Humification Process 31
Robert Wershaw and Igor Goljer

Carbon Group Chemistry of Humic and Fulvic Acid: A Comparison of C-1s NEXAFS and ^{13}C-NMR Spectroscopies 39
A.C. Scheinost, R. Kretzschmar, I. Christl and Ch. Jacobsen

Data, Mobility and Stability

Aspects of Measurement of the Hydrodynamic Size and Molecular Mass Distribution of Humic and Fulvic Acids 51
Manfred Wolf, Gunnar Buckau, Horst Gekeis, Ngo Man Thang, Enamul Hoque, Wilfried Szymczak and Jae-Il Kim

Solid State NMR Evidence for Multiple Domains in Humic Substances 63
Amrith S. Gunasekara, L. Charles Dickinson and Baoshan Xing

Investigation of Molecular Motion of Humic Acids with 1-D and 2-D Solution NMR 73
Kaijun Wang, L. Charles Dickinson, Elham A. Ghabbour, Geoffrey Davies and Baoshan Xing

Variation of Free Radical Concentration of Peat Humic Acid by Methylations 83
T. Shinozuka, Y. Enomoto, H. Hayashi, H. Andoh and T. Yamaguchi

Studies of the Structure of Humic Substances by Electrospray Ionization Coupled to a Quadrupole-Time of Flight (Qq-TOF) Mass Spectrometer 95
Robert W. Kramer, Elizabeth B. Kujawinski, Xu Zang, Kari B. Green-Church, R. Benjamin Jones, Michael A. Freitas and Patrick G. Hatcher

Capillary Electrophoresis of Humic Substances in Physical Gels 109
M. De Nobili, G. Bragato and A. Mori

Are There Humic Acids in Antarctica? 121
D. Gajdošová, L. Pokorná, P. Prošek, K. Láska and J. Havel

The Stability of Humic Acids in Alkaline Media 133
L. Pokorná, D. Gajdošová, S. Mikeska, P. Homoláč and J. Havel

Masses, Similarities and Properties

Proton and Metal Cation Binding to Humic Substances in Relation to Chemical Composition and Molecular Size 153
Ruben Kretzschmar and Iso Christl

Characterization of Trace Metals Complexed to Humic Acids Derived from Agricultural Soils, Annelid Composts and Sediments by Flow Field-Flow Fractionation-Inductively Coupled Plasma-Mass Spectrometry (Flow FFF-ICP-MS) 165
Thomas Anderson, Laura Shifley, Dula Amarasiriwardena, Atitaya Siripinyanond, Baoshan Xing and Ramon M. Barnes

Apparent Size Distribution and Spectral Properties of Natural Organic Matter Isolated from Six Rivers in Southeastern Georgia, USA 179
J.J. Alberts, M. Takács, M. McElvaine and K. Judge

Models and Theories

Molecular Modeling of Humic Structures 193
A.G. Bruccoleri, B.T. Sorenson and C.H. Langford

Modeling of Molecular Interactions of Soil Components with Organic Compounds 209
G. Haberhauer, A.J.A. Aquino, D. Tunega, M.H. Gerzabeck and H. Lischka

Binding of Hydrophobic Organic Compounds to Dissolved Humic Substances: A Predictive Approach Based on Computer Assisted Structure Elucidation, Atomistic Simulations and Flory-Huggins Solution Theory 221
Mamadou S. Diallo, Jean-Loup Faulon, William A. Goddard III and James H. Johnson, Jr.

Images, Oxidation and Humification

Atomic Force Microscopy (AFM) Study of the Adsorption of Soil HA and Soil FA at the Mica-Water Interface 241
A.-M. Tugulea, D.R. Oliver, D.J. Thompson and F.C. Hawthorne

The Influence of Catechol Humification on Surface Properties of Metal Oxides 253
C. Liu and P.M. Huang

Spectroscopic Evaluation of Humin Changes in Response to Soil Managements 271
Guangwei Ding, Jingdong Mao, Stephen Herbert, Dula Amarasiriwardena and Baoshan Xing

Effects of Clear Cutting on Structure and Chemistry of Soil Humic Substances of the Hubbard Brook Experimental Forest, New Hampshire, USA 281
David A. Ussiri and Chris E. Johnson

Significance of Burning Vegetation on the Formation of Black Humic Acids in Japanese Volcanic Ash Soils 297
H. Shindo and H. Honma

Organic Ores and Analysis of Commercial HSs

Leonardite and Humified Organic Matter 309
D.M. Ozdoba, J.C. Blyth, R.F. Engler, H. Dinel and M. Schnitzer

Some Chemical and Spectroscopic Characteristics of Six Organic Ores 315
M. Schnitzer, H. Dinel, T. Paré, H.-R. Schulten and D. Ozdoba

Interpretation by Principal Component Analysis of Pyrolysis-Field Ionization Mass Spectra of Lignite Ores 329
H. Dinel, M. Schnitzer, T. Paré, H.-R. Schulten, D. Ozdoba and T. Marche

A Comparative Evaluation of Known Liquid Humic Acid Analysis Methods 337
A.K. Fataftah, D.S. Walia, B. Gains and S.I. Kotob

Plant Growth Stimulation and Antimutagenesis

Response of Alfalfa to Calcium Lignite Fertilizer 345
T. Paré, M. Saharinen, M.J. Tudoret, H. Dinel, M. Schnitzer and D. Ozdoba

Effects of Humic Acids and Nitrogen Mineralization on Crop Production in Field Trials 355
Mir-M Seyedbagheri and James M. Torell

Antimutagenic and Antitoxic Actions of Humic Substances on Seedlings of Monocotyledon and Dicotyledon Plants 361
Guiseppe Ferrara, Elisabetta Loffredo and Nicola Senesi

Subject Index 373

History, Philosophy and Spectroscopy

AN ORGANIC MATTER TRAIL: POLYSACCHARIDES TO WASTE MANAGEMENT TO NITROGEN/CARBON TO HUMIC SUBSTANCES

C. Edward Clapp

USDA-ARS, Department of Soil, Water & Climate, University of Minnesota, St. Paul, MN 55108

1 INTRODUCTION

J. E. Dawson
Cornell
University

A farm boy from Princeton, MA went to the University of Massachusetts to study chemistry and learn how to make dynamite. After 4 years with a chemistry major and agronomy minor he proceeded to Cornell University to become a soil biochemist. This paper represents a summary of the research and experiences initiated in 1952 under the guidance and direction of Professor Jeffrey Earl Dawson, who started me on the way along the Organic Matter Trail. Jeff was an outstanding teacher and researcher who thought thermodynamically and later became a true friend and colleague. His life and career were cut short due to health complications, and he passed away at age 49. This paper is dedicated to his memory. Along with the highlights of research on organic matter, many references will be made to the people, colleagues and friends who influenced the Trail throughout the years.

2 AN ORGANIC MATTER TRAIL

The Organic Matter Trail started for the author in the Organic Soils Laboratory of the Department of Agronomy at Cornell University in Ithaca, NY in June 1952. After M.S. and Ph.D. degrees were completed, the trail led to Beltsville, MD in October 1956 with the Soil and Water Conservation Research Division (SWCRD) of USDA-ARS. Work on soil and rhizobial polysaccharides was carried out to study aggregate stabilization. After 5 years, a transfer of the organic matter/soil structure project to the Cornbelt Branch of the SWCRD in the Department of Soil Science at the University of Minnesota was completed in May 1961. In August, a visiting scientist arrived from Australia to spend a year investigating organic matter and soil crumbs. This work continued with students and other support for 10 years. When funding for basic research became difficult to justify, a call was issued for more practical and field-oriented projects. This led to waste management experiments involving application of municipal sewage sludges and wastewaters to agricultural land.

After ten more years, in 1980, a series of field and laboratory experiments was started involving nitrogen-tillage-residue management (NTRM) including ^{15}N and later ^{13}C isotope studies. Many graduate students, ARS Soil and Water Research Unit colleagues and Soils Department cooperators were involved in research and publication activities. During the late 1980s, the Organic Matter Trail was side-tracked but later developed into a super highway! Humic substances were re-introduced during an educational leave in 1988-89 at the University of Birmingham in England, and continued in 1989 at the Hebrew University of Jerusalem in Rehovot, Israel. Many facets of humic substances research have continued through the 1990s and into the 21st century.

Involvements with the International Humic Substances Society (IHSS), including many former associates and friends, led to an IHSS Board position and a relationship with the IHSS Standard Sample Collection. Together, we have set out on a combination of research projects, Society programs, publication endeavors (some controversial) and other exciting activities. With cooperating humic scientists we are looking at new isolation and characterization techniques, the sizes and shapes of humic macromolecules, at organic contaminant studies, at metal-humic interactions, at water treatment and at humic-plant stimulation, among other topics. The research areas in this version of the Organic Matter Trail will serve to summarize the career of one soil biochemist who has chosen to spend his time in the laboratory and field.

2.1 Polysaccharides

An agricultural muck soil from Orange County, NY was extracted with hot water in a Soxhlet system to produce several kilograms of lyophilized organic matter. After concentration, electrodialysis and ultrafiltration, a low-charged polysaccharide was isolated using convection as well as continuous flow paper curtain electrophoresis. Electrophoretic mobilities in phosphate and borate buffers were determined for the isolated polysaccharide using a column electrophoresis apparatus.[1,2] The polysaccharide was hydrolyzed and component sugars identified by comparison of their relative electrophoretic mobilities with the relative mobilities of pure sugars (Figure 1). Earlier experiments by Dawson[3] (in cooperation with Arne Tiselius) laid the groundwork for the electrophoresis studies.

The first real job started in October 1956 to study the effects of organic matter on soil structure at the USDA-ARS Soils Laboratory at Beltsville, MD under Hub Allaway. Administrative guidance was provided by Frank Allison and cooperative projects on rhizobial and soil polysaccharide effects on aggregate stability[4] were carried out with Bob Davis.

A transfer in 1961 to the USDA-ARS Cornbelt Branch under C.A. Van Doren in the Soil Science Department (Bill Martin, Head) of the University of Minnesota, St. Paul, led to a continuation of the organic matter/soil structure work. Administratively, the St. Paul Unit was directed by Bob Holt of the USDA Morris, Minnesota Lab. The one-year sabbatical of Bill Emerson, CSIRO soil physicist from Adelaide looking at soil crumb stability[5] and clay-polysaccharide complexes[6] provided a cooperative and highly rewarding venture. Tables 1 and 2 show the effects of organic matter extraction coupled with periodate oxidation of polysaccharide components. Al Olness joined the group as a graduate student and we carried on the polysaccharide-clay interaction studies.[7,8] X-ray diffractograms of clay-Polytran complexes (Figure 2) show two first-order (001) peaks and shoulders of the 3.6, 7.3 and 16.6% dextran complexes, indicating more than one basal spacing. No evidence of spacings greater than 15 Å was found in the diffractograms of the Polytran complexes dried at 105°C.

2.2 Waste Management

The Organic Matter Trail continued when Bill Larson came to St. Paul in 1968 as Research

Leader of the ARS Soils Unit. With the advent of the waste management studies in 1972, field and laboratory experiments were initiated at St. Paul, Rosemount and Elk River, MN. Other chief players were Soil Scientists Bob Dowdy and Dennis Linden, along with graduate students Steve Stark and Scott Harding. A 40-acre terraced sewage sludge watershed was constructed at Rosemount[9] where crop yields, nutrient uptake and soil and water quality were investigated for over 20 years, with corn (*Zea mays* L.) and reed canarygrass (*Phalaris arundinacea* L.) as crops. Nitrogen mineralization and availability from application of different types of sewage sludge was also a featured topic (Figure 3). Results indicated that the sludge application rate was more important than sludge type.[10] An ^{15}N-labeled fertilizer experiment was superimposed on sewage sludge field plots to further study N availability.[11]

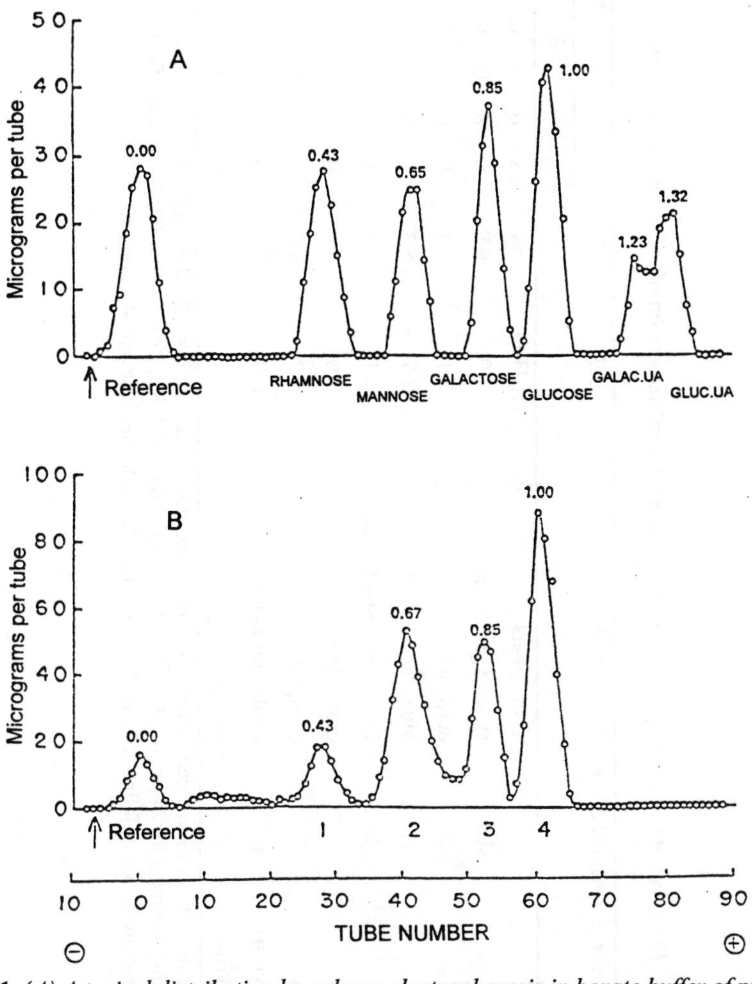

Figure 1 *(A) A typical distribution by column electrophoresis in borate buffer of pure sugars and uronic acids. Mobilities relative to glucose are indicated. The arrow denotes the upper end of the cellulose column. (B) Distribution by column electrophoresis in borate buffer of components of hydrolyzed organic soil polysaccharide*[1]

Table 1 *The effect of various treatments on the slaking (s) and dispersion (d) of crumbs from 4 Minnesota soils*[5]

Soil series	*Textural class*	*Cropping history*	*pH*	*C*	*Treatments*[a,b]									
					B		I_6B		$I_{24}B$		*PA*		$PAI_{24}B$	
		Crop, years	1:1,H_2O	%	s	d	s	d	s	d	s	d	s	d
Clyde	sicl	swamp grass, virgin	7.9	6.9	0	0	0	0	0	0	0	0	4	4
Clyde	sicl	corn, 10+	7.1	5.1	0	0	3	2	3	2	1	0	4	4
Fargo	cl	bluegrass, virgin	7.7	5.3	0	0	1	0	3	3	0	0	4	4
Fargo	sic	Continuous small grain, 5+	8.1	3.8	1	1	4	4	4	4	4	2	4	4
Judson	sil	grass, 15+	7.0	4.4	0	0	1	1	2	2	0	0	4	4
Judson	sil	corn, 10	6.2	4.0	0	0	4	4	4	4	2	1	4	4
Port Byron	sil	alfalfa-brome, 6	6.9	3.9	0	0	0	0	3	1	0	0	4	4
Port Byron	sil	corn, 6	5.9	3.3	0	0	2	2	3	2	0	0	4	4

[a] Treatments: All treatments preceded by 2 x 0.05 M NaCl (A) extraction; B = 0.25M $Na_2B_4O_7$; I_6B = 0.05 M $NaIO_4$ (6 hr) + 0.025 M $Na_2B_4O_7$; $I_{24}B$ = same treatment for 24 hr of I; PA = 0.1 M $Na_4P_2O_7$ + 0.05 M NaCl; $PAI_{24}B$ = PA + $I_{24}B$; [b] values of s and d represent various degrees of slaking and dispersion as follows: 0, no effect; 1, slight; 2, moderate; 3, severe; 4, complete.

Table 2 *The effect of various treatments on the slaking (s) and dispersion (d) of grassland or virgin soil crumbs*[5]

Soil series	*Textural class*	*Cropping history*	*pH*	*C*	*Treatments*[A,B]											
					p		*$I_{24}p$*		*B*		*$I_{24}B$*		*PA*		*$PAI_{24}p$*	
		Crop, years	1:1,H_2O	%	s	d	s	d	s	d	s	d	s	d	s	d
Fargo	cl	bluegrass, virgin	7.7	5.3	0	0	2	1	0	0	3	3	0	0	4	4
Houston	c	grass, 20+	5.9	2.7	0	0	3	2	1	1	3	3	0	0	4	4
Houston Black	c	native grass, virgin	7.8	3.6	0	0	0	0	0	0	0	0	1	0	4	4
Houston Black[c]	c	native grass, virgin	---	3.5	1	1	3	3	1	0	4	3	2	2	0	0
Mexico	sil	meadow, 2	7.2	1.6	1	1	4	2	0	0	3	2	1	1	4	4
Paulding	c	grass, 4	6.4	3.4	1	1	2	2	1	0	3	2	2	1	4	3
Paulding	c	forest, virgin	6.7	6.9	0	0	0	0	0	0	1	0	0	0	0	1

[a] Treatments: All treatments preceded by 2 x 0.05 M NaCl (A) extraction. p = 0.01M NaCl/Napnp; $I_{24}p$ = 0.05 M $NaIO_4$ (24 hr) + 0.01 M NaCl/Napnp; B = 0.025 M $Na_2B_4O_7$; $I_{24}B$ = 0.05 M $NaIO_4$ (24 hr) + 0.25 M $Na_2B_4O_7$; PA = 0.1*M* $Na_4P_2O_7$ + 0.05 M NaCl; $PAI_{24}P$ = PA + $I_{24}P$; [b] values of s and d represent various degrees of slaking and dispersion as follows: 0, no effect; 1, slight; 2, moderate; 3, severe; 4, complete. [c] $CaCO_3$ removed by treatments with 0.1 M HCl at 0°C.

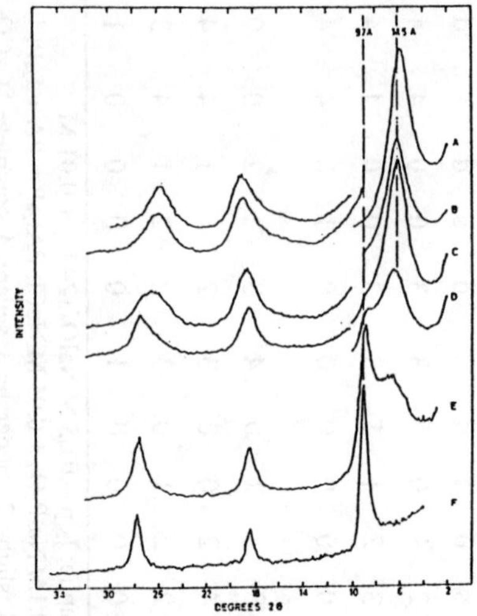

Figure 2 *X-ray diffractograms of oven-dry montmorillonite-Polytran complexes. Patterns: A, 37.2%; B, 34.8%; C, 16.6%; D, 7.3%; E, 3.6%; F, Na-montmorillonite*[7]

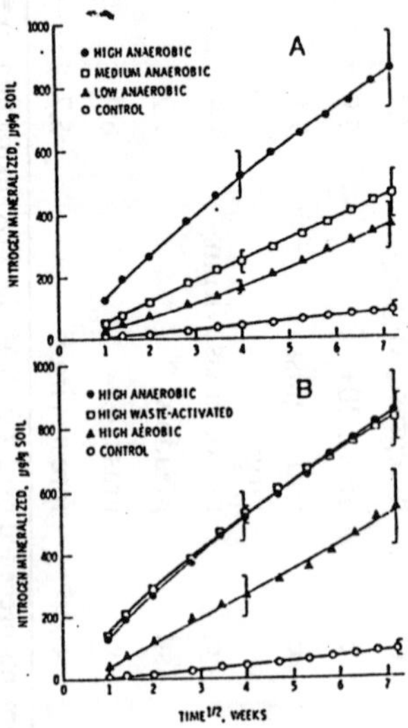

Figure 3 *Cumulative N mineralized during aerobic incubation in relation to the square root of time for (A) the anaerobic sludge treatments, and (B) the high sludge application rate treatments. Standard error intervals are shown for the 16- and 52-week periods*[10]

A field experiment[12] applying treated municipal wastewater (secondary effluent) to corn and 8 forage grasses was co-sponsored by the US Army Corps of Engineers (CRREL) and carried out at Apple Valley, MN. Main investigators, in addition to the ARS-Soils group, were Gordon Martin (ARS-Crops), Tony Palazzo (CRREL), Jim Nylund (Soils Dept.) and Karen Chopp (a graduate student). Results of the 6-year study showed that municipal wastewater could be applied effectively to agricultural crops with a reduction of nitrogen and phosphorus to acceptable safe levels for surface and ground water quality. A summary of nitrogen uptake by reed canarygrass (Figure 4) shows the large amount of nitrogen removed from wastewater compared with a fertilizer control.[13] Additional studies involving microbial activity and nitrate risk assessment[14,15] were carried out with Ed Schmidt (Soils Dept.).

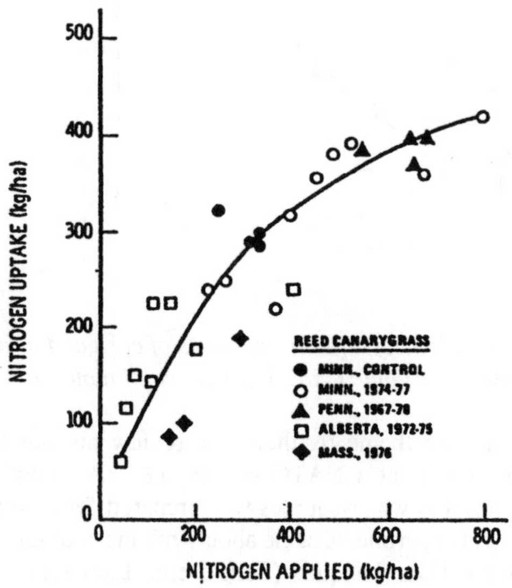

Figure 4 *Total seasonal nitrogen up-take by reed canarygrass irrigated with secondary municipal wastewater effluent*[13]

2.3 Nitrogen/Carbon

A series of field and laboratory experiments started in 1980 involved nitrogen-tillage-residue management (NTRM), including ^{15}N and ^{13}C isotope studies. An earlier experiment initiated by Bill Larson at Ames, IA was completed and the effects of residue additions on gain and loss of soil carbon[16] were compared (Figure 5). It would take approximately 6 Mg ha^{-1} yr^{-1} of corn stalks to prevent loss of organic matter in a soil containing about 1.8% C.

The NTRM experiment involved 8 large tractor-size blocks with three nitrogen levels (0-100-200 kg ha^{-1}), three tillage management systems (no-till, moldboard-plow, and chisel-plow) and two residue treatments (returned or harvested). A comparison ^{15}N-labeled fertilizer study on smaller plots (the most expensive 100 m^2 area in Dakota County!) was carried out during the same period under the initial inspiration and guidance of Art Edwards. Treatments included two fertilizer levels (2 and 20 g N m^{-2} with 40 and 4 atom-% ^{15}N, respectively) applied for 15 years, two tillage practices (no-till or rototill) and two residue choices (returned or harvested). Reports of the early years are given by Dave Clay[17,18] (graduate student), including predictions based on models NCSOIL and NCSWAP (Figure 6) by Jean Molina

(Soils Dept.) and Randy Deans[19] (graduate student). A recent modeling paper[20] based on data from the NTRM experiment reported the soil conditions necessary to reproduce the measured kinetics of seven variables over a 13-year period.

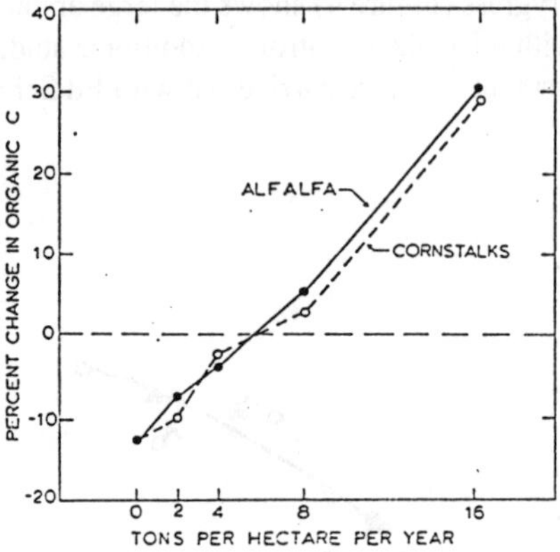

Figure 5 *Effect of annual additions of different amounts of corn and alfalfa residues on the gains and losses of organic C in soil under continuous corn culture*[16]

Continuing the organic matter theme resulted in a review on interactions of organic macromolecules with soil colloids at a NATO conference.[21] A summary review of ^{13}C abundance in world-wide soil and water sources was prepared for a Dublin conference.[22] Access to other long-term field experiments came about with the addition to the team of Ray Allmaras (ARS, St. Paul) and Dave Huggins (Soils Dept., Lamberton). A 10-year corn-soybean cropping sequence study[23] provided a unique opportunity to investigate C dynamics using natural ^{13}C abundance. Our results have shown that long-term field experiments are among the best means to predict soil management impacts on soil C storage. Soil organic C and natural abundance ^{13}C ($\delta^{13}C$) were sensitive to tillage, stover harvest and N management during 13 years of continuous corn in the NTRM study.[24] Figure 7 shows results where $\delta^{13}C$ values generally increased in both the 0-15 and 15-30 cm layers. All slopes except those in the NTh treatment in the 0-15 cm layer were greater than 0, irrespective of N rate.

2.4 Humic Substances

The re-introduction of humic substances research actually occurred with the IHSS Conferences in 1983 at Estes Park, CO and in 1984 at Birmingham, UK. A chapter on viscosity measurements of humics was presented and written for an IHSS book.[25] Another book[26] co-edited with Pat MacCarthy, Ron Malcolm and Paul Bloom on humic substances in soil and crop sciences resulted from a symposium at the SSSA meeting in 1985. The educational leave at Birmingham in 1988-89 brought together two old Cornell lab-mates and friends. Mike Hayes completed his M.S. with Jeff Dawson in 1955 and we relived many old memories. We carried out many experiments involving isolation and characterization of humic substances using techniques that Mike had been developing at Birmingham. Some of this work was reported at a conference in Lancaster, UK and published in 1993.[27] Another research paper introducing a new extraction method was presented and published as part of the proceedings of the IHSS Conference in Trinidad in 1994.[28]

A research project started during the extended educational leave in Israel (with Uri Mingelgrin and Yona Chen) was completed and published in 1997.[29] Plots in Figure 8 show significant differences in complexation coefficients (K_c) between the two herbicides napropamide and atrazine. The dialysis experiments carried out demonstrated a procedure by which the extent of complexation of small organic molecules with stable humic acids could be quantitatively estimated.[29]

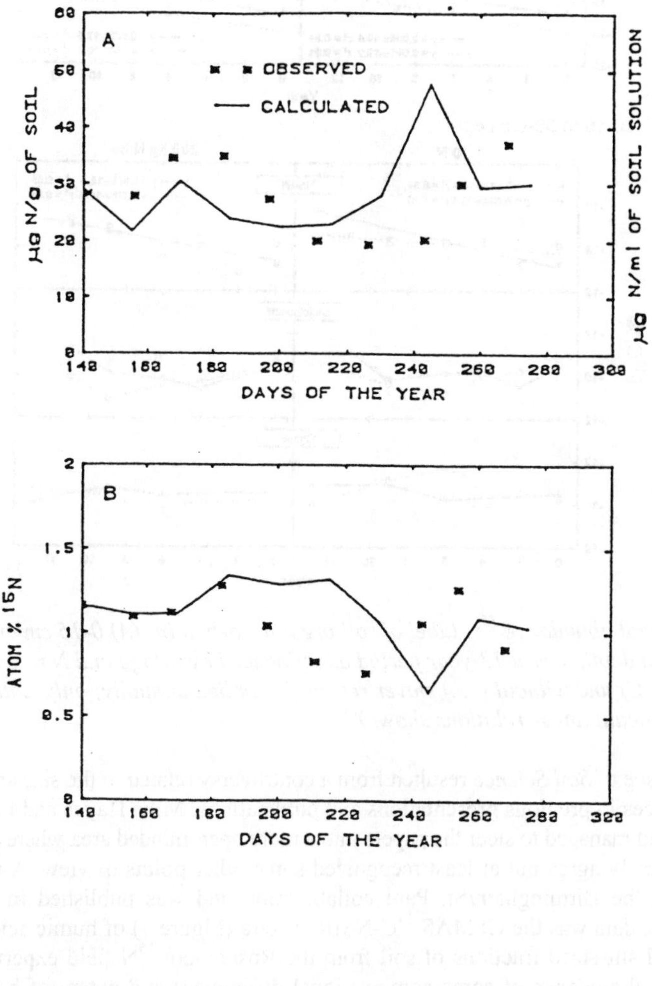

Figure 6 *Calculated and observed nitrate-N movement to the 100-cm soil depth for the 20 g N m^{-2} tilled-no residue treatment*[17]

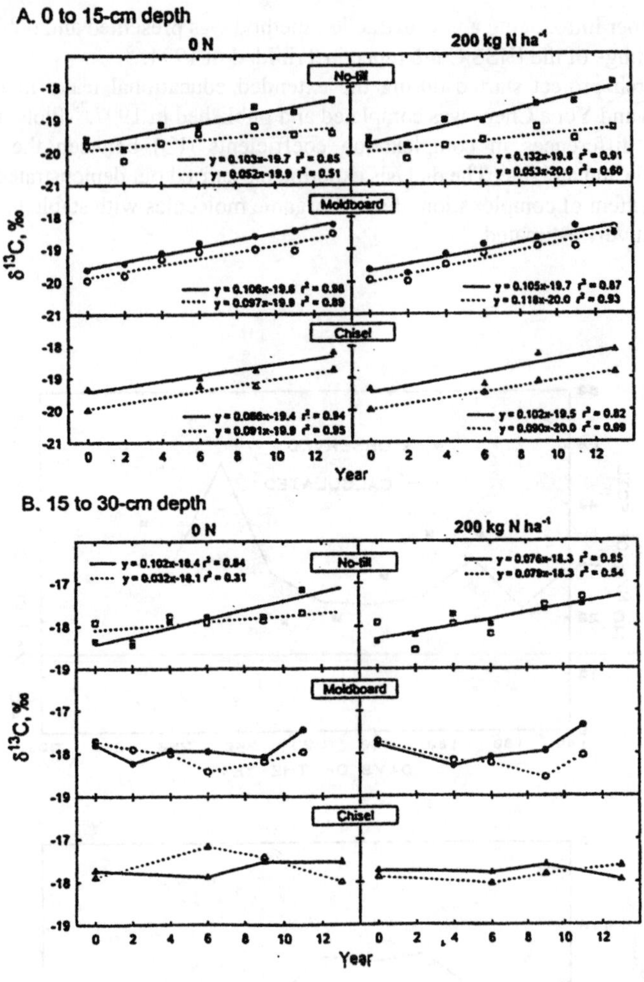

Figure 7 *Natural abundance ^{13}C label of soil organic carbon in: (A) 0-15 cm and (B) 15-30 cm depth over a 13-year period as influenced by tillage and N rate treatments with (C) and without (. . .) stover return (N applied annually; only statistically significant linear relations shown)*[24]

A special issue of *Soil Science* resulted from a controversy related to the size and shape of humic substances in previous presentations and publications. Mike Hayes and I introduced the subject[30] and managed to steer the papers into a more open-minded area where the authors did not completely agree but at least recognized some other points of view. A final paper surfaced from the Birmingham/St. Paul collaboration and was published in 1999.[31] A highlight of the data was the CPMAS ^{13}C-NMR spectra (Figure 9) of humic acids isolated from clay- and silt-sized fractions of soil from the Rosemount ^{15}N field experiment. The spectra provided evidence of some compositional differences and extent of humification between the humic acids from the two soil fractions.

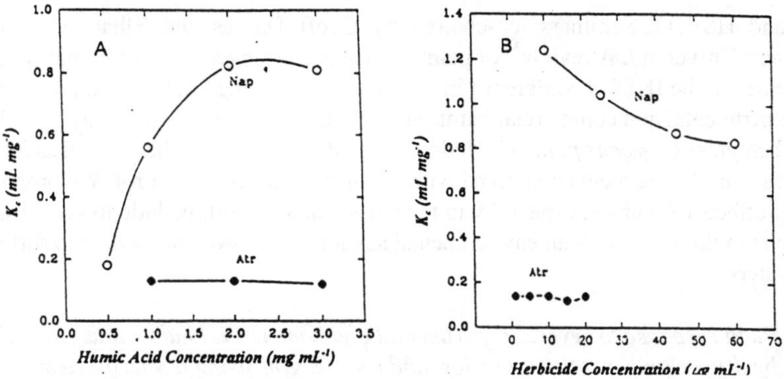

Figure 8 *A. Complexation coefficients (K_c) for different concentrations of a soil humic acid reacted with napropamide (Nap), 35 µg mL^{-1}; and atrazine (Atr), 20 µgmL^{-1}. B. Complexation coefficients (K_c) for a soil humic acid (2 mg mL^{-1}) reacted with napropamide (Nap) and atrazine (Atr)*[29]

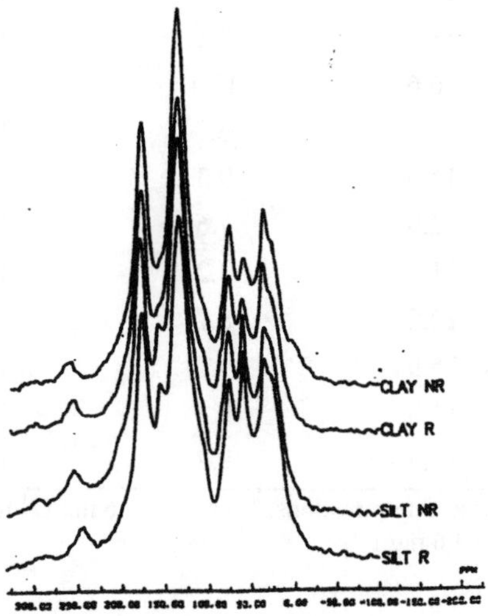

Figure 9 *CPMAS ^{13}C-NMR spectra of humic acids isolated from the clay- and silt-sized fractions of a Mollisol soil to which corn residues had been added (R) or not added (NR) over a period of 8 years*[31]

An area of increasing interest and importance is the stimulation of plant growth by humic substances, both in agricultural and horticultural usage. Turfgrass management, especially for golf courses, has been a major area of application for humic substances. Although a complete answer to the effects of enhanced growth is not yet known, we believe plant growth stimulation by humic substances is the result of improved iron and possibly zinc nutrition. Colleagues in this research effort were Yona Chen (Hebrew Univ.), Van Cline (Agronomist, TORO Co.), and Ruilong Liu (Research Associate, Soils). Preliminary papers reporting results of the humic/plant growth response have been presented and published as part of the

HS-II[32] and HS-III[33] Seminars, co-chaired by Geoff Davies and Elham Ghabbour at Northeastern University. A review[34] of plant growth promoting activity of humic substances was presented at the IHSS-9 conference in Adelaide and is being published in 2001. Data for several experimental and commercial humic and fulvic acids (Table 3) demonstrate enhanced creeping bentgrass (*Agrostis palustris* Huds.) root and shoot dry weights as well as increased root lengths for all treatments compared with the nutrient solution control. We propose that beneficial effects of humics, especially in turfgrass management, include lower fertilizer N application requirements with an environmental advantage in lower N leaching to surface and ground waters.

Table 3 *Plant response to several experimental and commercial humic and fulvic acids by dry weights, root/shoot ratios, and root lengths using a Microsystem Method*[a][32]

Treatment[b]	*Dry weight (mg well^{-1})*		*Root/shoot*	*Root length*
	Root	*Shoot*		*(cm well^{-1})*
Control	11.4	7.3	1.6	135
Hemic peat HA	15.0	8.3	1.8	180
Sapric peat HA	13.2	7.1	1.9	222
Fibric peat HA	16.6	12.1	1.4	207
Norfolk peat HA	9.7	6.1	1.6	172
Kerry soil HA	14.6	9.3	1.6	193
Leonardite HA	12.9	7.6	1.7	172
Moss HA	11.0	7.2	1.5	156
Norfolk peat FA	12.2	8.7	1.4	234
Kerry soil FA	15.9	9.7	1.6	160
Moss FA	11.3	8.1	1.4	174
Leonardite HA-FA	13.4	8.2	1.6	195

[a] Values are means of triplicate determinations: [b] N rate of 3.6 mg L^{-1} for control and all treatments. HA and FA rates 3.6 mg L^{-1}.

3 CONCLUDING REMARKS

The Organic Matter Trail is still on-going. Affiliation with the IHSS at the biennial meetings, in associations and correspondence with members around the world, and especially with those I have been privileged to know well: The Honorary Members, Past Officers, Present and Past Board Members, Co-authors and Co-editors, The Collection Committee and other special long-time associates.

A final editing effort of a book Humic Substances and Chemical Contaminants[35] to be published in 2001 by the SSSA and IHSS was the outcome of a workshop/symposium held in 1997. Many thanks are extended to the co-editors Michael Hayes, Nicola Senesi, Paul Bloom and Phil Jardine.

In my 45+ years on the Organic Matter Trail I have worked with many researchers and published with approximately 100 different scientists. I have advised about 20 graduate students. I treasure the associations very much and continue contact with most of these colleagues. While I take limited credit, I believe we have made some major advances along the Trail during my career, especially in the areas of organic matter/soil structure, waste management, nitrogen/carbon dynamics and humic substances characterization.

Many memories will survive. It was a special privilege to be chosen as Honorary Chair of HS Seminar V in March, 2001. It was a great meeting with good presentations, a rousing debate or two, lots of old friends and a memorable occasion. In addition, my wife and family deserve thanks, gratitude and love for sharing me with the USDA and University positions and the IHSS, putting up with weird schedules, editing, attending meetings and travel requirements. We have not solved all of the problems still out there but in my own case, after some 45 years on the Organic Matter Trail, we trust that the younger and wiser scientists will carry on for the next generations to come.

References

1. C.E. Clapp, *High Molecular Weight Water-soluble Muck: Isolation and Determination of Constituent Sugars of a Borate Complex-forming Polysaccharide Employing Electrophoretic Techniques*, Ph.D. Thesis, Cornell University, 1957.
2. C.E. Clapp, J.E. Dawson and M.H.B. Hayes, 'Composition and Properties of a Purified Polysaccharide Isolated from an Organic Soil', *Proc. Peat in Agric. and Hort.*, Special Publication No. 205, 1980, p. 153.
3. J.E. Dawson, C.E. Clapp and M.H.B. Hayes, 'Studies on the Chemical and Physico-chemical Properties of Extracts of Organic Soils: Electrophoretic Characteristics of Water-soluble Extracts', *Proc. Peat in Agric. and Hort.*, Special Publication No. 205, 1980, p. 278.
4. C.E. Clapp, R.J. Davis and S.H. Waugaman, 'The Effect of Rhizobial Polysaccharides on Aggregate Stability', *Soil Sci. Soc. Am. Proc.*, 1962, **26**, 466.
5. C.E. Clapp and W.W. Emerson, 'The Effect of Periodate Oxidation on the Strength of Soil Crumbs: I. Qualitative Studies. II. Quantitative Studies', *Soil Sci. Soc. Am. Proc.*, 1965, **29**,127 and 130.
6. C.E. Clapp and W.W. Emerson, 'Reactions between Ca-Montmorillonite and Polysaccharides', *Soil Sci.*, 1972, **114**, 210.
7. A.E. Olness and C.E. Clapp, 'Occurrence of Collapsed and Expanded Crystals in Montmorillonite-Dextran Complexes', *Clays Clay Minerals,* 1973, **21**, 289.
8. A.E. Olness and C.E. Clapp, 'Influence of Polysaccharide Structure on Dextran Adsorption by Montmorillonite', *Soil Biol. Biochem.*, 1975, **7**, 113.
9. C.E. Clapp, R.H. Dowdy, D.R. Linden, W.E. Larson, C.M. Hormann, K.E. Smith, T.R. Halbach, H.H. Cheng and R.C. Polta, 'Crop yields, Nutrient Uptake, Soil and Water Quality during 20 Years on the Rosemount Sewage Sludge Watershed', in *Sewage Sludge: Land Utilization and the Environment*, eds. C.E. Clapp, et al., SSSA Misc. Publ., Soil Sci. Soc. Am., Madison, WI, 1994, p. 137.
10. S.A. Stark and C.E. Clapp, 'Residual nitrogen availability from soils treated with sewage sludge in a field experiment', *J. Environ. Qual.*, 1980, **9**, 505.
11. S.A. Harding, C.E. Clapp and W.E. Larson, 'Nitrogen Availability and Uptake from Field Soils Five Years after Addition of Sewage Sludge', *J. Environ. Qual.*, 1985, **14**, 95.
12. C.E. Clapp, D.R. Linden, W.E. Larson, G.C. Marten and J.R. Nylund, 'Nitrogen Removal from Municipal Wastewater Effluent by a Crop Irrigation System', in

Land as a Waste Management Alternative, ed. R.C. Loehr, Cornell University Press, Ithaca, NY, 1977, p. 139.
13. C.E. Clapp, A.J. Palazzo, W.E. Larson, G.C. Marten and D.R. Linden, *Uptake of Nutrients by Plants Irrigated with Municipal Wastewater Effluent*, Proc. State of Knowledge in Land Treatment of Wastewater, International Symposium, 1977, **1**, 395.
14. K.M. Chopp, C.E. Clapp and E.L. Schmidt, 'Ammonia-Oxidizing Bacteria Populations and Activities in Soil Irrigated with Municipal Wastewater Effluent', *J. Environ. Qual.*, 1982, **11**, 221.
15. C.E. Clapp, R. Liu, D.R. Linden, W.E. Larson and R.H. Dowdy, 'Nitrates in Soils and Waters from Sewage Wastes on Land', in *Managing Risks of Nitrates to Humans and the Environment*, eds. W.S. Wilson et al., Royal Society of Chemistry, Cambridge, 1999, p. 139.
16. W.E. Larson, C.E. Clapp, W.H. Pierre and Y.B. Morachan, 'Effects of Increasing Amounts of Organic Residues on Continuous Corn: II. Organic Carbon, Nitrogen, Phosphorus and Sulfur', *Agron. J.*, 1972, **64**, 204.
17. D.E. Clay, C.E. Clapp, J.A.E. Molina and D.R. Linden, 'Nitrogen-tillage-residue Management: 1. Simulating Soil and Plant Behavior by the Model NCSWAP', *Plant Soil,* 1985, **84**, 67.
18. D.E. Clay, J.A.E. Molina, C.E. Clapp and D.R. Linden, 'Nitrogen-tillage-residue management: 2. Calibration of Potential Rate of Nitrification by Model Simulation', *Soil Sci. Soc. Am. J.*, 1985, **49**, 322.
19. J.R. Deans, J.A.E. Molina and C.E. Clapp, 'Models for Predicting Potentially Mineralizable Nitrogen and Decomposition Rate Constants', *Soil Sci. Soc. Am. J.*, 1986, **50**, 323.
20. J.A.E. Molina, C.E. Clapp, D.R. Linden, R.R. Allmaras, M.F. Layese, R.H. Dowdy and H.H. Cheng, 'Modeling the Incorporation of Corn (*Zea mays* L.) Carbon from Roots and Rhizodeposition into Soil Organic Matter', *Soil Biol. Biochem.*, 2001, **33**, 83.
21. C.E. Clapp, R. Harrison and M.H.B. Hayes, 'Interactions between Organic Macromolecules and Soil Inorganic Colloids and Soils', in *Interactions at the Soil Colloid-Soil Solution Interface*, eds. G.H. Bolt et al., Proc. 1986 NATO Conf., Kluwer, Dordrecht. 1991, p. 409.
22. C.E. Clapp, M.F. Layese, M.H.B. Hayes, D.R. Huggins and R.R. Allmaras, 'Natural Abundance of ^{13}C in Soils and Water', *in Humic Substances in Soils, Peats and Sludges*, eds. M.H.B. Hayes and W.S. Wilson, Royal Society of Chemistry, Cambridge, 1997, p. 158.
23. D.R. Huggins, C.E. Clapp, R.R. Allmaras, J.A. Lamb and M.F. Layese, 'Carbon Dynamics in Corn-Soybean Sequences as Estimated from Natural Carbon-13 Abundance', *Soil Sci,. Soc. Am. J.*, 1998, **62**, 195.
24. C.E. Clapp, R.R. Allmaras, M.F. Layese, D.R. Linden and R.H. Dowdy, 'Soil Organic Carbon and ^{13}C Abundance as Related to Tillage, Crop Residue and Nitrogen Fertilization under Continuous Corn Management in Minnesota', *Soil Tillage Res.*, 2000, **55**, 127.
25. C.E. Clapp, W.W. Emerson and A.E. Olness, 'Sizes and Shapes of Humic Substances by Viscosity Measurements', in *Humic Substances, Vol. 2: In Search of Structure*, eds. M.H.B. Hayes et al., Wiley, Chichester, 1989, p. 497.
26. P. MacCarthy, C.E. Clapp, R.L. Malcolm and P.R. Bloom, eds., *Humic Substances in Soil and Crop Sciences: Selected Readings*, Soil Sci. Soc. Am., Madison, WI,

1990.
27. C.E. Clapp, M.H.B. Hayes and R.S. Swift, 'Isolation, Fractionation, Functionalities, and Concepts of Structures of Soil Organic Macromolecules', in *Organic Substances in Soil and Water: Natural Constituents and Their Influences on Contaminant Behaviour*, eds. A.J. Beck et al., Royal Society of Chemistry, Cambridge, 1993, p. 31.
28. C.E. Clapp and M.H.B. Hayes, 'Isolation of Humic Substances from an Agricultural Soil Using a Sequential and Exhaustive Extraction Process', in *Humic Substances and Organic Matter in Soil and Water Environments*, eds. C.E. Clapp et al., IHSS, St. Paul, MN, 1996, p. 3.
29. C.E. Clapp, U. Mingelgrin, R. Liu, H. Zhang and M.H.B. Hayes, 'A Quantitative Estimation of the Complexation of Small Organic Molecules with Soluble Humic Acids', *J. Environ. Qual.*, 1997, **26**, 1277.
30. C.E. Clapp and M.H.B. Hayes, 'Sizes and Shapes of Humic Substances', *Soil Sci.*, 1999, **164**, 777.
31. C.E. Clapp and M.H.B. Hayes, 'Characterization of Humic Substances Isolated from Clay-and Silt-Sized Fractions of a Corn Residue-Amended Agricultural Soil', *Soil Sci.*, 1999, **164**, 899.
32. C.E. Clapp, R. Liu, V.W. Cline, Y. Chen and M.H.B. Hayes, 'Humic Substances for Enhancing Turfgrass Growth', in *Humic Substances: Structures, Properties, and Uses*, eds. G. Davies and E.A. Ghabbour, Royal Society of Chemistry, Cambridge, 1998, p. 227.
33. Y. Chen, C.E. Clapp, H. Magen and V.W. Cline, 'Stimulation of Plant Growth by Humic Substances: Effects on Iron Availability', in *Understanding Humic Substances: Advanced Methods, Properties and Applications*, eds. E.A. Ghabbour and G. Davies, Royal Society of Chemistry, Cambridge, 1999, p. 255.
34. C.E. Clapp, Y. Chen, M.H.B. Hayes and H.H. Cheng, 'Plant Growth Promoting Activity of Humic Substances', in *Understanding and Managing Organic Matter in Soils, Sediments, and Waters*, eds. R.S. Swift and K.M. Spark, IHSS, Adelaide, 2001, p. 243.
35. C.E. Clapp, M.H.B. Hayes, N. Senesi, P.R. Bloom and P.M. Jardine, eds., *Humic Substances and Chemical Contaminants*, SSSA Misc. Publ., Soil Sci. Soc. Am., Madison, WI, (in press) 2001.

THE PRINCIPLES OF HUMIC SUBSTANCES: AN INTRODUCTION TO THE FIRST PRINCIPLE

Patrick MacCarthy

Department of Chemistry and Geochemistry, Colorado School of Mines, Golden, CO 80401 USA

1 INTRODUCTION

A vast body of empirical data has accumulated on the nature and properties of humic substances. What is lacking is an overall set of rules or guiding principles that account for the chemical nature and environmental roles of these substances and would serve as a conceptual model for further research on humic materials. The objective of this paper is to examine critically the extensive body of published data, prior observations and ideas on humic substances in order to identify those features that are intrinsic and unique to humic materials. The results of this endeavor are expressed in the form of two principles. The First Principle is introduced in this paper and the Second Principle will be presented elsewhere.

2 THE FIRST PRINCIPLE OF HUMIC SUBSTANCES

The chemical nature of humic substances can be rationalized on the basis of the following statement, which will be referred to as the First Principle of humic substances.

Humic substances comprise an extraordinarily complex, amorphous mixture of highly heterogeneous, chemically reactive yet refractory molecules, produced during early diagenesis in the decay of biomatter, and formed ubiquitously in the environment via processes involving chemical reaction of species randomly chosen from a pool of diverse molecules and through random chemical alteration of precursor molecules.

This principle describes the fundamental molecular nature and origin of humic substances by addressing the questions: *what* are humic substances, and *when, where* and *how* are they formed? It represents an attempt to look beyond the accidental and incidental in an effort to view the big picture and grasp the true essence of humic substances. The overriding truth of this principle supersedes the myriad details of the chemical and microbial processes that participate in the formation, reactions and interactions of humic substances in nature and in the laboratory. Five important corollaries are derived from the First Principle and expose its consequences in more practical terms:

Corollary A. *Humic substances are devoid of a regularly recurring, extended skeletal entity.*

Corollary B. *Humic substances cannot be purified in the conventional meaning of purity.*

Corollary C. *The essence of humic substances resides in the combination of their extreme molecular heterogeneity and pronounced chemical reactivity.*

Corollary D. *Humic substances from different sources display a remarkable uniformity in their gross properties.*

Corollary E. *It is not possible to write a molecular structure, or set of structures, that fully describes the connectivity within molecules of a humic substance.*

Corollary A states that there is no long-range chemical order recognizable in humic substances, and that there is no identifiable backbone or skeletal structure that could be regarded as uniquely characteristic of these materials. Corollary B is self-explanatory and will be elaborated on later. Corollary C recognizes those unique features that are essential to humic substances from both a chemical and an ecological point of view. Corollary D provides the basis for considering humic substances as a unique class of materials. Corollary E addresses the nebulous nature of the molecular structure concept when applied to humic substances. In this paper, data and arguments are presented to justify this principle and then some implications of the principle are discussed.

3 DEFINITIONS OF TERMS

The term *humic substances* is intended to designate that class of organic material occurring in, or extracted from, decayed or decaying biomatter in soil, sediment or natural waters, and that does not fall into any of the discrete classes of organic substances. Organic matter extracted from plants, senescent leaves, recently fallen leaves and so on is not considered to constitute humic substances in the context of this paper. The word *biomatter* refers to biologically synthesized matter that is no longer living or part of a living cell. The terms *heterogeneous* and *heterogeneity*, as used in the principle and throughout this paper, include the lack of structural regularity within and among the humic molecules. This topic is addressed more rigorously in Section 5.2 of this paper. The term *regularly recurring, extended, skeletal entity* refers to a structural component that contains more than six carbon atoms and that occurs in an orderly, repeating manner within a molecule. In the context of this paper, *chemical reactivity* refers to the diverse reactions and interactions exhibited by humic substances in the environment. These reactions and interactions underlie the multiple functions of humic substances in nature and include acid dissociation, metal complexation, ion exchange, sorption on minerals and redox reactions with metal ions and organic species. Chemical reactivity of humic substances is discussed further in Section 4.2 below. The term *refractory* means that the substrate resists decomposition by microorganisms. That is, microbial degradation of humic substances is considerably slower than that of discrete biopolymers. It is not intended to imply that humic substances are absolutely recalcitrant and that they are not degraded by microorganisms.

4 JUSTIFICATION FOR THE FIRST PRINCIPLE

4.1 On the Origin and Ubiquity of Humic Substances

Referring to the First Principle, humic substances are indeed *ubiquitous* and are *produced*

... *in the decay of biomatter* in all terrestrial and aquatic environments throughout the world. The formation of humic substances occurs during *early diagenesis* of biomatter in contrast to the formation of coal and petroleum that are produced over geologic periods and under more severe metamorphic conditions. Materials that do not originate from the decay of biomatter in soil, sediment or natural waters should not be considered as humic substances even though they may display the same acid/base solubility behavior and color as humic substances.

4.2 On the Chemical Reactivity of Humic Substances

One of the more prominent characteristics of humic and fulvic acids is their pronounced chemical reactivity. Humic substances have an abundance of carboxyl groups and also possess weakly acidic phenolic groups. In addition to conferring acidity on humic substances, these groups also contribute to the complexation and ion-exchange properties of humic materials. Humic substances are also known to be redox active, and this activity manifests itself in geochemically and environmentally important processes.[1,2] Humic substances possess free-radicals[3-5] and can bind small molecules through both hydrogen bonding and nonpolar interactions. Humic substances exhibit both hydrophobic and hydrophilic characteristics and can bind to mineral surfaces. Being a complex mixture, it is unlikely that all constituent molecules in a humic sample possess each of these chemical characteristics.

4.3 On the Refractory Nature of Humic Substances

Despite the prominence and diversity of the chemical reactivity of humic substances, as discussed above, these materials are also known to be particularly unreactive in terms of their susceptibility to breakdown by microorganisms; that is, humic materials are refractory.[6,7] The refractory nature of humic substances has been well documented throughout the literature. For example, Schnitzer has stated that "all humic fractions [exhibit] resistance to microbial degradation";[8] according to Stevenson, "The resistance of humus to biological decomposition has long been known",[6] and the long lifetimes of humic substances in soil "attest to the high resistance of humus to microbial attack".[6] Malcolm has stated that humic substances from all environments display "a refractory nature to microbial decay".[9] Such observations are the basis for inclusion in the First Principle of the qualification *chemically reactive yet refractory.*

4.4 On the Inability to Separate/Purify Humic Substances

There is an overwhelming body of evidence from separation studies indicating that humic substances comprise an enormously complex mixture. In attempts to isolate pure humic substances, virtually every method of separation has been applied, ranging from classical methods such as fractional precipitation[10] to the more modern separation techniques including chromatography of all types, electrophoresis and field-flow fractionation. In all separation studies, the fractions obtained still consisted of very complex mixtures and no material has ever been isolated in significant amounts that could be described as a pure, or reasonably pure, humic substance. However, it is misleading and counterproductive to focus on the apparent negative aspects of these observations. It is more insightful to view the results as providing a substantial body of data in support of the extraordinary complexity of humic materials. Many published statements alluding to this highly heterogeneous nature of humic substances, as suggested by separation studies, can be found in the literature from the early 1960's to recent times. For example, as early as 1963 Dubach and Mehta[11] stated "... in spite of intensive efforts involving most diverse methods, no discrete fractions have ever been isolated from humic substances". In 1985, Perdue[12] stated that "... humic substances are indisputably a highly complex mixture that has thus far been essentially unresolvable into significant amounts of pure components".

It appears that molecules in a humic system differ gradually from one another in their chemical and physical properties in a continuous manner, thereby accounting for the inability to separate these materials into pure components. For example, in 1964 Dubach et al.[13] pointed out that "the methods used to characterize the fractions ... show rather a continuous variation", and in 1965 Felbeck[14] stated that "at the present time there has been no evidence presented to indicate that a definite fraction has ever been isolated from a humic substance. Each fraction can be refractionated into subfractions by some other technique, and this can be repeated, apparently, *ad infinitum.*" Swift, in 1985, stated that humic substances consist of "a broad spectrum of related molecules, each one differing almost imperceptibly from the next in terms of one or other of its properties".[15] These many observations relating to the inability to separate humic substances into pure or reasonably pure fractions, with no reports to the contrary, testify to the highly complex and heterogeneous nature of humic substances. Such observations support the statement in the First Principle that *Humic substances comprise an extraordinarily complex, amorphous mixture of highly heterogeneous ... molecules.* The issue of purity in the context of humic substances is addressed in more detail in Section 5.2.

4.5 On the Molecular Nature of Humic Substances

The compositions of naturally occurring organic molecules are dominated by a relatively small number of structural moieties (such as benzene rings, aliphatic segments, hexose and pentose units, amino acids), functional groups (such as carboxyl, hydroxyl, amine), and linkages (such as ester, amide, ether). Obviously, such entities are found in humic substances and to that extent there are recurring structural units in humic structures. However, numerous studies on higher molecular weight humic substances have failed to identify a *regularly* repeating structural unit or set of units that could be considered characteristic of these materials. In 1989 Hayes et al. stated that, "there is no evidence for regularity in the structures of humic macromolecules".[16] According to Stevenson "...each fraction (humic acid, fulvic acid, etc.) came to be regarded as being made up of a series of molecules of different sizes, few having precisely the same structural configuration or array of reactive functional groups".[17] One of the more intriguing statements in this context dates back to 1963 when Dubach and Mehta[11] speculated that "perhaps no two molecules of humic substances are exactly alike". There have been many similar statements in the intervening literature. For example, in 1976 Gjessing[18] commented that "every molecule of humus could be different" and in 1985 Stevenson stated that it "is probably safe to say that few, if any, humic molecules will be precisely the same".[6]

There have been many investigations of humic substances by chemical and thermal degradative methods[19] as well as by chemical derivatization.[20] While such studies have provided much information about the composition of humic substances (e.g. functional group contents), as well as identifying aromatic and aliphatic components among the degradation products, they have not revealed any extended molecular unit that could be considered an essential building block that is uniquely characteristic of these materials. Similarly, no molecular entity has been identified as constituting a molecular backbone of the molecules in humic substances. The refractory nature of humic substances is also consistent with, and can be used as supporting evidence for, the highly heterogeneous character of these materials. Additional evidence for the molecular complexity of humic substances comes from a wide variety of spectroscopic methods.[21,22] Such chemical and instrumental data add further support to Corollary A and to the statement in the First Principle that *humic substances comprise an extraordinarily complex, amorphous mixture of highly heterogeneous ... molecules*. These observations are also the basis for the statement in Corollary E.

4.6 Concerted Investigations of Humic Substances

Because humic substances are not unique compounds and can vary from source to source and with the method of extraction and treatment, it was seldom that researchers worked with common samples of humic materials. It was proposed in 1976 that an international reference collection of humic substances be established.[23] That proposal was implemented[24] and led to the formation of the International Humic Substances Society (IHSS) in 1982. The IHSS maintains and distributes the reference humic materials. Even with the availability of these reference materials since the mid-1980's and the consequent ability of researchers from all over the world to work on humic substances disbursed from physically homogenized stockpiles, researchers have not been able to identify regularly recurring, extended skeletal units in humic substances or any structural moiety that could be considered uniquely characteristic of humic substances. In the 1980's, researchers from the United States Geological Survey and several other laboratories embarked on a concerted study of a single aquatic fulvic acid. The study produced some interesting results. However, it did not isolate or identify specific substances within the fulvic acid or a regularly occurring molecular fragment that was characteristic of fulvic acid.[25] A major review of structural aspects of humic substances was published in 1989, wherein it was pointed out that "the term *structure* cannot have a precise meaning for such complicated mixtures of macromolecular substances, and it may well be pointless to try to establish accurate structures for anything other than the smallest and simplest humic molecules".[16]

Consequently, the large bodies of data that are generally interpreted as failures to separate humic substances and failures to determine molecular structure or a molecular backbone for humic substances must be reassessed. Such data and observations should instead be regarded as a unified and compelling body of evidence supporting the view that humic substances comprise a complex mixture of highly heterogeneous molecules formed through processes involving reactions of randomly chosen species as asserted in the First Principle. These observations are consistent with the amorphous character of all humic substances.

4.7 On Random Features in the Formation of Humic Substances

Biomolecules with uniquely defined structures are synthesized under biological and enzymatic control and through discrete formation pathways. This control is necessary in order to produce the specific molecules that are uniquely required for each of thousands of distinct biological tasks. Because of the complexity and indefiniteness of the experimental data, many diverse models for the humic formation process have been proposed.[6,17] The functions of humic substances in the environment do not require the participation of unique molecular structures and there is no need or experimental evidence for specific molecular structure control in the formation of humic substances. In that respect, the formation of humic substances may represent the antithesis of biological synthesis.

One prominent model for the formation of humic substances incorporates an initial degradative step involving microbial decomposition of the biomatter into smaller molecules. These small molecules are then consumed as an energy and carbon source by microbes with the release of other small molecules. This is followed by a formative step: some of the initial degradation products and the microbially-produced diverse small molecules, chosen randomly from the mixture, undergo condensation reactions to generate substances that, unlike the initial biological matter, resist breakdown by the organisms themselves. That is, refractory materials are produced. These are humic substances. Other proposed models for the formation of humic substances involve chemical alteration at sites along polymer chains of biomolecules such as lignin or partially degraded lignin. These alterations could include condensation, hydrolytic or oxidative processes. Such reactions, occurring randomly, would contribute to increased heterogeneity within a system.

The opportunity for molecular diversity in a system derived from a pool of varied reactive components resulting from microbial decomposition of plant and animal tissue is enormous. As stated by Orlov in 1974, "With a multitude of sources, a diversity of reactions and a continuous exchange of the components of humus substances, the latter are formed out of various products whose composition is governed by random factors. This is the big difference between the synthesis of humus substances and the synthesis of organic compounds in living organisms".[26] Stevenson made a similar statement in 1982, "The number of precursor molecules is large and the number of ways in which they combine is astronomical, thereby accounting for the heterogeneous nature of the humic material in any given soil." Swaby and Ladd[7] proposed that humus "is not synthesized enzymatically, but by heterogeneous chemical catalysis ..." involving "... chaotic chemical combination of monomers produced enzymatically". Such ideas are the basis for the statement in the First Principle that humic substances are *formed ... via processes involving chemical reaction of species randomly chosen from a pool of diverse molecules and through random chemical alteration of precursor molecules*. This random feature in the humic formation processes contributes to the molecular heterogeneity that is intrinsic to humic substances. The specific nature of the reactants and reactions involved in these proposed processes has not been established and this paper does not address that deficiency.

4.8 On the Uniformity and Variability Among Humic Substances

While there are variations among humic substances from different sites and between humic and fulvic acids from a given source,[27] the overall similarities in many of their properties are more striking than the differences.[8] Malcolm listed features that are common among all humic substances and pointed out specific differences among humic materials isolated from the same and different types of environments.[9] The *uniformity* is evident in many humic properties despite the diversity in geographic location, botanic origin and climatic conditions from which the humic substances arise: elemental compositions of humic substances from all over the world are remarkably similar[28,29] (Table 1); all humic substances exhibit the same general array of functional groups and there are no major compositional differences between old and recent humic materials.[7,30] As a consequence, all humic substances exhibit the same types of reactions and interactions. The uniformity of humic substances is consistent with and can be considered as supporting evidence for the First Principle -- the formation processes involving reactions of randomly chosen molecules resulting from the initial microbial degradation of biomatter lead to a molecularly heterogeneous but macroscopically homogenized mixture regardless of the source material. The nature of the source material is generally not strongly evident in the humified products. It is this uniformity among humic substances as a whole and the uniformity among the humic acids and fulvic acids individually[28,29] that give these materials an identity that justifies their study as unique classes of substances. These ideas are embodied in Corollary D of the First Principle.

Table 1 *Elemental compositions for humic substances. Group A: average values for humic acid, fulvic acid and humin from all over the world and not segregated by source. Group B: average values for humic acids from soil, freshwater and peat sources. Group C: average values for fulvic acids from soil, freshwater and peat sources*[28,a]

Group A	*C*	*H*	*O*	*N*
Humic Acid (410)	55.1±5.0	5.0±1.1	35.6±5.8	3.5±1.5
Fulvic Acid (214)	46.2±5.4	4.9±1.0	45.6±5.5	2.5±1.6
Humin (26)[b]	56.1±2.6	5.5±1.0	34.7±3.4	3.7±1.3
Group B				
Soil Humic Acids (215)	55.4±3.8	4.8±1.0	36.0±3.7	3.6±1.3
Freshwater Humic Acids (56)	51.2±3.0	4.7±0.6	40.4±3.8	2.6±1.6
Peat Humic Acids (23)[c]	57.1±2.5	5.0±0.8	35.2±2.7	2.8±1.0
Group C				
Soil Fulvic Acids (127)	55.4±3.8	4.8±1.0	36.0±3.7	3.6±1.3
Freshwater Fulvic Acids (63)	51.2±3.0	4.7±0.6	40.4±3.8	2.6±1.6
Peat Fulvic Acids (12)	54.2±4.3	5.0±0.8	35.2±2.7	2.8±1.0

[a] The uncertainties with each value are absolute standard deviations; the numbers in parenthesis following the name of each sample give the number of samples used in calculating the averages and standard deviations; [b] only 24 samples used for the nitrogen value in this row; [c] only 21 samples used for the nitrogen value in this row

4.9 On Humic Substances as an Equilibrium Mixture

Because of the reactive nature of the constituent molecules from which humic substances are considered to form and because of the wide variety of reactive humic species differing gradually from one another, it is quite possible that molecules of humic substances in an aqueous medium could be in a state of dynamic equilibrium. In addition to the expected acid-base dissociation reactions there could also be continuous, random compound formation and dissociation. A given humic sample in the aqueous state would then represent an equilibrium condition that could be disturbed by a fractionation, after which a new equilibrium would be established in the resulting fractions. Actually, there have been allusions in the literature to this type of humic equilibrium.[10] Such an effect could be an additional factor contributing to the inability to purify and to establish detailed structural features of humic substances. These associative interactions could also include various types of noncovalent forces such as hydrogen bonding. Thus, while humic substances as a whole are persistent materials, some of the individual molecules may have a more transitory existence. This is a topic that needs to be examined experimentally.

5 SOME CONSEQUENCES AND IMPLICATIONS OF THE FIRST PRINCIPLE

There are many consequences and implications of the First Principle, three of which are presented here.

5.1 Molecular Heterogeneity-cum-Chemical Reactivity -- The Essence of Humic Substances

The First Principle offers a simple rationale for what appear to be disparate and frustrating observations relating to the fractionation, structural nature and formation of humic substances. The inability to purify humic substances is a direct consequence of their intrinsic molecular heterogeneity, which in turn results from the molecular nonspecificity in the processes that generate these materials. Thus, it is not surprising in view of the First Principle that: (i) no unique molecular structure, discrete skeletal entity or repeating, extended molecular unit has been identified in humic substances, and (ii) no one has succeeded in purifying these materials. In fact, it becomes evident that the conventional concepts of purity and unique structures are foreign to humic substances. Not only do humic substances comprise an extraordinarily complex mixture of structurally randomized molecules as outlined in the First Principle, but such molecular heterogeneity, randomness and complexity, in combination with their chemical reactivity, in fact constitute the very essence of these materials. It is these features that distinguish humic substances from all other natural materials, as stated in Corollary C.

5.2 Humic Substances Defined as a *Supermixture*

Referring to Corollary B, the ultimate meaning of the term *chemically pure substance* could be exemplified by an ensemble of molecules all having the same molecular structure (without addressing issues of isotopic purity). Few, if any substances are absolutely pure in that strictly literal sense. Generally, a material is considered pure when the sample is comprised largely (98%, 99%, 99.9%, 99.99% and so on) of the stated substance and where the acceptable assay level depends on the specific circumstances. When dealing with a mixture of polymeric molecules, purification can refer to the isolation of fractions all having the same repeating sequence of monomers, even though the molecules may display a range of molecular weights. While such systems are not pure in the ultimate sense defined above, they evidently represent a high degree of purity in that the constituent molecules possess a common backbone. The difficulty of applying the term *purity* to humic substances was addressed by Hurst and Burges who stated that "The concept of purity, so easy to apply to simple molecules, has a less precise meaning when applied to mixed polymers and is especially difficult to apply to humic acid".[31] More recently, Clapp and Hayes stated that "all attempts to isolate a humic fraction that satisfy the criteria of purity have failed".[32]

Four features contribute to the *heterogeneity* of a mixture:
(i) the *multiplicity* of the system, i.e. the number of structurally different molecules (per mole or per stated number of moles) in the system;
(ii) the *molecular diversity,* i.e. the extent to which those molecules differ from each other;
(iii) polydispersity, i.e., the range of molecular weights displayed by the molecules, and
(iv) *chemical equilibrium* in which molecules constantly associate and dissociate as already discussed.

In the absence of detailed data to confirm the dynamic aspect of humic substances, this discussion will focus on their multiplicity, diversity and polydispersity aspects. A macroscopic system comprising an ensemble of molecules where no two molecules have identical molecular structures would have an extremely high multiplicity, but would not necessarily represent a very high degree of complexity. For example, it has been shown that for a mole of partially demethylated pectinic acid, the probability of two molecules being identical is virtually zero.[33] Nevertheless, there is still a high degree of simplicity or homogeneity in such a system since all the molecules possess a common type of backbone and have the same, limited types of functional groups. The molecular diversity in such a system would be small and consequently the heterogeneity of that mixture would also be relativity small. Near the other extreme of such a high-multiplicity system would be an ensemble of molecules lacking a common, regularly recurring, identifiable skeletal structure or other long-range order. In such a system the molecular diversity and multiplicity would both be very large, leading to a vastly more complex mixture. The larger the constituent molecules, the greater the statistical opportunity for increased molecular diversity. Clearly, variations in molecular weight among the species (polydispersity) would add another dimension to the complexity. The term *mixture* does not adequately convey the enormity of the challenges posed by these systems and such extremely complex mixtures will be referred to herein as *supermixtures.* A supermixture is defined in conceptual terms as *a mixture having a degree of complexity and heterogeneity equivalent to that of a large ensemble of molecules where no two molecules are identical, where the molecules are essentially devoid of a regularly recurring, extended skeletal entity and display a high degree of molecular diversity.* A supermixture has both a high multiplicity and a high diversity, and it generally would be polydisperse.

The formation of large molecules by the combination of species randomly chosen from a pool of diverse reactants would lead to a highly heterogeneous mixture of high statistical probability and thus high entropy of mixing -- a *supermixture.* In the case of highly complex mixtures as described above, purification in the ultimate sense would necessitate a molecule-by-molecule separation, which is certainly not within the normal domain of what is meant by purification. The conventional meaning of purity implies that macroscopic amounts of the purified materials are isolated. Two reasonably sized samples of a given supermixture would not share a common identical molecule although the gross properties of both samples are identical.

The combination of data from fractionation and structural investigations on humic substances indicates that these substances are, or approach, supermixtures in their complexity. It is thus clear why *humic substances cannot be purified in the conventional meaning of purity,* as stated in Corollary B. The wording in Corollary B could be replaced with the following: *Humic substances comprise a supermixture.* Considering humic substances as a supermixture focuses attention on their extreme multicomponent and structurally chaotic nature. With this view of humic substances it is possible to concentrate on the true nature of these materials during structural studies and while conducting experimental investigations on these materials. A humic substance (that is, a sample free of nonhumic materials) may be considered "pure" in the context of humic substances; however, the meaning of purity in that context must be distinguished from the conventional chemical definition of the term. The supermixture character of humic substances imposes constraints on the qualitative and quantitative analysis of these materials that will be examined in more detail elsewhere.

5.2.1 The Mixture-Supermixture Boundary. The boundary between a supermixture and a regular mixture is vague and one type of mixture may gradually merge into the other in the lower molecular weight region. Consider again the generation of a highly heterogeneous mixture in the manner described above, and where the distribution of molecular weights extends from quite low values (say 300 - 400 Da) to very high values. Such a system may be analogous to a humic-fulvic mixture. Consider further the lower MW fraction of that mixture, for example with molecular weights less than about 500 Da. The opportunity for molecular diversity in this fraction is diminished and the fraction may contain multiple identical molecules. While such a low MW fraction would still comprise an extremely complex mixture, it may not constitute a supermixture in the ultimate sense.

In a recent electrospray ionization/tandem mass spectrometric (MS^n) investigation of a fulvic acid, Leenheer et al.[34] isolated the 329 Da peak and subjected it to multiple MS experiments. They then proposed hypothetical fragmentation pathways that could account for the observed mass spectral peaks. While the isolated 329 Da peak is still likely to contain a mixture of different molecules, the study does indicate the relative simplicity of the very low molecular weight components. Another recent study of aquatic fulvic acids using electrospray ionization coupled with quadrupole time-of-flight mass spectrometry was conducted by Plancque et al.[35] They report that "for the first time to our knowledge, a possible structure was found for fulvic acid." In another recent mass spectrometric study, Hatcher et al.[36] proposed molecular building blocks based on lignin structures for a "humic acid" extracted from a degraded wood sample from which the cellulosic components had been removed by fungal/bacterial degradation.

5.3 Humic Substances – A Distinct Class of "Natural Product"

The term *natural product* has long been used in organic chemistry to classify materials having common structural characteristics into specific categories such as alkaloids, terpenoids, steroids and so on. Based upon the unique molecular nature of humic substances; their recognition by researchers as materials worthy of study for hundreds of years,[37,38] the ubiquity and general uniformity of these substances; and their established importance in soil fertility, environmental and geochemical processes,[6,39] it is proposed that they be considered as a class of "natural product". Humic substances, then, constitute that unique class of "natural product" comprising a mixture of molecularly heterogeneous molecules that are both chemically reactive and refractory that originate in decayed biomatter. This class of "natural product" is real and is distinguished from the conventional classes of natural product in its intrinsic molecular heterogeneity and randomness. Because of their fundamentally different type of molecular constitution, humic substances must be studied as inseparable, complex mixtures of structurally-heterogeneous molecules. That is, these materials must be studied as a supermixture and the limitations imposed by this complexity on the applicability of various experimental techniques and on the interpretability of the experimental data needs to be appreciated.

ACKNOWLEDGEMENTS

Critical input from the following individuals contributed to the technical and editorial refinement of this paper: Drs. R. W. Klusman, D. L. Macalady, K. Mandernack and P. Ross all of the Colorado School of Mines; Drs. J. A. Leenheer and R. L. Wershaw of the U.S. Geological Survey; Dr. J. A. Rice, South Dakota State University and Dr. J. Burdon, The University of Birmingham, U.K.

References

1. N.L. Wolfe and D.L. Macalady, *J. Contam. Hydrol.*, 1992, **9**, 17.
2. D.R. Lovley, J.D. Coates, E.L. Blunt-Harris, E.J.P. Phillips and J.C. Woodword, *Nature,* 1996, **382,** 445.
3. R.W. Rex, *Nature,* 1960, **188**, 1185.
4. C. Steelink and G. Tollin, in *Soil Biochemistry*, eds. A.D. McLaren and G.H. Peterson, Dekker, New York, 1967, Chap. 6.
5. N. Senesi, Y. Chen and M. Schnitzer, *Soil Biol. Biochem.*, 1997, **9**, 371.
6. F.J. Stevenson, *Humus Chemistry: Genesis, Composition, Reactions*, Wiley, New York, 1982.
7. R.J. Swaby and J.N. Ladd, *Transactions of Joint Meeting Comms*, IV & V, International Society of Soil Science, Palmerston North, New Zealand, 1962, p. 197.
8. M. Schnitzer and S.U. Khan, *Humic Substances in the Environment*, Dekker, New York, 1972.
9. R.L. Malcolm, in *Humic Substances in Soil and Crop Sciences: Selected Readings*, eds. P. MacCarthy, C.E. Clapp, R.L. Malcolm and P.R. Bloom, Soil Science Society of America, Madison, 1990, Chap. 2.
10. F.J. Sowden and H. Deuel, *Soil. Sci.,* 1961, **91**, 44.
11. P. Dubach and N.C. Mehta, *Soils Fert.,* 1963, **26,** 293.
12. E.M. Perdue, in *Humic Substances in Soil, Sediment, and Water: Geochemistry, Isolation, and Characterization*, eds. G.R. Aiken, D.M. McKnight, R.L. Wershaw and P. MacCarthy, Wiley, New York, 1985, Chap. 20.
13. P. Dubach, N.C. Mehta, T. Jakab, F. Martin and M. Roulet, *Geochim. Cosmochim. Acta,* 1964, **28**, 1567.
14. G.T. Felbeck, *Advan. Agron.*, 1965, **17**, 327.
15. R.S. Swift, in *Humic Substances in Soil, Sediment, and Water: Geochemistry, Isolation, and Characterization*, eds. G.R. Aiken, D.M. McKnight, R.L. Wershaw and P. MacCarthy, Wiley, New York, 1985, Chap. 15.
16. M. H. B. Hayes, P. MacCarthy, R. L. Malcolm and R. S. Swift, eds., *Humic Substances: II. In Search of Structure*, Wiley, Chichester, 1989, p. 729.
17. F. J. Stevenson, in *Humic Substances in Soil, Sediment, and Water: Geochemistry, Isolation, and Characterization*, eds. G.R. Aiken, D.M. McKnight, R.L. Wershaw and P. MacCarthy, Wiley, New York, 1985, Chap. 2.
18. E.J. Gjessing, *Characterization of Aquatic Humus*, Ann Arbor Science, Ann Arbor, Michigan, 1976.
19. M.H.B. Hayes, P. MacCarthy, R.L. Malcolm and R.S. Swift (eds.), *Humic Substances: II. In Search of Structure*, Wiley, Chichester, 1989, Chaps. 1-8.
20. J.A. Leenheer and T.I. Noyes, in *Humic Substances II: In Search of Structure*, eds. M.H.B. Hayes, P. MacCarthy, R.L. Malcolm and R.S. Swift, Wiley, Chichester, 1989, Chap. 9.
21. G.R. Aiken, D.M. McKnight, R.L. Wershaw and P. MacCarthy, eds., *Humic Substances in Soil, Sediment and Water: Geochemistry, Isolation and Characterization*, Wiley, New York, 1985, Chaps. 21, 22.
22. M.H.B. Hayes, P. MacCarthy, R.L. Malcolm and R.S. Swift, eds., *Humic Substances: II. In Search of Structure*, Wiley, Chichester, 1989, Chaps. 10-14.
23. P. MacCarthy, *Geoderma,* 1976, **16**, 179.
24. P. MacCarthy, R.L. Malcolm, M.H.B. Hayes, R.S. Swift, M. Schnitzer and W.L.

Campbell, in *Transactions of the XIII Congress of the International Society of Soil Science*, Hamburg, 1986, p. 378.
25. R.C. Averett, J.A. Leenheer, D.M. McKnight and K.A. Thorn, eds., *Humic Substances in the Suwannee River, Georgia: Interactions, Properties, and Proposed Structures*, U. S. Geological Survey, OFR 87-557, Washington, DC, 1989.
26. D.S. Orlov, *Humus Acids of Soil*, Moscow, 1985. (Translation of Humusovye Kisloty Pochv., Moscow University Publishers, Moscow, 1974).
27. R.L. Malcolm and P. MacCarthy, in *Advances in Soil Organic Matter Research and the Impact on Agriculture and the Environment*, ed. W.S. Wilson, Royal Society of Chemistry, London, 1991, p. 23.
28. J.A. Rice and P. MacCarthy, *Org. Geochem.*, 1991, **17**, 635.
29. M. Schnitzer, in *Proceedings of the Symposium on Soil Organic Matter Studies*, Braunsweig, 1977, p. 117.
30. G. Calderoni and M. Schnitzer, *Geochim. Cosmochim. Acta*, 1984, **48**, 2045.
31. H.M. Hurst and N.A. Burges, in *Soil Biochemistry*, eds. A.D. McLaren and G.H. Peterson, Dekker, New York, 1967, Chap. 11.
32. C.E. Clapp and M.H.B. Hayes, *Soil Sci.*, 1999, **164,** 777.
33. R. Speiser, C.H. Hills and C.R. Eddy, *J. Phys. Chem.*, 1945, **49**, 328.
34. J.A. Leenheer, C.E. Rostad, P.M. Gates, E.T. Furlong and I. Ferrer, *Anal. Chem.*, 2001, **73**, 1461.
35. G. Plancque, B. Amekraz, V. Moulin, P. Toulhoat and C. Moulin, *Rapid Commun. Mass Spectrom,* 2001, **15,** 827.
36. P.G. Hatcher, X. Zang and M.A. Freitas, in *Proceedings of 10th International Meeting of the International Humic Substances Society*, Toulouse, 2000, p. 1093.
37. F.K. Achard, *Crell's Chem. Ann.,* 1786, **2,** 391.
38. J.J. Berzelius, *Lehrbuch der Chemie*, 3rd Edn., translated by F. Wohler, Dresden & Leipzig, 1839.
39. M.H.B. Hayes and R.S. Swift, in *The Chemistry of Soil Constituents*, eds. D.J. Greenland and M.H.B. Hayes, Wiley, Chichester, 1978, p. 179.

NMR CHARACTERIZATION OF THE MOBILE COMPONENTS IN INTACT GREEN AND SENESCENT LEAVES AS A MEANS OF STUDYING THE HUMIFICATION PROCESS

R.L. Wershaw[1] and Igor Goljer[2]

[1] U.S. Geological Survey, Denver, CO 80225, USA
[2] Varian NMR Systems, Columbia, MD 21045, USA

1 INTRODUCTION

Non-living natural organic matter (NOM) in soils and natural waters has been studied for more than 200 years. NOM has most commonly been referred to as humic substances or humus. These terms were originally applied to soil NOM, but more recently they have been used for dissolved NOM as well. Early workers recognized that soil humus arises mainly from the degradation of dead plant tissue. Many of these early workers assumed that humus was composed of the end products of synthetic reactions that alter the structures of plant degradation products. Other workers, however, maintained that humus is a complex mixture of plant degradation products. This controversy has persisted to the present time.[1] In a recent book Stevenson[2] has articulated the position of those who maintain that humic substances arise from synthetic reactions; he defined soil humic substances as “A series of relatively high-molecular weight, yellow to black colored substances formed by secondary synthesis reactions,” and humus as the “Total of the organic compounds in soil exclusive of undecayed plant and animal tissues, their ‘partial decomposition’ products, and the soil biomass.” In contrast, a number of workers recently have proposed that humic substances consist mainly of the partial degradation products of plant polymers.[3-5] The preponderance of evidence supports the contention that humic substances are mostly composed of partially degraded plant components; however, the resulting degradation products can undergo subsequent condensation and coupling reactions.

Most workers have attempted to elucidate the structures of the chemical components of humic substances by applying the same procedures that have been used to elucidate the chemical structures of pure chemical compounds. These procedures are intrinsically unsuitable to this task because humic substance isolates from soils and natural waters are invariably composed of complex mixtures of compounds. Wershaw[1] has proposed that a more fruitful approach would be to study the chemical reactions that the chemical components of plant tissue undergo as they degrade to form NOM. We have chosen to study the degradation of leaf components because leaf fall from deciduous trees is a major source of NOM in forest soils.[6,7]

The molecular species of living leaf tissue can, in general, be divided into two categories as a function of their ability to undergo molecular motion. The molecules of some leaf components, such as lignin and cellulose, normally exist in rigid networks in which they can undergo only very limited molecular motion. The molecules of other components can move much more freely. These mobile species in leaves may be relatively low molecular weight molecules dissolved in cellular fluids, higher molecular weight components that are strongly hydrated, cell membrane components that exist in dynamically-ordered bilayer structures or molecules in waxy coatings. These mobile leaf components usually are much more susceptible to chemical reaction than the nonmobile components. As long as a leaf is alive, the chemical reactions that take place in the leaf must be very tightly controlled in order to maintain the physiological processes upon which the life of the organism depends. Much of this control is achieved by spatial segregation of the mobile components into specialized organelles of the leaf cells. However, during senescence, which is the first step in leaf death, the cell walls rupture and the component molecules from one organelle mix with molecules from other organelles.

The reactions that these molecules undergo during senescence constitute the first steps in the humification process that produces NOM. A senescent leaf is, in reality, a chemical reaction vessel that contains a number of highly reactive components. Leshem[8] pointed out that oxygen-containing free radical species such as superoxide, hydroxyl, peroxyl, alkoxyl, polyunsaturated fatty acid, and semiquinone free radicals are active during plant senescence. It is these free radicals that cause the catabolic breakdown of plant tissue components.[9,10] In addition to degradation by active-oxygen species, hydrolytic enzymes hydrolyze polysaccharides and proteins to soluble monosaccharides and amino acids that are transported to storage organs for subsequent use when the plants reawaken in the spring.[8,11,12]

Measurement of the liquid-state NMR spectra of the mobile components of intact leaves before and after senescence is a sensitive way to monitor chemical changes that take place prior to movement of the mobile components out of the leaves into soils or natural waters. In order to obtain high-resolution measurements in intact leaves, it is necessary to use probes that are designed to eliminate the effects of magnetic susceptibility non-uniformities that exist in heterogeneous systems. Magic-angle spinning (MAS) will eliminate line-broadening due to magnetic susceptibility differences in nonhomogeneous samples; however, in order to obtain the best possible resolution it is also necessary to reduce magnetic susceptibility mismatches in the body of the probe itself as much as possible.[13]

2 MATERIALS AND METHODS

2.1 Sample Collection

Attached green and senescent Aspen (*Populus tremuloides*) leaves were collected from trees near Pinecliffe, Colorado, at an elevation of approximately 2500 meters on September 1, 2000. Attached green and senescent *Malus* 'Manbeck Weeper' leaves were collected from the same tree at the Denver Botanic Garden on October 26, 2000.

2.2 NMR Analysis

The NMR analyses were performed with Varian Nano.NMR probes• on 500 MHz Varian spectrometers. Cut strips of leaf tissue were packed into the cavity of the rotor (40 μL

• The use of tradenames in this report is for identification purposes only and does not constitute endorsement by the U.S. Geological Survey.

capacity) and D_2O was added to fill the cavity. The two dimensional (2D) spectra were measured with a gradient Nano.NMR probe using gradient selected heteronuclear single quantum coherence (gHSQC) experiments.[14] The gHSQC experiment was performed in phase sensitive mode with 128 complex increments using 4 transients per increment with a 1 s delay. The total acquisition time was 21 minutes and the spinning rate of the rotor was 2140 Hz. The ^{13}C direct polarization (DP) measurements were made using a ^{13}C Nano.probe. A 45 μsec pulse width, 26 or 40 KHz sweep width, a 0 or 1 sec pulse delay, broad-band continuous WALTZ decoupling and a spinning rate of 2100 or 2400 Hz were used for these DP measurements.

3 RESULTS AND DISCUSSION

3.1 Direct Polarization (DP) NMR

The ^{13}C DP spectra of green and senescent Aspen and *Malus* leaves are shown in Figures 1 and 2. The aliphatic regions (15 to 50 ppm) of the two Aspen spectra are very similar, as are the two *Malus* spectra. However, the aliphatic band pattern of the Aspen leaves is very different from that of the *Malus* leaves. The most likely source of many of the bands between 15 and 40 ppm is the waxy material that coats the outer surfaces of the leaves.

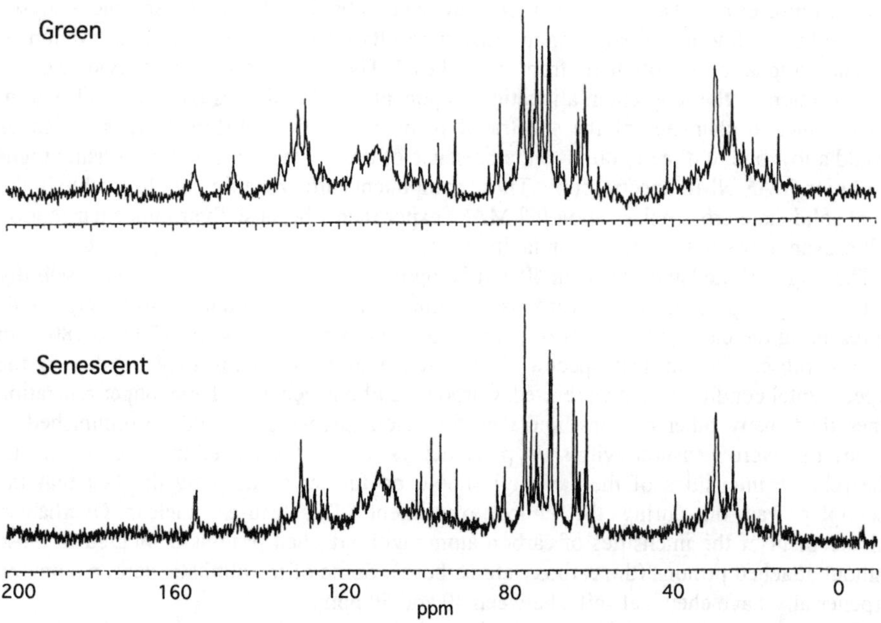

Figure 1 *^{13}C NMR spectra of green and senescent Aspen leaves*

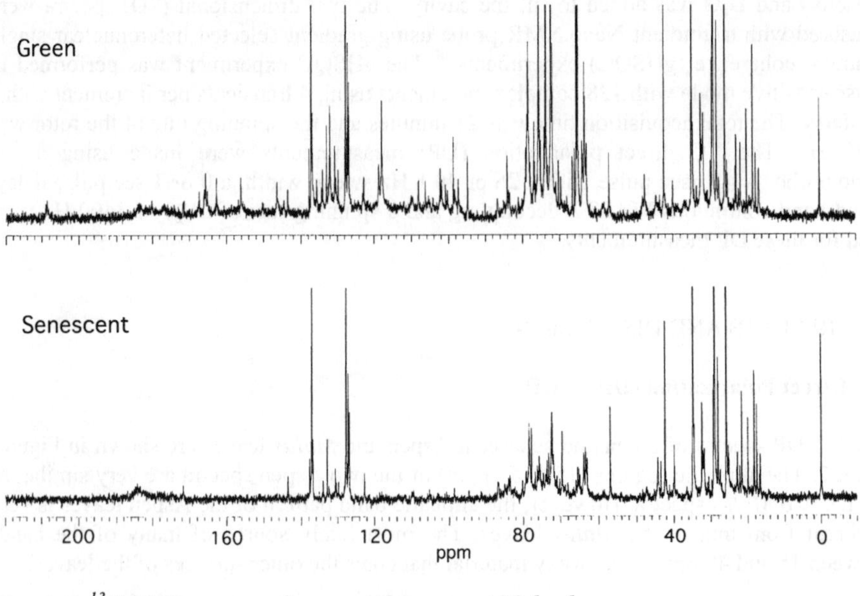

Figure 2 *^{13}C NMR spectra of green and senescent Malus leaves*

Lichtfouse et al.[15] have proposed from evidence obtained from ^{13}C isotopic analyses that the humin fraction of soil organic matter results from the "selective preservation of resistant aliphatic biopolymers from microbes." They have also found evidence for incorporation of the long-chain aliphatic components of the plant waxes into soil humin. The distinctive character of the spectra of plant waxes from different types of leaves should allow one to directly observe the presence of these waxes in soil humus using high-resolution MAS NMR spectroscopy. These components are probably not observable in the usual solid-state cross polarization (CPMAS) experiments because their molecular motion will prevent cross polarization from taking place.

The origin of the bands between 40 and 50 ppm is not clear at this time. One possibility is that they represent the μ carbons of amino acids or proteins. However, bands representing the carbonyl carbons of amino acids and peptides between 170 and 180 ppm are not present in our leaf spectra. The absence of these bands may be due to the experimental conditions that were used. Carbonyl carbons generally have longer relaxation times than many other carbons, and therefore their intensities should be diminished in spectra that were measured with short pulse delays such as those used in our experiments. The relative intensities of the carbonyl should be further reduced by the fact that the decoupler was on during the entire experiment. This causes nuclear Overhauser enhancement of the intensities of carbon atoms with attached protons compared to those without attached protons (the carboxylate carbons). Some of the aliphatic carbon atoms in terpenes also have chemical shifts between 40 and 50 ppm.

There are a number of differences in the carbohydrate regions between 60-105 ppm of the spectra (Figures 3 and 4). The relative intensities of all of the carbohydrate bands in the senescent *Malus* leaves are much less intense than in the green leaves. In the Aspen leaves some of the carbohydrate bands are less intense in the senescent leaves than in the green leaves and some of the bands are more intense. There is practically a complete absence of anomeric carbon bands between 90 and 105 ppm in the senescent *Malus* leaves; in the nonsenescent Aspen leaves there are two major bands at 92 and 104 ppm in the anomeric region. The intensity of the 92 ppm band is diminished in the senescent leaves and the 104 ppm band is completely absent.

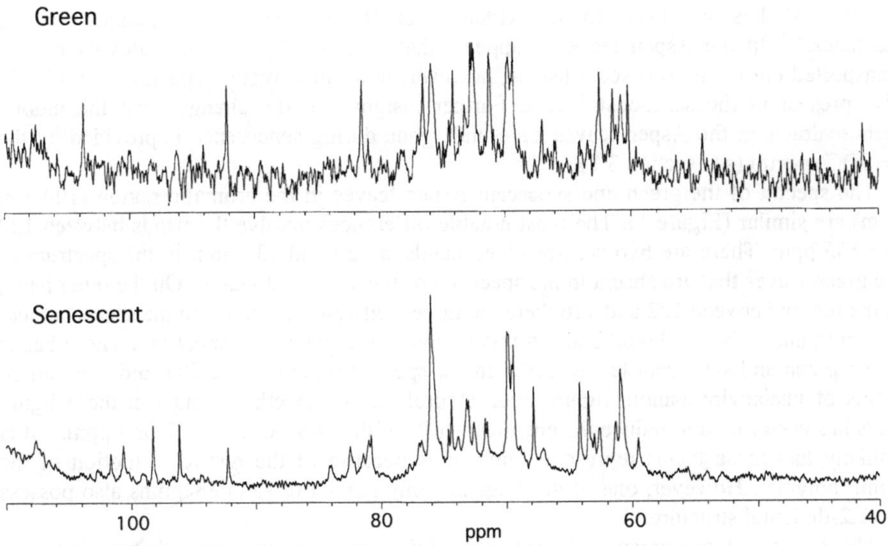

Figure 3 *Carbohydrate regions of the ^{13}C NMR spectra of green and senescent Aspen leaves*

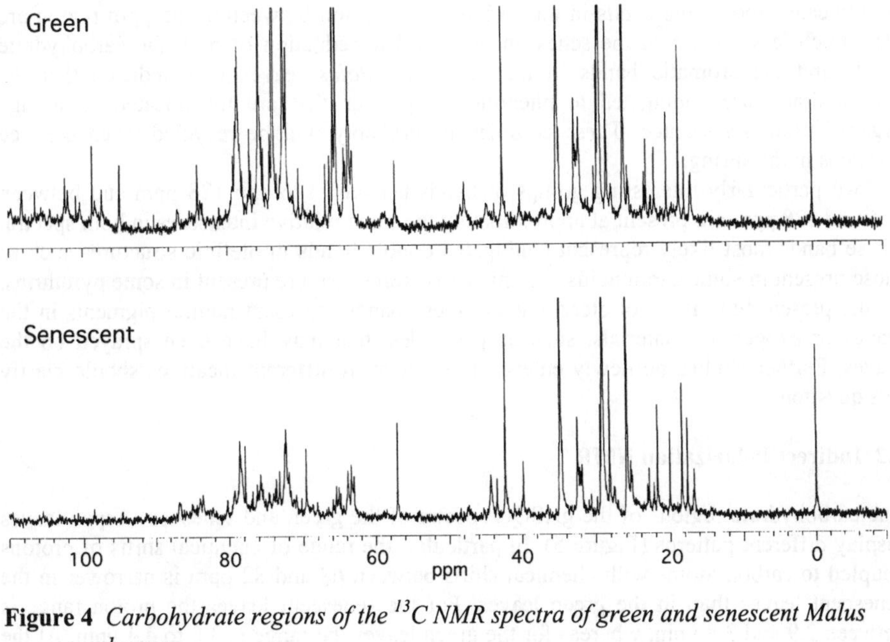

Figure 4 *Carbohydrate regions of the ^{13}C NMR spectra of green and senescent Malus leaves*

The reduction in intensities of the carbohydrate bands of the senescent *Malus* leaves indicates that many of the soluble carbohydrates have moved out of the leaves into storage organs. The soluble carbohydrates that have remained in the senescent leaves appear to be those that do not have anomeric carbon atoms such as sugar alcohols or inositols. In previous studies we have found evidence for these types of compounds in leaf leachates.[6,16] In the Aspen leaves it appears that some of the carbohydrates have been transported out of the senescent leaves; however, new carbohydrate species are probably also present in the senescent leaves. Further insight into the changes that the mobile carbohydrates in the Aspen leaves have undergone during senescence is provided by the gHSQC spectra (see section 3.2).

The spectra of the green and senescent Aspen leaves in the aromatic region (100-160 ppm) are similar (Figure 1). The most notable differences involve the bands between 120 and 135 ppm. There are two well-resolved bands at 129 and 131 ppm in the spectrum of the green leaves that are absent in the spectrum of the senescent leaves. On the other hand, in the region between 122 and 126 there are three well-resolved bands in the senescent leaf spectrum and only one broad band in this region in the green leaf spectrum. These bands in the green and senescent leaves occur in the spectral region of the C-1 and C-6 carbon atoms of guaiacylpropanoid lignin units. Hydrolysis of the ether groups in these lignin units has been found to reduce the chemical shifts of the two carbons by 2 or 3 ppm.[17] It is unlikely that these bands represent lignin units because of the restricted motion of the lignin polymer. However, one of the rings in many nonhydrolyzable tannins also possess the 1,2-diphenol structure.

The spectra of the green and senescent *Malus* leaves show more change than was observed in the Aspen spectra. Almost all of the bands between 95 and 180 ppm in the green leaves have disappeared in the senescent leaves with the exception of the bands between 126 and 138 ppm and weak bands at 123 and 144 ppm (Figure 2). A few other very weak bands may also be present in the 95-180 ppm region of the senescent spectrum. As indicated above, the bands in the carbohydrate region between 60-95 ppm region are also much less intense in the senescent leaves. The reduction of both the carbohydrate bands and the aromatic bands in the senescent *Malus* leaves may indicate that the carbohydrates are conjugated to phenolic compounds that are translocated to storage organs during senescence. These components probably will be recycled when the tree awakens in the spring.

Two particularly interesting groups of bands between 126 and 128 ppm and between 136 and 138 ppm are present at approximately the same relative intensities in both spectra. These bands most likely represent conjugated double bonds in olefinic structures such as those present in some carotenoids.[18] Similar structures also are present in some pyrethrins. At the present time it is not clear whether these bands represent natural pigments in the leaves or exogenous materials, such as pesticides, that may have been sprayed on the leaves. Further studies on newly emergent leaves from different locations should clarify this question.

3.2 Indirect Polarization NMR

The carbohydrate regions of the gHSQC spectra of the green and senescent Aspen leaves display different patterns (Figure 5). In particular, the range of chemical shifts of protons coupled to carbon atoms with chemical shifts between 68 and 82 ppm is narrower in the senescent leaves than in the green leaves. For the senescent leaves the proton range is between 2.9 and 3.8 ppm, whereas for the green leaves the range is 3.0 to 4.4 ppm. At the present time it is not clear why these differences in the range of proton chemical shifts arise. One possibility is that the wider range of proton shifts in the green leaves results from the presence of more complex polymeric structures in the green leaves than in the senescent leaves. Another possibility is that a wider variety of carbohydrate polymers is present in the green leaves than in the senescent leaves. We intend to continue this work with Aspen leaves from other environments and with leaves of other tree species in order to determine if similar differences are observed in their 2-D spectra. Experiments with known compounds will also be undertaken.

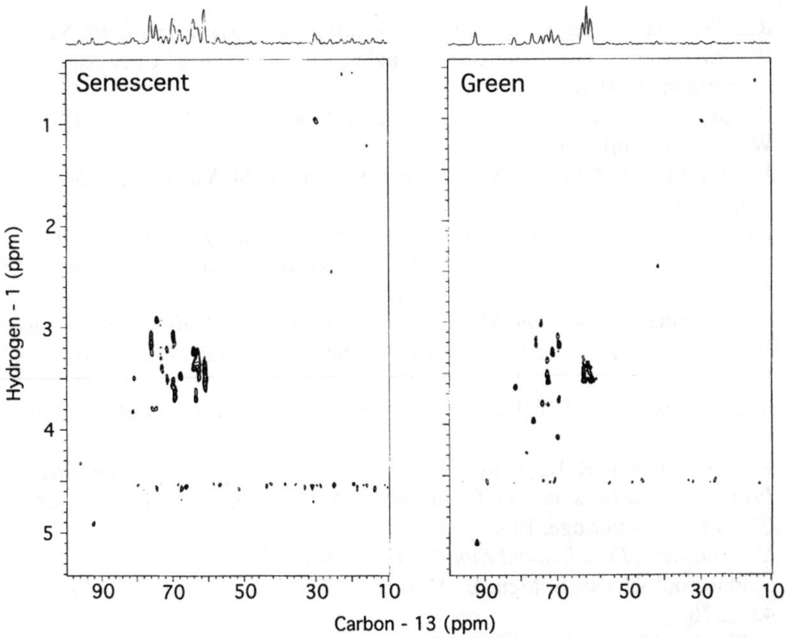

Figure 5 *Carbohydrate regions of the gHSQC spectra of green and senescent Aspen leaves*

4 CONCLUSIONS

High-resolution NMR spectra of the mobile components of leaf tissue before and after senescence were measured *in situ* with a new type of magic-angle-spinning probe. The NMR spectra of green and senescent Aspen (*Populus tremuloides*) and *Malus* 'Manbeck Weeper' leaves were compared. Fewer differences were observed between the spectra of the green and senescent leaves of both species in the aliphatic regions (15 to 50 ppm) of the spectra than in any of the other regions. These results indicate that the waxy material that coats the outer surfaces of the leaves undergoes little change during senescence. The carbohydrate region (60 to 105 ppm) and aromatic region (100 to 160 ppm) of both sets of spectra showed more pronounced differences between the green and senescent leaves. Further work will be necessary in order to understand the significance of these spectral differences. The results of these new studies should provide insight into the first reactions that the mobile leaf components undergo during the humification process.

ACKNOWLEDGEMENTS

Some of the spectra used in this study were measured by Professor Charles Mayne of the Chemistry Department at the University of Utah.

References

1. R.L. Wershaw, in *Humic Substances-Versatile Components of Plants, Soils and Waters*, eds. E.A. Ghabbour and G. Davies, Royal Society of Chemistry, Cambridge, 2000, p. 1.
2. F.J. Stevenson, *Humus Chemistry. Genesis, Composition, Reactions*, 2nd Edn., Wiley, New York, 1994.
3. J.A. Baldock, J.M. Oades, A.G. Waters, X. Peng, A.M. Vassallo and M.A.Wilson, *Biogeochem.*, 1992, **16**, 1.
4. P.G. Hatcher and E.C. Spiker, in *Humic Substances and their Role in the Environment*, eds. F.H. Frimmel and R.F. Christman, Wiley, Chichester, 1988, p. 59.
5. R.L. Wershaw, *Membrane-Micelle Model for Humus in Soils and Sediments and Its Relation to Humification*, U.S. Geological Survey Water-Supply paper 2410, 1994.
6. R.L. Wershaw, J.A. Leenheer, K.R. Kennedy and T.I. Noyes, *Soil Sci.*, 1996, **161**, 667.
7. R.L. Wershaw, K.R. Kennedy and J.E. Henrich, in *Humic Substances-Structures, Properties and Uses*, eds. G. Davies and E.A. Ghabbour, Royal Society of Chemistry, Cambridge, 1998, p. 29.
8. Y.Y. Leshem, *Free Radical Biol. Medic.*, 1988, **5**, 39.
9. S. Philosoph-Hadas, S. Meir, B. Akiri and J. Kanner, *J. Agric. Food Chem.*, 1994, **42**, 2376.
10. S. Strother, *Gerontology*, 1988, **34**, 151.
11. R. Aerts, *J. Ecology*, 1996, **84**, 597.
12. K.T. Killingbeck, *Ecology*, 1996, **77**, 1727.
13. P.A. Keifer, *Drugs of the Future*, 1998, **23**, 301.
14. A.L. Davis, J. Keeler, E.D. Laue and D. Moskau, *J. Magnetic Reson.*, 1992, **98**, 202.
15. E. Lichtfouse, C. Chenu, F. Baudin, C. LeBlond, M. da Silva, F. Behar, S. Derenne, C. Largeau, P. Wehrung and P. Albrecht, *Org. Geochem.*, 1998, **28**, 411.
16. R.L. Wershaw, J.A. Leenheer and K.R. Kennedy, in *Humic Substances-Structures, Properties and Uses*, eds. G. Davies and E.A. Ghabbour, Royal Society of Chemistry, Cambridge, 1998, p. 47.
17. H.Y. Hassi, M. Aoyama, D. Tai, C.-L. Chen and J.S. Gratzl, *J. Wood Chem. Tech.*, 1987, **7**, 555.
18. E. Breitmaier and W. Voelter, *Carbon-13 NMR Spectroscopy*, VCH, Weinheim, 1987, p. 335.

CARBON GROUP CHEMISTRY OF HUMIC AND FULVIC ACID: A COMPARISON OF C-1s NEXAFS AND ¹³C-NMR SPECTROSCOPIES

A.C. Scheinost,[1] R. Kretzschmar,[1] I. Christl[1] and Ch. Jacobsen[2]
[1] Department of Environmental Sciences, ETH Zurich, 8952 Schlieren, Switzerland
[2] Department of Physics and Astronomy, SUNY, Stony Brook, NY 11794, USA

1 INTRODUCTION

Bulk analysis of natural organic matter (NOM) may be performed with methods such as ^{13}C NMR, fluorescence and FTIR spectroscopies, and pyrolysis coupled with mass spectrometry.[1,2] However, it is increasingly recognized that the role of NOM in natural systems is not sufficiently explained by bulk chemical properties, but requires knowledge of chemical and physical structures at the microscopic level.[3] Therefore, attempts have been made to investigate the internal structure of organic colloids in marine, fresh and waste water by so-called correlative microscopy, a combination of optical, scanning confocal laser, transmission electron and atomic force microscopies.[4-6] Spatially resolved spectroscopic investigations of coal and colloids have been performed with infrared and fluorescence micro-spectroscopies, and with laser ionisation/desorption coupled with pyrolysis mass spectrometry at resolutions ≥ 1 μm.[7-11]

An alternative method, scanning transmission X-ray microscopy (STXM), may be ideally suited to study the physical and chemical structure of NOM.[12] Firstly, the theoretical optical resolution is between that of light and electron microscopes due to the use of X-rays with wavelengths around 4 Å.[13] The zone plates (Fresnel lenses) used to focus the X-ray beam currently limit the effective resolution to >30 nm.[14] This optical resolution covers a major part of the particle size range of colloids (10^{-9} to 10^{-6}m). Secondly, X-rays with energies between the oxygen 1s and the C 1s edges are able to penetrate a thin film of water without substantial attenuation. Hence, soil particles can be investigated without dehydration, which leaves the chemical and physical structure of soil organic matter (SOM) unaltered.[15] Thirdly, by connecting a tuneable monochromator to the STXM, near-edge X-ray absorption fine structure (NEXAFS) spectra at the C1s-edge can be collected with spatial resolution. Electronic transitions from 1s to π* antibonding orbitals and to mixed Rydberg/valence states give rise to strong bands in the range from 285 to 291 eV that provide information on the covalent bonds of carbon and the nearest neighbors.[16]

C-NEXAFS has been employed to study biopolymers like amino acids and peptides,[17,18] coal[19,20] and humic substances.[21] In these studies, the functional group chemistry of samples with unknown composition has been inferred from reference spectra in a more or less qualitative way by comparing band heights and shapes. Therefore, the intention of our study was to investigate if NEXAFS spectra of NOM can be interpreted more quantitatively. We used a fulvic acid and humic acid size fractions whose functional group composition had been determined by CPMAS ^{13}C NMR spectroscopy.[22]

2 MATERIALS AND METHODS

2.1 Humic and Fulvic Acid Samples

Humic acid (PUHA) and fulvic acid (PUFA) were isolated from a Humic Gleysol in Northern Switzerland as recommended by the International Humic Substances Society[23] and purified by dialysis. Subsequently, the humic acid was separated into four size fractions using a cross-flow hollow-fibre ultrafiltration technique.[22] For quantification of chemical carbon groups, solid-state ^{13}C NMR spectra of the freeze-dried samples were recorded on a Bruker DSX 200 NMR spectrometer at a resonance frequency of 50.3 MHz and with a cross-polarization magic-angle spinning rate of 6.8 kHz.[22]

2.2 NEXAFS Spectroscopy

C1s-NEXAFS was conducted at beamline X-1A of the National Synchrotron Light Source, Brookhaven National Laboratory, Upton, NY using the STXM endstation.[13] The essential components of the STXM are a tuneable undulator, which is inserted in the 2.5 GeV electron storage ring to generate a high flux of photons (10^9 photons/s) in the soft X-ray region, a spherical grating monochromator with an energy resolution of 0.3 eV, a 160-μm Fresnel zone plate with a nominal spatial resolution of 30 nm and a proportional counter to detect the transmitted photons.[14] The monochromator was calibrated using the second-strongest absorption band of CO_2.

Films of FA and HAs were prepared by drying a 1-μL droplet of aqueous suspension on a silicon nitride window (100 nm thick). Sample spectra were recorded through these films and the silicon nitride window from 277 to 310 eV in steps of 0.3 eV. I_o spectra were taken through a sample-free region of the same silicon nitride window. Only spectra with at least 1 % transmittance of the incoming beam were used to avoid spectral distortions.

As references, acetic, malonic, phthalic, salicylic, gallic acids and catechol (all ACS reagent grade) were sorbed at pH 5 on freshly precipitated ferrihydrite gel using 2g/L ferrihydrite suspensions and initial organic ligand concentrations of 10 mM. After 1 hour, > 99% of the organic ligands were removed from solution (Shimadzu TOC analyser), forming aggregates > 1 μm in diameter (optical microscope, STXM). About 1 μL of this suspension was inserted between two silicon nitride windows to achieve a capillary water film of about 1 μm in thickness.[24] Sample spectra were then collected through the C-rich ferrihydrite aggregates, while I_0 spectra were taken through the C-poor solution between the ferrihydrite aggregates.

3 RESULTS AND DISCUSSION

3.1 Qualitative Analysis of Carbon Groups

Depending on their composition, the NEXAFS spectra of the investigated reference compounds showed bands indicative of aromatic, phenolic and carboxyl C groups (Table 1). The positions of these bands correspond well with band positions derived from electron energy-loss spectroscopy (EELS) (Table 1). Because of the consistency of the two methods, published EELS spectra were used to establish energy ranges for alkyl and O-alkyl groups (Table 1). With the exception of a slight overlap of band positions for alkyl and carboxyl groups, characteristic band positions are sufficiently separated to allow for a discrimination of the carbon groups.

Table 1 *Energies (eV) of C1s π* and Rydberg transitions (C-H) of reference compounds and their assignment*

	aromat.	*phenol*	*C-H*	*COOH*	*COH*	*Method*	*Ref*
Benzene	285.2					EELS	25
Phenol	285.2	287.1				EELS	26
Hydroquinone	285.2	287.2				EELS	26
Ethane			287.9			EELS	27
Propane			287.7			EELS	28
Butane			287.8			EELS	28
Pentane			287.6			EELS	28
Hexane			287.4			EELS	28
Formic acid				288.2		EELS	29
Acetic acid				288.6		EELS	29
Propionic acid				288.6		EELS	30
Methanol			287.9		289.4	EELS	30
Ethanol			288.3		289.5	EELS	31
Propanol			288.1		289.5	EELS	30
Phthalic acid	285.0			288.4		XAFS	this work
Salicylic acid	285.0			288.3		XAFS	this work
Gallic acid	285.0	287.2		288.5		XAFS	this work
Catechol	285.3	287.0				XAFS	this work
Acetic acid				289.0		XAFS	this work
Malonic acid				288.6		XAFS	this work
Mean	285.1	287.1	287.8	288.5	289.5		
Min	285.0	287.0	287.4	288.2	289.4		
Max	285.3	287.2	288.3	289.0	289.5		

NEXAFS spectra of PUFA and PUHA samples and their size fractions are shown in Figure 1 together with band regions derived from Table 1. The spectra show three well resolved peaks between 284 and 290 eV.

The spectra were deconvoluted using an arctangent function for the ionisation step at 290 eV, four Gaussian functions for the π* transitions below the ionisation energy and two Gaussian functions for the σ* transitions above the ionisation energy (Figure 2). Bands 1, 3 and 4 match the regions of aromatic, carboxyl and O-alkyl groups, respectively, indicating the existence of these groups in the samples (Table 1). Band 2 is ca. 0.4 eV

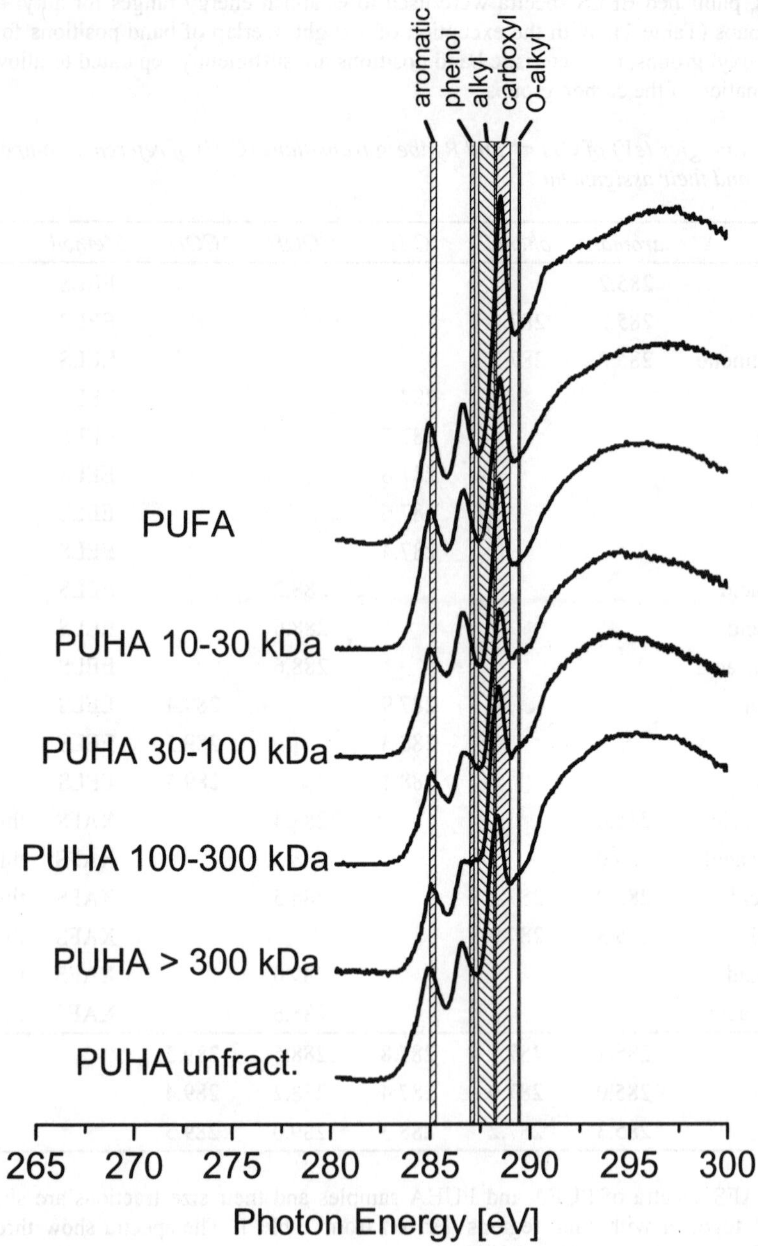

Figure 1 *NEXAFS spectra of thin films of PUFA and PUHA and of PUHA size fractions*

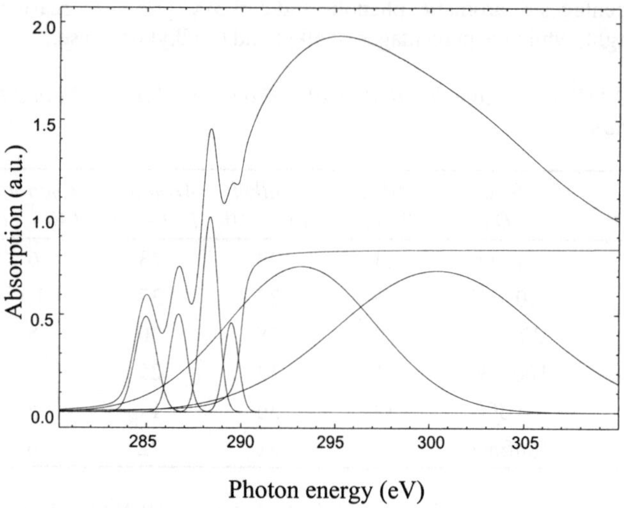

Figure 2 *Deconvolution of NEXAFS spectra*

Table 2 *Band positions [eV] of Gaussian bands 1 to 6 (see Figure 2)*

	1	*2*	*3*	*4*	*5*	*6*
PUFA (unfract.)	285.0	286.6	288.5	289.6	295.1	302.2
PUHA 1 (10 – 30 kDa)	285.0	286.7	288.5	289.6	294.0	301.9
PUHA 2 (30 – 100 kDa)	285.0	286.7	288.5	289.5	293.8	301.8
PUHA 3 (100 – 300 kDa)	285.1	286.8	288.4	289.5	293.2	300.8
PUHA 4 (> 300 kDa)	285.1	286.8	288.4	289.4	292.8	300.5
PUHA 0 (unfract.)	285.0	286.7	288.3	289.5	293.2	300.4
Mean	285.0	286.7	288.4	289.5	293.7	301.3
Min	285.0	286.6	288.3	289.4	292.8	300.4
Max	285.1	286.8	288.5	289.6	295.1	302.2

below the energy range of phenolic groups, indicating a slightly different chemical environment. Although NMR spectra indicated the presence of alkyl groups, the corresponding Rydberg transition in the energy range 282.4 to 288.3 eV could not be reliably fitted. This may be due to a relatively weak band intensity as compared to the carboxyl π^* transition. The two σ^* transitions (bands 5 and 6) show a systematic shift towards higher energy with decreasing molecular size (Table 2).

3.2 Quantitative Analysis of Carbon Groups

Table 3 shows that the unfractionated PUFA and PUHA samples have surprisingly similar contents of aromatic, phenolic, carboxyl and O-alkyl groups, while PUHA has more alkyl groups. However, there is a substantial difference between the size fractions of the humic acid. The percentages of aromatic, phenolic and carboxyl groups increase with decreasing molecular weight, while the percentages of alkyl and O-alkyl decrease.[22]

Table 3 *^{13}C-NMR percentage distribution of carbon in PUFA, PUHA and PUHA size fractions*[22]

	Size kDa	*Alkyl 0 - 45*	*O-alkyl 45 – 110*	*Aromatic 110 – 160*	*Phenolic 140 – 160*	*Carboxyl 160 – 185*
PUFA	unfract.	24	29	23	6	18
PUHA 1	10 – 30	20	22	37	10	16
PUHA 2	30 - 100	20	28	30	8	17
PUHA 3	100 - 300	21	33	25	7	14
PUHA 4	> 300	33	36	17	5	12
PUHA 0	unfract.	28	31	22	6	15

Table 4 shows the correlation coefficients between NEXAFS band areas and the percentage of carbon in the different groups as determined by NMR. The areas of bands 1 and 2 are significantly ($p<0.05$) correlated with both aromatic and phenolic groups (Figure 3). The cross-correlations (area of band 1 to phenolic groups; area of band 2 to aromatic groups) are due to the high correlation between aromatic and phenolic groups ($r=0.994$) which, in turn, indicates a small variation of the number of phenolic groups per aromatic ring structure.

Furthermore, positive correlations exist between the area of band 3 and the percentage of carboxyl groups, and between the area of band 4 and the percentage of O-alkyl groups ($p<0.15$). As a substitute for the not fittable Rydberg/alkyl band, we used the total area of σ^* bands (bands 5 and 6), which showed a positive correlation ($p<0.12$) with alkyl groups.

In spite of a relatively large scattering of the correlation plots, the systematic changes of aromatic, phenolic, carboxyl and O-alkyl groups with the molecular size of the humic acid fractions are properly detected by NEXAFS, revealing that the method is well suited for the quantitative analysis of carbon groups in NOM (Figure 3).

Table 4 *Correlation coefficients between percentage chemical groups from ^{13}C-NMR (horizontal) and areas of NEXAFS bands (vertical)*

	Aromatic	*Phenolic*	*Carboxyl*	*O-alkyl*	*Alkyl*
1	0.89	0.87	0.55	-0.94	-0.63
2	0.91	0.91	0.44	-0.91	-0.61
3	0.32	0.24	0.71	-0.50	-0.27
4	-0.56	-0.55	-0.24	0.57	0.30
5+6	-0.50	-0.49	-0.40	0.40	0.71

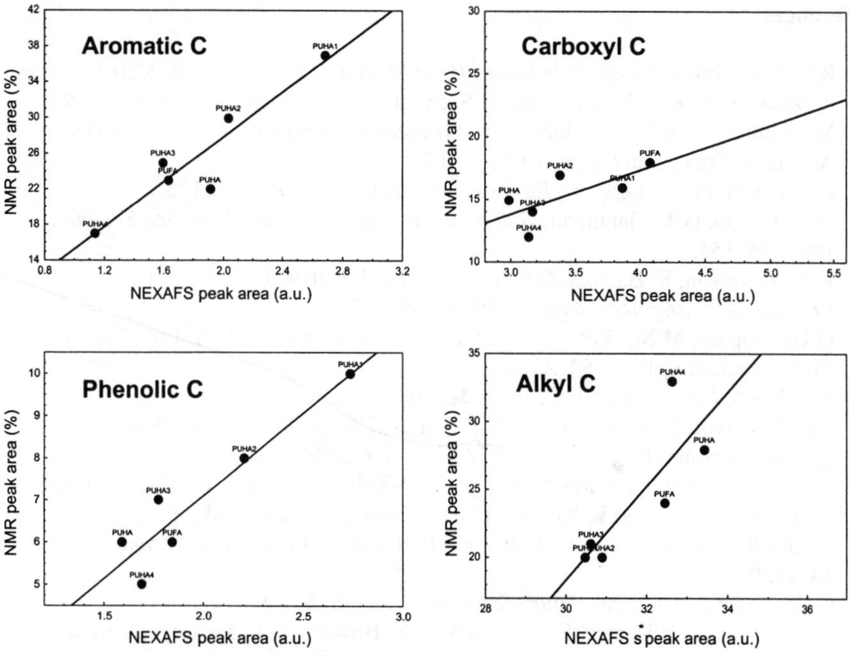

Figure 3 *Correlations between NMR and NEXAFS peak areas*

4 CONCLUSIONS

NEXAFS can be used to quantify the amounts of aromatic, phenolic, carboxyl and O-alkyl groups in NOM. The quantification of alkyl groups appears to be less reliable. Further efforts have to be made to improve the deconvolution, for instance by including a mixed Rydberg/valence band. Alternative approaches, for example by performing first principles calculations of NEXAFS spectra to study the influence of compositional changes will be tested.[32]

ACKNOWLEDGEMENTS

CPMAS ^{13}C NMR spectra were measured by H. Knicker, Technische Universität München. The group of Janos Kirz and Chris Jacobsen at SUNY Stony Brook developed the STXM at beamline X-1A, Brookhaven National Laboratory, with support from the Office of Biological and Environmental Research, US Department of Energy under contract DE-FG02-89ER60858 and the US National Science Foundation under grant DBI-9605045. Steve Spector and C. Jacobsen of Stony Brook and D. Tennant of Lucent Technologies, Bell Laboratories developed the zone plates with support from NSF under grant ECS-9510499.

References

1. R.L. Malcolm, in *Humic Substances II: In Search of Structure*, eds. M.H.B. Hayes, P. MacCarthy, R.L. Malcolm and R.S. Swift, Wiley, Chichester, 1989, p. 339.
2. M. Schnitzer and H.-R. Schulten, in *Advances in Agronomy*, ed. D.L. Sparks, Academic Press, San Diego, 1995, p. 167.
3. J. Buffle and G.G. Leppard, *Environ. Sci. Technol.*, 1995, **29**, 2176.
4. I.G. Droppo, D.T. Flannigan, G.G. Leppard and S.N. Liss, *Water Sci. Technol.*, 1996, **34**, 155.
5. K.J. Wilkinson, E. Balnois, G.G. Leppard and J. Buffle, *Coll. Surf. A - Physicochem. Engineer. Aspects*, 1999, **155**, 287.
6. G.G. Leppard, M.M. West, D.T. Flannigan, J. Carson and J.N.A. Lott, *Can. J. Fish. Aquat. Sci.*, 1997, **54**, 2334.
7. S.A. Stout, *Int. J. Coal Geol.*, 1993, **24**, 309.
8. P.J. Greenwood, H.F. Shaw and A.E. Fallick, *Clay Miner.*, 1994, **29**, 637.
9. M. Mastalerz and R.M. Bustin, *Int. J. Coal Geol.*, 1996, **32**, 55.
10. D. Srzic, S. Kazazic, S. Martinovic, L. Pasa-Tolic, N. Kezele, D. Vikic-Topic, S. Pecur, A. Vrancic and L. Klasinc, *Croat. Chem. Acta*, 2000, **73**, 69.
11. U. Ghosh, J.S. Gillette, R.G. Luthy and R.N. Zare, *Environ. Sci. Technol.*, 2000, **34**, 1729.
12. C. Jacobsen and J. Kirz, *Nature Struct. Biol.*, 1998, **5**, 650.
13. C. Jacobsen, S. Williams, E. Anderson, M.T. Browne, C.J. Buckley, D. Kern, J. Kirz, M. Rivers and X. Zhang, *Optics Commun.*, 1991, **86**, 351.
14. S. Spector, C. Jacobsen and D. Tennant, *J. Vac. Sci. Technol. B*, 1997, **15**, 2872.
15. S.C.B. Myneni, J.T. Brown, G.A. Martinez and W. Meyer-Ilse, *Science*, 1999, **286**, 1335.
16. J. Stöhr, in *NEXAFS Spectroscopy*, eds. G. Ertl, R. Gomer and D.L. Mills, Springer-Verlag, 1992.
17. H. Ade, X. Zhang, S. Cameron, C. Costello, J. Kirz, and S. Williams, *Science*, 1992, **258**, 972.
18. J. Boese, A. Osanna, C. Jacobsen and J. Kirz, *J. Electr. Spectrosc. Rel. Phenom.*, 1997, **85**, 9.
19. G.D. Cody, R.E. Botto, H. Ade, S. Behal, M. Disko and S. Wirick, *Energy Fuels*, 1995, **9**, 75.
20. G.D. Cody, R.E. Botto, H. Ade, S. Behal, M. Disko and S. Wirick, *Energy Fuels*, 1995, **9**, 525.
21. J. Rothe, M.A. Denecke and K. Dardenne, *J. Coll. Interface Sci.*, 2000, **231**, 91.
22. I. Christl, H. Knicker, I. Kögel-Knabner and R. Kretzschmar, *Eur. J. Soil Sci.*, 2000, **51**, 617.
23. R.S. Swift, in *Methods of Soil Analysis: Part 3, Chemical Methods*, eds. D.L. Sparks, J.M. Bartels and J.M. Bigham, SSSA, Madison, WI, 1996, p. 1018.
24. U. Neuhäusler, PhD Thesis, Georg-August-Universität, Göttingen, 1999.
25. A.P. Hitchcock, P. Fischer, A. Gedanken and M.B. Robin, *J. Phys. Chem.*, 1987, **91**, 531.
26. J.T. Francis and A.P. Hitchcock, *J. Phys. Chem.*, 1992, **96**, 6598.
27. I. Ishii, R. McLaren, A.P. Hitchcock, K.D. Jordan, Y. Choi and M.B. Robin, *Can. J. Chem. - Rev. Canad. Chimie*, 1988, **66**, 2104.
28. A.P. Hitchcock and I. Ishii, *J. Electr. Spectrosc. Rel. Phenom.*, 1987, **42**, 11.
29. I. Ishii and A.P. Hitchcock, *J. Chem. Phys.*, 1987, **87**, 830.
30. I. Ishii and A.P. Hitchcock, *J. Electr. Spectrosc. Rel. Phenom.*, 1988, **46**, 55.

31. A.P. Hitchcock and D.C. Mancini, *J. Electr. Spectrosc. Rel. Phenom.*, 1994, **67**, 1.
32. V. Caravetta, O. Plashkevych and H. Ågren, *J. Chem. Phys.*, 1998, **109**, 1456.

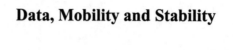

Data, Mobility and Stability

ASPECTS OF MEASUREMENT OF THE HYDRODYNAMIC SIZE AND MOLECULAR MASS DISTRIBUTION OF HUMIC AND FULVIC ACIDS

Manfred Wolf,[1a] Gunnar Buckau,[2] Horst Geckeis,[2] Ngo Manh Thang,[2] Enamul Hoque,[1a] Wilfried Szymczak[1b] and Jae-Il Kim[2]

[1] GSF- National Research Center for Environment and Health [a]Institute of Hydrology, [b]Institut of Radiation Protection, 85764 Neuherberg, Germany
[2] Forschungszentrum Karlsruhe, Institut für Nukleare Entsorgung, 76021 Karlsruhe, Germany

1 INTRODUCTION

Humic and fulvic acids represent operationally defined fractions of humic substances that are ubiquitous in soils, sediments and natural waters.[1] Chemically, they consist of a complex heterogeneous mixture of mainly polycarboxylic/polyhydroxycarboxylic acids of different aromaticity with unknown or only partly known structures and high polydispersity. Humic substances (HSs) are able to complex radionuclides and heavy metals and can also interact with toxic organic substances like pesticides.[1] These properties of humic and fulvic acids are important with respect to the migration potential of pollutants in groundwater. To get insight into chemical processes responsible for the formation and stability of metal complexes with HSs, determination of molecular mass and size distributions of humic and fulvic acids is essential.

Molecular size or mass distribution of humic substances can be measured by various methods that are based on different principles. Methods based on the measurement of hydrodynamic size are, for example, size exclusion chromatography (SEC),[2-5] high performance size exclusion chromatography (HPSEC)[6-11] and flow field-flow fractionation (FFFF).[12-16] Methods for determination of mass distribution are, for example laser desorption ionization - Fourier-transform mass spectrometry (LDI-FTMS),[17] LDI FT-ion cyclotron resonance (ICR) MS,[18] infrared LD MS,[19] electrospray-ionization (ESI) MS,[20] FT-ICRMS,[18,21] ESI quadrupole time-of-flight (Q-TOF) MS,[22] matrix assisted LDI-TOF MS[23] and time-of-flight secondary ion mass spectrometry (TOF-SIMS).[24]

A large number of data on molecular masses/sizes of humic substances have been published. These results range from about some hundred to more than a million Da.[1] HSs from different origins and/or geochemical environments may have different molecular mass/size distributions. However, published values of molecular mass/size for identical humic or fulvic acids vary considerably. Therefore, methodological problems are involved in generating such contradicting data. Some discrepancies may result from different experimental conditions and the calibration standards used, differences in data evaluation and HSs association, agglomeration or micelle generation.[4,8,9,25-27] The aim of this paper is to discuss various aspects of measurement of hydrodynamic size and mass distributions of HSs in connection with our recent results measured by HPSEC, organic-aqueous SEC, FFFF and TOF-SIMS to get a better understanding of molecular mass/size distributions of humic and fulvic acids.

2 MATERIALS AND METHODS

2.1 Reagents

All chemicals and organic solvents were purchased from commercial sources (Merck, Aldrich, Fluka) with the highest purity available and were used without further purification. Aqueous solutions were prepared with high purity water (Milli-Q_{PLUS}, Millipore).

2.2 Humic Substances

The Suwannee River fulvic acid standard (SR(FA)) was purchased from the International Humic Substances Society (IHSS). Aldrich humic acid (Aldrich(HA)) was purchased from Aldrich as the sodium salt and used after purification.[28] Fulvic acid (Dach(FA)) from a peat bog in the Dachauer Moos (Bavaria, Germany) was extracted from water.[10,11] Two fulvic acids (Gohy-73(FA) and Gohy-573(FA)) and two humic acids (Gohy-73(HA) and Gohy-573(HA)) were extracted from groundwaters. The original groundwater (Gohy-2227 Groundwater) of the Gorleben aquifer (Lower Saxony, Germany) was used after filtration through a 0.45 µm membrane filter.[2,29] Fulvic acid (Kranichsee(FA)) and humic acid (Kranichsee(HA)) were extracted from a boggy lake (Kleiner Kranichsee) in southern Saxony, Germany.[30] Fulvic acid (Derwent(FA)) was extracted from groundwater of the Derwent Reservoir (Derwent, UK).[31] Original porewater (Boom Clay Porewater) from the Boom clay formation (Mol, Belgium) was used after filtration through a 0.45 µm membrane filter.

2.3 HPSEC and Organic-Aqueous SEC

Size exclusion chromatography and high performance size exclusion chromatography are methods for the measurement of molecular size distribution. The analyte, dissolved in an eluent, flows or is pumped through a column filled with a gel or porous matrix with a well defined pore size depending on the fractionation range. In the ideal case (no effects other than size exclusion) analyte molecules with hydrodynamic diameters larger than the pore diameter can not diffuse into the pores (they are excluded) and are transported faster than analyte molecules that have hydrodynamic diameters smaller than the pores. These smaller molecules can diffuse into the pores of the gel and are therefore retarded depending on their hydrodynamic size. If all the analyte molecules fall in the fractionation range of the porous material, then the space in the gel for the individual analyte molecules is inversely proportional to the respective molecular volume and therefore the smallest molecules elute the latest. It is very important in SEC or HPSEC to suppress non-size exclusion effects like ion exclusion (faster transport resulting in overestimation of the molecular size) or sorption (retarded transport resulting in underestimation of the molecular size). This requires using sufficient ionic strength and/or the use of additives such as methanol.

HPSEC analysis of FAs was carried out with a GFC 300-8 column (hydrophilic gel surfaces, length 300 mm, inner diameter 7.7 mm, particle size 8 µm, pore size 30 nm, Macherey-Nagel, Germany) connected to a Hewlett-Packard HPLC system (Model HP 1090). The mobile phase was 5 mM Na_2HPO_4 + 0.1% MeOH (pH 7.0; final ionic strength 13 mM). The void volume (V_0) was determined with blue dextran (M_r ca. 2 x 10^6 Da) (M_r: molecular mass) or polystyrene sulfonate (PSS) standard (M_p 780 kDa) (M_p: molecular mass distribution maximum) and the total permeation volume (V_p) with KNO_3.[11,32] For calibration of the column the polycarboxylic acids maleic acid (M_r 134.09), benzene-1,3-

dicarboxylic acid (166.13), citric acid (192.13), benzene-1,2,4,5-tetracarboxylic acid (254.15), EDTA (292.25), DETPA (393.35) and polyacrylic acid (PAA) standards M_p 1250 and 2925 Da were used.[10,11] The UV absorption (λ = 240 nm) was acquired on-line using a Hewlett-Packard photodiode array detection (DAD) system. For analysis a 20 µL sample equivalent to 0.4 - 1 µg humic substance was injected.

Organic-aqueous SEC analysis of methylated or non-methylated FAs and HAs was carried out on Sephadex-L-60-120 gel using 50% and 90% N-methylformamide (NMF) in 0.1 M $NaClO_4$ as a mobile phase. The SEC column was calibrated with globular protein standards albumin (M_r 66 kDa), carbonic anhydrase (29 kDa), cytochrome c (12.4 kDa) and aprotinin (6.5 kDa). The void volume (V_0) was determined with blue dextran (M_r ca. 2 x 10^6 Da) and the total permeation volume (V_p) with benzyl alcohol (108.14 Da). Quantitative methylation of carboxylic groups in fulvic and humic acids was confirmed by IR and ^{13}C-NMR spectroscopy.[2]

2. 4 Flow Field-Flow Fractionation

Field-flow fractionation (FFF), first proposed in 1966 by Giddings,[12] is a technique for size fractionation of macromolecules and colloids. In this method the analyte, dissolved in a carrier medium, is pumped with constant velocity through a thin ribbon-like channel equipped with a membrane permeable for the carrier. The flow-field is established by a cross-flow perpendicular to the channel flow and the size fractionation is a function of the diffusion coefficient.

Flow FFF (FFFF) was carried out with the model F-1000 (FFFractionation Inc.) using a regenerated cellulose membrane (Schleicher and Schuell) with a cut-off of 5 kDa (related to globular proteins). As the mobile phase, 5 mM Tris-buffer (pH 9.1) was used at a constant channel flow of 1 mL min^{-1}. The cross-flow was in the case of polystyrene sulfonate (PSS) standards and humic substances 5 mL min^{-1}; in the case of globular protein standards it was 2 mL min^{-1}. Calibration of the FFFF channel was carried out with PSS standards (Polysciences) with nominal molecular masses 1.37 to 30.9 kDa and with globular protein standards (Sigma-Aldrich) carbonic anhydrase (M_r 29 kDa), bovine albumin (66 kDa), β-amylase (200 kDa), apoferritin (443 kDa) and bovine thyroglobulin (669 kDa). The absorbance of the effluent from the FFFF channel was recorded with a UV/vis detector (Lambda-Max LC Model 481, Waters) at 254 nm for HSs and 225 nm for PSS standards.[16]

2.5 Time-of-Flight Secondary Ion Mass Spectrometry

TOF-SIMS in combination with oblique 30 keV SF_5^+ ion bombardment[33] produces highly reproducible mass spectra of FA that cover more than five orders of magnitude in dynamic range without background subtraction. Spectra in the negative ionization mode are less affected by fragmentation than in the positive mode. Therefore spectra of FAs in the negative mode are presented.[24]

Solid samples for TOF-SIMS analysis were prepared by spray-deposition of FA in pure water (100 to 200 mg/L) onto cleaned and etched silicon substrates until complete coverage of the silicon. The samples were bombarded with a pulsed 25 to 30 keV SF_5^+ ion beam and the sample bias was ± (3.5 to 6) keV. Spectra were recorded over 0.5 to 1 x 10^7 primary ion pulses (beam current 1 to 1.5 nA, pulse width 7 ns, channel width 2 ns/bin, repetition rate 19 kHz), well within the limit of low-damage SIMS. The corresponding upper limit of detectable mass was m/z 7000.

3 RESULTS AND DISCUSSION

3.1 TOF-SIMS

TOF-SIMS spectra in the negative ion mode of three different fulvic acids (SR(FA), Dach(FA) and Derwent(FA)) are shown in Figure 1. The corrected yield vs. molecular mass calculated from the raw TOF-SIMS spectra[24] was converted into mass concentration and normalized to the mass integral (m/z 0 to 7000). The spectra are smoothed for better visualization of the general shape. The range 350 - 450 mass units is shown for the SR(FA) in the inset. Various fine structures can be seen in this unsmoothed original spectrum.

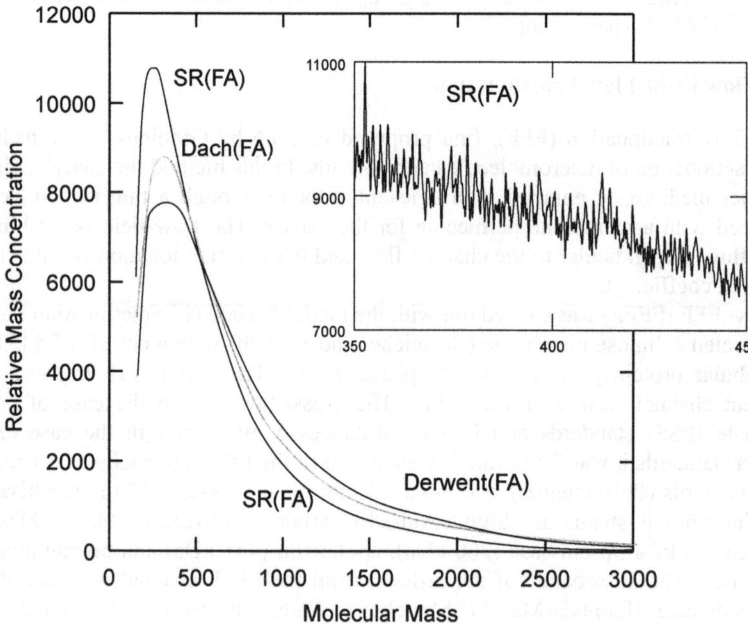

Figure 1 *Heavily smoothed TOF-SIMS spectra in the negative ion mode in the range of molecular mass 150 to 3000 for SR(FA), Dach(FA) and Derwent(FA). Part of original unsmoothed spectrum of SR(FA) is shown in the inset*

The molecular mass distributions of SR(FA), Dach(FA) and Derwent(FA) are slightly different from each other. The mass spectra show a broad maximum around molecular mass 300 to 350 Da with tailing up to at least a molecular mass of 3 kDa. In the lower mass region the spectra are characterized by high-yield atomic ions and fragments (Figure 2).[24] In order to evaluate the number and weight-averaged molecular masses (M_n and M_w), the background from fragmentation needs to be corrected for. This is done by two different approaches, namely (i) subtracting a spectrum from a high coverage sample target with very low coverage of the fulvic acid, and (ii) setting a mass-threshold that discriminates against small fragments. These methods are approximations and lead to slightly different results.

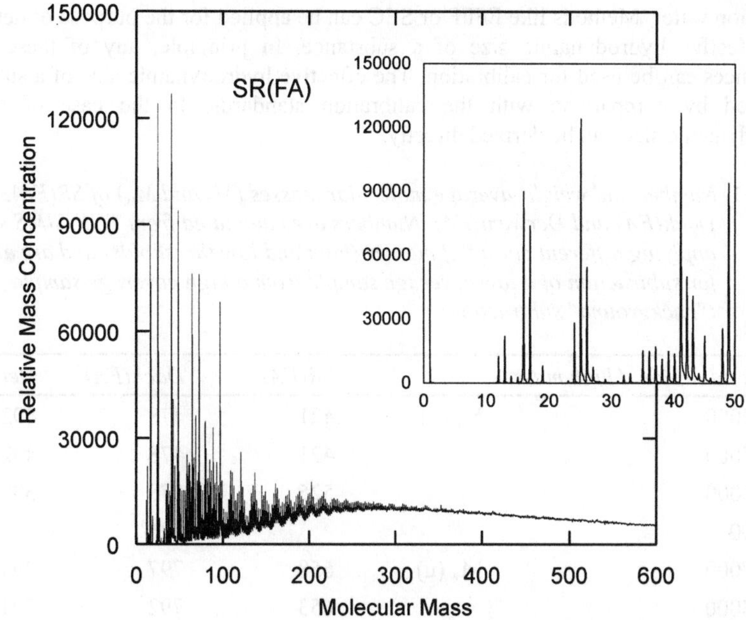

Figure 2 *TOF-SIMS spectrum in the negative ion mode of SR(FA) in the molecular mass range up to 600 Da. The range up to 50 is shown in the inset*

The results in Table 1 show that the contribution of masses between 5 and 7 kDa is negligible. Selection of a lower threshold value for the original spectra and subtraction of a "low coverage background" is not straightforward and the impact of fragmentation cannot be corrected for without some uncertainty.[24] Agreement within some 100 mass units is found for a threshold value of between m/z 150 and 250 from the original spectra and subtraction of "low coverage spectra." Dach(FA) and Derwent(FA) show very similar values, whereas the molecular mass distribution of SR(FA) shows lower values. Number and weight average molecular masses cluster around approximately 500 to 600 and 700 to 800 Da, respectively. These numbers are in good agreement with estimates from vapor pressure osmometry (VPO),[34-37] results of other mass spectrometrical methods such as infrared LD MS[19] and ESI quadrupole time-of-flight (Q-TOF) MS.[23] Similar values can also be obtained by HPSEC and organic-aqueous SEC measurements under certain conditions (see below).

3.2 Hydrodynamic Size

The effective hydrodynamic size of a molecule is a function of the mass, associated entities (especially water) and the overall geometry. The relationship between molecular mass and molecular size for globular proteins, dextran and polystyrene sulfonate is shown in Figure 3. A slope of 3 would result if the effective density (the effective size versus mass) does not change with increasing mass. As expected, globular proteins show a behavior close to such an ideal globular structure. The other standard curves have lower slopes and thus the effective size relative to mass increases with increasing mass. This shows deviation from globular shape, possibly in conjunction with increasing amounts of hydration water. Methods like FFFF or SEC can be applied for the purpose of determining the effective hydrodynamic size of a substance. In principle, any of these types of substances can be used for calibration. The effective hydrodynamic size of a substance is obtained by comparison with the calibration standards. In the case of FFFF the hydrodynamic size can be derived directly.

Table 1 *Number and weight-average molecular masses (M_n and M_w) of SR(FA), Dach(FA) and Derwent(FA). Numbers are calculated from TOF-SIMS spectra applying different cut-off of masses (high and low thresholds) and are also given for subtraction of a low coverage sample from a high coverage sample ("background" subtraction)*

Threshold (u) low / high mass		*SR(FA)*	*Dach(FA)*	*Derwent(FA)*
150 / 7000	M_n (u)	421	478	492
150 / 5000		421	478	492
250 / 7000		525	578	597
0 / 5000			606[a]	
150 / 7000	M_w (u)	658	797	794
150 / 5000		653	792	791
250 / 7000		738	870	864
0 / 5000			774[a]	

[a] Subtraction of a low coverage sample from a high coverage sample.

In Figure 3, the molecular mass and effective hydrodynamic size of Gohy-573(FA) in aqueous medium (pH 8.5, I=0.1 M) is also shown. Significantly different numbers are obtained depending on the experimental set-up. The effective hydrodynamic size of the Gohy-573 by HPSEC[42] is equivalent to the effective hydrodynamic size of PSS with a molecular mass of 1.4 kDa, dextran with a molecular mass of 3.1kDa and 5 kDa for the globular protein.

3.3 FFFF

Molecular size distributions derived from flow field-flow fractionation (FFFF) analysis[16] of Aldrich(HA), aqueous humic and fulvic acids and original ground- and porewater samples containing humic substances are presented in Table 2. The molecular sizes are related to polystyrene sulfonate (PSS) standards. The size distributions of these humic substances vary somewhat. The size distribution maxima for Aldrich(HA), Kranichsee(HA) and (FA) vary between 1.3 and 1.9 kDa_{pss} whereas Derwent(FA), the Gorleben-2227 groundwater and the Boom Clay porewater cluster around 1 kDa_{pss}, with tailing up to 5-10 kDa_{pss}. The higher numbers for M_n and M_w molecular size reflect the tailing towards larger molecules. The hydrodynamic size correlates with the molecular mass. The numbers obtained for size distribution maxima are very similar to the 1.4 kDa_{pss} found for Gohy-573(FA) by HPSEC. This shows that HPSEC and FFFF give comparable results.

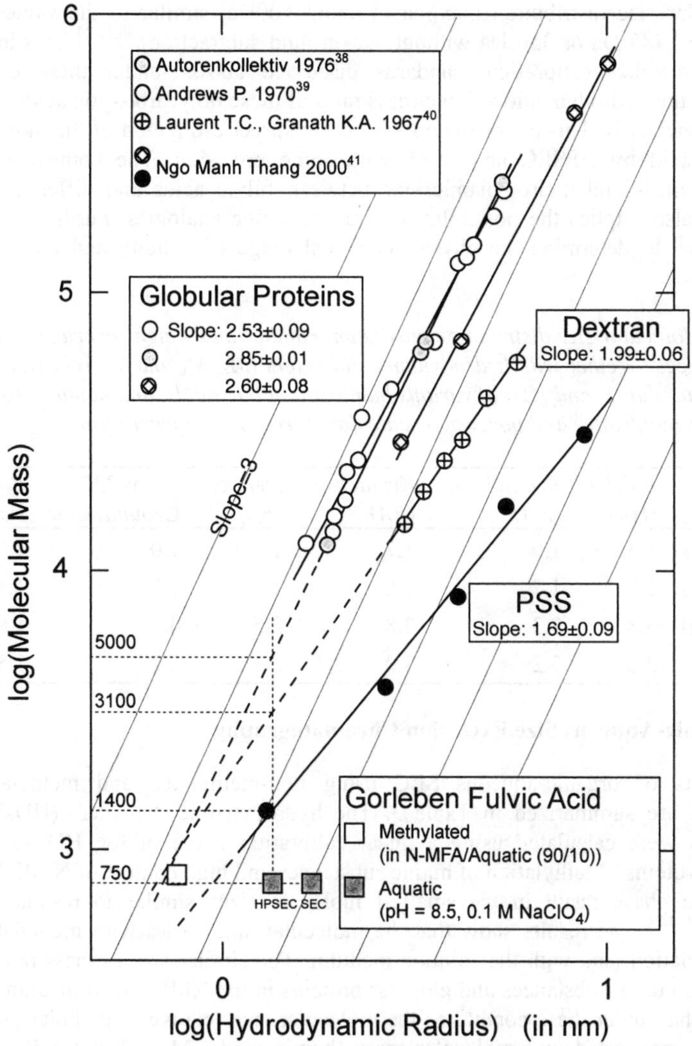

Figure 3 *Log (molecular mass) vs. log (hydrodynamic radius) for various frequently used calibration standards*[38-41] *and the Gorleben fulvic acid Gohy-573(FA) by SEC with Sephadex 100-120 and by HPSEC with Fractogel TSK HW-50. Numbers used for the fulvic acid under aquatic conditions are 750 Da for molecular mass and 1.7 nm (SEC) and 1.3 nm (HPSEC) for hydrodynamic radius. For the methylated fulvic acid in organic-aqueous medium a molecular mass of 820 Da and hydrodynamic radius of 0.75 nm is used (cf. Table 3, refs. 2 and 42, and Table 1 for comparison)*

3.4 HPSEC Calibrated with PAA and Simple Organic Molecules

The HPSEC calibration curve using various polycarboxylic acids and polyacrylic acid (PAA) standards is shown in Figure 4. Application of these standards to Dach(FA) results in molecular size distribution maxima of about 500Da, similar to the values found by TOF-SIMS (485 Da or 350 Da without background subtraction).[10,11,24] This implies that, contrary to other calibration standards discussed above, under these experimental conditions the hydrodynamic volume/mass ratio of these polycarboxylic acids is similar to that of fulvic acids. This opens the possibility for direct estimation of the molecular mass of fulvic acid by HPSEC and FFFF with some precision. The comparison between molecular mass and hydrodynamic size between fulvic acids and different calibration standards also implies the possibility of drawing some analogies. Further investigations are required to determine the physico-chemical range of validity and precision of this approach.

Table 2 *Molecular size distribution maximum, number and weight-average molecular sizes together with hydrodynamic diameters (M_p, M_n and M_w related to PSS standards, and HD) of various humic and fulvic acids, and humic substances of groundwater and porewater, calculated from FFFF fractograms*[16]

	Aldrich (HA)	*Kranichsee (HA)*	*Kranichsee (FA)*	*Derwent (FA)*	*Gohy-2227 Groundwater*	*Boom Clay Porewater*
M_p, (kDa)	1.3	1.9	1.7	1.0	1.0	1.0
M_n, (kDa)	1.7	1.7	1.6	1.4	1.8	1.1
M_w, (kDa)	3.3	3.3	2.8	2.5	4.1	1.8
HD (nm)	2.6	3.2	3.0	2.1	2.2	1.8

3.5 Organic-Aqueous Size Exclusion Chromatography

The results of organic-aqueous SEC using non-methylated and methylated humic substances are summarized in Table 3. The hydrodynamic diameters (HDs) of humic substances were calculated using a linear calibration curve of log HD vs. log M_r of globular proteins.[2] Methylation of humic substances and high content of N-MFA (90 %) in the mobile phase result in the smallest molecule sizes similar to results from other methods.[10,11,24] The results show that the molecular size varies with methylation versus non-methylation, and with the solution medium. The similar size to mass ratios between methylated humic substances and globular proteins in the NMF (90%) medium (Figure 3) indicates that under these conditions the molecular size relative to globular proteins may be directly converted into molecular mass (humic acids: M_p 1.2 and 1.38 kDa; fulvic acids: M_p 640 and 820 Da). The proton-exchange capacity shows that there are about 8 to 9 carbon atoms per carboxyl group, resulting in an average of 5 carboxyl groups per humic substance molecule.[2] After subtraction of 5 methyl groups per humic substance molecule, peak maxima of about M_p 570 and 750 Da for fulvic acids or 1130 and 1310 Da for humic acids are obtained.

3.6 Association

Regarding published results of high molecular sizes/masses of HSs, it is important to note that humic and fulvic acids can associate, agglomerate or can generate micelles under different conditions.[4,8,9,25-27] This is more pronounced for humic acid because it normally has a higher hydrophobicity than fulvic acids. Binding forces may be hydrophobic interactions, hydrogen bonding, van der Waals bonding or bridging via complexed metal

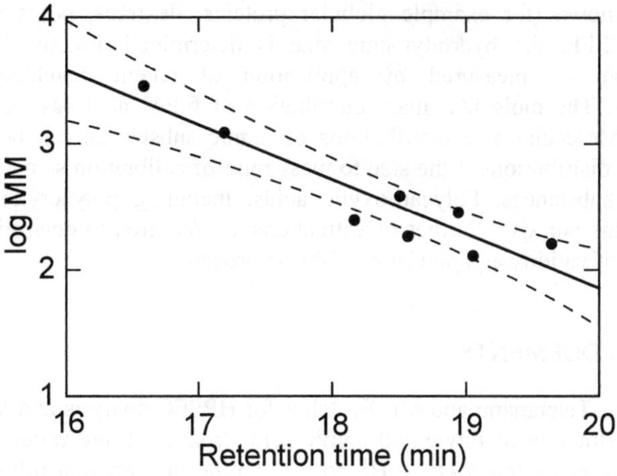

Figure 4 *HPSEC calibration curve using various polycarboxylic acids and polyacrylic acid (PAA) standards (confidence limit 95 %; $r^2 = 0.86$)*

Table 3 *Molecular sizes (M_p: equivalent to the size of globular proteins of the given mass) and hydrodynamic diameters (HD) of non-methylated and methylated humic and fulvic acids determined by organic-aqueous SEC (N-methylformamide (NMF) in 0.1 M $NaClO_4$; calibration with globular proteins)*[2]

	Mobile phase: 50% NMF				*Mobile phase: 90% NMF*			
Sample	*Non-methylated*		*Methylated*		*Non-methylated*		*Methylated*	
	M_p, Da	HD, nm	M_p, Da	HD, nm	M_p, Da	HD, nm	M_p, Da	HD, nm
Gohy-73 (HA)	11000	3.6	-	-	6900	3.1	1200 (1130)[a]	1.7
Gohy-73 (FA)	4300	2.6	1600	1.9	1500	1.9	640 (570)[a]	1.4
Gohy-573 (HA)	11000	3.6	-	-	8300	3.3	1380 (1310)[a]	1.8
Gohy-573 (FA)	4400	2.7	1300	1.8	1650	1.9	820 (750)[a]	1.5

[a] After subtraction of five methyl groups per molecule (see text)

ions. These processes can generate products with larger sizes and as a consequence measurement of their molecular size distribution results in higher values than those of the primary units (non-associated molecules). Also, the hydrodynamic size of the primary units of humic substances can vary due to changes in pH,[14] ionic strength,[25] type of metal loading[14] or concentration. Consequently, for the purpose of detecting primary units of humic substances it is essential to keep their concentration sufficiently low and to avoid high concentrations of protons, electrolytes and multivalent metal ions.

4 CONCLUSIONS

The hydrodynamic size of humic and fulvic acids can be measured by SEC, HPSEC and FFFF. For SEC and HPSEC, a consistent set of standards is needed for which the size to mass ratio is known (for example globular proteins, dextrans, polystyrene sulfonates (PSS)). With FFFF, the hydrodynamic size is determined directly. Molecular mass distributions can be measured by application of recent developments in mass spectroscometry. The molecular mass distribution of fulvic acid has its maximum well below 1 kDa. Molecular size distributions of humic substances can be converted into molecular mass distributions if the size to mass ratio of calibration standards is similar to that of humic substances. Polycarboxylic acids, including polyacrylic acids (PAAs), appear to suit this purpose. Further investigations are required to determine the physico-chemical range of validity and precision of this approach.

ACKNOWLEDGEMENTS

We thank Mr. G. Teichmann and Mr. E. Peller for HPSEC analysis and Mr. D. Jurrat for sampling and extraction of fulvic and humic acids from peat bog water in the Dachauer Moos. Furthermore, we thank Dr. J. Higgo for making the Derwent fulvic acid available within a cooperation with BGS, UK, and Dr. K. Schmeide (Research Center Rossendorf, Germany) for making the Kranichsee humic and fulvic acids available.

References

1. F.J. Stevenson, *Humus Chemistry: Genesis, Composition, and Reactions*, 2nd Edn., Wiley, New York, 1994.
2. G. Buckau, PhD Thesis, Faculty of Chemistry, FU Berlin, 1991.
3. J.I. Kim, G. Buckau, G.H. Li, H. Duschner and N. Psarros, *Fresenius J. Anal. Chem.*, 1990, **338**, 245.
4. A. Piccolo, S. Nardi and G. Concheri, *Chemosphere*, 1996, **33**, 595.
5. I.V. Perminova, F.H. Frimmel, D.V. Kovalevskii, G. Abbt-Braun, A.V. Kudryavtsev and S. Hesse, *Water Res.*, 1998, **32**, 872.
6. Y.-P. Chin, G.R. Aiken and E. O'Loughlin, *Environ. Sci. Technol.*, 1994, **28**, 1853.
7. S.E. Cabaniss, Q. Zhou, P.A. Maurice, Y.-P. Chin and G.R. Aiken, *Environ. Sci. Technol.*, 2000, **34**, 1103.
8. P. Conte and A. Piccolo, *Environ. Sci. Technol.*, 1999, **33**, 1682.
9. A. Piccolo, P. Conte and A. Cozzolino, *Soil Sci.*, 2001, **166**, 174.
10. M. Wolf, G. Teichmann, E. Hoque, W. Szymczak and W. Schimmack, *Fresenius J. Anal. Chem.*, 1999, **363**, 596.
11. E. Hoque, M. Wolf, G. Teichmann, E. Peller, W. Szymczak, W. Schimmack and G. Buckau, *J. Chromatogr. A*, in review.
12. J.C. Giddings, *Sep. Sci.*, 1966, **1**, 123.
13. R. Beckett, Z. Jue and J.C. Giddings, *Environ. Sci. Technol.*, 1987, **21**, 289.
14. M.E. Schimpf and M.P. Petteys, *Colloids Surf. A*, 1997, **120**, 87.
15. D. Amarasiriwardena, A. Siripinyanond and R.M. Barnes, in *Humic Substances: Versatile Components of Plants, Soils and Water*, eds. E.A. Ghabbour and G. Davies, Royal Society of Chemistry, Cambridge, 2000, p. 215.
16. N.M. Thang, H. Geckeis, J.I. Kim and H.P. Beck, *Colloids Surf. A*, 2001, **181**, 289.
17. J.A. Rice and D.A. Weil, in *Humic Substances in the Global Environment and Implications on Human Health*, eds. N. Senesi and T.M. Miano, Elsevier, Amsterdam, 1994, p. 355.
18. A. Fievre, T. Solouki, A.G. Marshall and W.T. Cooper, *Energy Fuels*, 1997, **11**,

554.
19. T.L. Brown and J.A. Rice, *Org. Geochem.*, 2000, **31**, 627.
20. U. Klaus, T. Pfeifer and M. Spiteller, *Environ. Sci. Technol.*, 2000, **34**, 3514.
21. T.L. Brown and J.A. Rice, *Anal. Chem.*, 2000, **72**, 384.
22. G. Plancque, B. Amekraz, V. Moulin, P. Toulhoat and C. Moulin, *Rapid Commun. Mass Spectrom.*, 2001, **15**, 827.
23. G. Haberhauer, W. Bednar, M.H. Gerzabek and E. Rosenberg, in *Humic Substances: Versatile Components of Plants, Soils and Water*, eds. E.A. Ghabbour and G. Davies, Royal Society of Chemistry, Cambridge, 2000, p. 143.
24. W. Szymczak, M. Wolf and K. Wittmaack, *Acta hydrochim. hydrobiol.*, 2000, **28**, 350.
25. Y. Kaneko, W. Agui, M. Abe and K. Ogino, *Yukagaku*, 1988, **37**, 108.
26. R.L. Wershaw, *Environ. Sci. Technol.*, 1993, **5**, 814.
27. J.A. Rice, T.F. Guetzloff and E. Tombacz, in *Humic Substances: Versatile Components of Plants, Soils and Water*, eds. E.A. Ghabbour and G. Davies, Royal Society of Chemistry, Cambridge, 2000, p. 135.
28. Th. Rabung, H. Geckeis, J.I. Kim and H.P. Beck, *Radiochim. Acta.*, 1998, **82**, 243.
29. R. Artinger, B. Kienzler, W. Schüssler and J.I. Kim, in *Effects of Humic Substances on the Migration of Radionuclides: Complexation and Transport of Actinides*, ed. G. Buckau, 1st Technical Progress Report, Forschungszentrum Karlsruhe, 1998, FZKA 6124, p. 23.
30. K. Schmeide, H. Zänker, K.H. Heise and H. Nitsche, in *Effects of Humic Substances on the Migration of Radionuclides: Complexation and Transport of Actinides*, ed. G. Buckau, 1st Technical Progress Report, Forschungszentrum Karlsruhe, 1998, FZKA 6124, p. 161.
31. J.V. Higgo, J.R. Davies, B. Smith and C. Milne, in *Effects of Humic Substances on the Migration of Radionuclides: Complexation and Transport of Actinides*, ed. G. Buckau, 1st Technical Progress Report, Forschungszentrum Karlsruhe, 1998, FZKA 6124, p. 103.
32. E. Hoque, *J. Chromatogr. A*, 1995, **708**, 273.
33. W. Szymczak and K. Wittmaack, *Nucl. Instrum. Methods B*, 1994, **88**, 149.
34. J.H. Reuter and E.M. Perdue, *Geochim. Cosmochim. Acta*, 1981, **45**, 2017.
35. G.R. Aiken and R.L. Malcolm, *Geochim. Cosmochim. Acta*, 1987, **51**, 2177.
36. F.J. Novotny, PhD Dissertation, South Dakota State University, Brookings, SD, 1993.
37. G.R. Aiken, P.A. Brown, T.I. Noyes and D. Pickney, in *Humic substances in the Suwannee River, Georgia: Interactions, Properties, and Proposed Structures* IX, eds. R.C. Averett, J.A. Leenheer, D.M. McKnight and K.A. Thorn, U. S. Geological Survey Water Supply Paper 2373, 1994, p. 167.
38. Autorenkollektiv, *Strukturveränderungen an Biopolymeren mit spektroskopischen und hydrodynamischen Methoden*, Akademie-Verlag, Berlin, 1976, p. 132.
39. P. Andrews, in *Methods of Biochemical Analysis*, Vol. 18, ed. D. Glick, Wiley, New York, 1970.
40. T.C. Laurent and K.A. Granath, *Biochim. Biophys. Acta*, 1967, **136**, 191.
41. N.M. Thang, PhD Thesis, University of Saarland, Saarbrücken, 2000.
42. R. Artinger, G. Buckau, J.I. Kim and S. Geyer, *Fresenius J. Anal. Chem.*, 1999, **364**, 737.

SOLID STATE NMR EVIDENCE FOR MULTIPLE DOMAINS IN HUMIC SUBSTANCES

Amrith S. Gunasekara,[1] L. Charles Dickinson[2] and Baoshan Xing[1]

[1] Department of Plant and Soil Sciences, University of Massachusetts, Amherst, MA 01003, USA
[2] Department of Polymer Science and Engineering, University of Massachusetts, Amherst, MA 01003, USA

1 INTRODUCTION

The continued sustenance of life on the earth is highly dependent on soils. Soils support plant growth because they contain a major sorbent, soil organic matter (SOM). Soil organic matter controls the sorption/release of water and essential plant nutrients, although their abundance in mineral soils range from only 1-5% on average.[1,2] A major fraction of SOM is amorphous, dark colored materials known as humic substances (HSs), which are produced from highly transformed plant and animal residues.[2] Humic substances are high molecular weight compounds that are found in soils, sediments and waters. Due to their sorption properties and presence in many environments, they have been extensively studied for their potential to sorb hydrophobic organic contaminants (HOCs).[3-5] Recent research has shown that HOCs have a high affinity for HSs in soils and sediments.[6] Xing and Pignetello[7] have studied sorption/desorption isotherms of HOCs by HSs and proposed the dual-mode sorption model (DMM). The DMM hypothesizes that the sorption of HOCs into SOM occurs by sorbing first into an expanded, rubbery (mobile) domain and then into a condensed, glassy (rigid) domain that contains nm size holes (micropores). According to the DMM, sorption of HOCs into the rubbery (mobile) domain occurs through partition or dissolution. Sorption into the glassy (rigid) domain occurs through a competitive process between HOCs for a limited number of high affinity binding sites. Evidence for the DMM is based on isotherm non-linearity and competitive sorption that cannot be explained by other sorption models such as the partition model.[6] A study by Xing and Chen[8] using X-ray diffraction showed the presence of expanded and condensed domains in HSs. Further, the study found that the distribution of these domains was different for HSs from mineral and organic horizons of a soil profile. This type of study exemplifies the importance of using spectroscopy methods to obtain evidence that supports or refutes the DMM. There are only a few spectroscopic studies that support the existence of multiple domains within HSs. Nuclear magnetic resonance (NMR) is one of the most advanced spectroscopic tools for the determination of molecular structures and dynamics in complex organic polymers. In this study we used NMR to observe the spin-spin relaxation (T_2) of protons (^1H) in coal and different HSs. The objective of the experiments was to identify multiple domains within humic acid (HA) samples. Variable temperature studies were also performed to determine the effect of heat on the domain distribution.

2 MATERIALS AND METHODS

2.1 Sample Source and Treatment

Two HA samples and a coal sample were used in this study. The HA samples are representative of HSs and were extracted from different soils. Amherst humic acid (AHA) was extracted from a mineral soil collected in Amherst, MA. The International Humic Substances Society humic acid (IHSSHA) was extracted from a well-humified Pahokee peat purchased from the International Humic Substances Society. The HA samples were extracted and purified using the procedures outlined by Chen and Pawluk.[9] The bituminous coal sample (AC) was obtained from Massey Coal Company. After extraction, the two HA samples were treated with dilute hydrofluoric and hydrochloric acids to reduce the ash content and remove paramagnetic elements from the samples. Prior to NMR analysis the samples were oven dried overnight at 105°C to remove moisture. The samples were slowly cooled to room temperature by placing them in a vacuum desiccator containing Drierite absorbent purchased from Fisher Scientific Company. This experimental step was necessary because initial NMR studies showed that the ^1H of the water molecules greatly enhanced the overall proton intensity in the sample.

2.2 NMR Spectroscopy

All NMR experiments were conducted on a Bruker DSX 300 MHz instrument. The proton relaxation of each sample was measured using a ^1H channel of a Bruker 7 mm CP/MAS probe. This solid state probe is routinely used for ^{13}C cross polarization studies. Samples were tightly packed into 7 mm/18 zirconia rotors and sealed with Kel-F caps. They were analyzed in a static mode because spinning induced sidebands that complicated the T_2 analysis. The pulse program used was the Carr-Purcell pulse sequence followed by Gill-Meibohm phase cycling. This pulse program included a 6 μsec 90° pulse with an incremented defocusing delay. Observation of the relaxing ^1H signal took place during refocusing of the proton magnetization back to equilibrium, perpendicular to the main magnetic field. The ^1H T_2 decay curve contained many data points that represented proton intensities with increasing time. The relaxation intensities of the proton signal with increasing time decayed monotonically.

2.3 Data Treatment

It is well understood that ^1H T_2 relaxation is related to a correlation time that reflects molecular motion.[10] In a sample, protons in condensed or rigid regions will relax rapidly, corresponding to short T_2 values, while slowly relaxing protons, having long T_2 values, correspond to expanded or mobile regions.[11] Figure 1a shows the T_2 relaxation of the overall ^1H population with time for sample AC. The graph shows the monotonic decay of the protons that approached complete relaxation at approximately 58 μsec. This curve could be well fitted to two decay functions. A Gaussian decay function relates protons in a rigid region to short T_2 relaxation times. This region is indicative of a tightly configured structure. Exponential decay is related to protons having longer relaxation times. These protons have greater mobility and could be found in a loosely configured structure. The Gaussian portion of the T_2 decay curve with rapid ^1H relaxation is expressed in Equn. (1),

$$I_F(t) = I_F(O) \exp\left[-\frac{1}{2}\left(\frac{t}{T_{2F}}\right)^2\right] \qquad (1)$$

where $I_F(t)$ is the total intensity for the Gaussian portion of the T_2 decay curve, $I_F(O)$ is the initial intensity at time = 0, t is time in μs and T_{2F} is the T_2 value for the rigid domain in μs. The exponential portion of the T_2 decay curve with slow ^1H relaxation is expressed in Equn. (2),

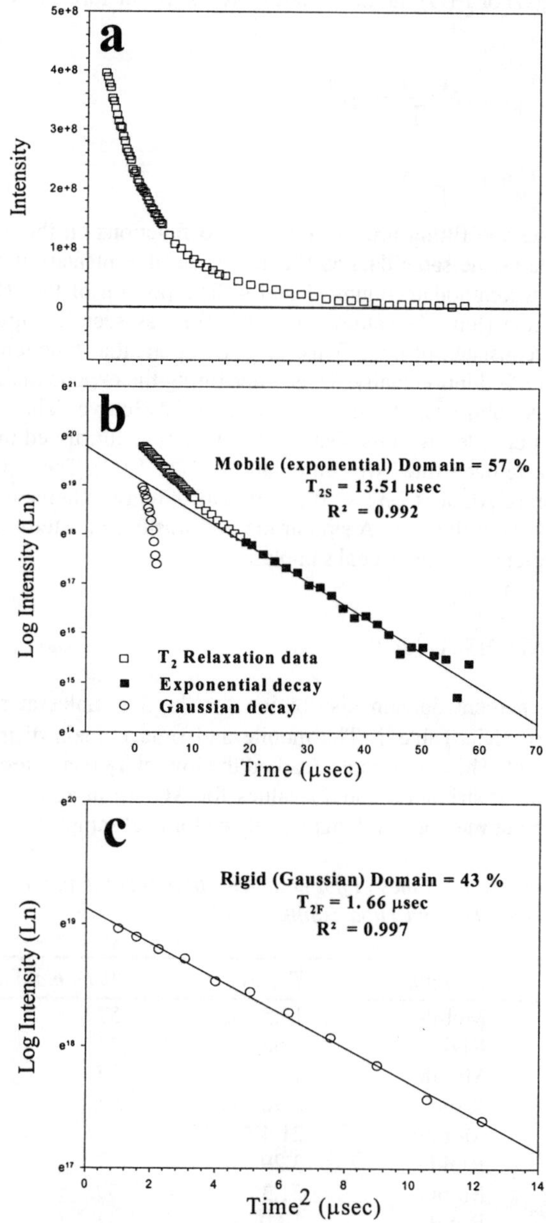

Figure 1 *Spin-spin (T_2) relaxation decay curve of sample AC (a). Fitting of the exponential function (b) and the Gaussian function (c) is used to obtain the percentage of mobile and rigid domains, respectively*

$$I_S(t) = I_S(O)\exp\left(-\frac{t}{T_{2S}}\right) \quad (2)$$

where $I_S(t)$, $I_S(O)$, t and T_{2S} are the parameters analogous to the Gaussian function. By fitting the data to the linear form (natural logarithm) of Equns. 1 and 2, we can obtain the values for T_2 and $I_S(O)$ or $I_F(O)$. Equns. 3 and 4 are the linear forms of equations 1 and 2, respectively.

$$\mathrm{Ln}\, I_F(t) = \mathrm{Ln}\, I_F(O) - \left[\frac{1}{2}\left(\frac{1}{T_{2F}}\right)^2\right]t^2 \quad (3)$$

$$\mathrm{Ln}\, I_S(t) = \mathrm{Ln}\, I_S(O) - \left(\frac{1}{T_{2S}}\right)t \quad (4)$$

Figure 1 illustrates the fitting process for the two functions on the T_2 decay curve of AC. Figure 1b contains the same data as Figure 1a, but the intensity is now on a natural logarithm scale to accommodate Equns. 3 and 4. The portion of the ^1H T_2 decay curve having slow relaxation (long T_2 value) was fitted first, as seen in Figure 1b. Since the decay curve of AC consisted of both Gaussian and exponential functions, they had to be separated to be quantified individually. Thus, after fitting the exponential function for long retention times it is subtracted from the total T_2 decay curve. The residual after the subtraction can then be fitted with the Gaussian function, as illustrated in Figure 1c. Once the two functions have been fitted we can obtain $I_S(O)$ or $I_F(O)$. The $I_S(O)$, representative of the mobile region of AC was 2.32×10^8, while $I_F(O)$, representative of the rigid region, was 1.77×10^8, or 43% of the total. A similar method was used to study the percentages of mobile and rigid regions in various coal samples.[10-13]

3 RESULTS AND DISCUSSION

The T_2 values and percent domain distribution for the 3 samples at room temperature (19°C) are summarized in Table 1. The mobile and rigid domain distributions of the 3 samples were different. The coal sample AC had the lowest T_2 relaxation time for both the domains. The domain distribution and T_2 values for AC are in agreement with previous studies that used similar methods of domain analysis for coal samples.[12-14]

Table 1 *Spin-spin relaxation times (T_2) and distribution of rigid and mobile domains for AC, AHA, IHSSHA and Yubarishinko Coal*

Sample	*Domain*	*T_2 (μsec)*	*% Domain distribution*
AC	Mobile	13.5	57
	Rigid	1.66	43
AHA	Mobile	19.1	37
	Rigid	2.96	63
IHSSHA	Mobile	21.4	43
	Rigid	3.29	57
Yubarishinko Coal[a]	Mobile	21.3	59
	Rigid	10.0	41

[a] Data from Yokono and Sanada[10]

The two separate decays for AC can be seen in Figure 1b. If this graph contained only one domain, then only one decay would be present. For AC, two distinct decays (i.e. functions) were well fitted, as indicated by the R^2 values. The ability to fit the T_2 decay curve of each sample with two different line shapes is a clear indication of the heterogeneous nature of the samples of this study.

Amherst humic acid from the mineral soil had a large fraction (63%) of ^1H in its rigid (glassy) domain. Although AHA has a larger percent rigid domain than AC, it is less tightly structured, as seen by the increased T_2 value (2.96 μs for AHA compared to 1.66 μs for AC). The same was true when comparing the mobile domains of AHA and AC. Larger T_2 values for both domains were observed for IHSSHA when compared to AC and AHA. The IHSSHA sample had a lower percentage of rigid domain than AHA. These differences may be attributed to the source from which the respective HA was extracted: AHA was extracted from a mineral soil whereas IHSSHA was extracted from a well-humified Pahokee peat, which is an organic soil.

The experimental applications thus far illustrate the capabilities of applying Gaussian and exponential functions to the ^1H T_2 decay curves to approximate rigid and mobile domain distributions for heterogeneous organic samples. Using other analytical techniques, recent studies have applied polymer concepts such as the glass transition temperature (T_g) to support the presence of multiple domains within HSs.[15] The glass transition is a process that changes the glassy phase to a mobile (rubbery) one with increasing temperature. Thus far, our results show that HA protons are localized into two domains. Based on these results and a theoretical understanding that ^1H in the two domains may undergo reorganization as a result of heating, variable temperature T_2 relaxation studies were initiated. Although this study did not investigate the T_g directly, we observed different ^1H T_2 relaxations for the HA samples over a range of different temperatures (19-80°C). The samples were maintained at a given temperature for one hour before excitation and T_2 observation. This provided adequate time for the sample to equilibrate with the applied temperature. Figures 2a and 2b illustrate the logarithm transformed T_2 decay curves for IHSSHA and AHA, respectively, at different temperatures. It is evident from Figures 2a and 2b that two functions can be fitted for every T_2 decay curve. Each curve at a given fixed temperature was fitted with exponential and Gaussian functions to obtain approximate domain distributions.

The variable temperature results for IHSSHA and AHA are shown in Figures 3a and 3b, respectively. As the temperature increased, the percentage distribution decreased for the rigid domain and correspondingly increased for the mobile domain. This trend was observed for both IHSSHA and AHA. There is a relationship, though not proportional, between increased temperature and increasing mobile domain percentage. These results are consistent with glass transition temperature (T_g) studies.[15] Using differential scanning calorimetry, LeBoeuf and Weber[16] found that the T_g of a dry HA was 62.2°C. As shown in Figure 3a, the domain distribution of IHSSHA was strongly influenced by temperatures between 19°C and 70°C: we observed a decrease from 57 % to 45 % of the rigid domain that inversely corresponded to a mobile domain increase. At 70°C and 80°C the domain distribution was not strongly affected; the distribution was constant at 45 % rigid and 55 % mobile domains. A similar trend was observed for AHA, where there was a 10 % change in domain distribution between the temperatures of 19°C and 60°C. Only a 1 % phase change was observed between 70-80°C. Thus, these results correlate well to previous studies that used polymer concepts to explain and observe the presence of multiple domains within HA.[15,16] The large T_g range and domain distribution change could be attributed to the heterogeneous nature of the samples in terms of their structures and molecular weights.

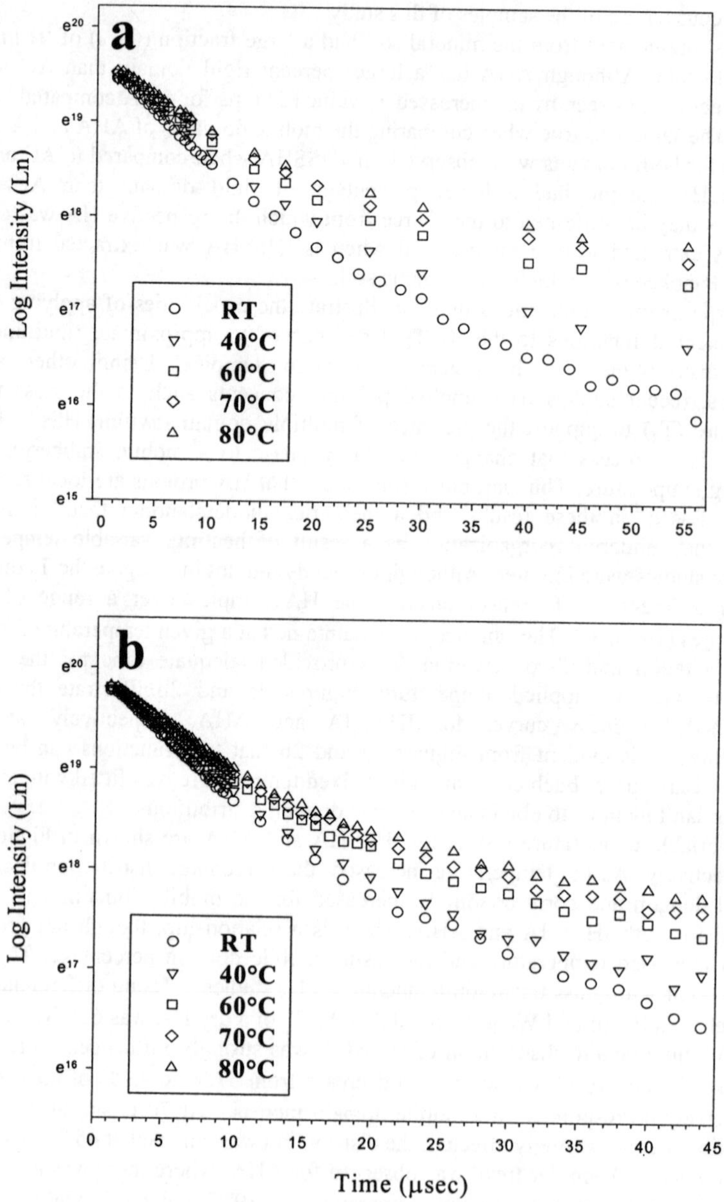

Figure 2 *The logarithm transformed T_2 relaxation decay curves of IHSSHA (a) and AHA (b) at different temperatures*

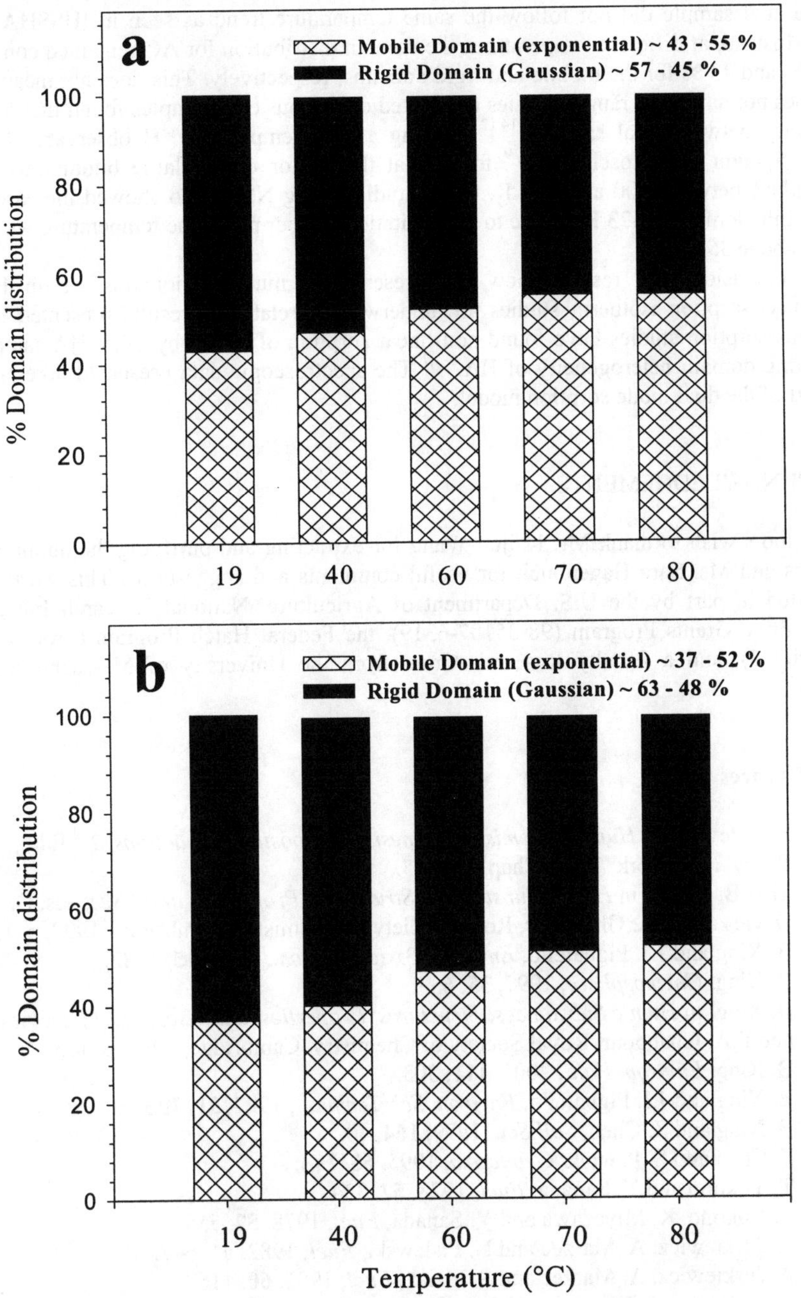

Figure 3 *Mobile and rigid domain distribution of IHSSHA (a) and AHA (b) at different temperatures*

The coal sample did not follow the same temperature trend as seen in IHSSHA and AHA. Instead, at different temperatures the domain distribution for AC remained constant at 57 % and 43 % for the mobile and rigid domains, respectively. This does not mean that AC does not have a T_g range. Studies conducted on various coal samples found that the T_g may vary between coal samples.[14,17,18] Using a high temperature ^1H observable NMR probe, Speight and Moschopedis[19] found that the T_g for high-volatile bituminous coal takes place between 600 and 800 K. Other studies using NMR also showed increases in the mobile domain at 423 K.[20] Due to the limitations of our probe the temperature was not raised above 353 K.

In conclusion, our results show the presence of multiple domains within HAs. Currently, sorption isotherm studies are underway to relate the results presented here. Previous sorption studies have found non-linear sorption of HOCs by other HA samples, indicating domain heterogeneity of HA.[3-5,7] The spectroscopic data presented here are in support of the dual mode sorption model.

ACKNOWLEDGEMENTS

The authors wish to thank Mr. Kaijun Wang for extracting and purifying the humic acid samples and Ms. Sara Hagenbuch for useful comments and suggestions. This work was supported in part by the U.S. Department of Agriculture, National Research Initiative Competitive Grants Program (98-35107-6319), the Federal Hatch Program (Project No. MAS00773), and a Faculty Research Grant from the University of Massachusetts at Amherst.

References

1. F.J. Stevenson, *Humus Chemistry: Genesis, Composition, Reactions*, 2nd Edn., Wiley, New York, 1994, Chap. 1, p. 15.
2. M.H.B. Hayes, in *Humic Substances; Structures, Properties, and Uses*, eds. G. Davies and E.A. Ghabbour, Royal Society of Chemistry, Cambridge, 1998, p. 1.
3. B. Xing and J.J. Pignatello, *Environ. Toxicol. Chem.*, 1996, **15**, 1282.
4. B. Xing, *Chemosphere*, 1997, **35**, 633.
5. B. Xing, in *Humic Substances; Structures, Properties, and Uses*, eds. G. Davies and E.A. Ghabbour, Royal Society of Chemistry, Cambridge, 1998, p. 173.
6. B. Xing, *Environ. Poll.*, 2001, **111**, 303.
7. B. Xing and J.J. Pignatello, *Environ. Sci. Technol.*, 1997, **31**, 792.
8. B. Xing and Z. Chen, *Soil Sci.*, 1999, **164**, 40.
9. Z. Chen and S. Pawluk, *Geoderma*, 1995, **65**, 173.
10. T. Yokono and Y. Sanada, *Fuel*, 1978, **57**, 334.
11. T. Yokono, K. Miyazawa and Y. Sanada, *Fuel*, 1978, **57**, 555.
12. A. Jurkiewicz, A. Marzec and N. Pislewski, *Fuel*, 1982, **61**, 647.
13. A. Jurkiewicz, A. Marzec and S. Idziak, *Fuel*, 1981, **60**, 1167.
14. W.A. Barton, L.J. Lynch and D.S. Webster, *Fuel*, 1984, **63**, 1262.
15. E.J. LeBoeuf and W.J. Weber, Jr., *Environ. Sci. Technol.*, 2000, **34**, 3623.
16. E.J. LeBoeuf and W.J. Weber, Jr., *Environ. Sci. Technol.*, 1997, **31**, 1697.
17. P.H. Given, A. Marzec, W.A. Barton, L.J. Lynch and B.C. Gerstein, *Fuel*, 1986, **65**, 155.
18. Y. Sanada and H. Honda, *J. App. Polym. Sci.*, 1962, **6**, 94.

19. J.G. Speight and S.E. Moschopedis, *Fuel*, 1979, **58**, 235.
20. W.J. Weber Jr., P.M. McGinley and L.E. Katz, *Environ. Sci. Technol.*, 1992, **26**, 1955.

INVESTIGATION OF MOLECULAR MOTION OF HUMIC ACIDS WITH 1D AND 2D SOLUTION NMR

Kaijun Wang,[1] L. Charles Dickinson,[2] Elham A. Ghabbour,[3] Geoffrey Davies[3] and Baoshan Xing[1]

[1] Department of Plant and Soil Science, University of Massachusetts, Amherst, MA 01003, USA
[2] Department of Polymer Science and Engineering, University of Massachusetts, Amherst, MA 01003, USA
[3] Chemistry Department and the Barnett Institute, Northeastern University, Boston, MA 02115, USA

1 INTRODUCTION

Humic substances (HSs) in soils, sediments, and waters are the result of the biological and chemical metabolism of organic matter.[1] They play significant roles in the formation of soil aggregates, control of soil acidity, cycling of nutrients, soil moisture retention, detoxification of hazardous compounds, sustainable agriculture and environmental quality.[2] Humic acids (HAs) are the fraction of HSs that are not soluble in acidic aqueous solutions but soluble in alkaline solutions. The structural and functional properties of HAs are critical in determining their reactivity with heavy metals and organic contaminants. In this respect, NMR has been shown to be a powerful tool for study of these materials. Although the heterogeneous nature of HAs presents a great challenge for the structural determination by NMR spectroscopy, valuable information on fragmental compositions and functionalities has been obtained.[1,3,4]

More recently, two-dimensional solution NMR spectroscopy has been incorporated in studies of HAs.[5-8] NOESY (Nuclear Overhauser Enhancement Spectroscopy) and ROESY (Rotating-frame Nuclear Overhauser Enhancement Spectroscopy) are 2D methods of determining the through-space dipolar interactions between nuclei. They can be used to establish through-space proximity between functionalities, which can give insight into molecular shape and structure.[9] ROESY has some advantages over NOESY. The NOE in the rotating-frame under spin-lock conditions is always positive. This enables ROESY to detect dipolar interactions for molecules of any size and even those of intermediate molecular weight where the laboratory frame NOE is near zero, so that the NOESY is not suitable.[10] The second advantage of ROESY is that the effects of spin diffusion for macromolecules are reported to be less marked than in the corresponding NOESY experiments, so that the ROESY spectra may be simpler and more certainly interpreted.[11] In this study, ROESY was used to demonstrate configurational changes of HAs in solution.

Among other solution NMR parameters, the spin-lattice relaxation time (T_1) is a sensitive and powerful probe of molecular dynamics of chemical systems. Interest in NMR relaxation parameters has shown a dramatic increase in the past two decades.[12] T_1^{-1} ($1/T_1$) is the rate constant for a nuclear spin to return to its equilibrium state after excitation by radio frequency. The process of spin-lattice relaxation involves the transfer of magnetization between the magnetic nuclei (spins) and their environment (the lattice).[13]

Spin-lattice relaxation is induced by field fluctuations at the NMR frequency due to molecular motion. Therefore, T_1 is an indicator of how rapidly a nuclear spin moves in its environment. The theoretical description of dipolar relaxation for an assembly of molecules under the motional narrowing limit leads to the proportionality relationship $1/T_1 \propto \tau_c$. In the extreme narrowing limit ($\omega_o\tau_c \ll 1$, ω_o is Larmor frequency), this can be expressed in the form of a useful rule: the faster a molecule moves, the greater is T_1.[14] Here τ_c is the correlation time, which, roughly, is the time for a molecule to reorient by one radian. There are many spin-lattice relaxation mechanisms. Important mechanisms are dipole-dipole interactions, spin-rotation, scalar interactions and chemical shift anisotropy. For protons in solutions, the dominant relaxation mechanism for energy transfer is usually the intramolecular dipole-dipole interaction.[15]

The proton T_1 is a well-established NMR parameter for structural and configurational determination of organic molecules in solution. Although its use for investigation was limited by the misconception that proton dipolar relaxation rates for average molecules are too complicated for solution of stereochemical problems,[15] proton spin-lattice relaxation has been widely studied and such experiments have indicated great promise in structural and dynamic studies of organic molecules.[16-20] Despite this, proton spin-lattice relaxation has received little attention in the field of humic chemistry. Therefore, the objectives of this study were to directly determine the proton spin-lattice relaxation times of HAs in solution and to evaluate the effects of HA concentration and solvents on this proton spin-lattice parameter. Molecular motion and intramolecular interactions of HAs are also discussed.

2 MATERIALS AND METHODS

2.1 Materials

Two humic acids were used in this study: Amherst HA and New Hampshire HA. They were extracted using the standard procedures recommended by the International Humic Substances Society. Elemental compositions of humic acids were determined by the Microanalytical Laboratory at the University of Massachusetts, Amherst. The composition of Amherst HA is: 54.6% C, 4.16% H, 38.5% O, 2.46% N and 0.24% ash; New Hampshire HA contains 52.9% C, 5.40% H, 39.7% O, 2.00% N and 0.25% ash. The functional group compositions of the two HAs are shown in Table 1. To estimate the effects of HA concentrations and solvents, four concentration levels (1%, 3%, 5%, and 8% w/w) were prepared for Amherst HA, while three concentration levels (1%, 3%, and 5% w/w) for New Hampshire HA were investigated. 0.5 M deuterated sodium hydroxide (NaOD) in D_2O and deuterated dimethylsulfoxide (d_6-DMSO) were used as solvents. Solution pH was estimated for the samples in 0.5 M NaOD and the value changed from 12.5 to 12.8 depending on the HA concentration.

2.2 Methodology

T_1 was measured by the commonly used inversion recovery pulse sequence (PD - 180° - τ - 90° - AT)$_n$ (Figure 1A). The delay time τ between two pulses was the variable parameter, which was set at a different fixed value for each spectrum. Here, AT is the acquisition time, PD is the pulse delay, and n is the total number of scans required for an acceptable signal-to-noise ratio. AT+PD was set to be greater than $5T_1$.

Table 1 *Functional group composition of humic acids*[a]

Sample	*Aliphatic*	*Carbohydrate*	*Aromatic*	*Carboxylic*
Amherst HA	17.3	25.0	40.8	16.9
New Hampshire HA	17.0	31.6	39.0	12.4

[a]These data are from Mao, et al.[4]

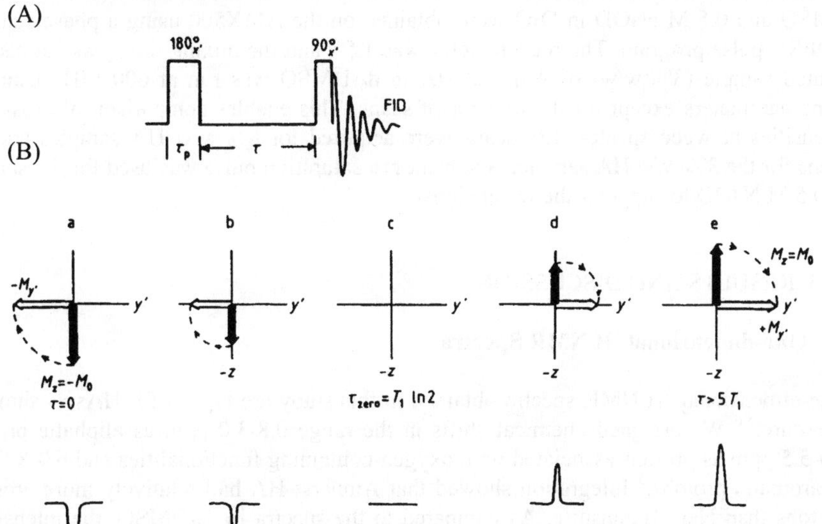

Figure 1 *Pulse sequence (A) and vector diagrams and signals for five different τ (B). Modified from Friebolin*[14]

Figure 1B shows the evolution stages of magnetization (M_z) through relaxation for five different delay times τ. The relationship between magnetization and τ is given by Equn. (1). Taking logarithms of both sides of the equation, Equn. 1 becomes Equn. (2).

$$Mo - Mz = 2Moe^{-\tau/T_1} \quad (1)$$

$$\ln(Mo - Mz) = \ln 2Mo - \frac{\tau}{T_1} \quad (2)$$

Because M_o and M_z are not directly measurable quantities, the intensities (I) of the signals obtained directly from experiments after Fourier transformation can be used to replace the magnetizations, as in Equn. (3),

$$\ln(Io - Iz) = \ln 2Io - \frac{\tau}{T_1} \quad (3)$$

where I_o is the maximum signal intensity at $\tau = 0$ and I_z is the intensity at $t = \tau$. We can calculate T_1 values by fitting the intensities of signals at different τ to Equn. (3).

All the T_1 experiments were carried out on a Bruker DPX300 spectrometer using the inversion recovery pulse sequence at room temperature. The ^{1}H 90° and 180° pulse lengths were 9.25 μs and 18.5μs, respectively and the recycle delay was 10 seconds. Proton spectra were recorded at 10 different τ: 10μs, 500μs, 10ms, 100ms, 250ms, 500ms, 1s, 2s, 5s and 10s. A preliminary test showed that this time range was suitable for proton T_1 measurement.

Two-dimensional ROESY spectra were recorded with Bruker AMX500 and AVANCE600 NMR spectrometers. The ROESY spectra for Amherst HA (8% w/w) in d_6-DMSO and 0.5 M NaOD in D_2O were obtained on the AMX500 using a phase sensitive ROESY pulse program. The recycle delay was 1.5 s and the mixing time was 150ms. The diluted sample (3% w/w) of Amherst HA in d_6-DMSO was run at 600 MHz using the same parameters except for the number of scans. This enables comparison of cross-peak intensities between spectra. 124 scans were acquired for 8% w/w HA samples and 324 scans for the 3% w/w HA sample. A solvent pre-saturation pulse was used for the samples in 0.5 M NaOD to suppress the water signal.

3 RESULTS AND DISCUSSION

3.1 One-dimensional ^{1}H NMR Spectra

One-dimensional ^{1}H NMR spectra obtained in this study are typical for HAs as shown in literature.[5,6] We assigned chemical shifts in the range 0.8-3.0 ppm as aliphatic protons, 3.0-5.5 ppm as protons associated with oxygen-containing functionalities and 6.0-8.5 ppm as aromatic protons.[3] Integration showed that Amherst HA had relatively more aromatic protons than New Hampshire. As compared to the spectra in d_6-DMSO, the intensity of protons in the range 6.0-8.5 ppm was lower in 0.5 M NaOD solution, probably due to proton exchange.[6] We did not observe any change between ^{1}H spectra of the same HA at different concentrations. To simplify the relaxation data analysis, we selected 5 chemical shifts from different regions as examples: 0.8 ppm and 2.0 ppm from the aliphatic region representing methyl and methylene/methine groups, 3.8 ppm in the oxygen-substituted aliphatic region, and 6.8 ppm and 7.5 ppm for aromatic protons.

3.2 Spin-lattice Relaxation Time (T_1)

The proton spectra for T_1 measurement consisted of a series of one-dimensional spectra recorded at different τ (Figure 2). The intensities of signals changed from negative at very short τ (10μs) to a positive maximum before τ = 10 s. Generally, the zero-crossing points (the τ at which the signal intensity reached zero) varied for protons of different chemical moieties. The signal intensities for most aliphatic protons became zero at τ of about 250 ms, while the zero-crossing points for aromatic protons were around 500 ms.

Figure 3 illustrates the proton T_1 of humic acids in d_6-DMSO. For all concentrations, proton T_1 values were less than 1 second, ranging from 0.34 s to 0.75 s. It is clear that the T_1 values decreased with increasing concentration. This result is consistent with the commonly held notion that molecular mobility is higher in more diluted solutions, perhaps, in this case, due to some level of aggregation. Similar results were obtained in studies of proton relaxation in agar and protein (lysozyme and Rnase) solutions.[21,22] One explanation is that with increasing concentration, rotational diffusion slows down due to the enhanced friction among solute molecules. As a result, the rotational correlation time

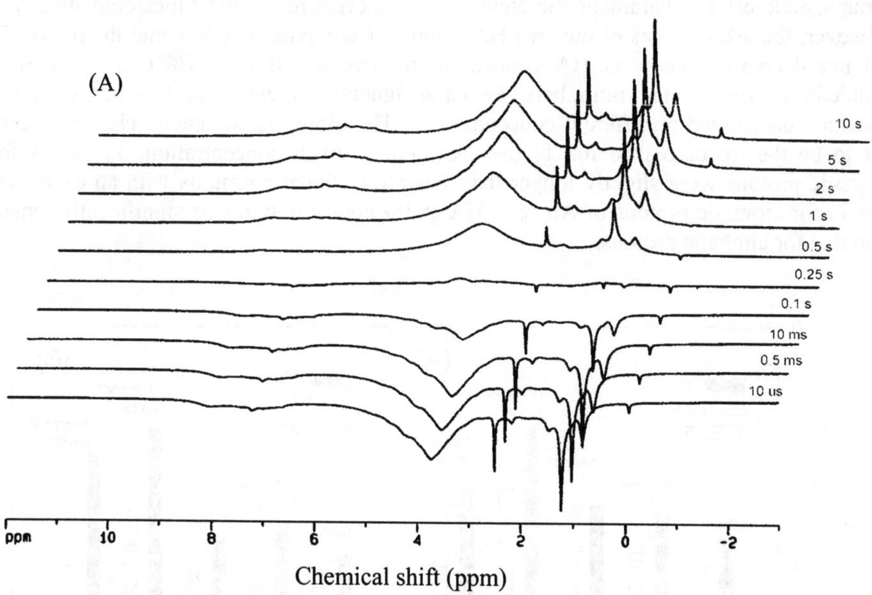

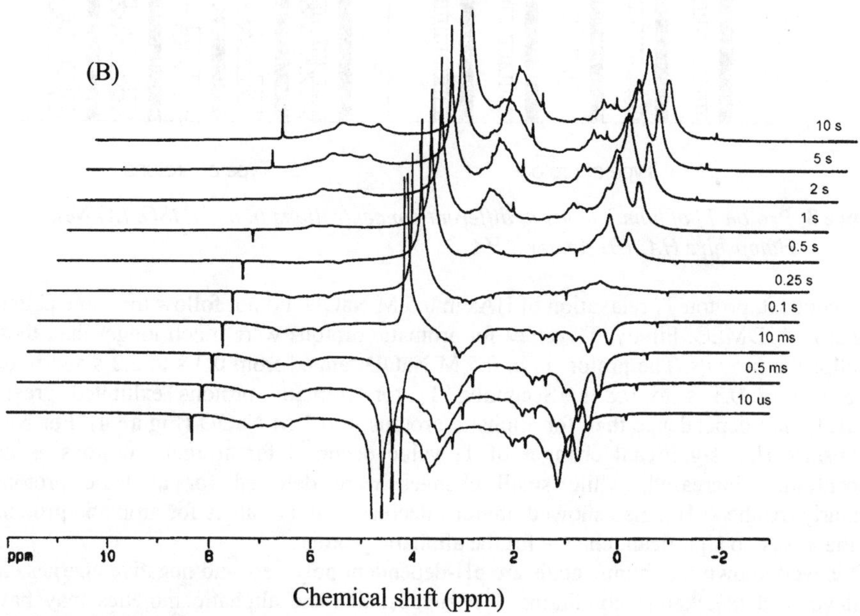

Figure 2 *Proton NMR spectra of New Hampshire HA in d_6-DMSO (A) and in 0.5 M NaOD (B)*

becomes longer and T_1 gets shorter. Molecules that exhibit hydrogen bonding are expected to show a particularly large concentration dependence.[23] HAs contain many functional groups that can form hydrogen bonds. Thus, the result showing that proton T_1 was sensitive to HA concentration is not surprising. Another possible explanation is paramagnetic effects. Paramagnetic elements can shorten relaxation times dramatically.[23] However, the ash contents of our two HA samples were extremely low and the proton T_1 did not decrease greatly as HA concentration increased from 3-8% (i.e., more than doubled). Moreover, the ratio between paramagnetic contents and HA concentration remained unchanged at different concentrations. Therefore, paramagnetic elements seem not to be the major reason for T_1 decrease. For a given concentration, T_1 values for aromatic protons were slightly longer than those for aliphatic protons with an exception that T_1 for aromatic protons of Amherst HA at 1% concentration was significantly longer than that for aliphatic protons.

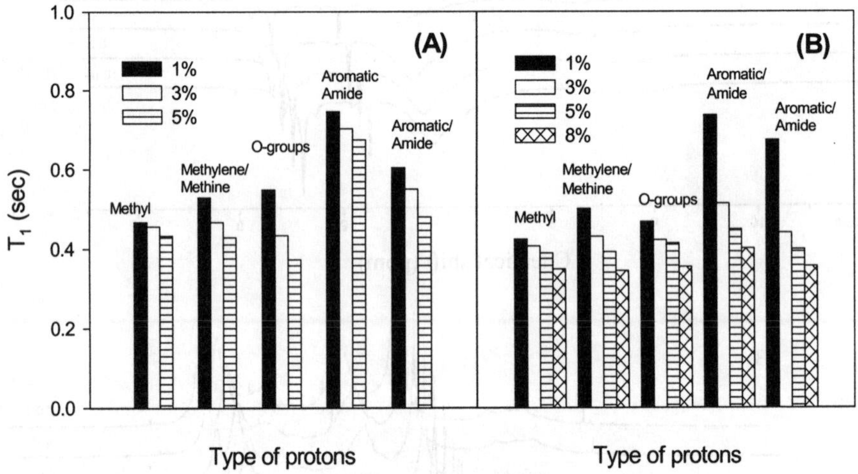

Figure 3 *Proton T_1 of humic acids at different concentrations in d_6-DMSO. (A) New Hampshire HA; (B) Amherst HA*

In contrast, proton T_1 relaxation of HAs in 0.5 M NaOD did not follow the same pattern as that in d_6-DMSO. Firstly, T_1 values for aromatic protons were much longer than those for aliphatic protons. The proton T_1 in 0.5 M NaOD ranged from 0.3 s to 2.2 s and in d_6-DMSO from 0.3 s to 0.8 s. Secondly, T_1 for aromatic protons exhibited greater concentration dependence than for aliphatic protons in 0.5 M NaOD (Figure 4). For New Hampshire HA, significant changes of T_1 value occurred for aromatic protons as the concentration increased, while small changes were detected for aliphatic protons. Similarly, Amherst HA also showed obvious decreases of T_1 values for aromatic protons, but there was no significant change for the aliphatic protons.

It is well known that humic acids are pH-dependent polymers and negative charges can be developed in alkaline conditions. We suggest that the aliphatic moieties may have relatively more functional groups that could be deprotonated at high pHs. Thus, more negative charges may be developed on aliphatic segments than on aromatic moieties. As a result, the aliphatic segments may repel each other more strongly. In basic solution, the repulsion force would still dominate between aliphatic regions even in relatively concentrated solutions. Thus, minimal interactions would occur between these regions. Cook and Langford[24,25] also suggested that there were relatively more functional groups associated with aliphatic moieties than with aromatic ones for a fulvic acid. However, aromatic segments may have lower negative charges due to fewer functional groups. A tendency towards aggregation may gradually surmount the repulsion as the HA concentration increases. This would result in T_1 decreases of aromatic protons because of more interactions at higher concentrations.

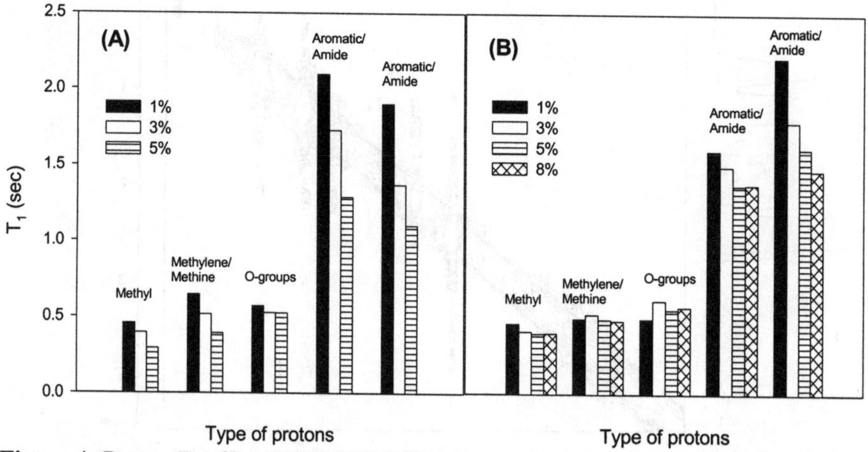

Figure 4 *Proton T_1 of humic acids at different concentrations in 0.5 M NaOD. (A) New Hampshire HA; (B) Amherst HA*

Spin-lattice relaxation involves magnetization transfer between nuclear spins and the lattice. The proton T_1 reflects the changes of environment around nuclear spins, usually in terms of fluctuations of interproton distance.[10] Therefore, the different proton spin-lattice relaxation behaviors in d_6-DMSO and 0.5 M NaOD might result from local configurational changes of HAs. Because T_1 is directly related to the molecular motion, general ideas on segmental mobility can be obtained from the comparison of T_1 values among functional groups within a molecule. Aromatic protons had longer T_1s than aliphatic protons, especially in 0.5 M NaOD. It is inferred that the aromatic protons may move faster than aliphatic protons and that protonated aromatic rings may have higher mobility. Similarly, Cook and Langford[24] reported that aromatic moieties appeared to be smaller than aliphatic segments for a HA and a fulvic acid. They concluded that aromatic moieties were more mobile for the HSs used in their study. It is here speculated that some protonated aromatic rings may be connected to the end of aliphatic chains, so that they can rotate relatively freely as a whole ring, which may lead to longer T_1.

3.3 Two-dimensional ^{1}H NMR Spectra

ROESY was employed to investigate HA configurational changes as concentration increased and the effects of different solvents. Cross peaks were observed in the aliphatic region (0-4ppm) for Amherst HA at 8% w/w in d_6-DMSO (Figure 5A). This indicates dipolar interactions between functional groups in the aliphatic region. By contrast, most of the cross peaks disappeared in the same region at 3% w/w (Figure 5B). This evidence supports the idea that folding in the aliphatic region may occur at high HA concentrations. These results are consistent with T_1 evidence that molecules are more mobile in diluted solutions (see the discussion above).

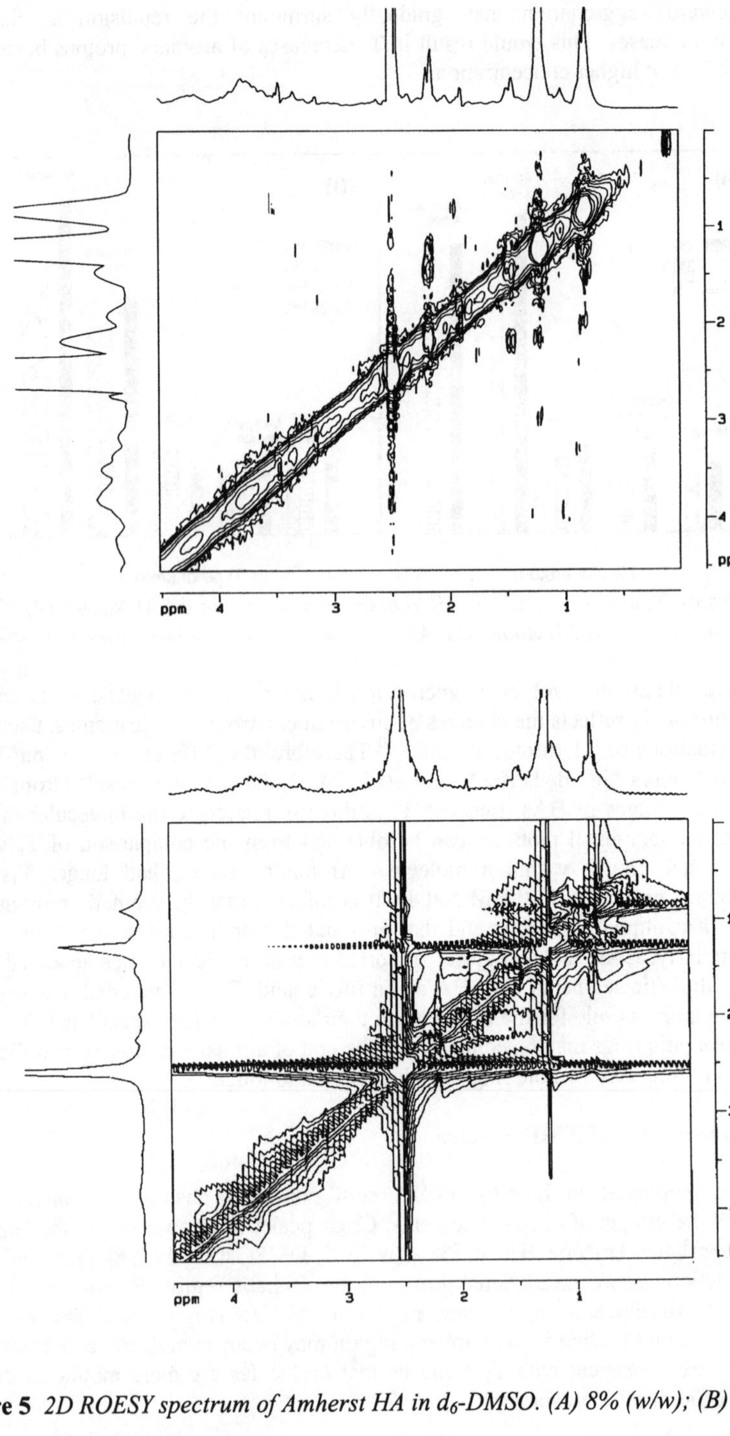

Figure 5 *2D ROESY spectrum of Amherst HA in d_6-DMSO. (A) 8% (w/w); (B) 3% (w/w)*

The ROESY spectrum for Amherst HA at 8% w/w (data not shown) in 0.5 M NaOD showed fewer dipolar interactions among functional groups. This result supports our speculation that aliphatic segments might carry more negative charges at high pH. They could repel one another strongly and it becomes difficult for one group to be close enough to another for dipolar interactions to occur. From these results it can be inferred that the configuration of HAs is different in these two solvents.

4 CONCLUSIONS

T_1 measurement is a useful approach for molecular dynamic studies. This study explored proton T_1 of HAs in different solvents. Proton T_1 ranged from 0.3 s to 0.8 s in d_6-DMSO and from 0.3 s to 2.2 s in 0.5 M NaOD. Generally, proton T_1 decreased with increasing concentration of HAs, probably due to reduced molecular mobility. In 0.5 M NaOD, aromatic protons had much longer T_1 values than aliphatic protons and appear to have higher mobility. In d_6-DMSO, T_1 for aromatic protons were slightly longer than that for aliphatic protons, suggesting that aromatic protons were only slightly more mobile than aliphatic protons. ROESY spectra demonstrated that dipolar interactions among aliphatic segments occurred for HAs at high concentrations in d_6-DMSO, indicative of folding in the aliphatic region. By contrast, far fewer cross peaks were observed in the same chemical shift region in 0.5 M NaOD. From T_1 measurements and ROESY spectra, we conclude that the configuration of HAs in 0.5 M NaOD is different from that in d_6-DMSO.

ACKNOWLEDGEMENTS

This work was supported in part by the U.S. Department of Agriculture, National Research Initiative Competitive Grant Program (98-35107-6319), the Federal Hatch Program (Project No. MAS00773) and a Faculty Research Grant from the University of Massachusetts at Amherst.

References

1. M.H.B. Hayes, in *Humic Substances: Structures, Properties and Uses*, eds. G. Davies and E.A. Ghabbour, Royal Society of Chemistry, Cambridge, 1998, p. 1.
2. B. Xing, in *Humic Substances: Structures, Properties and Uses*, eds. G. Davies and E.A. Ghabbour, Royal Society of Chemistry, Cambridge, 1998, p. 173.
3. C.M. Preston, *Soil Sci.*, 1996, **161**, 144.
4. J.-D. Mao, W.-G. Hu, K. Schmidt-Rohr, G. Davies, E.A. Ghabbour and B. Xing, *Soil Sci. Soc. Am. J.*, 2000, **64**, 873.
5. A.J. Simpson, R.E. Boersma, W.L. Kingery, R.P. Hicks and M.H.B. Hayes, in *Humic Substances, Peats and Sludges: Health and Environmental Aspects,* eds. M.H.B. Hayes and W.S. Wilson, Royal Society of Chemistry, Cambridge, 1997, p. 46.
6. T.W.M. Fan, R.M. Higashi and A.N. Lane, *Environ. Sci. Technol.*, 2000, **34**, 1636.
7. Y. Chien and W.F. Bleam, *Environ. Sci. Technol.*, 1998, **32**, 3653.
8. W.L. Kingery, A.J. Simpson, M.H.B. Hayes, M.A. Locke and R.P. Hicks, *Soil Sci.*, 2000, **165**, 483.

9. N. Chandrakumar and S. Subramanian, in *Modern Techniques in High Resolution FT-NMR*, Springer-Verlag, New York, 1987.
10. A.-D. Bax and D.G. Davis, *J. Magn. Reson.*, 1985, **63**, 207.
11. C.J. Bauer, T.A. Frenkiel and A.N. Lane, *J. Magn. Reson.*, 1990, **87**, 144.
12. R.E. Wasylishen, in *NMR Spectroscopy Techniques*, ed. M.D. Bruch, Dekker, New York, 1996, p. 105.
13. G.C. Levy, J.D. Cargioli and F.A.L. Anet, *J. Am. Chem. Soc.*, 1973, **95**, 1527.
14. H. Friebolin, *Basic One- and Two-Dimensional NMR Spectroscopy*, Wiley-VCH, Weinheim, 1998. p. 161.
15. P. Dais and A.S. Perlin, in *Advances in Carbohydrate Chemistry and Biochemistry*, eds. R.S. Tipson and D. Horton, Academic Press, New York, 1987, Vol. 45, p. 125.
16. M. Koike, D.-G. Vandervelde and J.R. Shapley, *Organometallics*, 1994, **13**, 1404.
17. S. Kaku, *Plant Cell Physiol.*, 1993, **34**, 535.
18. A. Martinez, S. Olafsdottir and T. Flatmark, *Eur. J. Biochem.*, 1993, **211**, 259.
19. S. Kiihne and R.-G. Bryant, *Biophysical J.*, 2000, **78**, 2163.
20. M. Effemy, J. Lang and J. Kowalewski, *Magn. Reson. Chem.*, 2000, **38**, 1012.
21. M. Askin, K. Yurdakock and Z. Gulsun, *Spect. Lett.*, 1993, **26**, 1039.
22. A.-G. Krushelnitsky and V.-D. Fedotov, *J. Biomol. Struct. & Dyn.*, 1993, **11**, 121.
23. F.W. Wehrli, A.P. Marchand and S. Wehrli, *Interpretation of Carbon-13 NMR Spectra*, Wiley, Chichester, 1988, p. 211.
24. R.L. Cook and C.H. Langford, *Environ. Sci. Technol.*, 1998, **32**, 719.
25. R.L. Cook and C.H. Langford, *Polymer News*, 1999, **24**, 6.

VARIATION OF FREE RADICAL CONCENTRATION OF PEAT HUMIC ACID BY METHYLATIONS

T. Shinozuka, Y. Enomoto, H. Hayashi, H. Andoh and T. Yamaguchi

Chiba Institute of Technology, Chiba, Japan

1 INTRODUCTION

It is well known that humic substances (HSs) contain high concentrations of stable free radicals. During the last twenty years, several studies have tried to characterize and quantify these functional groups in soil humic acids (HAs). It is widely understood that the radical stabilizing structure is of the semiquinone type,[1] which is formed during the process of humification. This free radical form is considered to be at least partly responsible for the dark color of HA.

Electron spin resonance (ESR) measurements[2-4] of HAs extracted from a tropical peat showed that the free radical concentration of this peat HA was somewhat higher than those from soil reported in the previous papers. The special feature of the tropical peat humic acid is that it has a higher carboxyl group content than soil humic acids.[5] This suggests that the carboxyl group might affect the stabilization of the free radicals in the peat humic acid.

To clarify the relation between the carboxyl group content and the free radical concentration, the methylations of the oxygen containing groups of the tropical peat HA with certain methods were investigated in this study. The following three methylation methods were applied; (1) permethylation by the Hakomori method,[6] (2) methoxylation of phenolic hydroxyl groups with diazomethane[7] and (3) esterification of carboxyl groups with trimethoxymethane.[8]

2 EXPERIMENTAL

2.1 Extraction

Powdered samples of HA were extracted with the prescribed method of the International Humic Substances Society (IHSS) from tropical peat (Kalimantan, Indonesia) and andosol (Japan). These source materials were rinsed in a 0.1M HCl solution and filtered. The insoluble component was extracted with 0.1M NaOH and precipitated by acidification of the alkaline supernatant with 0.1M HCl under a nitrogen atmosphere. The solution was centrifuged at 7000 rpm for 20 min. at 293K and the supernatant was freeze dried.

2.2 Methylation of Humic Acid

2.2.1 Methylation by the Hakomori Method.[6] Figure 1 illustrates the experimental procedure of the Hakomori method. A humic acid (0.5 g) solution in dimethyl sulfoxide (DMSO, 25 mL) was shaken for 0.5h at 293K, then a DMSO solution of excess methylsulifinyl carbanion was added to it under a nitrogen atmosphere and the mixture was stirred for 1h at room temperature. Methyl iodide (1.5 mL) was added to the solution with stirring for 30 min at room temperature and the supernatant was dialyzed against distilled water. The methylated humic acid was freeze dried.

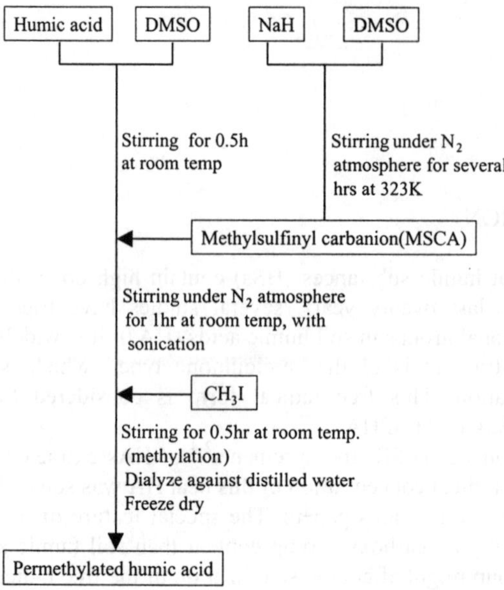

Figure 1 *Permethylation of humic substance by the Hakomori method*

2.2.2 Methylation with Diazomethane.[7] Figure 2 shows the procedure of methylation with diazomethane. Production of diazomethane: N-methyl-N-nitrosourea (5 g) was added to a mixture of dichloromethane (50 mL) and 40% KOH at 273K. To this mixture, humic acid (1.0 g) in DMSO solution (30 mL) was added and stirred for the prescribed time. The methylated humic substance was obtained by freeze drying.

2.2.3 Methylation with trimethoxymethane.[8] Figure 3 shows the pathway of methylesterification by trimethoxymethane used as a dehydrating agent. Trimethoxymethane (10 mL) was added to the humic acid (1.0 g) suspension in methanol (30 mL) at 333K with a small amount of concentrated sulfuric acid and shaken for 1 to 24 hours. The suspension was centrifuged and then the precipitate was washed three times with cold water and freeze dried.

2.3 Analysis of Fundamental Properties of Humic Acid

The contents of carboxyl and phenolic hydroxyl groups were measured by the ion exchange capacity with calcium acetate[9] and the Follin-Ciocalteu method,[10] respectively.

The E_4/E_6 ratio was determined by measuring absorbances of the humic acid samples dissolved in 0.1M NaOH at 220-700 nm on a Shimadzu Model UV-1700 spectrometer. The moisture content was determined by heating samples at 378K for 24h. The ash content was measured by ignition at 973K for 1h. FTIR measurements were performed with a JASCO FT/IR-410 spectrometer. Spectra were measured for KBr disk samples (KBr:sample = 500:1). The aromaticity *fa* was calculated by CPMAS ^{13}CNMR spectrometry (BRUKER AMX-400) using commonly assigned spectral regions for aliphatic and aromatic components.[5]

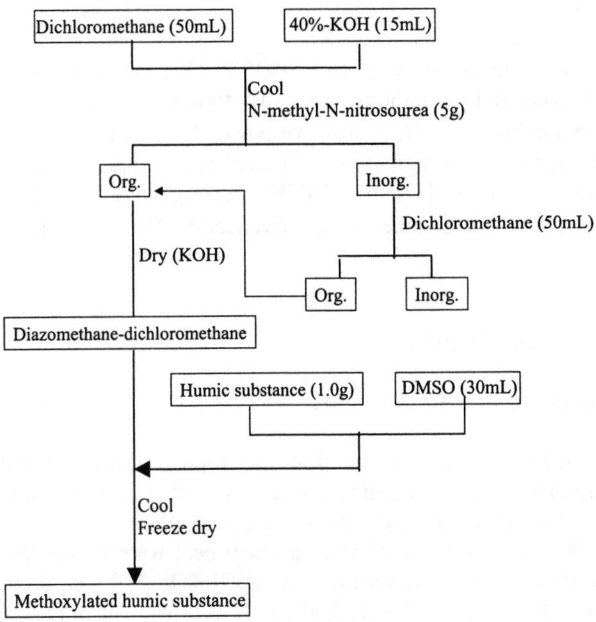

Figure 2 *Methoxylation of humic substance with diazomethane*

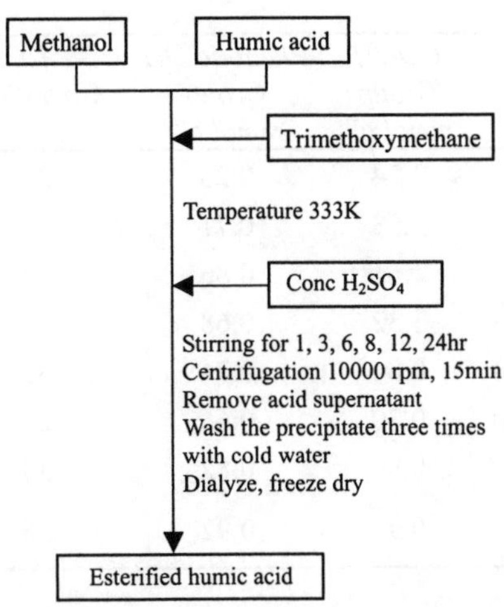

Figure 3 *Methylesterification of humic substance with trimethoxymethane*

2.4 MALDI-TOF MS Measurements[11]

A KOMPACT MALDI II mass spectrometer with a nitrogen laser (337 nm) was used for the measurements. The linear TOF mode was used with an accelerating voltage of 20kV. The full mass range of > 20 kDa was examined. The sample in DMSO solution was placed on the target and dried in a vacuum oven at 313K for 1h. Then, a solution of α-cyano-4-hydroxycinnamic acid (CHCA) in THF was applied over the sample as the selected matrix.

2.5 ESR Measurements

ESR measurements were performed with an ESP 300E (X-band 9GHz) spectrometer on solid samples. Experimental conditions were as follows; spectrometer modulation 100kHz, modulation amplitude 6.3 G, microwave power 1.0mW, gain range 3350±500 G. The relative concentration of ESR signals was obtained by the approximation based on the absolute signals for free radicals using DPPH (1,1-diphenyl-2-picrylhydrazyl; MW 394.24; 3.1×10^{20} spin g^{-1}). The g values were referenced to Mn^{2+}/MgO (g_3 = 1.981, g_4 = 2.034).

3 RESULTS AND DISCUSSION

3.1 Characteristics of the Peat Humic Acids

The analytical data of humic acid extracted from the tropical peat along with those from the other sources are given in Table 1. All results are expressed on a moisture and ash-free basis. It should be noted that both the content of carboxyl group and the concentration of free radicals of the humic acid extracted from tropical peat were higher than those from soils, including the IHSS standard HA sample. MALDI-TOF MS measurements showed that the molecular weight of the peat humic acid was lower than that from soil (Figure 4)

Table 1 *Characteristics of humic acids*

Sources ofhumic acid	*COOH Group mmol g^{-1}*	*Phenolic OH Group mmol g^{-1}*	*fa Aromaticity %*	*Free radical Concentration spin x 10^{18} g^{-1}*
Tropical peat (Indonesia)	3.24	0.25	23	6.81
Grass peat (Belarus)	1.73	0.71	24	1.14
Sphagnum peat (Canada)	1.61	0.86	26	3.01
Grass peat (Hokkaido)	1.32	0.68	22	0.37
Gray forest soil (Chiba, Japan)	0.84	0.74	32	0.47
Paddy field soil (Chiba, Japan)	0.70	0.66	26	0.10
Farm soil (Chiba, Japan)	0.86	0.82	30	0.28
IHSS standard soil	0.91	0.92	36	1.29

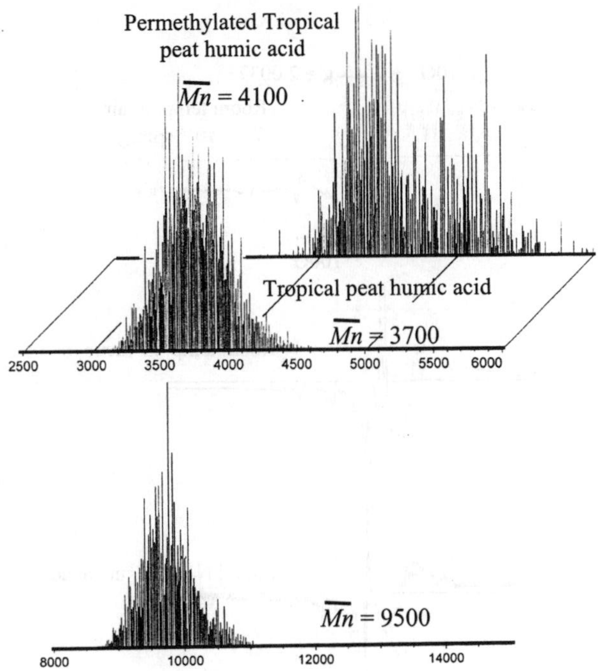

Figure 4 *MALDI-TOFMS spectra of peat and soil humic acids, sample:matrix=1:500*

3.2 Peat Humic Acid ESR Spectra

ESR spectra of humic acids usually consist of a single line that exhibits hyperfine splitting with g values ranging from 2.0031-2.0045 and line widths from 2.0G to 3.6G. The free radical concentrations vary from 1.4×10^{17} to 37.4×10^{17} spin/g.[12] The measured ESR spectra of the tropical peat humic acid are shown in Figure 5. The spectra consist of a single line with g value 2.0037, a line width of 3.6G and a free radical concentration of 6.8×10^{18} spin/g in air (Figure 5).

It was confirmed that although the molecular weight distribution and the aromaticity were lower, both the carboxyl group content and the free radical concentrations of the tropical peat humic acid were higher than for the other humic acids. Free radical concentrations increase with the carboxyl group content (Figure 6).

3.3 Permethylation of HA with the Hakomori Method

The Hakomori method can be applied to the permethylation of carboxyl and hydroxyl groups of HA, as indicated by the IR spectrum of the product, but the reaction rate is rather low under the prescribed reaction conditions used in this study. The FTIR spectrum of the methylated humic acid showed smaller R-C=O (1725 cm^{-1}) and -OH (3400 cm^{-1}) peaks than the unmethylated sample. The bands of phenolic hydroxyl and carboxyl groups were diminished by this methylation (Figure 7).

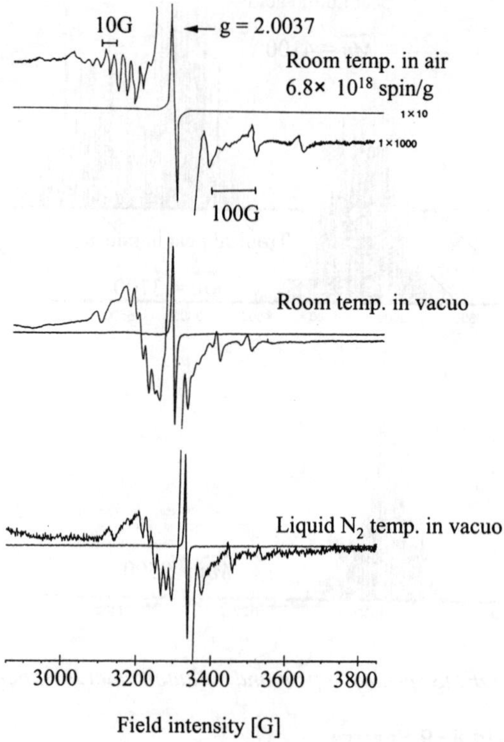

Figure 5 *ESR spectra of tropical peat humic acid*

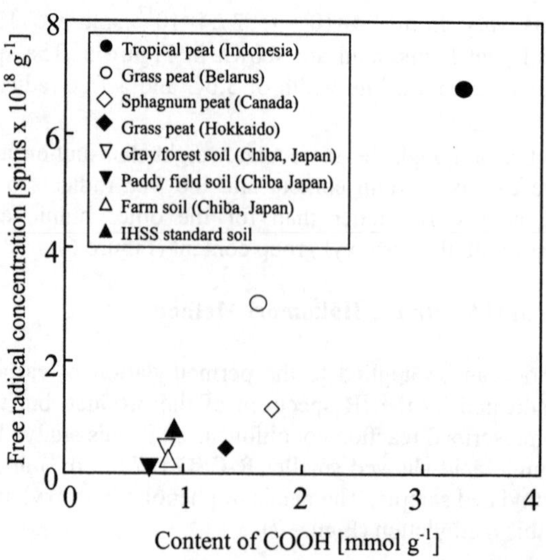

Figure 6 *Relationship of COOH contents and free radical concentrations in humic acids*

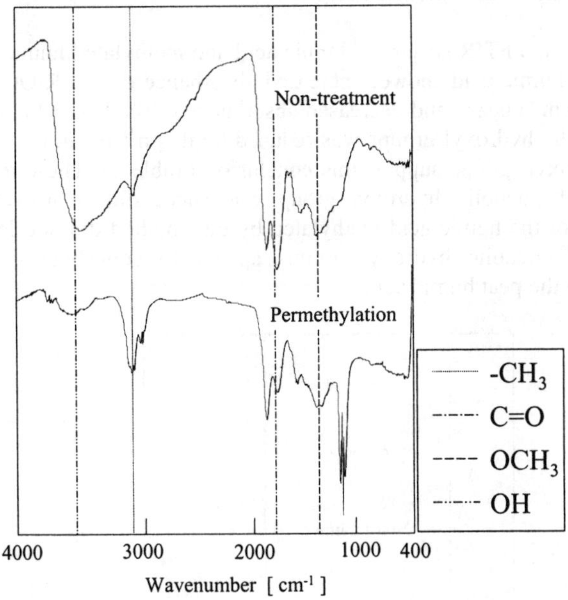

Figure 7 *FTIR spectra of permethylated and non-treated humic acid*

The free radical concentrations and the carboxyl group contents of the product decreased with the methylation time (Figure 8). Furthermore, a fairly good correlation was found between the free radical concentrations and the carboxyl group contents (Figure 6). This suggests that carboxylic groups contribute to free radical stabilization in the peat humic acid.

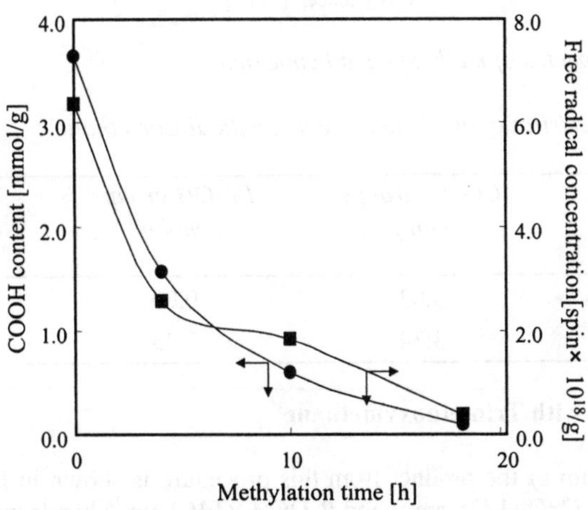

Figure 8 *Decrease of free radical concentrations with COOH contents as a function of methylation time*

3.4 Methylation with Diazomethane

Figure 9 shows the FTIR spectra of humic acid and methylated humic acid. The spectrum of methylated humic acid showed increased absorbance due to R-OCH_3 (1460 cm^{-1}) and R-CH_3 (2960 cm^{-1}) peaks and decreased absorbance of the R-C=O (1725 cm^{-1}) peak. The band of phenolic hydroxyl groups was reduced by this methylation. Chemical analysis of phenolic hydroxyl groups support this conclusion (Table 2). These results show that the methylation of phenolic hydroxyl group was successful. However, the free radical concentration of the humic acid methylated by this method did not decrease appreciably. Consequently, phenolic hydroxyl groups appear to contribute little to free radical stabilization in the peat humic acid.

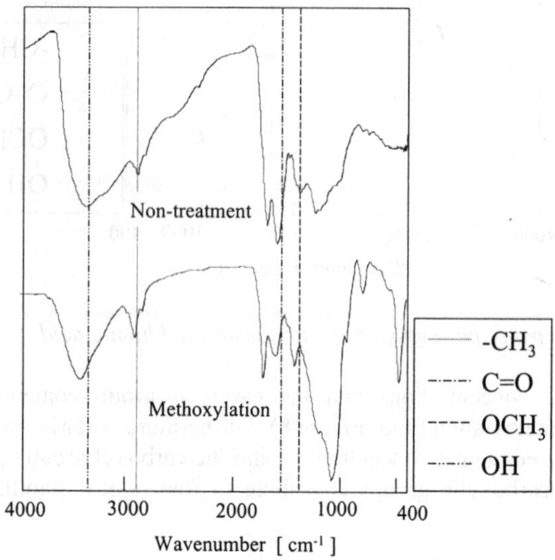

Figure 9 *FTIR spectra of methoxylated humic acid*

Table 2 *Characteristics of humic acid treated with diazomethane*

Sample	*COOH groups mmol g^{-1}*	*Ph-OH groups mmol g^{-1}*	*Free radical concentration Spin x 10^{18} g^{-1}*
Methoxylated HA	3.03	0.00	5.24
Non-treated HA	3.34	0.25	6.81

3.5 Methylation with Trimethoxymethane

The FTIR spectrum of the product from this procedure is shown in Figure 10. Increased intensity of the R-C=O (1725 cm^{-1}) and R-OCH_3 (1460 cm^{-1}) bands indicate esterification. Figure 11 shows the effect of the reaction time on the free radical concentration and the carboxyl group content in the humic acid sample. As the reaction proceeds, the carboxyl group content decreases. However, complete esterification of the carboxyl group was not achieved with this procedure even though the reaction time was prolonged. Furthermore, no substantial decrease of the free radical concentration was observed in the products.

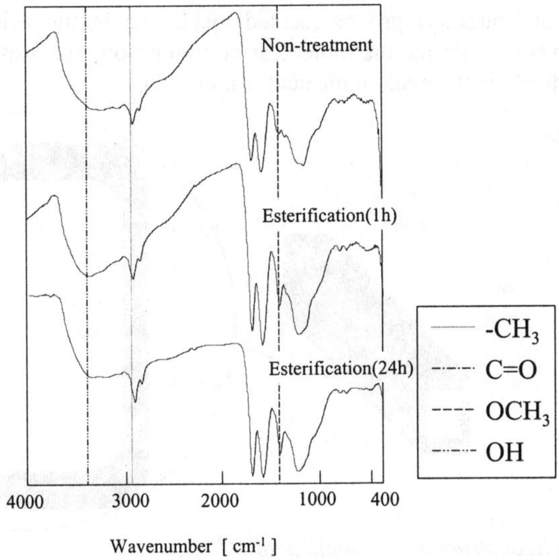

Figure10 *FTIR spectra of esterified humic acid*

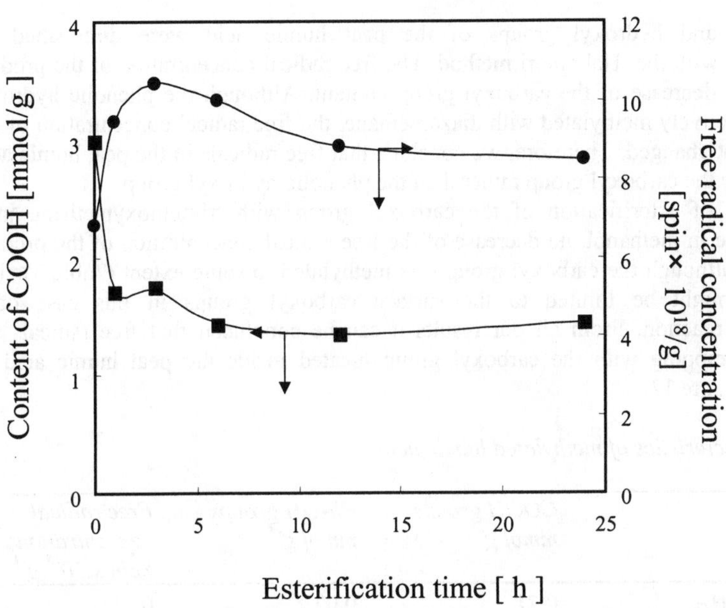

Figure 11 *Variations of COOH content and free radical concentration in humic acid with esterification time*

As mentioned in the experimental procedure, this reaction was carried out on a HA suspension. This means that the methylation would be limited to the carboxyl groups located on the surface of the humic acid in this heterogeneous reaction. These results suggest that the carboxyl groups located inside the humic acid, presumably forming hydrogen bonds to stabilize the molecular configuration, are responsible for stabilization of the free radicals in the peat humic acid (Figure 12).

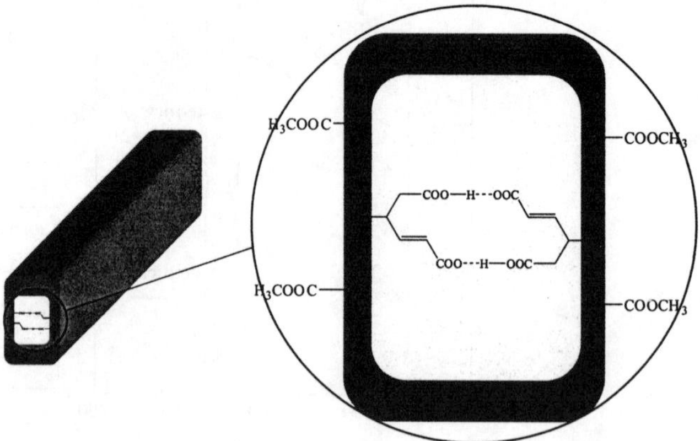

Figure 12 *Helical structure of humic acid*[13,14]

4 CONCLUSIONS

The carboxyl and hydroxyl groups of the peat humic acid were diminished by permethylation with the Hakomori method. The free radical concentration of the product decreased with decrease of the carboxyl group content. Although the phenolic hydroxyl group was selectively methylated with diazomethane, the free radical concentration of the product was not changed. Therefore, we conclude that free radicals in the peat humic acid are stablized by the carboxyl group rather than the phenolic hydroxyl group.

In the case of esterification of the carboxyl group with trimethoxymethane in a suspension state in methanol, no decrease of the free radical concentration of the product was observed although the carboxyl group was methylated to some extent (Table 3). The esterification might be limited to the surface carboxyl groups in this case of a heterogeneous reaction. From all our results it can be concluded that free radicals are stabilized by trapping with the carboxyl group located inside the peat humic acid as envisaged in Figure 12.

Table 3 *Characteristics of methylated humic acid*

Sample	*COOH groups mmol g^{-1}*	*Ph-OH groups mmol g^{-1}*	*Free radical concentration Spin x 10^{18} g^{-1}*
Permethylated HA	0.21	0.07	0.13
Methoxylated HA	3.03	0.00	5.24
Methylesterfied HA	1.44	0.23	8.50
Non-treated HA	3.34	0.25	6.81

References

1. N. Senesi and M. Schnitzer, *Soil Sci.*, 1977, **123**, 224.
2. T. Yamaguchi, N. Banbalov, H. Hayashi and M. Uomori, *13th Japan Humic Substance Society Symposium*, 1997, **13**, 20.
3. T. Yamaguchi, H. Hayashi and Y. Enomoto, *14th Japan Humic Substance Society Symposium*, 1999, **14**, 7.
4. T. Shinozuka, H. Hayashi, Y. Enomoto, H. Andoh, Y. Takiguchi and T. Yamaguchi, *16th Japan Humic Substance Society Symposium*, 2000, **16**, 11.
5. T. Yamaguchi, H. Hayashi, Y. Yazawa, M. Uomori, F. Yazaki and N.N. Bambalov, *Intl. Peat J.*, 1998, **8**, 87.
6. S. Hakomori, *J. Biochem.*, 1964, **55**, 205.
7. F.J. Stevenson, *Humus Chemistry*, Wiley, New York, 1982, p. 54.
8. J.E. Parker, C.A.F. Johnson, P. John, G.P. Smith and A.A. Herod, *Fuel*, 1993, **72**, 1381.
9. M. Schnizer and S.U. Khan, *Humic Substances in the Environment*, Elsevier, New York, 1972, p. 29.
10. K. Tsutsuki and S. Kuwatsuka, *Soil Sci. Plant Nutr.*, 1978, **24**, 547.
11. J.E. Parker, C.A.F. Johnson, P. John, G.P. Smith, A.A. Herod, B.J. Stokes and R. Kandiyoti, *Fuel*, 1993, **72**, 1381.
12. C. Saiz-Jimenez and F. Shafizadeh, *Soil Sci.*, 1985, **139**, 319.
13. G. Davies, A. Fataftah, A. Cherkasskiy, E.A. Ghabbour, A. Radwan, S.A. Jansen, S. Kolla, M.D. Paciolla, L.T. Sein, Jr., W. Buermann, M. Balasubramanian, J. Budnick and B. Xing, *J. Chem. Soc., Dalton Trans.*, 1997, 4047.
14. L.T. Sein, Jr., J.M. Varnum and S.A. Jansen, *Environ. Sci. Technol.*, 1999, **33**, 546.

STUDIES OF THE STRUCTURE OF HUMIC SUBSTANCES BY ELECTROSPRAY IONIZATION COUPLED TO A QUADRUPOLE-TIME OF FLIGHT (QQ-TOF) MASS SPECTROMETER

Robert W. Kramer,[1] Elizabeth B. Kujawinski,[1] Xu Zang,[1] Kari B. Green-Church,[2] R. Benjamin Jones,[2] Michael A. Freitas[1] and Patrick G. Hatcher[1]

[1] Department of Chemistry, The Ohio State University, Columbus, OH 43210, USA
[2] Campus Chemical Instrument Center, The Ohio State University, Columbus, OH 43210, USA

1 INTRODUCTION

Humic substances (HSs) are refractory materials that comprise the major organic portion of soils. They serve as a major reservoir of organic carbon in soils and marine sediments and play a key role in a number of geochemical cycles, including contaminant fate and transport. In order to understand these substances and their role in different processes, it is important that their structures be understood on a molecular level. This would allow specific interactions and structural characteristics of HSs to be better understood and applied to explain their behavior on a larger scale.

Several analytical techniques are used to study the general structural characteristics of humic substances. For example, solid state ramp cross polarization magic angle spinning ^{13}C nuclear magnetic resonance spectroscopy (CPMAS ^{13}C NMR) is a widely used technique that determines the relative abundance of different common functional groups in a sample.[1] Although a great deal of information can be obtained by this method, specific molecular structural information is elusive in complex samples such as humic acids. Other analytical techniques have been used to study humic acids,[2,3] but each has some disadvantage that limits precise structural characterization at the molecular level.

Electrospray ionization (ESI) mass spectrometry is a novel technique that has been applied recently to the characterization of humic substances.[4-10] ESI is a "soft" ionization technique in which ionizable compounds such as proteins, polar molecules and humic substances become charged by the action of a volatilizing nebulizer spray. This process has been shown not to fragment the components of similar molecules such as proteins.[11] It is thought that HSs will remain intact as well. This assumption is crucial in light of the debate on whether humic substances are high molecular weight macromolecules or aggregates of noncovalently linked molecules such as sugars, carbohydrates and fatty acids.[12]

This paper describes the application of ESI ionization coupled to a quadrupole time of flight mass analyzer to highlight differences among several HSs. We chose these Armadale and diluvial Japan humic substances as our analytes because of the large difference in structural composition as determined by CPMAS ^{13}C NMR. The ESI-QqTOF method is capable of achieving resolving powers in excess of 10,000, which is sufficient to resolve many of the peaks in the spectrum of HSs. The resolving power was estimated using the mass to charge ratio (m/z) of a peak relative to its full width at half height. Much higher resolving power can be attained for HSs with other techniques such as Fourier transform ion cyclotron resonance (FT ICR) mass spectrometry.[6,10] The QqTOF analyzer was chosen because of its robust and sensitive nature and its ability to show little mass discrimination over a relatively wide range of masses. Several types of adducts are possible, such as H^+, Na^+, K^+ and NH_4^+, but only H^+ and Na^+ were expected in these samples, as demonstrated previously.[10] The sodium ion would be expected as a result of extraction in sodium hydroxide. However, it appears that the peaks in the samples of this study consist mostly of hydrogen adducts.

Positive ion mode was chosen over negative ion mode to limit the number of multiply charged peaks in the spectra. It is well known that significant amounts of carboxyl groups are present in HSs and these could be deprotonated, resulting in multiply-charged ions. Several reports[6,8,9] have shown that while the positive ion mode is less sensitive than the negative ion mode it produces predominantly singly-charged species, whereas the negative ion mode is more conducive to production of multiply charged species that complicate interpretations.[8]

Using this technique together with the information about general structure obtained from other techniques such as NMR, one can assign molecular structures to peaks within the mass spectra. The negative mass defect was also used in determining the nature of molecules that have m/z ratios at positions well above or below the nominal mass. Since hydrogen has an exact mass of 1.0078 Da, the more hydrogen present in a molecule the higher above the nominal mass the peak will appear. Conversely oxygen, with an exact mass of 15.9949, lies slightly below the nominal mass. As a result, molecules with a high oxygen content and/or multiple bonds will appear at or slightly below the nominal mass. Use of this approach is particularly important in the case where resolution is limited and compounds having a high mass defect are present within the mixture.

While several recent reports have been published on the use of ESI coupled to various mass spectrometers as described above, we wish to make a distinction between the reports published to date and this current study. Most previous reports have focused on details of the method as applied to one or a limited set of samples, usually fulvic acids because of their high solubility and availability. The reported spectra all generally showed a broad distribution of peaks extending to 1-2 kDa, but most of the peaks were present in mass ranges less than 1000 Da. A peak is observed at every mass unit on instrumentation with limited mass resolving power. Using high resolution instrumentation, Brown and Rice[6] and Kujawinski et al.[10] have shown that the peaks observed in low resolution instruments actually are clusters of multiple peaks, often with more than 5-10 peaks per cluster. Aside from some differences in the mass ranges and peak maxima, minimal structural information can be gleaned from the massive data sets generated. Attempts by Plancque et al.[9] to obtain MS/MS data to aid structural assignments provided less than convincing information on possible structural entities, mainly because the limited resolution of the quadrupole portion of the QqTOF allows the multitude of peaks to enter the collision cell to be fragmented. As part of an ongoing investigation of humic substances by both high and medium resolution ESI/MS, we have observed significant differences among humic substances from varied sources.[10] This report focuses on a more detailed evaluation of the mass spectra than was possible before[10] and is an attempt to relate humic structures to the combined mass spectral and NMR data. We specifically focus on peaks that can be readily discerned because they have a mass discrimination that aids structural assignment.

2 METHODS

2.1 Sample Preparation

Several HSs were used in this study: the humic and fulvic acid fraction from the Armadale soil[13,14] and humic acids from a diluvial soil from Iwata, Japan.[14,15] These samples were supplied by M. Schnitzer (Agriculture Canada, Ottawa, Ontario). Each sample was obtained as dried, ash-free humic acid. The stock solutions were prepared by dissolving the samples to 1 mg mL^{-1} in aqueous ammonia at pH 8. These stock solutions were then diluted with methanol or isopropanol before analysis with the ESI QqTOF mass spectrometer.

2.2 Instrument Parameters

Solid-state ^{13}C NMR spectra were obtained by the method of cross polarization and magic angle spinning (CPMAS) using 4 mm rotors spun at 13 kHz frequency as described by Dria et al.[16] The ramp cross polarization method was used with two-pulse-phase modulation decoupling added. Spectra were referenced to tetramethysilane (0 ppm) using glycine as a surrogate chemical shift standard. Contact times were 2 ms and recycle delays of 1 s were used.

All of the mass spectra were obtained with a Micromass Q-TOF™ II mass spectrometer equipped with an orthogonal electrospray ion source operated in the positive ion mode. Positive ion mode was chosen in preference to negative ion mode to minimize multiply charged species. The HSs were analyzed in a 50:50 water: methanol or isopropanol solution and injected into the electrospray source at a rate of 5-10 μL min^{-1}. Optimum electrospray conditions were cone voltage 60 V, capillary voltage 3000 V and source temperature 100°C. The ESI gas was nitrogen. The initial quadrupole Q1 was set to pass ions from 100-2000 Da, and all ions transmitted into the pusher region of the TOF mass analyzer were scanned over m/z 100-3000 with an integration time of 1 s. Data were acquired in continuum mode until the averaged data were acceptable (10-15 min).

3 RESULTS AND DISCUSSION

3.1 NMR Data

Before beginning the interpretation of the ESI-QqTOF spectra of the two HAs it is useful to discuss the NMR spectra of the samples to determine what general types of compounds are likely to be present in each sample (Figure 1). The NMR spectrum of each HS provides the semi-quantitative distribution of functionalities present in the samples.

The difference in the general structural nature of these samples is evident from the NMR spectra. The CPMAS ^{13}C NMR spectrum of the Armadale humic acid shows that this sample is very rich in aliphatic structures, indicated by the intense signals between 0 and 110 ppm. Carboxyl and carbonyl (160-200 ppm), carbohydrate-like (60-110 ppm) and aromatic (110-160 ppm) structures are also prevalent in the sample. The CPMAS ^{13}C NMR spectrum for the Armadale fulvic acid is somewhat similar to that of its corresponding humic acid but is relatively richer in carboxyl (160-200 ppm) groups, as expected, and aromatic structures (110-160 ppm). The signals for carbohydrates or alkoxyl groups (60-110 ppm) and paraffinic carbons (0-60 ppm) are relatively less plentiful. From these spectra and the permanganate oxidation data of Ogner and Schnitzer[13] and Matsuda and Schnitzer,[14] we expect the Armadale samples to be rich in fatty acid type structures with some contributions from carboxylated aromatic compounds. In contrast, the dominant signals in the diluvial Japan humic acid spectrum occur in the aromatic region (100-160 ppm) and to a lesser extent in the carboxyl region (160-200 ppm). Previous studies have attributed these peaks to benzenecarboxylic acid-dominated structures.[15]

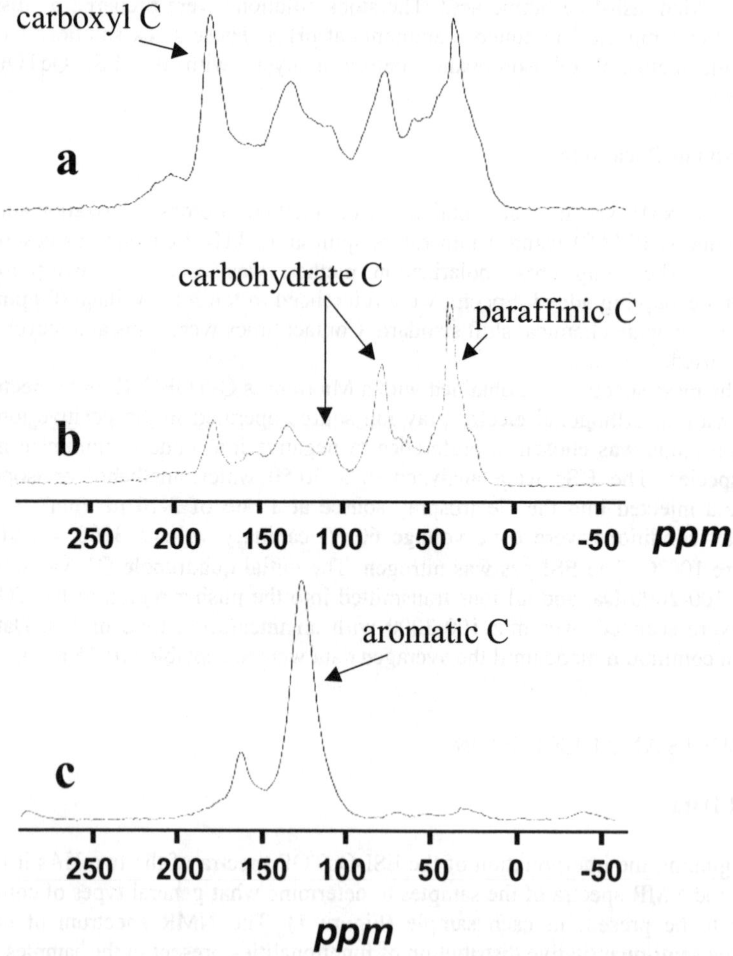

Figure 1 *Solid-state CPMAS ^{13}C NMR spectra of (a) Armadale FA (b) Armadale HA (c) diluvial Japan HA*

3.2 ESI-QqTOF Data

The NMR data established a structural foundation for further study using ESI-QqTOF mass spectrometry. Using this information together with the concepts of mass discrimination it will be possible to associate many peaks in the mass spectra with specific structures. While the MS/MS capabilities of the Qq-TOF instrument can be employed to assist in mass assignment as recently demonstrated,[9] we discovered that overlap of peaks in spectra of humic acids prevents appropriate use of this technology. This was readily apparent from studies of humic acids by ESI coupled to a high resolution Fourier transform mass spectrometer.[10] From these studies it is clear that a cluster of peaks exists at every nominal mass and that inability to select one of the multitude of peaks for collision induced dissociation leads to ambiguous results.

3.3 The Full Mass Range

The full mass range of data in the ESI-QqTOF mass spectra of the Armadale and the diluvial Japan humic acids extends to 3000 Da, but peaks are mainly observed in the mass range extending to 1000 Da. The data in Figure 2 show how two very different humic acids compare. Some important differences can be noted in the spectra. First, the Armadale humic acid exhibits a major cluster of peaks in distinct mass ranges 75-200 Da, 250-400 Da and 550-750 Da. The diluvial humic acid exhibits clusters in the same first two mass ranges but also contains two more clusters that are different. One exists between 500-600 Da and the other extends from 750-900 Da. Of the two mass ranges common to both humic acids (75-100 Da and 250-400 Da), the distributions of peaks are noticeably different between the two samples. Thus, we conclude that the ESI-QqTOF mass spectra reflect the observed fact from NMR measurements that these two humic acids structurally are very different.

The mass spectrum for the Armadale fulvic acid is shown plotted from 0-1000 Da in Figure 3. Comparing this spectrum with that of the corresponding humic acids reveals some great similarities. In fact the spectra are nearly identical, with peaks and regions of peaks matching precisely. Such a great similarity might be expected based on the sample origins, but is surprising considering the differences observed by NMR. At this time we do not have an explanation for this observation except to say that a more detailed assessment of the peaks might explain the apparent discrepancies.

A general observation also made in previous studies[8,9] is that the majority of intense peaks observed for humic substances are mostly below an m/z of 1000 Da. This suggests that humic substances generally are small molecules. Piccolo and Conte[12] have proposed that humic substances are supramolecular associations of small molecules and that the high average molecular weight determinations previously made by gel permeation methods were in error because the small molecules have tendencies to form aggregates. If this is true then ESI probably destroys these aggregates, thus explaining the low molecular weight range observed in ESI spectra. However, it is too soon to use this argument to explain the data because closer examination of the spectra indicates that broad, unresolved peaks at nearly every mass unit extend to high masses of 2-3 kDa. These broad peaks contribute substantially to the total area of intensity in each spectrum, but their intensity on the vertical axis is suppressed relative to the intense signals from pure components because they consist of a series of overlapping peaks. The pure component peaks are relatively more intense, but they contribute a smaller proportion of the total area. Thus, when the spectra are scaled to peak heights the pure components dominate the vertical axis. This effect can be seen clearly in Figure 4, which shows an expanded region of the spectrum for the two humic acids. As discussed below, most of the intense peaks can be assigned to fatty acids, pure compounds whose peak would only be broadened by processes that depend on the resolving power of the instrumentation. Other peaks in the region appear to be broadened by the fact that a multitude of peaks with unresolved masses are overlapping. These give rise to the envelope of peaks discussed above and appear to account for most of the signal intensity when one considers the entire mass range.

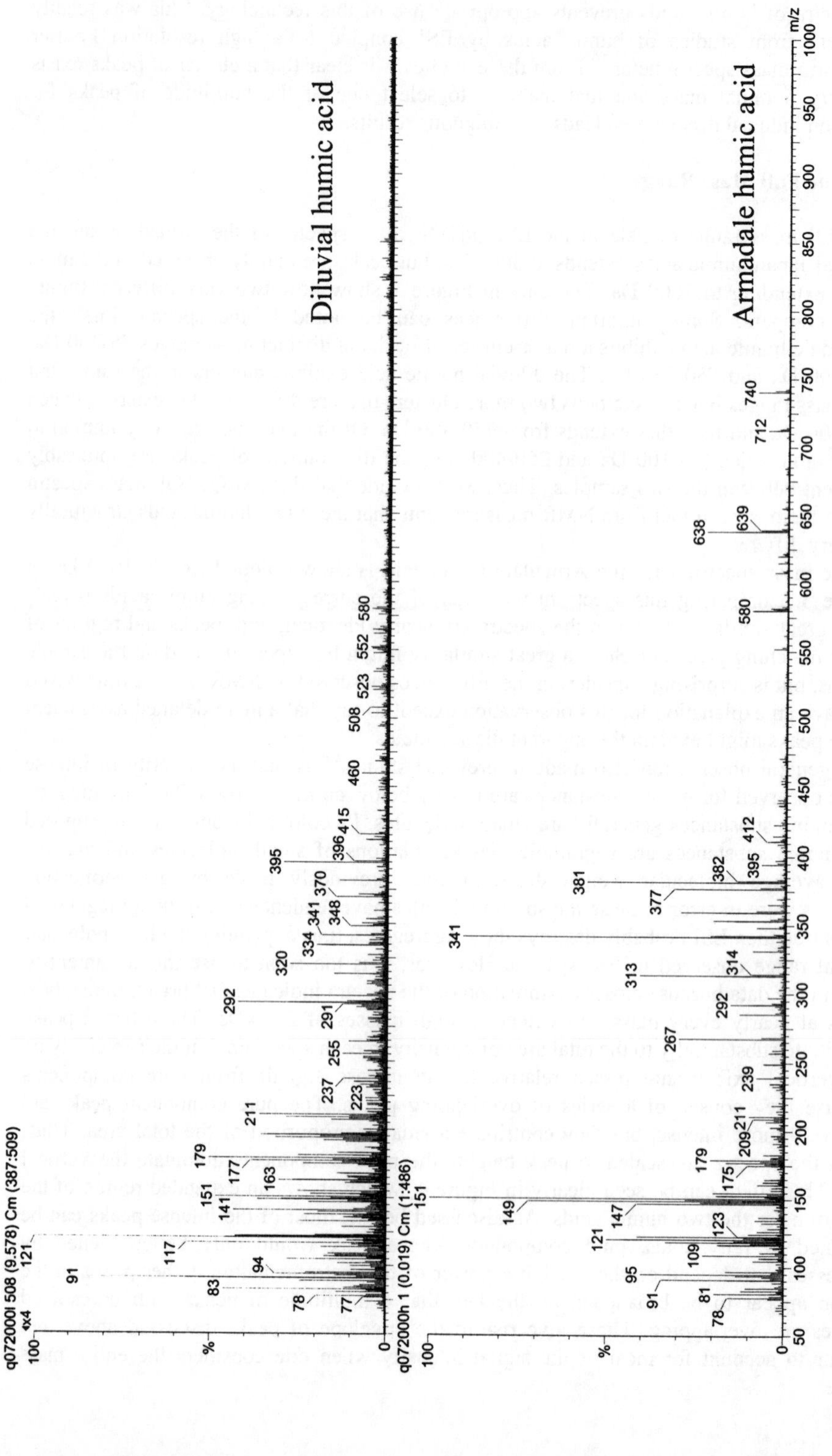

Figure 2 *ESI-Qq-TOF mass spectra of humic acids plotted over the mass range 50-1000 Da. Mass values above peaks have been rounded to the nominal mass*

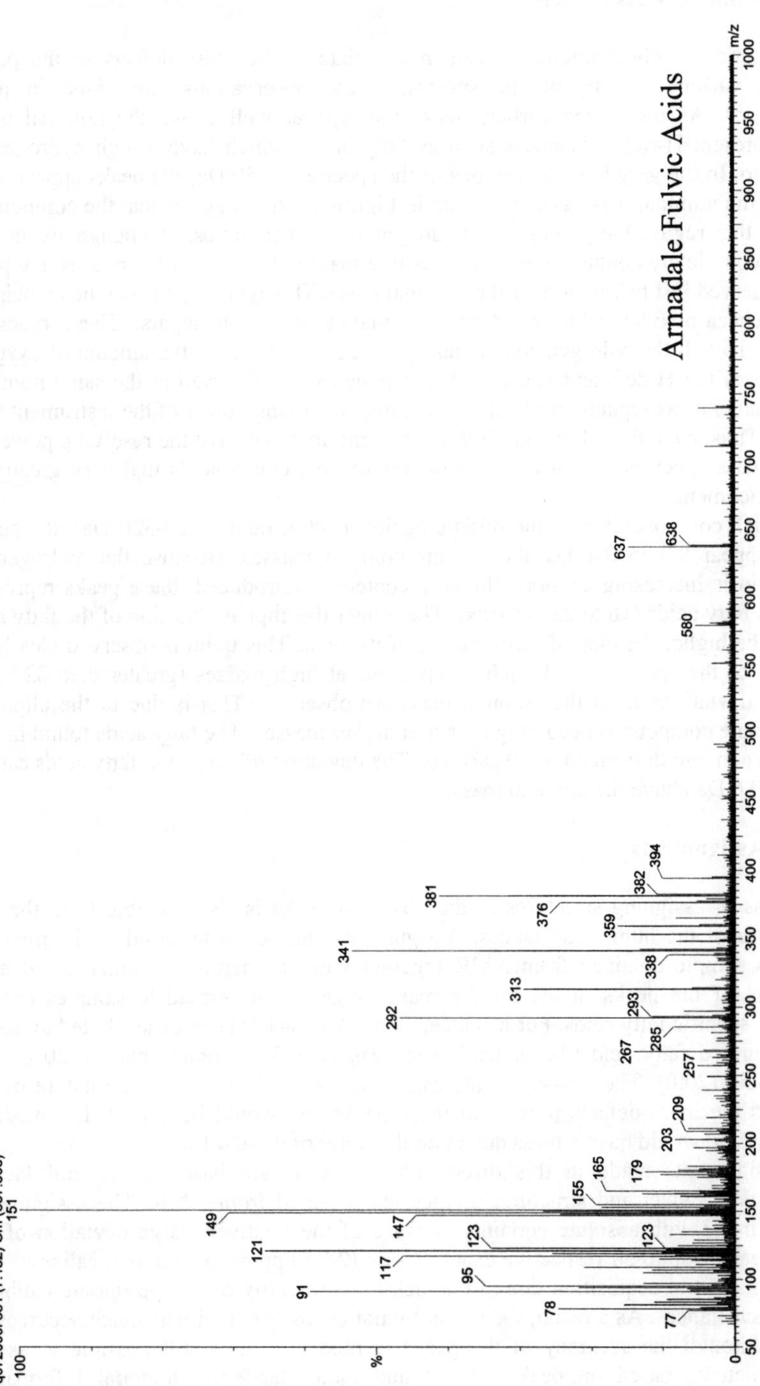

Figure 3 *ESI-Qq-TOF mass spectrum of the Armadale fulvic acids plotted over the mass range 50-1000 Da. Only nominal masses are given*

3.4 Effects due to Mass Defects

The next series of observations we can make relate to the mass defects of the peaks present in various regions of the spectra. These observations can assist in peak identifications. As mentioned earlier, peaks that appear well above the nominal mass usually represent H-rich substances such as fatty acids, which have a high hydrogen to oxygen ratio. In the very low m/z regions of the spectra (0-350 Da) the peaks appear at or just below the nominal mass as can be seen in Figure 4. This suggests that the compounds present in this region have very low hydrogen to oxygen ratios. If enough oxygen is present to completely counterbalance the positive mass defect due to hydrogens, the peak mass is observed just below or near the nominal mass. This type of peak can be associated with benzenecarboxylic acids, condensed aromatics and even sugars. These types of compounds have little hydrogen content and often contain a substantial amount of oxygen. In some cases the H-deficient and H-rich components are observed at the same nominal mass but can exist as separate peaks due to the high resolving power of the instrument (see Figure 4). Thus, even though the QqTOF instrument does not have the resolving power of a FTICR mass spectrometer, it can nonetheless resolve components that vary greatly in their H-enrichment.

For H-rich components near the middle of the spectral range (220-320 Da), the peaks begin to appear 0.3 to 0.4 Da above their nominal masses. Because the hydrogen to oxygen ratio is increasing as more aliphatic content is introduced, these peaks represent mid-length fatty acids (16 to 22 carbons). The longer the aliphatic portion of the fatty acid becomes, the higher the mass discrimination of its peak. This trend is observed clearly in this region of the spectra. For H-rich components at high masses (greater than 320 Da) very large deviations from the nominal mass are observed. This is due to the aliphatic chains in some compounds becoming longer at higher masses. The fatty acids found in this region range in length from 24 to 40 carbons. The deviation of very long fatty acids can be as high as 0.6 Da above the nominal mass.

3.5 Peak Assignments

The process of assigning structures to the observed peaks is also possible from the full mass spectra of the humic substances. Assignments can be made based on information about each sample obtained from NMR together with the mass discrimination of each peak. Some of the peaks in the middle mass range of the Armadale samples can be assigned to specific fatty acids. For instance, the peak at 341 Da can be attributed to the 22 carbon saturated fatty acid (22:0), while the peak at 313 Da represents the 20 carbon saturated acid (20:0). These assignments can be made solely based on the nature of the sample and the mass defects present in these peaks. As would be expected, a medium length fatty acid would have a mass defect on the order of 0.3-0.4 Da.

The assignments made in this discussion (Table 1) are based on several factors including mass defect and structural predictions obtained from NMR. The assignments cannot be made with absolute certainty because of the relatively large deviation of the assigned peaks from their respective exact masses (2-250 ppm). An internal calibrant was not used in spectral acquisition due to the lack of availability of an appropriate calibrant for humic substances. As a result, the peaks cannot be analyzed with as much accuracy as desired. Although the accuracy of the peaks is moderate, it is still possible to assign several structures based on peak position and molecular-level structural information provided by previous studies.[13-15]

Figure 5 shows the mass spectra of the two humic acids from 0-500 Da. Many of the

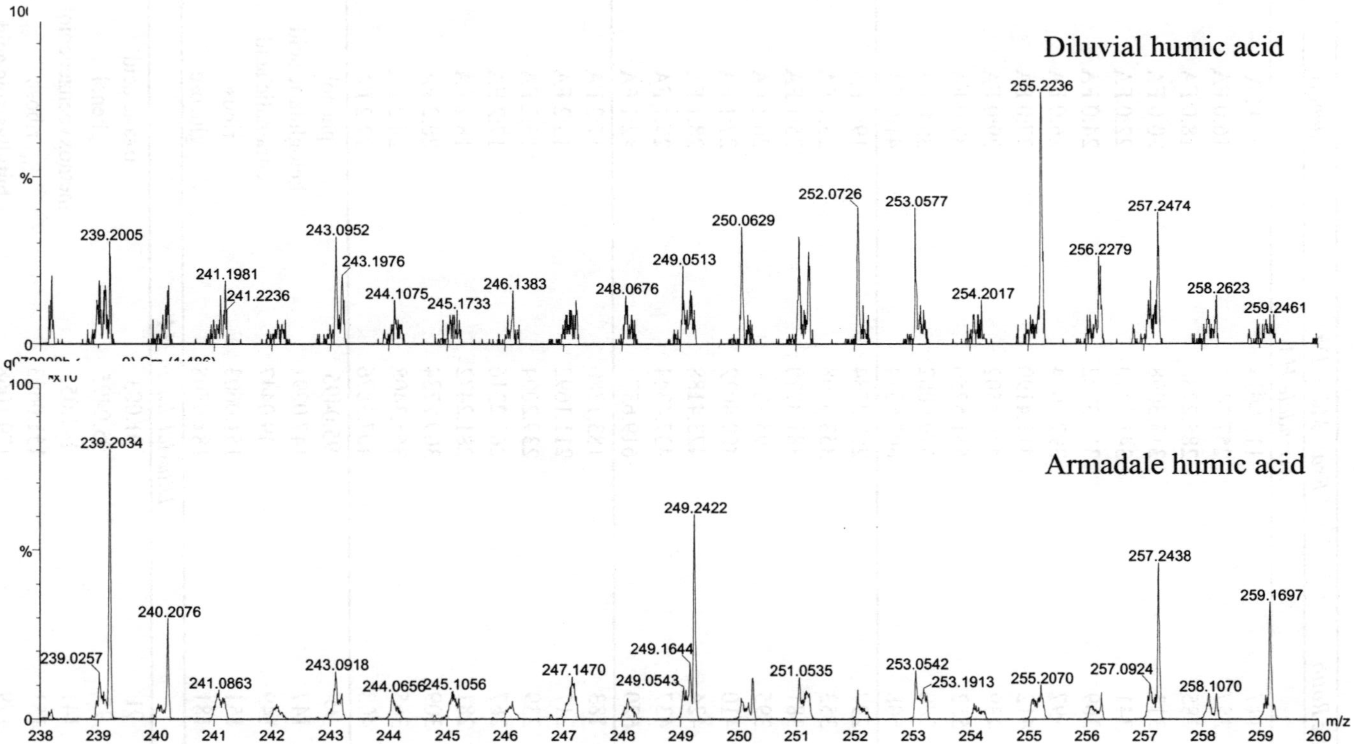

Figure 4 *Expanded region (238-260 Da) of the ESI-Qq-TOF mass spectra of humic acids. Exact masses are given above the peaks*

Table 1 *Tentative peak identifications for humic acids*

M/z Ratio	*Exact Mass, Da*	*Identity*
	Armadale HA	
117	117.0914	6:0 FA
257	257.2474	16:0 FA
285	285.2786	18:0 FA
313	313.3098	20:0 FA
341	341.3410	22:0 FA
369	369.3722	24:0 FA
397	397.4034	26:0 FA
412	411.4190	27:0 FA
439	439.4502	29:0 FA
552	551.5750	37:0 FA
580	579.6062	39:0 FA
608	607.6374	41:0 FA
297	297.2784	19:1 FA
353	353.3408	23:1 FA
381	381.3720	25:1 FA
395	395.3876	26:1 FA
410	409.4032	27:1 FA
424	423.4188	28:1 FA
437	437.4344	29:1 FA
620	619.6372	42:1 FA
183	183.1380	11:2 FA
211	211.1692	13:2 FA
239	239.2004	15:2 FA
267	267.2316	17:2 FA
281	281.2472	18:2 FA
309	309.2784	20:2 FA
365	365.3408	24:2 FA
408	407.3876	27:2 FA
95	95.0495	phenol
147	147.0291	ketoglutaric acid
149	149.0447	citramalic acid
151	151.0603	ribose
181	181.0708	glucose
	Diluvial Japan HA	
91	91.003	oxalic acid
95	95.0495	phenol
141	141.055	methoxybenzenediol
151	151.0603	ribose
179	179.1068	butylbenzoic acid
211	211.024	benzenetricarboxylic acid

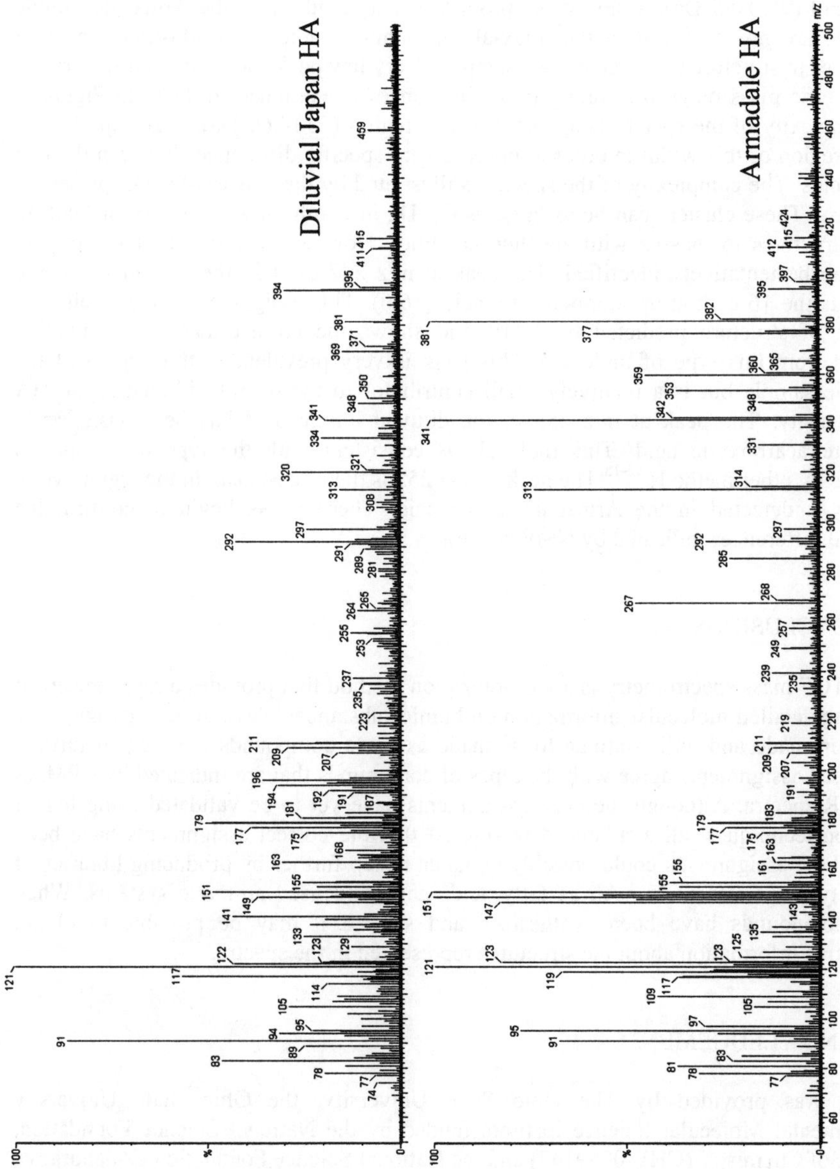

Figure 5 *ESI-Qq-TOF mass spectra of humic acids plotted over the mass range 50-500 Da*

peaks in this range can be more readily observed and assigned to specific molecular structures. Several fatty acids in the Armadale sample such as the 18 carbon saturated (18:0) and 25 carbon monounsaturated (25:1) fatty acids can be identified (Table 1). As mentioned previously, many peaks occur at or below the nominal mass, showing little or no mass defect. These peaks can be assigned to structures such as ribose (151.0603 Da) and phenol (95.0495 Da). Many of the prominent fatty acids from the Armadale sample (Table 1) are not observed in the diluvial Japan humic acid. This illustrates the vast differences in structure between the two samples. Very few peaks occur in both spectra.

A specific mass range illustrating these differences is examined in detail in Figure 4. The complexity of the spectra is apparent in this region (238-260 Da). It is important to study a region of this width in order to observe very specific differences between the two humic acids. The complexity of the spectra is illustrated by the cluster of peaks present at each mass. These clusters can be as large as 0.5 Da in width, making it very difficult to assign structures to masses with absolute certainty. However, a number of the intense peaks can be tentatively identified. The peak at m/z 257.2474 in the Armadale sample represents the 16 carbon unsaturated fatty acid (16:0). This assignment agrees well with the type of structure predicted by NMR and shows the correct mass discrimination expected from this type of molecule. This peak is very prevalent in this region of the Armadale sample but is a relatively small contributor to the overall diluvial Japan HA peak intensity. The peak at m/z 255 in the diluvial humic acid has been assigned to benzenetetracarboxylic acid. This molecule is consistent with the type of compound predicted by other methods.[13-15] The peak at m/z 255 is the largest peak in the region, yet it is nearly undetected in the Armadale humic acid. These peaks begin to confirm the structural differences indicated by NMR spectra.

4 CONCLUSIONS

ESI-QqTOF mass spectrometry is a soft ionization method that provides a rapid means of extracting detailed molecular information on humic substances. Several peak assignments have been made and will continue to be made as these compounds are studied further. These peak assignments agree with the types of compounds that are indicated by CPMAS ^{13}C NMR spectra. Although the peak assignments have yet to be validated using higher resolution techniques, all available data suggest that the correct assignments have been made. These assignments could possibly be taken a step further by producing libraries of various types of structures such as fatty acids and condensed aromatic systems. When model compounds have been synthesized and studied it may be possible to obtain quantitative information about the structures represented in the spectra.

ACKNOWLEDGEMENTS

Funding was provided by The Ohio State University, the Ohio State University Environmental Molecular Science Institute funded by the National Science Foundation, Division of Chemistry (CHE-0089147) and the National Science Foundation Collaborative Research in Environmental Molecular Science (CHE-0089172). The Ohio Board of Regents provided funds through the Hayes Investment Fund Competition for purchase of the Qq-TOF mass spectrometer. We thank Dr. Morris Schnitzer for providing the samples.

References

1. M.A. Wilson, *NMR Techniques and Applications in Geochemistry and Soil Chemistry*, Pergamon, Sydney, 1987.
2. M. Schnitzer and S.U. Khan, *Soil Organic Matter*, Elsevier, Amsterdam, 1978.
3. F.J. Stevenson, *Humus Chemistry*, 2nd Edn., Wiley, New York, 1994.
4. C. McIntyre, B.D. Batts and D.R. Jardine, *J. Mass Spectrom.*, 1997, **32**, 328.
5. A. Fievre, T. Solouki, A.G. Marshall and W.T. Cooper, *Energy Fuels*, 1997, **11**, 554.
6. T.L. Brown and J.A. Rice, *Anal. Chem.*, 2000, **72**, 384.
7. T. Solouki, M.A. Freitas and A. Alomary, *Anal. Chem.*, 1999, **71**, 4719.
8. J.A. Leenheer, C.E. Rostad, P.M. Gates, E.T. Furlong and I. Ferrer, *Anal. Chem.*, 2001, **73**, 1461.
9. G. Plancque, B. Amekraz, V. Moulin, P. Toulhoat and C. Moulin, *Rapid Comm. Mass Spectrom.*, 2001, **15**, 827.
10. E.B. Kujawinski, M.A. Freitas, X. Zang, P.G. Hatcher, K.B. Green-Church and R.B. Jones, *Org. Geochem.*, 2001, in press.
11. S.J. Gaskell, *J. Mass Spectrom.*, 1997, **32**, 677.
12. A. Piccolo and P. Conte, *Adv. Env. Res.*, 1999, **3**, 511.
13. G. Ogner and M. Schnitzer, *Can. J. Chem.*, 1971, **49**, 1053.
14. K. Matsuda and M. Schnitzer, *Soil Sci.*, 1972, **114**, 185.
15. P.G. Hatcher, M. Schnitzer, A.M. Vassallo and M.A. Wilson, *Geochim. Cosmochim. Acta*, 1989, **53**, 125.
16. K.J. Dria, J.R. Sachleben and P.G. Hatcher, *J. Environ. Qual.*, 2001 (in press).

CAPILLARY ELECTROPHORESIS OF HUMIC SUBSTANCES IN PHYSICAL GELS

M. De Nobili,[1] G. Bragato[2] and A. Mori[2]

[1] Dipartimento di Produzione Vegetale e Tecnologie Agrarie, University of Udine, via delle Scienze 208, 33100 Udine, Italy.
[2] Istituto Sperimentale per la Nutrizione delle Piante, Sezione di Gorizia, via Trieste 23, 34170 Gorizia, Italy

1 INTRODUCTION

Size exclusion chromatography can separate HSs into homogeneous molecular size fractions.[1,2] However, the estimated molecular size from the exclusion range given by manufacturers is not correct for HSs because their hydrodynamic size is different from those of the proteins or polysaccharides used for the calibration, and ion exclusion effects can frequently cause artifactual separations. An alternative method for fractionation of humic substances according to molecular size or for the analysis of their molecular weight distribution can reasonably be sought by the application of electrophoretic techniques. Conventional polyacrylamide gel electrophoresis (PAGE) was applied to HSs with the aim of separating them into distinct fractions,[3] but HSs showed bell-shaped electropherograms with a featureless appearance. PAGE did not achieve fractionation of HSs into defined bands. Nevertheless, a meaningful separation was obtained and HSs taken from the front of the migrating band displayed significantly different structural characteristics.[4] The separation into four distinct bands acheived by Troubetskoy et al.[5] is probably some kind of isotachophoretic mechanism. Conventional PAGE applied to HS fractions of reduced molecular weight polydispersity[3] showed that the electrophoretic mobility of humic substances is linearly related to the apparent molecular size as calculated from the cut-off limits of the ultrafiltration membranes from which the fractions were prepared. Moreover, extrapolation of electrophoretic mobilities to zero acrylamide concentration showed that all fractions had the same mobility in free solution.

Thanks to its versatility, low management costs and effectiveness, capillary zone electrophoresis (CZE) has spread quickly in the last ten years[6] and among many different applications is often adopted to characterise nucleic acids and proteins. Apart from the strength of the electric field and the molecular charge-to-size ratio of solutes, ion migration in CZE is determined by factors such as the capillary radius and the presence of the electroendoosmotic flow (EOF), which is activated by the ionisation of silanol groups at the inner capillary surface at pH above 2.5. The EOF causes solutes to migrate towards the cathode in spite of the direction of their electrophoretic velocity.

Separations are very poor when the charge-to-mass ratios of macromolecules are similar. In such a case, CZE in free solution is inapplicable and the use of sieving matrices, as in gel-filled capillaries, might be advantageous. However, gel shrinking, bubble formation and matrix hydrolysis shorten the lifetime of gel matrices[7] and make capillary gel electrophoresis unsuitable for macromolecules like HSs, which, because of their limited solubility at acid pH, need to be separated in neutral to alkaline buffers. An alternative strategy to CZE in free solution is the use of physical gels, i.e., entangled polymer solutions capable of forming a dynamic porous matrix with sieving properties. Physical gels are formed by hydrophilic, uncross-linked polymers - polyacrylamide (PAA), polyethylene glycol (PEG), polyvinyl alcohol (PVA), and so on - that overlap above a critical concentration called the entanglement threshold.[8,9] Some polymers - like PEG and PVA - can also interact with hydrogen-donor analytes by formation of hydrogen bonds.[10-12] This sometimes can even improve the separation of analytes, but it must be avoided if a separation based on molecular sizes is needed.

A few researchers have so far used CZE to characterize HSs.[13,14] They mainly sought a fingerprint characterisation of HSs, whereas the dynamic sieving of HSs in entangled polymers and the possible interaction of humic molecules, as hydrogen donors rich in phenolic and carboxylic groups, with physical gels was not considered. This is a brief review of studies made in our labotatories to investigate the applicability of CZE in physical gels to separate humic substances according to molecular size differences. This represents a first step towards the determination of the molecular size distribution of humic substances by capillary electrophoresis. The investigations were carried out on fractions of reduced molecular weight polydispersity obtained by ultrafiltration of HSs extracted from soil and peat samples, in coated and uncoated capillaries filled with solutions of PEGs of different chain lenghts and at different pH and ionic strength.

2 MATERIAL AND METHODS

2.1 Reagents

Polyethyleneglycol (PEG) of 400, 4,000, 20,000 and 35,000 relative molecular mass (M_r) and tris-hydroxymethylaminomethane (Tris) were from BDH (Poole, U.K.). Polyvinyl alcohol (PVA) 49,000 M_r and mesytil oxide (MSO) were from Fluka (Buchs, Switzerland). All reagents were of analytical grade.

2.2 HSs Extraction and Fractionation

HSs were extracted from a commercial sample of *Sphagnum* peat (Novobalt, Lithuania), dried and milled to pass a 0.5 mm sieve, and from the A1 horizon of a Spodosol, sieved at 2.0 mm after air drying. Extraction was done for 1 h at room temperature with 0.5 M NaOH under nitrogen flux. Extracts were filtered through a 0.2-µm cellulose nitrate filter (Whatman) and treated with Amberlite IR 120 H^+ cation exchange resin (Carlo Erba) in order to lower the pH to 7 and to remove excess sodium.

HSs extracts were then ultrafiltered on Diaflo YM and XM membranes (Amicon) with 0.1 M sodium pyrophosphate adjusted to pH 7.1. Five fractions of the following apparent M_r ranges, as deduced from the 95% cut-off limits of the membranes, were obtained: 1-5, 5-10, 10-30, 50-100 kDa. Ultrafiltration of each fraction ended when the solution flowing out of the cell was colourless. Fractions were exhaustively dialyzed with distilled water to eliminate pyrophosphate, concentrated on the same membrane to a final concentration of about 1 mg organic C per mL of water and stored at -18°C.

2.3 Capillary Zone Electrophoresis

CZE was performed with an ABI 270A-HT unit (Applied Biosystems) in three sets of 55-cm long fused-silica capillaries: uncoated capillaries (75 μm i.d.) (Composite Metal Services); μSil DB-WAX polyether-coated capillaries (100 μm i.d.) (J&W Scientific); and PAA-coated capillaries (50 μm i.d.) prepared in the laboratory as described by Hjerten.[15] Capillaries were 55 cm long (30 cm to the detector). CZE runs were carried out in a Tris-phosphate buffer solution at pH 8.3. Running conditions were as follows: temperature, 30°C; voltage, ±14 kV; detection wavelength, 360 nm. Samples were injected by hydrodynamic cathodic injection for coated capillaries and anodic injection for uncoated capillaries.

PEGs were dissolved in the buffer in different concentrations, whereas PVA 49,000 Da was added at a concentration of 2.5% (w/v). The effect of pH on EOF in uncoated capillaries was tested in the running buffer adjusted with either H_3PO_4 (pH 6.3) or Na_2HPO_4 (pH 10.3). The effect of ionic strength was investigated with 25 mM and 75 mM Tris-phosphate buffers. The EOF was determined using MSO as a neutral marker. MSO was dissolved in the running buffer to a final concentration of 1% v/v, submitted to CZE at the end of each series of measurements and detected at 210 nm. In uncoated capillaries, electropherograms show a baseline disturbance in correspondence with the MSO peak.[16] such disturbance was routinely used to measure the EOF in each run.

The effective mobility (μ_{eff}) of HS fractions was determined by Equn. (1),

$$\mu_{eff} = | \mu_{app} - \mu_{EOF} | \qquad (1)$$

where μ_{app} is the apparent electrophoretic mobility of fractions and μ_{EOF} is the mobility of EOF.[8,17] The latter is negligible in coated capillaries.

3 RESULTS AND DISCUSSION

3.1 Effect of EOF in Free Solution

Humic molecules are negatively charged at pH > 6 and in the presence of an electric field they migrate as a single symmetric broad band towards the anode.[2-4] In CZE, however, the EOF drives HSs towards the cathode when the capillary is uncoated and their separation is therefore modulated by a more or less effective anode attraction.

The different HSs fractions displayed very close mobilities in free solution, both in uncoated and in polyether-coated capillaries (Figure 1), confirming results previously obtained by conventional PAGE.[2] These findings suggest that relatively small differences in charge-to-mass ratios occur in the 1–100 M_r range, taking into account that differences in charge might also be partially masked by an enhanced hydration of molecules with higher charge densities. A different behaviour was observed in PAA-coated capillaries, where the apparent mass showed an opposite relationship with HSs migration times.[17] This fact may be a consequence of the partitioning of HS between the electrophoretic buffer and PAA.

The fact that mobilities appear to be independent of the molecular size should be regarded as an approximation due to the relatively low efficiency of the separation under the experimental conditions used. In fact, the same fractions analysed under similar experimental conditions, but in polyether DB-WAX capillaries 100-cm long (75 cm to the detector) showed an inverse relationship between size and migration time: the smaller the size, the faster was the migration.[17] However, the lack of observable charge effects is a perequisite for obtaining separations according to the size of molecules and subsequent work was carried out in 55 cm long uncoated capillaries.

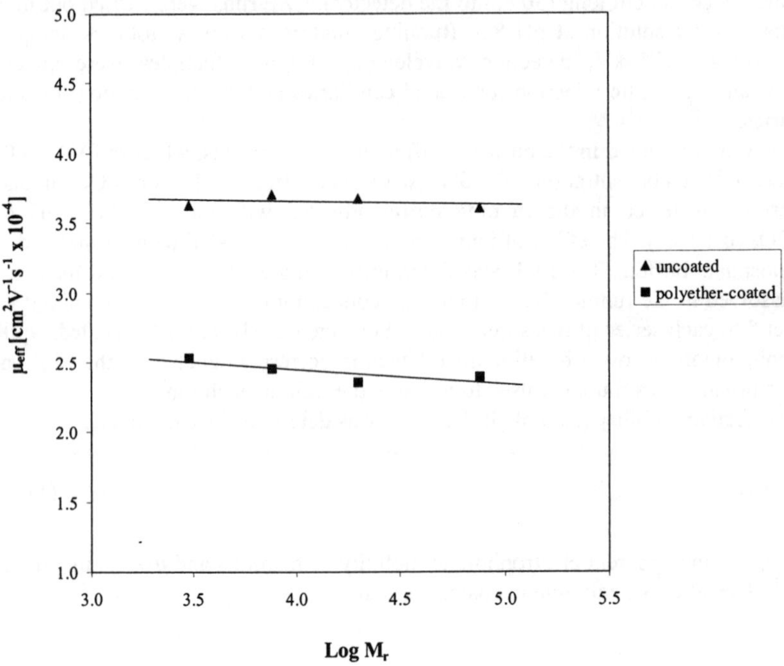

Figure 1 *Effective mobility (μ_{eff}) of HSs fractions from a peat versus the logarithm of their apparent molecular mass (M_r) in an uncoated and in a polyether-coated (DB-WAX) capillary (total length 55 cm, length to the detector 30 cm, 75 μm i.d.) filled with 50 mM Tris-phosphate buffer, pH 8.3. Running conditions: ±14 kV; 16 μA; hydrodynamic injection, 3 s*

3.2 HS Behaviour in Physical Gels

When CZE runs are carried out in buffer solutions containing a long chain hydrophilic polymer like PEG at a concentration above the entanglement threshold, the migration times of humic substances fractions increase (Figure 2).

Consequentially, mobilities of HSs fractions decrease, and an inverse relationship was found between effective mobility and the logarithm of mean apparent molecular mass. Figure 3 shows the effect of PEG 4,000 on the effective mobility of humic substances fractions at concentrations ranging from 2.5% (the entanglement threshold) to 15%. This decrease is probably attributable to the translational frictional coefficient that, in the presence of PEG, could amplify the effect of molecular size differences on hydrodynamic mobility. However, higher concentrations of PEG 4,000 did not improve the efficiency of separation. By contrast, the selectivity coefficient, on the contrary, decreased from 0.45

(2.5% PEG) to 0.29 (15% PEG). As the pore size of the polymeric matrix decreases with increasing polymer concentrations, the fact that separations were not improved suggests that humic molecules also are excluded from gel pores in 2.5% PEG and reptate in the space between entangled polymer chains.

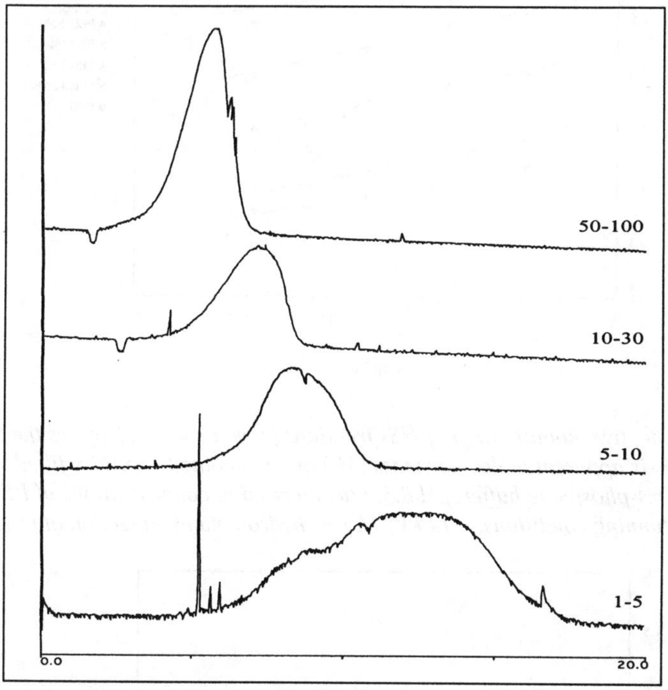

Figure 2 *Electropherograms of four HSs fractions (1-5, 5-10, 10-30 and 50-100 kDa) from a spodosol in an uncoated capillary filled with 50 mM Tris-phosphate buffer, pH 8.3, 5% PEG 4000. Running conditions: +14 kV; 16 μA; hydrodynamic injection at the anode, 3 s*

Similar behaviour was observed for PEGs of longer chain lengths (PEG 20,000 and PEG 35,000). However, the concentration range that could be tested was much more limited, because of the high viscosity of the solutions (Figure 4). The mobility of HSs increased with PEG 35,000, but the slopes of regression lines between mobility and log M_r were not significantly different from those obtained with PEG 4000.

Besides seiving, an alternative mechanism of separation of humic substances in entangled PEG solutions may arise from hydrogen-bonding interactions between hydrogen-donors and PEG.[12] However, the mobilities of HSs fractions were only marginally affected by the molecular size of PEG, suggesting that the separation is mainly governed by a sieving mechanism. This hypothesis was tested with PEG 400 solutions. Thanks to its reduced chain length, PEG 400 does not form physical gels at concentrations lower than 16%.

The addition to the buffer of PEG 400 at concentrations from 2.5% to 12.5% decreased effective mobilities of all fractions proportionally to the polymer concentration, but at any given concentration all fractions migrated at about the same velocity (Figure 5).

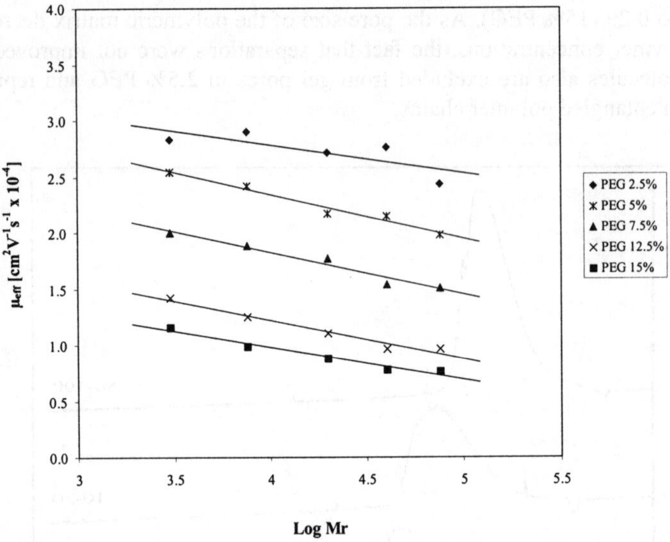

Figure 3 *Effective mobility (μ_{eff}) of HSs fractions from a spodosol versus the logarithm of their apparent molecular mass (M_r) in an uncoated capillary filled with 50 mM Tris-phosphate buffer, pH 8.3, plus increasing concentrations of PEG 4000. Running conditions: +14 kV; 16 μA; hydrodynamic injection at the anode, 3 s*

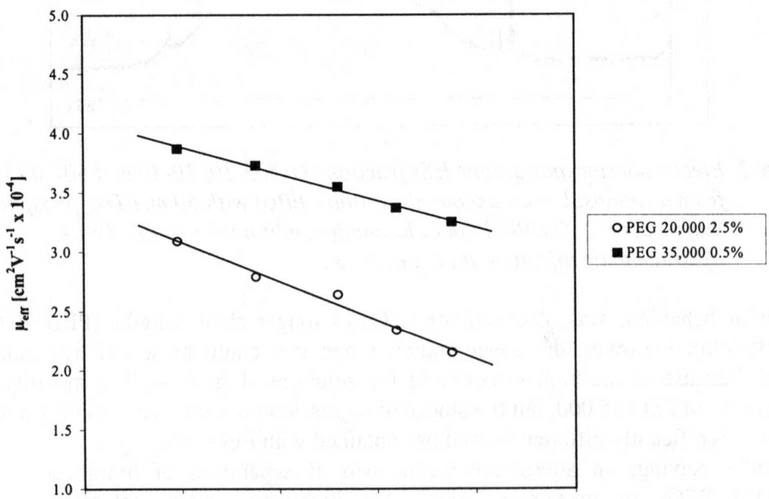

Figure 4 *Effect of increasing chain length of PEG on the effective mobility (μ_{eff}) of HSs fractions from a spodosol in 50 mM Tris-phosphate buffer, pH 8.3, 2.5% PEG 20,000 or 0.5% PEG 35,000. Running conditions: +14 kV; 16 μA; hydrodynamic injection at the anode, 3 s*

Such behaviour seems to exclude the possibility of describing the separation of HSs in PEG matrices in terms of hydrogen-bonding interactions. Instead, the effect of entangled PEG solutions on HSs migration appears to be the result of the following two factors: a decrease in the EOF caused by the corresponding increase in viscosity upon addition of PEG, and a sieving effect due to the formation of a physical gel. The effect is evident at low gel concentrations (2.5-5%) and for molecules of larger apparent molecular size, which are more strongly retarded in their migration to the cathode because of the larger probability of contacting the gel network.

Another UV transparent and low viscosity polymer that has been found to exhibit efficient sieving of SDS-protein complexes is PVA.[10] In previous work[18] we observed that contrary to PEG matrices, HS electropherograms in 2.5% PVA 49,000 showed two unresolved peaks whose relative aboundance depended on the PVA/humic substances ratios. This suggests that PVA interacts with HSs. The large number of hydroxo groups on the PVA chain as compared to PEG was considered a possible cause for the interaction via a hydrogen-bonding mechanism. In any event, PVA was not suitable for separating HSs in CZE.

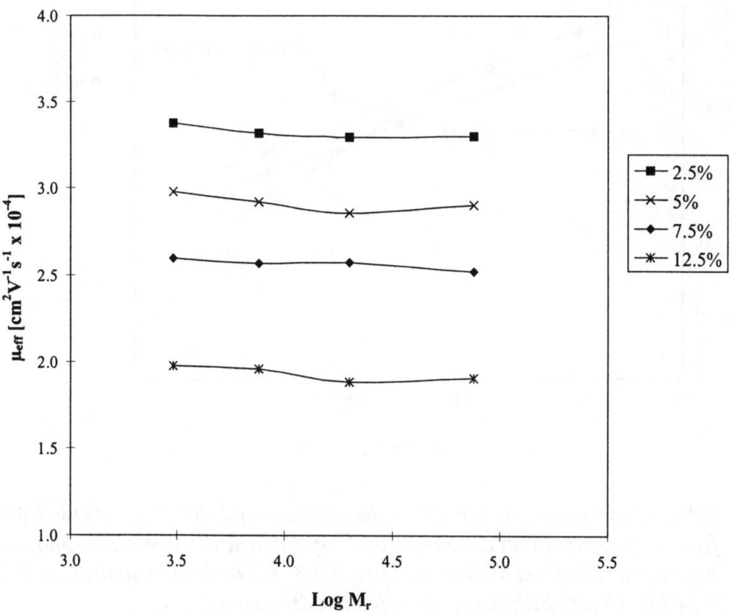

Figure 5 *Effective mobility (μ_{eff}) of HSs fractions from a peat in an uncoated capillary filled with 50 mM Tris-phosphate buffer, pH 8.3, plus increasing concentrations of PEG 400. Running conditions: +14 kV; 16 μA; hydrodynamic injection at the anode, 3 s*

3.3 Influence of pH and Ionic Strength

The effect of pH on the capillary electrophoretic behaviour of humic substances in physical gels is the result of two contrasting actions. The increase of buffer pH causes a further ionisation of weak acid groups of humic molecules, particularly of phenolic groups, which have an average pKa around 9.5, thereby increasing the negative charge on humic molecules. At the same time, the further ionisation of silanol groups on the capillary wall enhances the EOF. Positively charged amines sometimes are used as EOF modifiers as they interact with ionised silanol groups. The combination of these effects produced rather small changes in the migration times of humic molecules, whereas it increased the efficiency of separation at pH 10.3. The effective mobility of HSs is not greatly affected by a change of pH from 6.3 to 8.3. This probably is due to the fact that most carboxyl groups are already ionised at pH 6.3,[19] whereas phenolic groups dissociate above pH 8.5. A stronger phenolic character of low molecular weight fractions[18] could be the cause of the increased slope of the regression line exhibited at pH 10.3. Figure 6 shows the impact of pH on the mobility of HSs.

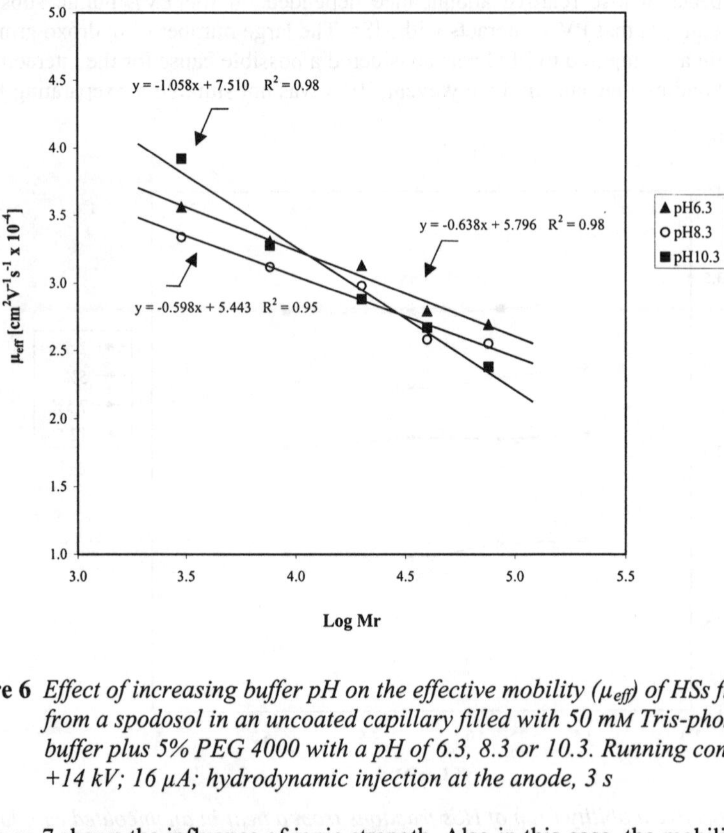

Figure 6 *Effect of increasing buffer pH on the effective mobility (μ_{eff}) of HSs fractions from a spodosol in an uncoated capillary filled with 50 mM Tris-phosphate buffer plus 5% PEG 4000 with a pH of 6.3, 8.3 or 10.3. Running conditions: +14 kV; 16 μA; hydrodynamic injection at the anode, 3 s*

Figure 7 shows the influence of ionic strength. Also in this case, the mobility of HSs is the combined result of changes in EOF and of the net charge of humic molecules. The latter originates from the screening due to increased concentration of counterions and also from possible conformational changes, and therefore acts again in the opposite direction from the EOF. In fact, the EOF decreases with the square root of the concentration of the buffer. The increase in ionic strength corresponding to a twofold increase in the buffer concentration actually reduces the effective mobility of HSs, but in a more concentrated buffer the correlation between mobility and apparent mass is effective only for medium and small-sized HSs.

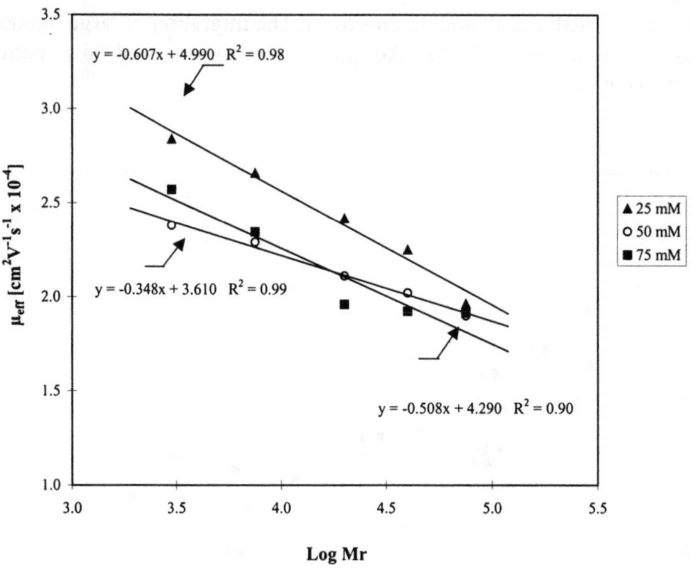

Figure 7 *Effect of increasing buffer concentration on the effective mobility (μ_{eff}) of HSs fractions from a spodosol in an uncoated capillary filled with 25, 50 or 75 mM Tris-phosphate buffer, pH 8.3, plus 5% PEG 4000. Running conditions: +14 kV; 16 μA; hydrodynamic injection at the anode, 3 s*

3.4 A Practical Application

An attempt to apply capillary electrophoresis in physical gels to characterise HSs extracted from *Sphagnum* peats in order to establish the geographical origin of a commercial sample has given encouraging results.[20] The characteristics of *Sphagnum* peats are related to the botanical origin and the degree of decomposition of the original plant material, which is commonly expressed by way of their *r* value. The mobilities of HSs extracted from peats of different geographical origin are plotted in Figure 8 against the corresponding *r* values. The bivariate plot displays a decrease of mobility from poorly decomposed peats of Finland (sample N) to the most decomposed Irish peat (sample A). The relationship between electrophoretic mobility of humic substances and the degree of decomposition of peats suggests that peat decomposition is accompanied by an increase in the size of humic molecules.

4 CONCLUSIONS

The linear relationship between electrophoretic mobility of HSs in entangled PEG solutions and their molecular size could certainly be exploited to determine the molecular size distribution of humic molecules by CZE. However, some problems still have to be solved. The main problem, which is common to other analytical techniques that cannot give absolute molecular weights, remains that of the calibration, as molecular weight standards of humic substances are not available at present. Other standards are specific to the application of CE in physical gels to humic substances: in particular, the separation efficiency is still limited and should be improved. The migration of large humic molecules in entangled PEG solutions probably takes place by reptation[21] and this would offer little room for improvement.

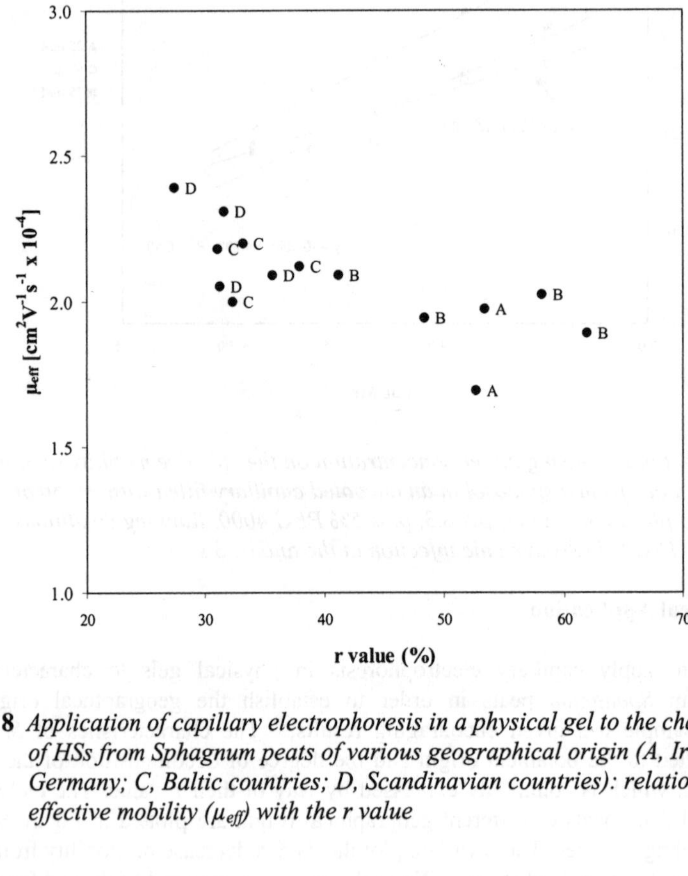

Figure 8 *Application of capillary electrophoresis in a physical gel to the characterisation of HSs from Sphagnum peats of various geographical origin (A, Ireland; B, Germany; C, Baltic countries; D, Scandinavian countries): relationship of effective mobility (μ_{eff}) with the r value*

References

1. M. De Nobili, E. Gjessing and P. Sequi, in *Humic Substances II: In Search of Structure*, eds. M.H.B. Hayes, P. MacCarthy, R.L. Malcolm and R.S. Swift, Wiley, New York, 1989, p. 561.
2. M. De Nobili and F. Fornasier, *Europ. J. Soil Sci.*, 1996, **47**, 223.
3. J.M. Duxbury, in *Humic Substances II: In Search of Structure*, eds. M.H.B. Hayes, P. MacCarthy, R.L. Malcolm and R.S. Swift, Wiley, New York, 1989, p. 593.
4. M. De Nobili, G. Bragato, J.M. Alcaniz, A. Puigbo and L. Comellas, *Soil Sci.*,1990, **150**, 763.
5. O.A. Trubetskoj, O.E. Trubetskaya, G.V. Afanas'eva, O.I. Reznikova and C. Saiz-Jimenez, *J. Chromatogr. A*, 1997, **767**, 285.
6. R. Weinberger, *Practical Capillary Electrophoresis*, Academic Press, New York, 1993.

7. T. Wehr, R. Rodriguez-Diaz and M. Zhu, *Capillary Electrophoresis of Proteins*, Dekker, New York, 1999, p. 239.
8. P.D. Grossman, in *Capillary Electrophoresis: Theory and Practice*, eds. P.D. Grossman and J.C. Coburn, Academic Press, Boston, 1992, p. 215.
9. C. Heller, *J. Chromatogr. A*, 1995, **698**, 19.
10. E. Simò-Alfonso, M. Conti, C. Gelfi, P.G. Righetti and C. Heller, *J. Chromatogr. A*, 1995, **689**, 85.
11. Y. Esaka, Y. Yamaguchi, K. Kano, M. Goto, H. Haraguchi and J. Takahashi, *Anal. Chem.*, 1994, **66**, 2441.
12. Y. Esaka, M. Goto, H. Haraguchi, T. Ikeda and K. Kano, *J. Chromatogr. A*, 1995, **771**, 305.
13. A. Rigol, J.F. López-Sánchez and G. Rauret, *J. Chromatogr. A*, 1994, **664**, 301.
14. C. Ciavatta, M. Govi, L. Sitti and C. Gessa, *Commun. Soil Sci. Plant Anal.*, 1995, **26**, 3305.
15. S. Hjerten, *J. Chromatogr. A*, 1985, **347**, 191.
16. K. Kenndler-Blachkolm, S. Popelka, B. Gas and E. Kenndler, *J. Chromatogr. A*, 1996, **734**, 351.
17. M. De Nobili, G. Bragato and A. Mori, in *Humic Substances-Structures, Properties and Uses*, eds. G. Davies and E.A. Ghabbour, Royal Society of Chemistry, Cambridge, 1998, p. 109.
18. M. De Nobili, G. Bragato and A. Mori, *J. Chromatogr. A*, 1999, **863**, 195.
19. E.M. Perdue, J.H. Reuter and M. Ghosal, *Geochim. Cosmochim. Acta*, 1980, **44**, 1841.
20. G. Bragato, A. Mori and M. De Nobili, *Europ. J. Soil Sci.*, 1998, **49**, 589.
21. M. De Nobili, G. Bragato and A. Mori, in *Understanding Humic Substances: Advanced Methods, Properties and Applications*, eds. E.A. Ghabbour and G. Davies, Royal Society of Chemistry, Cambridge, 1999, p. 101.

ARE THERE HUMIC ACIDS IN ANTARCTICA?

D. Gajdošová,[1] L. Pokorná,[1] P. Prošek,[2] K. Láska[2] and J. Havel[1]

[1] Department of Analytical Chemistry, Faculty of Science, Masaryk University, 611 37 Brno, Czech Republic
[2] Department of Geography, Faculty of Science, Masaryk University, 611 37 Brno, Czech Republic

1 INTRODUCTION

Soil is the earth's thin coat. It is a reservoir of water and nutrients and a participant in the cycling of carbon and other elements through the global ecosystem. Soil formation processes are influenced by such factors as topography, climate, organisms and time. Humic substances (HSs) and particularly humic acids (HAs) represent a major part of organic matter in soil. HSs are widespread and can be found in soils of all continents.[1-6]

Soil humification in Antarctica started millions of years ago. The process of humification depends on plants and organisms living there.[1] Because of extreme environmental conditions, the continent supports only a few cold-adapted land plants and animals. Plant and vegetation residues (algae, lichens and moss) fall on the soil where they are then transformed by microorganisms.[7]

The aim of this work was to monitor the occurrence and content of organic matter (OM) in Antarctic soil in different locations (different distance from the coast and different altitudes above sea level) of the vegetation oasis at Crepin Point in Admiralty Bay on King George Island (South Shetland Archipelago, West Antarctica).

King George Island, the largest island of the South Shetlands, lies between 61°50′S and 62°14′S and between 57°30′W and 59°02′W. The length of the island along the NE – SW axis is 75 km while from NW to SE it is only 30 km. The coastline is strongly differentiated, especially in the southern part where bays were formed as a consequence of tectonic predisposition and glacier activity.[8]

The island surface (about 1310 km^2) is 95 % covered by an ice cap created by more glacial domes. Numerous glacier tongues are present. Ice-free areas occur in the form of nunataks or coastal oases that are localized in several bays on the southern part of the island.[9]

The substrata samples of this study were collected at Crepin Point oasis. Organic matter was isolated and studied by chemical analysis, capillary zone electrophoresis (CZE), matrix assisted laser desorption/ionization time-of-flight mass spectrometry (MALDI TOF MS) and UV-vis spectroscopy.

2 MATERIALS AND METHODS

2.1 Antarctic Soil

The samples were taken in two profiles localized at the Peruvian Machu Picchu station (profile SLOPE and profile MORAINE) at the localities indicated in Figure 1.

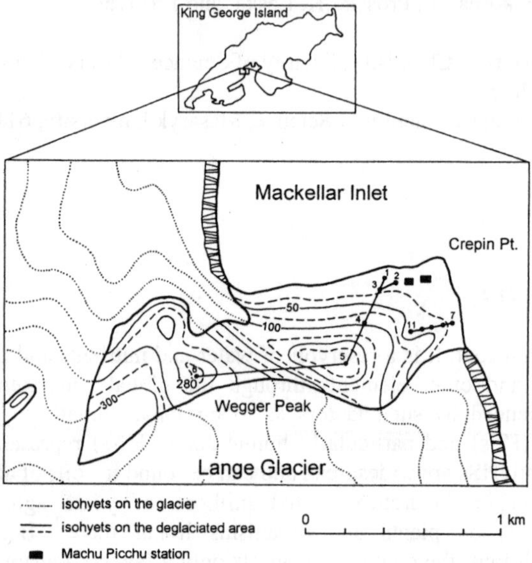

Figure 1 *Schematic map of the coastal oasis at Crepin Point as profiles SLOPE and MORAINE*

Profile SLOPE. Locality 1 and 2 – plane of the Machu Picchu Base about 10 m above sea level, partly inundated by melt water of the snow patches. Genetically uplifted and stabilized beach terrace built of beach pebbles with sand filler covered by mosses and insignificantly by *Deschampsia antarctica* (about 25 % of the whole surface). Locality 3 – slope floor about 20 m above sea level, built by regoliths filled by fine-grained material. Locality 4 – slope position with altitude about 100 m above sea level, situated to NE by inclination 30 – 35°. Slope surface built by debris filled by fine material. Locality 5 – top altitude 200 m above sea level. Surface built by weathered rock material (rock fragments with sand and gravel filling, solitary growth of lichens genus *Deschampsia*). Locality 6 – top altitude 300 m above sea level on a small plateau (surface about 25 m^2) inclined by 5 – 10° to the SSW. Surface partly covered by fine-grained material. Weathered rock material of the plateau change over sharp edge to rock walls.

Profile MORAINE. Localities 7 – 11 – the altitude of these points (located on N marginal moraine of the Lange Glacier) varies between 15 – 40 m above sea level. The moraine surface is built by typical unsorted mechanically weathered material with prevalence of regolith stones with filling of a fine-grained fraction between 0.6 – 2.0 mm.

Mineralogical classification of Antarctic soil is given in Table 1 (grain-size distributions were determined by sieving).

Table 1 *Grain-size distribution*

Sampling site	*Gravel %* (>2.0 mm)	*Sand %* (2.0-0.063 mm)	*Dust + Clay %* (<0.063 mm)
10 m	4.2	94.6	1.2
20 m	16.2	79.0	4.8
100 m	57.2	37.2	5.6
200 m	22.5	54.0	23.5
300 m	4.5	48.8	46.7
Moraine 1	44.2	38.1	17.7
Moraine 2	40.1	44.1	15.8
Moraine 3	36.0	44.2	19.8
Moraine 4	33.8	50.7	15.5
Moraine 5	50.2	40.5	9.3

2.2 Humic Substances

Soil HA Standard (1S102) was purchased from the International Humic Substances Society (IHSS). Chemapex HA Standard was obtained from Chemapex s.r.o. Prague, Czech Republic.[10]

Antarctic Organic Matter (OM). OM were extracted from Antarctic soil at 10, 20, 100, 200 and 300 m from the coast and from Antarctic moraine in different altitudes above sea level. The latter samples are labeled Moraine 1, 2, 3, 4 and 5.

2.1.1 Isolation of OM from Antarctic Soil. From several procedures described in the literature for extraction of HSs from soil,[11-15] we have applied sodium hydroxide because it is relatively fast and very efficient. The OM isolation procedure consisted of extraction of the soil (30 g) with 150 mL of 0.5 M NaOH for 10 hours and centrifuging for 10 min. Supernatant A was separated, 150 mL 0.5 M NaOH was added to the residue and the mixture was shaken (1 h). After centrifugation for 10 min, supernatant B was separated, 150 mL water was added to the residue and the mixture was shaken for 15 min. Supernatant C was separated and to added supernatants A and B in a glass bottle and adjusted to pH 1 with 1 M HCl at room temperature. The solution was filtered and the solid obtained was dried (60°C). The yield of OM was less than 1 % of the 30 g of soil used for extraction.

2.3 Characterization of Antarctic Soil and Organic Matter

2.2.1 Elemental Analysis. The total carbon, nitrogen and hydrogen contents of the Antarctic OM were determinated with a Perkin Elmer 2400 CHNS/O analyzer. Ash contents were determined by burning ~30 mg of OM in an oven at 900°C for 6 h. Silica, alumina and iron oxides content were measured with a Jobin-Yvon JY ULTRACE ICP spectrometer.

2.2.2 UV-vis Spectroscopy. Spectrophotometric measurements were made with a Unicam UV2 spectrometer. The samples were dissolved in 35 mM NaOH and diluted to concentration between 50-1000 mg/L.

2.2.3 MALDI TOF MS. Mass spectra were measured with a Kratos Kompact MALDI III in the linear positive mode. The instrument was equipped with a nitrogen laser (wavelength 337 nm, pulse duration τ = 10 ns, energy pulse 200 μJ). The energy of the laser applied was in the range of 0-180 units. For each sample 100 laser shots were used, the signals of which were averaged and smoothed. Insulin was used for mass calibration.

2.2.4 Capillary Zone Electrophoresis. A Beckman CZE (Model P/ACE) System 5500 equipped with a diode array detection (DAD) system, an automatic injector, a fluid-cooled column cartridge and a System Gold Data station was used. Fused silica capillary tubing of 47 cm (40.3 cm length to the detector) x 75 μm was used and the normal polarity mode of the CZE system was applied. Optimal conditions were found to be hydrodynamic injection 20 s, applied voltage 20 kV, detection at 210 nm and temperature 20°C. A mixture containing 90 mM boric acid, 90 mM TRIS and 1mM EDTA (pH 8.5) was used as a background electrolyte.

3 RESULTS AND DISCUSSION

3.1 Elemental Analysis

The results for total carbon, nitrogen and hydrogen content and ash (wt.%) analysis in Antarctic soil, OM and HSs are listed in Tables 2 and 3. Table 3 shows SiO_2, Al_2O_3 and Fe_2O_3 contents of OM extracted from Antarctic soil samples 10, 100, 300 m from the coast and from Moraine 4. The content of carbon in Antarctic soils and Antarctic OM is very low.[16]

Table 2 *Elemental analyses of Antarctic soils*

Sampling site	*C %*	*H %*	*N %*	*H/C ratio*
10 m	0.07	0.08	0.00	13.7
100 m	0.36	1.17	0.03	39.0
200 m	0.18	1.32	0.03	88.0
300 m	0.85	1.08	0.08	15.2
Moraine 1	0.09	0.25	0.01	33.4
Moraine 5	0.62	0.26	0.06	5.03

The H/C atomic ratio of one OM isolated from Antarctic soil is comparable with the ratio of the IHSS Soil standard. From the low H/C ratios of two of the samples it seems that some OM of Antarctic soil are aromatic.

The ash from Antarctic OM was analyzed and the results are given in Table 3. Ash contents of Antarctic OM are quite high. They are due to high contents of SiO_2, Al_2O_3, Fe_2O_3 or other oxides. More than ~70 % of the ash represents silica, alumina and iron oxides.

3.2 UV-vis Absorption Spectra

The absorption spectra of OM extracted from Antarctic soil sampled at various distances from the coast are given in Figure 2. Spectra are compared with absorption spectra of the

IHSS Soil HA Standard and the coal derived Chemapex HA Standard. The absorbances of solutions of OM extracted from Antarctic soil in the wavelength range 200-600 nm increase with increasing distance of the sampling site from the coast. On the other hand, the absorption spectra of solutions of OM extracted from Antarctic moraine (Figure 3) show some differences. Samples of soils from Antarctic moraine were taken at different altitudes above sea level. The composition of moraine soil differs most probably due to influence of the glacier movement.

Table 3 *Elemental analyses of humic substances*

Samples[a]	*C %*	*H %*	*N %*	*H/C ratio*	*Ash %*	*Ash analysis* SiO_2 *%*	Al_2O_3 *%*	Fe_2O_3 *%*
OM (10 m) from Antarctic soil	5.32 (26.0)	0.35 (1.7)	0.55 (2.7)	0.79	79.5	36.9	25.9	7.6
OM (100 m) from Antarctic soil	4.04 (17.0)	0.61 (2.6)	0.48 (2.0)	1.81	76.3	31.4	28.7	9.1
OM (300 m) from Antarctic soil	11.28 (19.2)	0.92 (1.6)	1.19 (2.0)	0.97	41.2	40.1	22.1	7.7
OM from Antarctic moraine (Moraine 4)	2.43 (4.6)	1.29 (2.4)	0.34 (0.6)	6.37	47.0	43.5	24.4	13.1
IHSS Soil HA standard	58.1	3.68	4.14	0.76	0.88	-	-	-

[a] valus in parantheses are the corresponding ash-free contents.

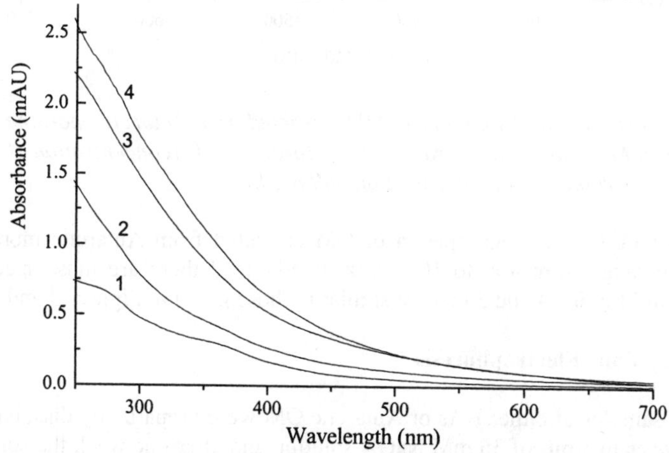

Figure 2 *UV-Vis spectra of solutions of IHSS Soil standard (3) (concentration 100 mg/L) and OM extracted from Antarctic soil and sampling distance 20 m (1) (concentration 1000 mg/L), 200 m (2) (concentration 1000 mg/L) and 300 m (4) (concentration 500 mg/L) from the coast*

3.3 MALDI TOF MS

All OM samples were dissolved in 36 mM NaOH solution and were measured in the linear positive mode (without matrix). Organic matter is ionized and positively charged ions are obtained.[17,18]

The mass spectrum of the IHSS Soil HA standard is presented in Figure 4 and compared with the coal derived Chemapex HA Standard and with the OM sample extracted from Antarctic soil taken near the coast. All spectra show some similarity. Peaks appear mainly in the range from 700 to 1000 m/z with differences between peaks equal to about 14, 26 or 28 m/z. This comparison shows that there are not great differences between soil and coal HAs. Mass spectra of OM extracted from Antarctic soil at sampling distance 100, 200 and 300 m are given in Figure 5. The most informative and similar to the mass spectra of HAs standards is the mass spectrum of OM extracted from soil taken 300 m from the coast.

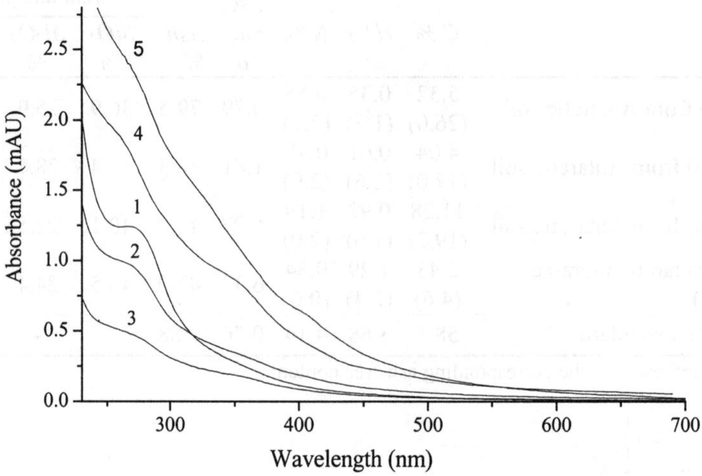

Figure 3 *UV-Vis spectra of solutions of OM extracted from Antarctic moraine – Moraine 1 (1), Moraine 2 (2), Moraine 3 (3), Moraine 4 (4) (concentration 1000 mg/L) and Moraine 5 (5) (concentration 500 mg/L)*

Figures 6 and 7 show mass spectra of OM extracted from Antarctic moraine. Peaks appear in the range from 400 to 1000 m/z. In Figure 7 there are mass spectra of OM extracted from Moraine 4 and 5 that are similar to those given in Figures 3 and 4.

3.4 Capillary Zone Electrophoresis

Solutions of samples of either HAs or Antarctic OM were prepared by dissolving 5 mg of dried substances in 5 mL of 36 mM NaOH solution and after one week the solutions were analyzed. It was found recently[17] that fresh HSs solutions do not give informative electropherograms. "Optimal" separation patterns are obtained after several days of ageing. Part of the electropherogram of the IHSS Soil HA standard is shown in Figure 8. It is possible to observe about 4 separated peaks. Electropherograms of OM extracted from Antarctic soil sampled at several distances from the coast (Figure 8) are compared here with the electropherogram obtained for the IHSS Soil HA standard. Similar features are seen in the electropherograms of OM extracted from Antarctic soil 300 m distance from the coast and OM extracted from Antarctic Moraine 5 (Figure 9). A total of 4 or 5 peaks (or groups of peaks) are observed. All the electropherograms have 4 or 5 main peaks that appear in the migration time range from 9 to 12 min.

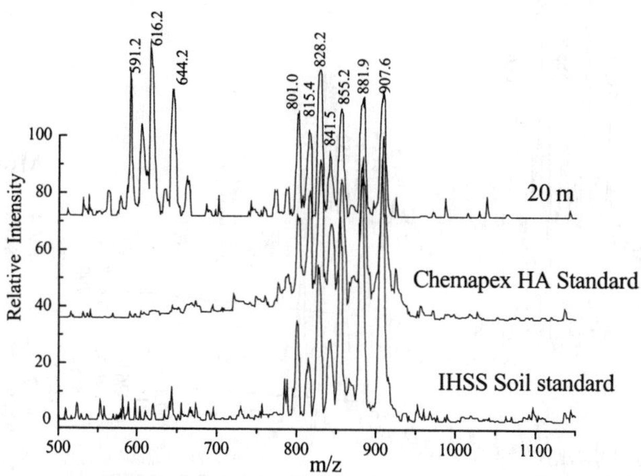

Figure 4 *Mass spectra of IHSS Soil standard, Chemapex HA Standard and OM extracted from Antarctic soil (sample distance 20 m from the coast); samples were dissolved in 36 mM NaOH solution (HA and OM concentration 1000 μg/mL) and spectra were measured in LDI linear positive mode*

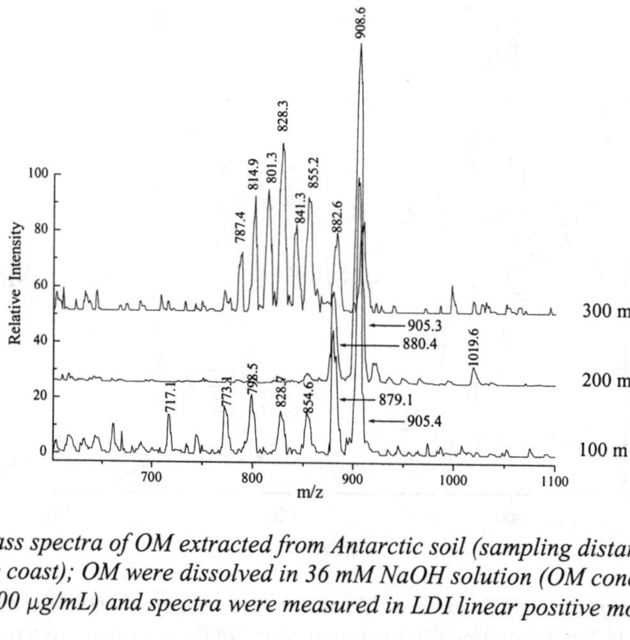

Figure 5 *Mass spectra of OM extracted from Antarctic soil (sampling distance 20 m from the coast); OM were dissolved in 36 mM NaOH solution (OM concentration 1000 μg/mL) and spectra were measured in LDI linear positive mode*

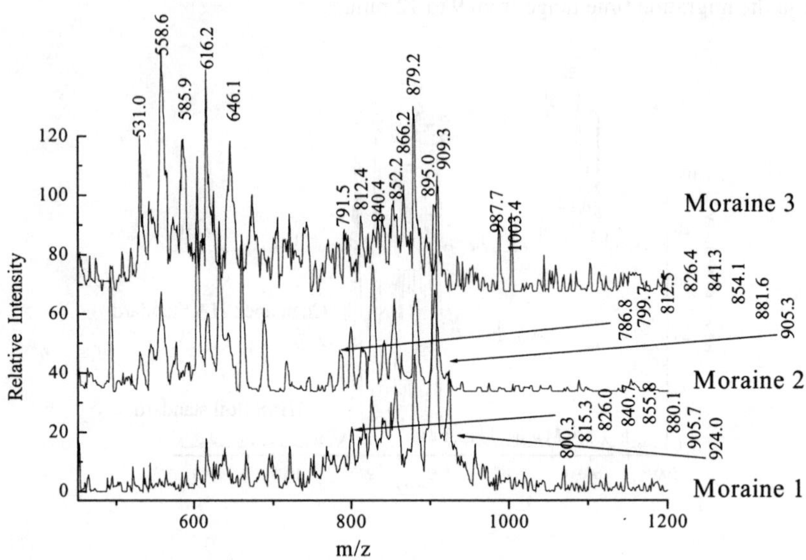

Figure 6 *Mass spectra of OM extracted from Antarctic moraine; OM were dissolved in 36 mM NaOH solution (OM concentration 1000 μg/mL) and spectra were measured in LDI linear positive mode*

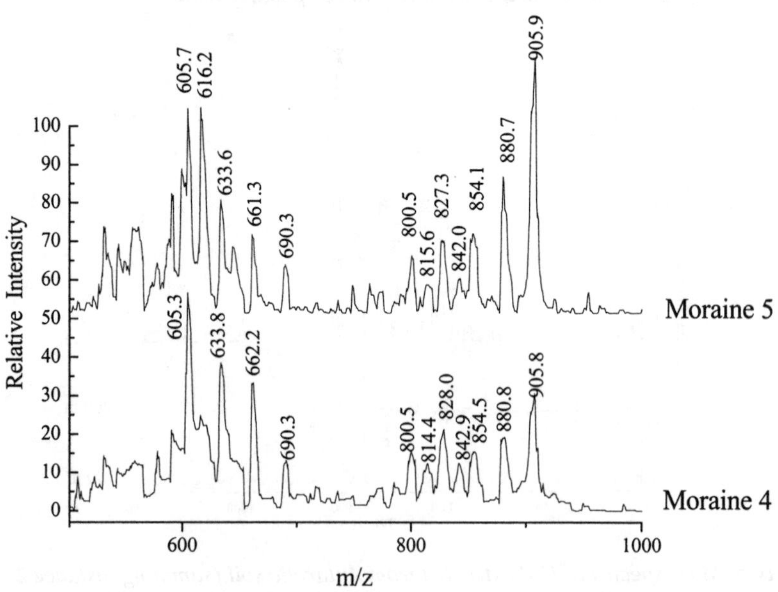

Figure 7 *Mass spectra of OM extracted from Antarctic moraine; OM were dissolved in 36 mM NaOH solution (OM concentration 1000 μg/mL) and spectra were measured in LDI linear positive mode*

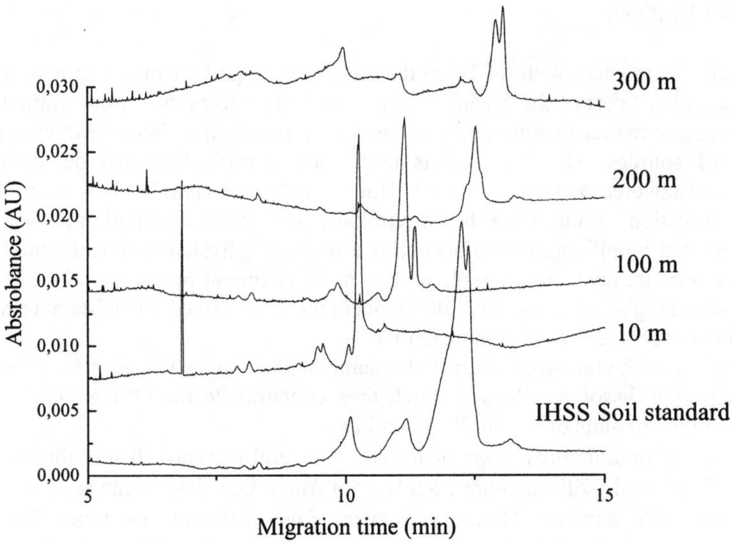

Figure 8 *Electropherograms of IHSS Soil standard (HA concentration 500 μg/mL) and OM extracted from Antarctic soil (sampling distance 10, 100, 200 and 300 m from the coast) dissolved in 36 mM NaOH solution (OM concentration 500 μg/mL)*

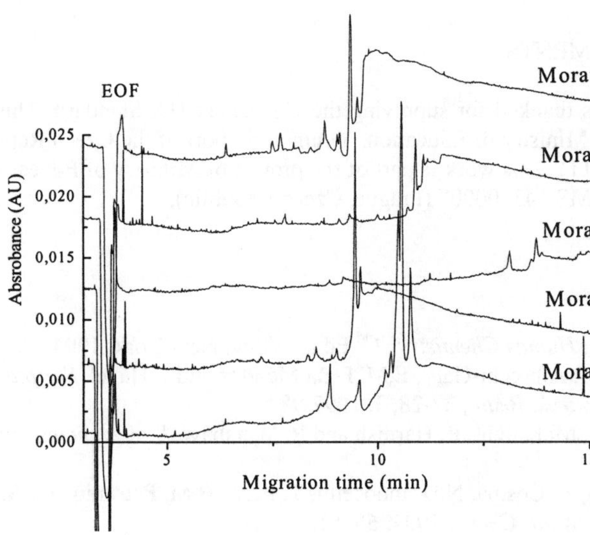

Figure 9 *Electropherograms of OM extracted from Antarctic moraine dissolved in 36 mM NaOH solution (OM concentration 500 μg/mL)*

4 CONCLUSIONS

It was found in agreement with ref 1 that the content of organic matter in Antarctic soil is low and that at least part of the organic matter may be humic acids.[19] Concerning the low content of organic matter in Antarctic soil, one has to realize that there are at least several different OM sources. The first one is prehistoric humification that occurred when Antarctica was covered by rich vegetation. Humification is a continuum of processes that include humification coming from the current biosphere on the land and in the sea. It can be suggested that humification also occurs as a result of physico-chemical conditions in this extreme environment, for example by means of chemical reactions induced by high pressure under the glacier. Generally, all the phenomena described above seem to result in observed low concentrations of organic matter.

We found high ash contents in extracted organic matter. The explanations are that either sodium hydroxide dissolves silicates which then contaminate the OM or that Antarctic OM contain high amounts of bound Si, Al, and Fe.

The extracted "organic ores" were analyzed by several methods. It was found both by MALDI TOF MS and CZE that humic acids from Antarctica show some similarities with the IHSS Soil HA standard. Humic substances from different soil types from other continents are similar to each other and it was found here that humic substances from Antarctic soil are also are somewhat similar to soil humic acids, for example from North America and even to coal derived humic acids. Several m/z peaks observed in mass spectra are the same as those in soil, peat and even in coal derived humics from other continents, which is further evidence that humic acids around the world have similar structures or that they contain the same chemical entities. In summary, organic ores extracted from Antarctic soils contain humic acids.

ACKNOWLEDGEMENTS

Chemapex Prague[10] is thanked for supplying the Chemapex HA Standard. This research was supported by the Ministry of Education, Youth and Sport of the Czech Republic (No. CEZ J07-98:143100011). This work is part of the project of Ministry of Education, Youth and Sport Project no. MS 143100007 (Prague, Czech Republic).

References

1. F.J. Stevenson, *Humus Chemistry*, 2nd Edn., Wiley, New York, 1994.
2. D. Fetsch, A.M. Albrecht-Gary, E.M. Peña Méndez and J. Havel, *Scripta Fac. Sci. Nat. Univ. Masaryk. Brun.*, **27-28**, 3, 1997-98.
3. G.R. Aiken, D. McKnight, R. Harnish and R. Wershaw, *Biogeochem.*, 1996, **34,** 157.
4. L. Campanella, B. Cosma, N.D. Innocenti, T. Ferri, B.M. Petronio and A. Pupella, *Int. J. Environ. Anal. Chem.*, 1994, **55,** 61.
5. L. Campanella, B.M. Petronio, C. Braguglia, R. Cini and N.D. Innocenti, *Int. J. Environ. Anal. Chem.*, 1995, **60,** 49.
6. C.M. Braguglia, L. Campanella, B.M. Petronio and R. Scerbo, *Int. J. Environ. Anal. Chem.*, 1995, **60,** 61.
7. G.J. Retallack and E.S. Krull, *Aust. J. Earth Sci.*, 1999, **46,** 785.

8. K. Birkenmajer, Geology of Admiralty Bay, King George Island (South Shetland Islands) - An outline, *Polish Polar Res.*, 1980, **1**, 29.
9. A.A. Marsz, in *Ecology of the Antarctic Coastal Oasis*, eds. P. Prosek and K. Láska, Folia Facultatis Scientiarum Naturalium Universitatis Masarykianae Brunensis, *Geographia*, 2001, **25**, 7.
10. www.chemapex.cz, Chemapex Praha s.r.o., V Háji 15, 170 00 Praha 7, Czech Republic.
11. M.D. Landgraf, R.C. Alonso Javaroni and M.O. Oliveira Rezende, *J. Cap. Elec.*, 1998, **5**, 6.
12. D. Garcia, J. Cegarra, M. Abad and F. Fornes, *Bioresource Technol.,* 1994, **47**, 103.
13. D. Tonelli, R. Seeber, C. Ciavatta and C. Gessa, *Fresen. J. Anal. Chem.*, 1997, **359**, 555.
14. C.M. Braguglia, L. Campanella, B.M. Petronio and R. Scerbo, *Int. J. Environ. Anal. Chem.,* 1995, **60**, 61.
15. E. Magi, T. Giusto and R. Frache, *Anal. Proc.,* 1995, **32**, 267.
16. N. Calace, L. Campanella, F. Depaolis and B.M. Petronio, *Int. J. Environ. Anal. Chem.,* 1995, **60,** 71.
17. L. Pokorná, D. Gajdošová, S. Mikeska and J. Havel, in *Humic Substances: Versatile Components of Plants, Soils and Water*, eds. E.A. Ghabbour and G. Davies, Royal Society of Chemistry, Cambridge, 2000, p. 299.
18. Y.P. Chin, G.R. Aiken and E. O'Loughlin, *Environ. Sci. Technol.,* 1994, **28**, 1853.
19. L. Beyer, H.-P. Blume, C. Sorge, H.-R. Schulten, H. Erlenkeuser and D. Schneider, *Arctic Alpine Res.*, 1997, **29**, 358.

THE STABILITY OF HUMIC ACIDS IN ALKALINE MEDIA

L. Pokorná,[1] D. Gajdošová,[1] S. Mikeska,[2] P. Homoláč[2] and J. Havel[1]

[1] Department of Analytical Chemistry, Faculty of Science, Masaryk University, 611 37 Brno, Czech Republic
[2] Chemapex Praha s.r.o., 170 00 Praha 7, Czech Republic

1 INTRODUCTION

Humic materials play an important role in nature and therefore they are intensively studied. Humic substances (HSs) are a complex mixture of many relatively low molecular weight compounds.[1-6]

Humic substances have been found to be negatively charged polymers containing various functionalities attached to an aliphatic-aromatic backbone structure.[7] As a result of both hydrophobic and hydrophilic moieties, different intra- and intermolecular interactions can be expected in aqueous systems of HSs, depending on solution conditions. In aqueous solutions humic substances show a tendency to self-assemble, and they accumulate at interfaces and solubilize organic compounds.[8]

Aggregation changes in humic substances structures caused by pH and ionic strength effects are described as well. The aggregation of humic acids decreases with increasing pH, but with increasing ionic strength aggregation is more likely. Aggregation was also demonstrated by capillary zone electrophoresis.[3]

The formation of HSs is assumed to involve the autooxidation of polyphenolic compounds in alkaline media in the presence of oxygen. This process is attributed to radical reactions, while the degradation of humic materials in alkaline solutions is usually related to the splitting of esters.[9] Changes in the behavior of HSs sample solutions were observed and the action of oxygen dissolved in aqueous solution was suggested as a possible explanation.[10]

Recently it was observed that quinones play very important roles in both biological and abiotic reduction and oxidation of humic substances.[11] Quinones in humic structures originate from the decomposition of lignin or microbial organic matter. They act as electron transmitters. The transfer of electrons to humic compounds supported by oxygen presence results in increased content of semiquinone radical intermediates as well as hydroquinones as shown in Scheme 1.

A direct correlation between the electron-accepting capacity of humic substances and their aromaticity was observed.[12] Except for oxidative degradation of humic compounds that occurs via *o*- and *p*-quinone ring-opening reactions followed by formation of two carboxylic groups, photodecarboxylation and homolytic cleavage generating O=C-OH• radicals take place in the degradation of humic substances. The efficiency of the latter degradation reactions depends strongly on the structure of the reactant and mainly on the presence or absence of oxygen.[13]

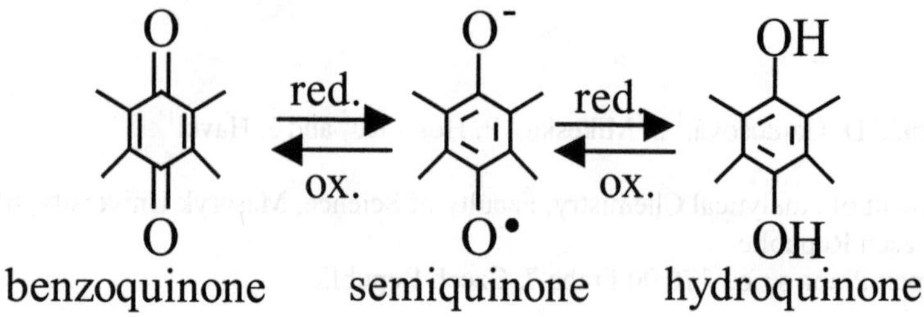

Scheme 1 *Quinone model compound. The semiquinone species contains an unpaired electron*[13]

Ideas about HSs molecular character and the concept of random coil solution conformations are useful in understanding reactions and interactions of HSs in aqueous media.[14,15]

Humic acids are not very soluble in water and more soluble in alkaline media. The aim of this work was to study the stability of humic substances in solutions of different bases such as hydrazine and sodium or tetramethylammonium hydroxide (TMAH). Therefore, a short review concerning interaction of humic acids with these reagents follows.

1.1 Sodium Hydroxide and its Effect on Humic Substances

Sodium hydroxide is a base that is known to dissolve HSs easily. The influence of sodium hydroxide on the behavior of humic substances in solutions was noted.[16-19] It is known that in aqueous media, especially in dilute NaOH solution, HSs undergo dissociation of their acidic functional groups. This process results in the spontaneous formation of an electrical double layer on the macromolecules of humic acids. Hence, electrostatic interactions definitely affect the conformational alterations of humic macromolecules, the colloidal stability of humic solutions and also the aggregation of the various humic components.[16]

In consequence of the interaction of humic charges, agglomerates are created in solutions of high pH, according to increases of solution viscosity.[19] Agglomeration was observed also in slightly acidic solutions of humic substances using CZE.[3]

1.2 Hydrazine Interactions with Humic Substances

Hydrazine (N_2H_4) is a strong base ($pK_a = 7.95$ at 25°C),[20] which in the form of the free base acts as a strong nucleophile in substitution reactions. It also is proposed to participate in condensation reactions with the carboxyl groups in humic structures.[21]

Quinone moieties are very reactive parts of humic substances whose presence has been confirmed, for example, by NMR.[22] Their interactions with hydrazine give pyrazole derivatives in isoflavone structures.[23] In like manner, activation by carbonyl groups in quinone structures in humic compounds would provide substitution products as shown in Scheme 2a.

There is strong reason to expect that condensation reactions occur between hydrazine and carbonyl groups in humic structures.[21] It is generally known that hydrazine interacts with carbonyl groups, aldehydes and ketones in aqueous media. The interactions of hydrazine with carbonyl groups of HSs can result in the formation of imines, as shown in Scheme 2b. The other possibility of interaction between carbonyl groups and hydrazine is the formation of intermediate hydrazones during the Wolff-Kishner reduction of carbonyl compounds. Other possible interactions of hydrazine and humic carbonyl components are shown in Scheme 2c.

Hydrazine incorporated from the interactions with quinone structures of humic substances would be less susceptible to hydrolysis than products of condensation with carbonyl groups.[21]

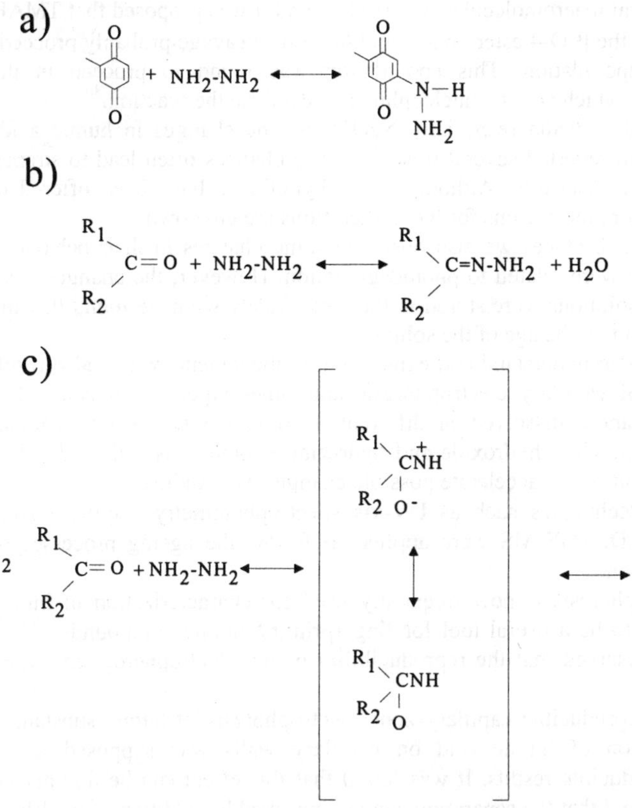

Scheme 2 *Interactions between quinone structures (a) and carbonyl groups (b,c) in humic substances with hydrazine (adapted from refs13,21,31)*

We used hydrazine hydrate as an HSs solubilizing agent because of the presumption that after the dissolution of HSs in hydroxide (sodium hydroxide or TMAH) oxygen dissolved in aqueous solution might cause humic acids decomposition. The presence of oxygen can amplify the ability of humic material to undergo radical reactions and also might induce the oxidation processes. Therefore, we have tried to eliminate these processes using hydrazine. This idea is supported by the fact that in the presence of hydrazine, phenols are not oxidized by dissolved oxygen in solution.[24]

1.3 Tetramethylammonium Hydroxide as a Dissolving Agent for Humic Acids

TMAH also easily dissolves humic acids. TMAH is a strong base that also can act as a strong oxidizing and methylating agent. The utilization of TMAH for oxidation and methylation of humic compounds was investigated, mainly using TMAH as an agent for thermochemolysis of humic materials that is usually followed by pyrolysis-gas chromatography/mass spectrometry.[25-28]

The interaction between TMAH and carboxyl groups of HSs can result in prevention of decarboxylation of preexisting carboxylic moieties. Carboxyl groups present are converted to their methyl esters by reaction with TMAH. Recently, the mechanism of interactions between TMAH and natural macromolecules was studied and it was proposed that TMAH is effective in cleavage of the β-O-4 ester bond[29] and that this cleavage probably proceeds through intermolecular epoxidation. This epoxidation was found to proceed in the presence of hydroxyl ions, which act as a nucleophile and catalyze the reaction.[30]

The influence of alkaline media (especially NaOH) on the changes in humic acids behavior was observed and reported several times.[16-18] The changes often lead to strange results of analysis, as also discussed. Although some hypotheses have been offered to understand this phenomenon, the reasons for these alterations are unknown.

When studying humic substances we also observed some changes in their behavior.[5] First of all, the alterations were related to photodegradation. However, the changes were still observed while HSs solutions were stored in the dark. Lately we have found that the alterations primarily depend on the age of the solutions.

The aim of this work is to understand and explain these phenomena, which also result in the irreproducibility of capillary electrophoresis and other types of analyses. The behavior of humic substances dissolved in different alkaline media, such as sodium hydroxide, tetramethylammonium hydroxide and hydrazine hydrate was followed. Also, the effect of heating of solutions to accelerate possible changes was studied.

Various instrumental techniques such as UV-vis spectrophotometry, capillary zone electrophoresis and MALDI-TOF MS were applied to follow the ageing processes of humic substances in solution.

Capillary zone electrophoresis is now frequently used for characterization of humic substances. It was found to be a useful tool for fingerprinting humic compounds.[2-6,31-33] However, it was often observed that the reproducibility of the electropherograms when analyzing HSs was low.

Recently, the highly reproducible capillary zone electrophoresis of humic substances was reported.[31] Adsorption of humic acid on capillary walls was supposed to be responsible for non-reproducible results. It was found that this effect can be diminished using a capillary coating and that the separation can be improved by addition of modifiers to the background electrolyte. However, it was observed that after elimination of sorption of humic acids on the capillary wall it is possible to reproduce electropherograms only if the solutions are of the same age.

Coal-derived humic acid Chemapex Standard prepared from capucine, a special form of oxidized bohemian brown coal from western Bohemia (which is an equivalent of Leonardite), was chosen as a representative HA to study the ageing effects. This humic acid was recently analyzed in detail.[6] It was concluded that its properties are very similar to other humic substances independent of their origin and that this humic acid is suitable for comparative studies. IHSS Peat standard HA and Aldrich HA were also used to follow HSs changes in solutions with the aim of comparing the ageing process of HSs solutions with HSs samples of different origin.

2 MATERIALS AND METHODS

2.1 Chemicals

All reagents used were of analytical grade. Sodium hydroxide, boric acid, TRIS (1,1,1-tris-(hydroxymethyl)-aminomethane) and 30% hydrazine aqueous solution were from Lachema a.s. (Brno, Czech Republic). EDTA was from ONEX (Rožnov pod Radhoštěm, Czech Republic). Tetramethylammonium hydroxide, H_3PO_4 for conditioning of the capillary and mesityl oxide used as an EOF marker were from Merck (Schuchardt, Germany). Deionized water used to prepare all solutions was double-distilled from a quartz Heraeus Quartzschmelze apparatus (Hanau, Germany).

2.2 Humic Substances

The humic substances used in present work were: humic acid sodium salt (Aldrich, Cat. No: H1,675-2), coal-derived Chemapex Standard humic acid and the IHSS Peat standard HA (1S103H). Stock solutions were prepared by dissolving 5 mg sample in 5 mL of the relevant reagent solution (1 mg/mL). For further analysis these solutions were diluted to 500, 250 and 100 μg/mL. The concentration of sodium hydroxide in the stock solution was 36 mM, the concentration of hydrazine hydrate in the stock solution was 0.15 M and the concentration of tetramethylammonium hydroxide was 0.6 % w/v.

2.3 UV-vis Spectrophotometry

UV-vis absorption spectra were measured with a Unicam Model UV2 UV/vis spectrophotometer (Cambridge, Great Britain). Spectrophotometry was used for measuring absorption spectra of humic substances and also to measure absorbances at 250, 272, 280, 350, 365, 465 and 665 nm. These data were used for determination of the E_4/E_6 ratio (absorbances at 465 and 665 nm), the E_2/E_3 ratio (absorbances at 250 and 365 nm), the absorptivity ε_{272} (absorbance at 272 nm) and the band half-width of the electronic transition absorbance bands Δ_{ET} (absorbance at 280 and 350 nm). All these data give information about the aromaticity of humic samples. The value of the E_4/E_6 ratio, the so called index of humification, correlates also with the average molecular weight and size and to the oxygen content of humic materials.[34] The specific absorptivity (L/g cm) can be calculated using Equn. (1),

$$\varepsilon_{272} = \frac{A_{272}}{c} \qquad (1)$$

where A_{272} is the absorbance at 272nm and c is the HA concentration (g/L). The value of the band half-width of the electronic transition absorbance bands is given by Equn (2).

$$\Delta_{ET} = 2.18\left(\ln\left(\frac{A_{280}}{A_{350}}\right)\right)^{-\frac{1}{2}} \qquad (2)$$

2.4 Capillary Zone Electrophoresis

A Beckman CZE (Model P/ACE) System 5500 (Palo Alto, CA, USA) equipped with a diode array detection system, an automatic injector, a fluid-cooled column cartridge and a System Gold Data station was utilized for all CZE experiments. Fused silica capillary tubing of 47 cm (40.3 cm length to the detector) x 75 μm I.D. was used. The normal polarity mode of the CZE system (cathode at the side of detection) was applied.

The CZE analyses were performed with a mixture of 90 mM boric acid + 90 mM TRIS + 1 mM EDTA, pH 8.3 as the background electrolyte. Separations were conducted in the hydrodynamic mode of injection. The time of injection was 20 s, the applied voltage was 20 kV and the working temperature was 20°C.

2.5 MALDI-TOF MS

Mass spectra were measured with a Kratos Kompact MALDI III mass spectrometer in the linear mode. The instrument was equipped with a nitrogen laser (wavelength 337 nm, pulse duration $\tau = 10$ ns, energy pulse 200 μJ). The applied energy of the laser was in the range 0-180 arbitrary units. For each sample we used 100 laser shots (positive mode), the signals of which were averaged and smoothed. Insulin was used for mass calibration. The laser energy of 120 arbitrary units was found optimal and was used for analyses.

3 RESULTS AND DISCUSSION

3.1 Characterization of Humic Substances Using UV-vis Spectrophotometry

UV-vis spectrophotometry was used to monitor time changes in humic substances dissolved in alkaline media. The ageing effects on the humic materials were followed by study of factors that characterize the aromaticity. Some of them also can determine approximate average molecular weights and the oxygen content of the samples. Last but not least, the influence of different alkaline agents on these alterations was also examined.

First, UV-vis spectra of HSs samples were measured. It was observed that the initial spectra of HSs in all media used are very similar. Decrease of absorbance with increasing age of the solution was observed for all of them. Factors related to aromaticity were calculated from UV-vis spectrometric data. It was observed that the highest values were always obtained for humics dissolved in NaOH and the lowest values were found for HSs dissolved in diluted aqueous hydrazine. The results are given in Table 1.

Table 1 *Influence of the dissolving agent on the values of factors indicating aromaticity*

HA sample	E_2/E_3	E_4/E_6	ε_{272}	Δ_{ET}
Chemapex Standard - NaOH	2.70	7.17	0.0350	2.68
Peat - NaOH	2.82	7.06	0.0310	2.68
Chemapex Standard - hydrazine	1.20	5.41	0.0150	1.58
Peat - hydrazine	1.20	5.25	0.0105	1.21
Peat - TMAH	1.36	6.65	0.0310	2.74

The dependence of individual factors characterizing aromaticity on the age of solutions is shown in Figures 1-3. Aromaticity of HSs solutions decreases with age.

This decrease is rather rapid, and then the values of individual parameters seem to be more or less constant, especially for samples dissolved in solutions containing hydrazine. This decrease of aromaticity can be explained by suggesting that hydrazine weakens hydrogen bonds and in this way leads to the decomposition of HSs supramolecules into their components.

Based on the fact that quinone moities play a significant role in HSs redox reactions as mentioned above, some other notions can be deduced. The reduction of quinone functional groups leads to the formation of semiquinone radicals whose content has a strong relationship with the electron-accepting capacity and in this way with the aromaticity. Because semiquinone radicals were found to be transient and their concentration decreases with time,[13] this can explain the observed decline of aromaticity.

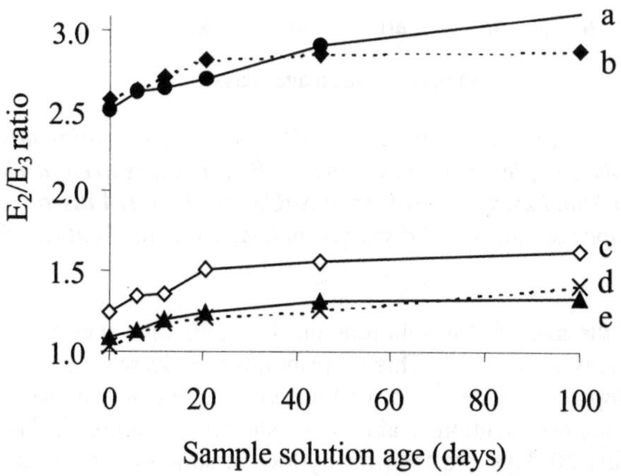

Figure 1 *Dependence of the E_2/E_3 ratio on sample solution age for Chemapex Standard HA dissolved in NaOH, (b) Peat humic acid (HA) dissolved in NaOH, (c) Peat HA dissolved in TMAH, (d) Chemapex Standard HA dissolved in hydrazine, (e) Peat HA dissolved in hydrazine*

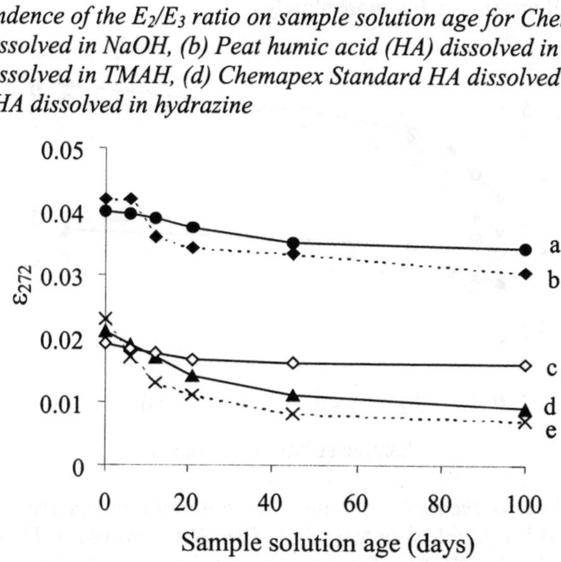

Figure 2 *Dependence of specific absorptivity ε_{272} on the sample solution age for Chemapex Standard HA dissolved in NaOH, (b) Peat humic acid (HA) dissolved in NaOH, (c) Peat HA dissolved in TMAH, (d) Chemapex Standard HA dissolved in hydrazine (e) Peat HA dissolved in hydrazine*

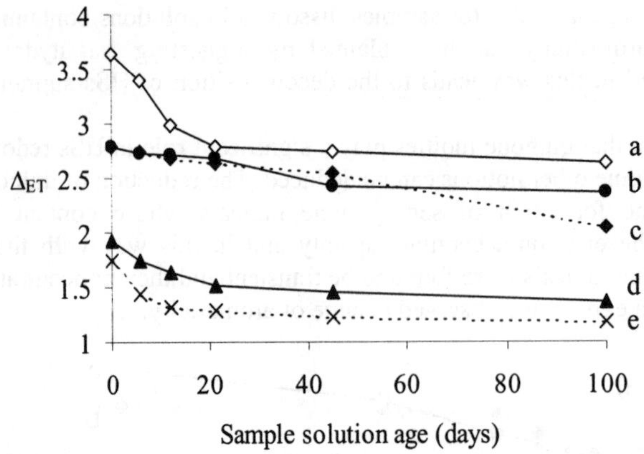

Figure 3 *Dependence of the band half-width of the electronic transition absorbance bands on the sample solution age for: (a) Peat HA dissolved in TMAH, (b) Chemapex Standard HA dissolved in NaOH, (c) Peat HA dissolved in NaOH, (d) Chemapex Standard HA dissolved in hydrazine, (e) Peat HA dissolved in hydrazine*

The influence of the age of HSs solutions on the E_4/E_6 ratio was also studied. It was found that E_4/E_6 increases with time. This corresponds to a decrease of average molecular weight and to the apparent enhancement of the oxygen content. The same situation was observed for other factors mentioned above, as shown in Figure 4. The increase was largest during the first 20 days and then for hydrazine solutions it seems to be constant. For NaOH it is still increasing, but more slowly.

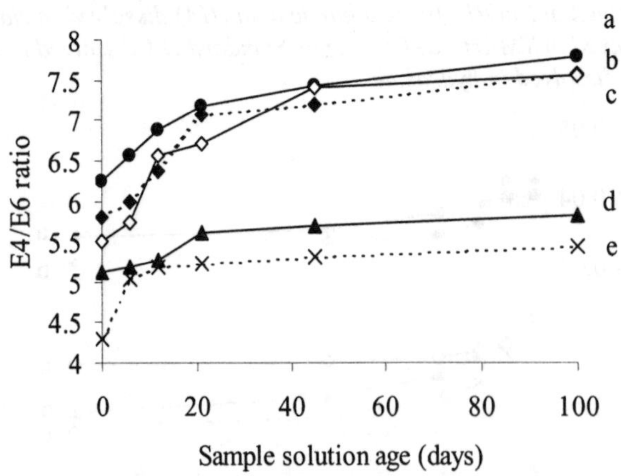

Figure 4 *Dependence of the E_4/E_6 ratio on the sample solution age for (a) Chemapex Standard HA dissolved in NaOH, (b) Peat HA dissolved in TMAH, (c) Peat HA dissolved in NaOH, (d) Chemapex Standard HA dissolved in hydrazine, (e) Peat HA dissolved in hydrazine*

The effect of elevated temperature (90°C) on the alteration of HSs solutions and accelerated ageing was also followed. Practically the same tendency of lowering the aromaticity was found but the differences in the values of the factors were not so pronounced (Table 2).

Table 2 *Influence of time of heating on factors characterizing aromaticity of peat humic acid dissolved in NaOH*

Age of heated solutions (hrs)	E_2/E_3	E_4/E_6	ε_{272}	Δ_{ET}
0	1.21	5.41	0.015	1.96
1	1.25	5.51	0.013	1.80
2	1.26	5.59	0.009	1.73
4	1.28	5.59	0.011	1.71
6	1.37	5.62	0.008	1.61

3.2 Capillary Zone Electrophoresis and MALDI-TOF MS

3.2.1 Capillary Zone Electrophoresis. Electropherograms obtained for different ages of the solutions when using sodium hydroxide for dissolution of humic acids are given in Figure 5. It was found that analysis of fresh solutions does not yield any peaks and an electropherogram containing four peaks was only obtained 3 weeks afterwards. The larger number of peaks and improvement of the separation corresponding to increasing age of the solution was observed. The "best" separation was found for the solutions that were 5 weeks old. For older solutions the same number of peaks was found but deformation of peak shapes was observed. This effect was observed for all HSs samples analyzed, independent of their origin. The electropherograms of various humic substances studied (30 day old solutions) are very similar (Figure 6).

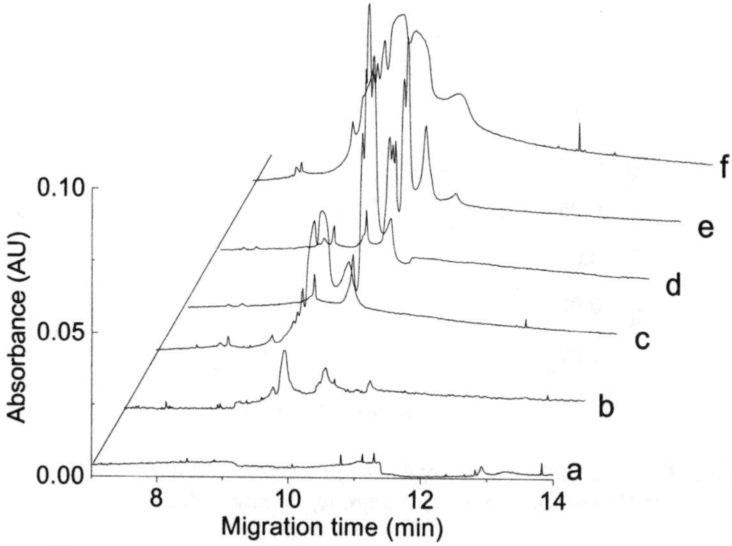

Figure 5 *Electropherograms of Chemapex Standard HA dissolved in 18 mM NaOH. Solution age: (a) 1 day, (b) 3 days, (c) 22 days, (d) 30 days, (e) 38 days, (f) 60 days*

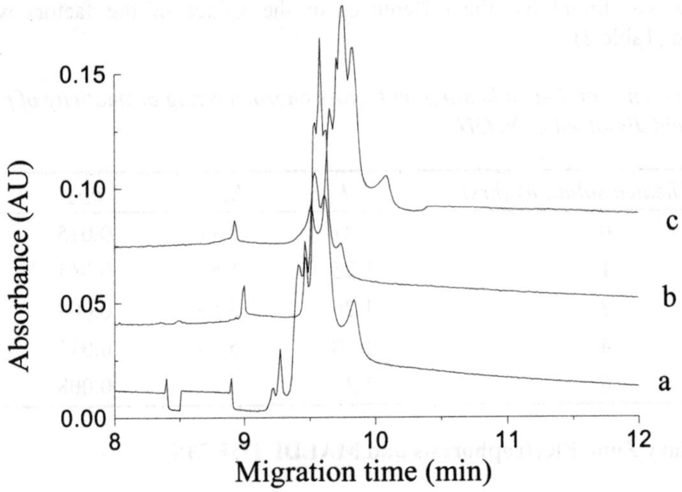

Figure 6 *Comparison of electropherograms of various humic acids dissolved in 18 mM NaOH (30 days old): (a) Chemapex Standard HA (b) Aldrich humic sample (c) IHSS Peat HA*

Electropherograms for HSs dissolved in TMAH are given in Figure 7. It was found that after 5 days of ageing separation into 4 peaks was obtained. The "best" separation was reached for ≈ 3 weeks old solutions and for 1 month old solutions some deformation of peaks on the electropherograms was observed.

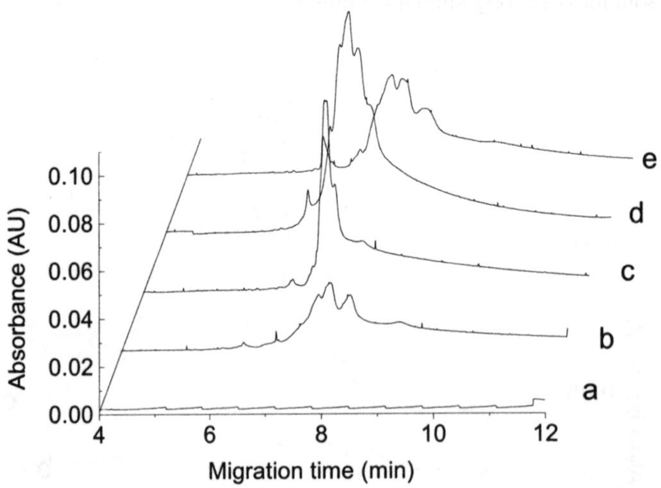

Figure 7 *The electropherograms of Chemapex Standard HA dissolved in 0.3 % w/v TMAH Age: (a) 1 day, (b) 5 days, (c) 16 days, (d) 22 days, (e) 30 days*

Analyzing solutions of HSs dissolved in diluted hydrazine, some differences in the electropherograms were also found (Figure 8). In this case the changes in electrophoretic behavior of the solutions were seen after 1 day of preparation of the solution while three peaks were detected. After 6 days the number of peaks increased to four and during 2 weeks electropherograms with five separated peaks were obtained. For two months old solution the number of peaks was the same.

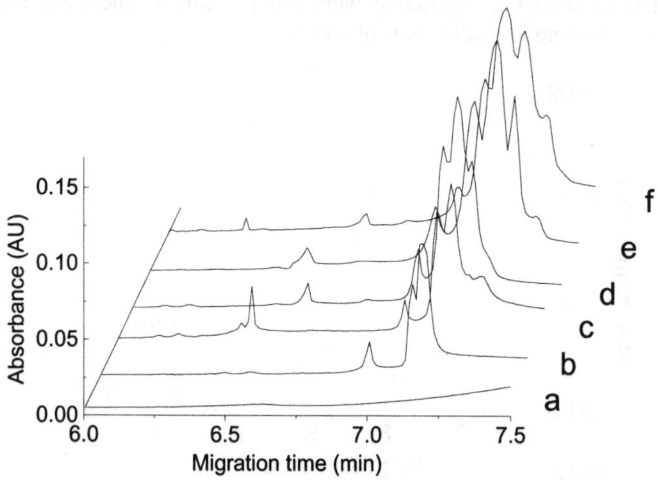

Figure 8 *The electropherograms of Chemapex Standard HA dissolved in 75 mM hydrazine. Age: (a) 4 hours, (b) 1 day, (c) 5 days, (d) 8 days, (e) 15 days, (f) 53 days*

A possible explanation of these alterations is that the presence of oxygen leads to radical reactions. It is suggested that in alkaline media the oxidation of hydroquinones is catalyzed and can result in peak deformations. When hydrazine is present, the influence of oxygen is eliminated and after the fast reduction process observed during the first 8 days the electropherograms for older solutions were practically identical and no further alterations occurred.

Ageing was also studied at elevated temperature. Solutions were heated at 90°C and similarities between results of analysis of samples of different age and heated solutions were found (Figure 9). For HSs dissolved in 18 mM NaOH solution, the electropherograms of samples measured after 6 hr of heating were practically identical to those for 38 day old solutions. Electropherograms for humics dissolved in hydrazine and collected after 6 hours of heating corresponded to those of a 6 days old solution at room temperature. This confirms that alterations in HSs are much faster when using hydrazine. Similar effects as for sodium hydroxide were observed for humics dissolved in TMAH. Electropherograms of solutions collected after 6 hours of heating at 90°C are similar to those for 30 day old solutions.

3.2.2 Matrix Assisted Laser Desorption/Ionization Time of Flight Mass Spectrometry. MALDI-TOF MS was used as another method of monitoring changes of HSs in alkaline solution. Mass spectra of HSs dissolved in diluted NaOH are given in Figure 10. For all HSs the set of peaks of the same m/z values in the range 700 to 900 m/z was found. Analyzing HSs dissolved in NaOH solutions of different age gave interesting results, as shown in Figure 11. The one month old solution shows a group of peaks in the range 800-1000 m/z. The difference between these peaks is about 14 m/z, which might indicate the CH_2 functional group. Some of these peaks diminish with ageing and only two of them remain for one year old solutions. Another difference is observed in the range of 1200-2000 m/z, where the spectrum for a one year old solution shows a new set of peaks.

However, the intensity of these peaks is relatively low. These alterations show that the original humic acid fractions (with m/z around 900) decompose and other fractions appear that can be ionized and visualized (m/z ≈ 1500-2000). It is interesting to note that mass spectra of HAs dissolved in hydrazine were not possible to measure, because in the presence of hydrazine no ionization was observed.

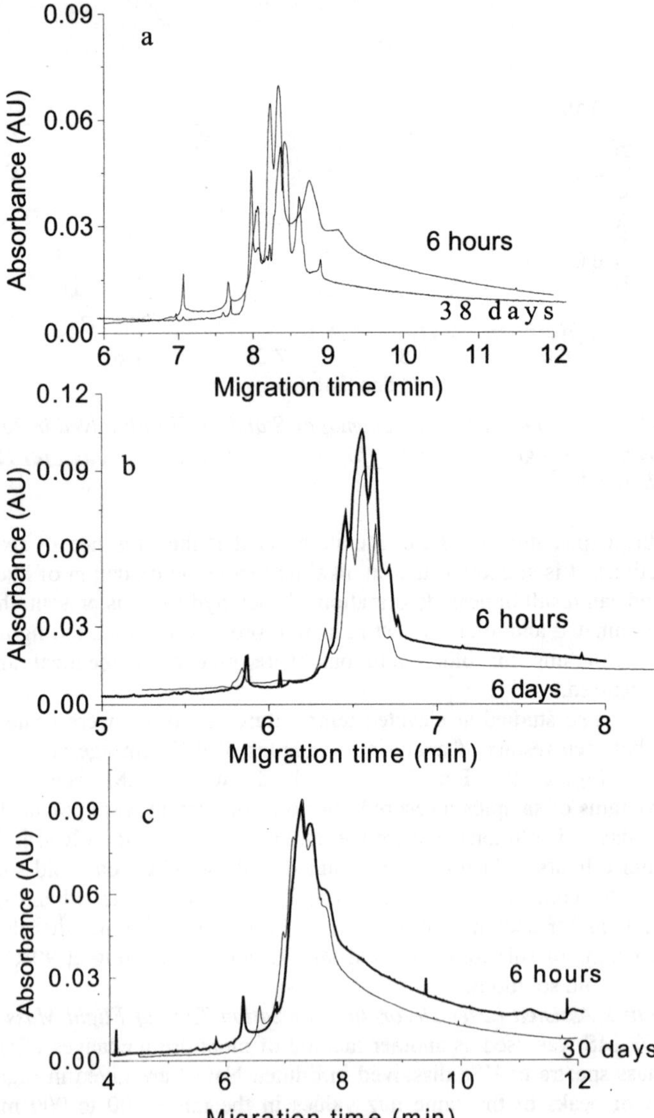

Figure 9 *Comparison of electropherograms of peat humic acid showing the effect of heating of samples dissolved in: (a) NaOH, (b) hydrazine, (c) TMAH (concentrations of NaOH, hydrazine hydrate and TMAH are the same as those in Figures 5,7,8)*

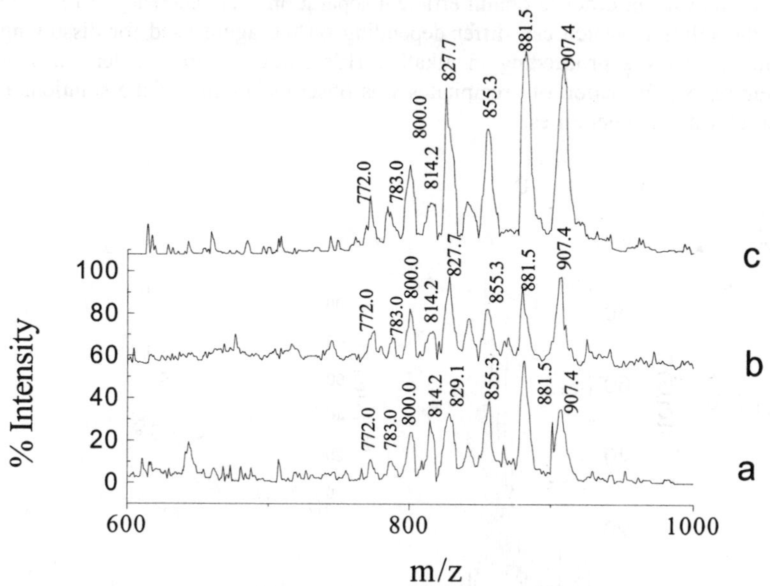

Figure 10 *Comparison of mass spectra of various humic substances dissolved in 36 mM NaOH (30 days old): (a) Chemapex Standard HA, (b) peat humic acid, (c) Aldrich humic sample*

Mass spectra of HSs samples dissolved in TMAH for 1 day and 1 month old solutions are given in Figure 12. The mass spectrum of a 1 day old solution shows quite low intensity. A set of peaks is observed in the ranges 500-600 m/z, 800-900 m/z and at 1700-1800 m/z. Most of the differences between the peaks in the set ≈ 830 m/z are about 14 m/z, which corresponds to the results obtained when using NaOH. The mass spectrum of a 1 month old solution shows the group of peaks in the range 800-900 m/z again differentiated by about 14 m/z. Some significant peaks in the range 1000-2000 m/z were observed with a new, high intensity peak at 1592.8 m/z. This observation confirms the results obtained when sodium hydroxide was used for dissolving the humic substances, as mentioned above.

4 CONCLUSIONS

MALDI-TOF MS and capillary zone electrophoresis were found to be suitable methods to follow HSs changes in different dissolving agents and especially due to processes connected with ageing of HSs solutions.

Humic substances are usually ranked among the most stable compounds, especially in soil solution, but considerable changes were observed in solutions of higher pH. These alterations are probably connected with radical reactions and redox processes in alkaline solution and they lead to decreasing aromaticity of samples. The difficulty in obtaining reproducible results of analysis during long-term measurements is explained by the changes due to the "ageing" process of humic substances in such solutions. It was found that to obtain reproducible results it is necessary to work with solutions of the same age.

For CZE analyses in order to obtain efficient separation it is necessary to find an optimal age of the solution, which can differ depending on the agent used for dissolving humic substances. Changes proceeding in alkaline HSs solutions are accelerated at elevated temperature. No formation of precipitates was observed in any of the solutions either at room or elevated temperatures.

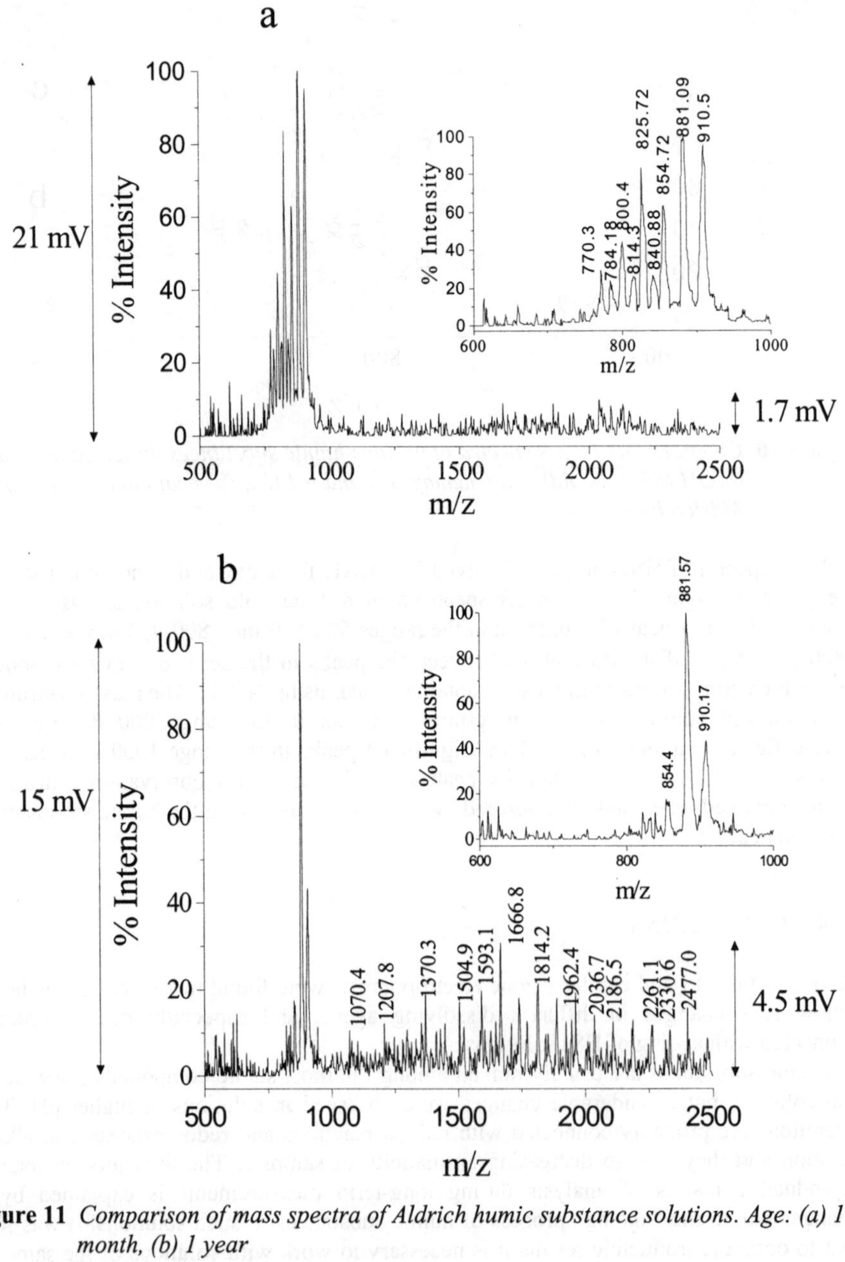

Figure 11 *Comparison of mass spectra of Aldrich humic substance solutions. Age: (a) 1 month, (b) 1 year*

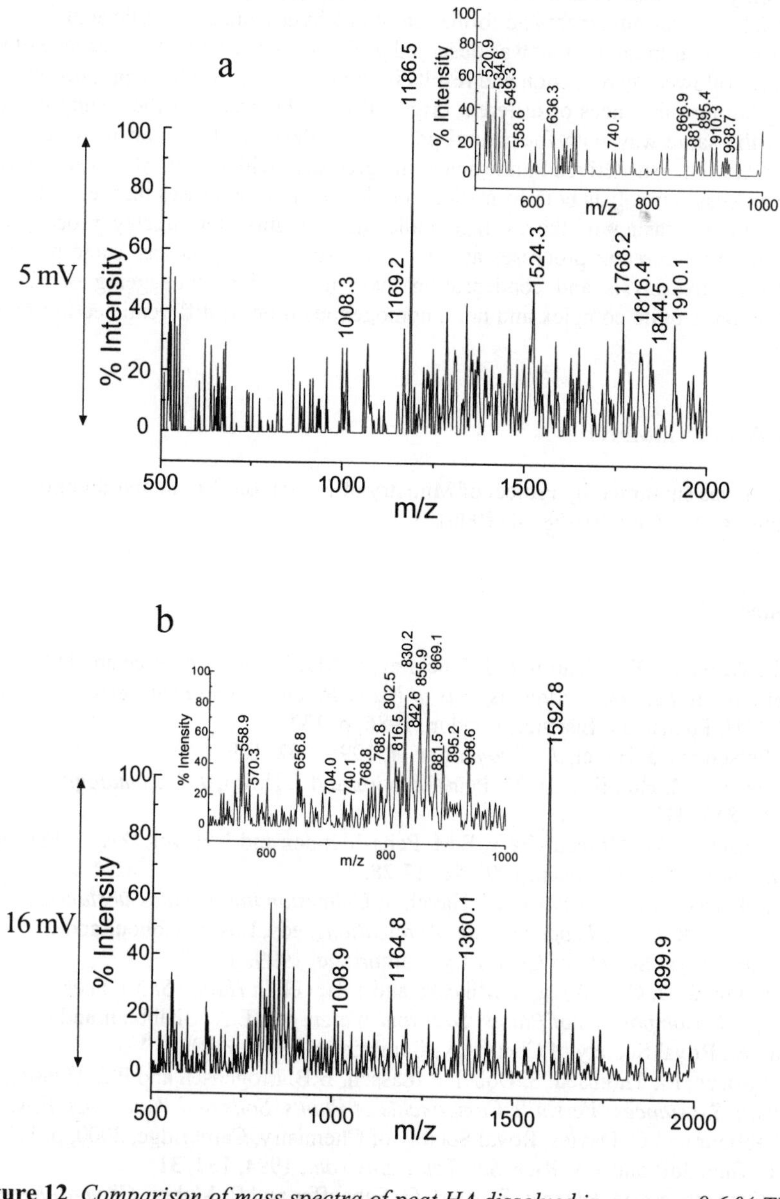

Figure 12 *Comparison of mass spectra of peat HA dissolved in aqueous 0.6 % TMAH. Age: (a) 1 day, (b) 1 month*

According to the idea that humic substances are products of lignin degradation and that aromatic rings with flexible long aliphatic chains are their main structural units, we can draw some conclusions. During degradation and/or decomposition oxidative ring opening processes take place that can result in differences occurring in mass spectra of differently aged solutions. Generally, it is known that oxidative processes in alkaline media are focused mainly on phenolic and enolic structures. The changes in these structures can result especially in rearrangement and formation of aldehydes and carboxylic acids.

The decrease of aromaticity that was observed probably is caused by the cleavage of the aromatic core followed by reduction and results in increase of the aliphatic proportion.

Studying humic substances of different origin, it was observed that the ageing process proceeds in the same way both for natural peat (IHSS Peat standard) and for Chemapex Standard or Aldrich coal-derived humic acids. In agreement with ref. 10, it was shown that the ageing processes involves both formation and degradation processes that result in both increasing and decreasing of the average molecular weights. The ageing process was found also to involve redox processes at some early stage but in the last stage probably mostly alkaline hydrolysis and condensation take place. Thus, the ageing of HSs in alkaline solution is quite complex and not a homogenous process differentiated in various time periods.

ACKNOWLEDGEMENT

This research was supported by Project of Ministry of Education, Youth and Sport of Czech Republic (No. CEZ J07-98:143100011).

References

1. R.L. Wershaw, K.A. Thorn, D.J. Pinckney, P. MacCarthy, J.A. Rice and H.F. Hemond, in *Peat and Water-Aspects of Water Retention and Dewatering in Peat*, ed. C.H. Fuchsman, Elsevier, London, 1986, p. 133.
2. D. Fetsch and J. Havel, *J. Chromatogr. A*, 1998, **802**, 189.
3. D. Fetsch, M. Hradilová, E.M. Peña-Méndez and J. Havel, *J. Chromatogr. A*, 1998, **817**, 313.
4. D. Fetsch, A.M. Albrecht-Gary, E.M. Peña-Méndez and J. Havel, *Scripta Fac. Sci. Nat. Univ. Masaryk. Brun.*, 1997-98, **27-28**, 3.
5. L. Pokorná, D. Gajdošová and J. Havel, in *Understanding Humic Substances: Advanced Methods, Properties and Applications*, eds. E.A. Ghabbour and G. Davies, Royal Society of Chemistry, Cambridge, 1999, p. 107.
6. L. Pokorná, D. Gajdošová, S. Mikeska and J. Havel, in *Humic Substances: Versatile Components of Plants, Soils and Water*, eds. E.A. Ghabbour and G. Davies, Royal Society of Chemistry, Cambridge, 2000, p. 299.
7. L. Carlsen, M. Thomsen, S. Dobel, P. Lassen, B.B. Mogensen and P.E. Hansen, in *Humic Substances: Versatile Components of Plants, Soils and Water*, eds. E.A. Ghabbour and G. Davies, Royal Society of Chemistry, Cambridge, 2000, p. 177.
8. T.F. Guetzloff and J.A. Rice, *Sci. Total. Environ.*, 1994, **152**, 31.
9. A. Eschenbach, M. Kästner, R. Bierl, G. Schaeffer and B. Mahro, *Chemosphere*, 1994, **28**, 683.
10. W. Ziechmann, M. Hübner, K.E.N. Jonassen, W. Batsberg, T. Nielsen, S. Hahner, P.E. Hansen and A.-L. Gudmundson, in *Humic Substances: Versatile Components of Plants, Soils and Water*, eds. E.A. Ghabbour and G. Davies, Royal Society of Chemistry, Cambridge, 2000, p. 9.
11. D.T. Scott, D.M. McKnight, E.L. Blunt-Harris, S.E. Kollesar and D.R. Lovley, *Environ. Sci. Technol.*, **32**, 2984, 1998.

12. R.L. Malcolm, in *Humic Substances II: In Search of Structure*, eds. M.H.B. Hayes, P. MacCarthy, R.L. Malcolm and R.S. Swift, Wiley, New York, 1989, p. 340.
13. T.M. Bockman, S.M. Hubig and J.K. Kochi, *J. Am. Chem. Soc.*, 1996, **118**, 4502.
14. M.H.B. Hayes and R.S. Swift, in *The Chemistry of Soil Constituents*, eds. D.J. Greenland and M.H.B. Hayes, Wiley, Chichester, 1978, p. 79.
15. R.S. Swift, in *Humic Substances II: In Search of Structure*, eds. M.H.B. Hayes, P. MacCarthy, R.L. Malcolm and R.S. Swift, Wiley, Chichester, 1989, p. 467.
16. E. Tombácz and J.A. Rice, in *Understanding Humic Substances: Advanced Methods, Properties and Applications*, eds. E.A. Ghabbour and G. Davies, Royal Society of Chemistry, Cambridge, 1999, p. 69.
17. F.J. Stevenson, *Humus Chemistry*, 2nd Edn., Wiley, New York, 1994.
18. M.H.B. Hayes and C.L. Graham, in *Humic Substances: Versatile Components of Plants, Soils and Water*, eds. E.A. Ghabbour and G. Davies, Royal Society of Chemistry, Cambridge, 2000, p. 91.
19. S.S. Gonet and K. Wegner, *Environ. Int.*, 1996, **22**, 485.
20. R.L. Hinman, *J. Org. Chem.*, 1958, **23**, 1587.
21. P.J. Isaacson and M.H.B. Hayes, *J. Soil Sci.*, 1984, **35**, 79.
22. V. Szabo, J. Borda and V. Vegh, *Acta Chim. Hung.*, 1978, **98**, 457.
23. K.A. Thorn, J. Arterburn and M. Mikita, *Environ. Sci. Technol.*, 1992, **26**, 107.
24. P.J. Kreuger, in *The Chemistry of Hydrazo, Azo and Azoxy Groups*, ed. S. Patai, Wiley-Interscience, London, 1975.
25. D. Fabbri and R. Helleur, *J. Anal. Appl. Pyrol.*, 1999, **49**, 277.
26. M. Grote, S. Klinnert and W. Bechmann, *J. Environ. Monit.*, 2000, **2**, 165.
27. J.C. del Rio, D.E. McKinney, H. Knicker, M.A. Nanny, R.D. Minard and P.G. Hatcher, *J. Chromatogr. A*, 1998, **823**, 433.
28. B. Chefetz, Y. Chen, C.E. Clapp and P.G. Hatcher, *Soil Sci. Soc. Am. J.*, 2000, **64**, 583.
29. P.G. Hatcher and R.D. Minard, *Org. Geochem.*, 1996, **24**, 593.
30. T.R. Filley, R.D. Minard and P.G. Hatcher, *Org. Geochem.*, 1999, **30**, 607.
31. L. Pokorná, M.L. Pacheco and J. Havel, *J. Chromatogr. A*, 2000, **895**, 345.
32. M.L. Pacheco and J. Havel, *J. Radioanal. Nucl. Chem.*, 2001, **248**, 565.
33. J. Havel and D. Fetsch, in *Encyclopedia of Separation Science*, ed. I. Wilson, Academic Press, London, 2000, p. 3018.
34. G.R. Aiken, D.M. McKnight, R.L. Wershaw and P. MacCarthy, eds., *Humic Substances in Soil, Sediment and Water*, Wiley, New York, 1985.

Masses, Similarities and Properties

PROTON AND METAL CATION BINDING TO HUMIC SUBSTANCES IN RELATION TO CHEMICAL COMPOSITION AND MOLECULAR SIZE

Ruben Kretzschmar and Iso Christl

Institute of Terrestrial Ecology, Swiss Federal Institute of Technology, CH-8952 Schlieren, Switzerland

1 INTRODUCTION

Humic substances (HSs) are ubiquitous in soil and aquatic environments and can strongly influence the mobility and bioavailability of trace metals that form complexes with humic acid (HA) functional groups. Therefore, geochemical modeling of metal cation behavior in natural systems is in many cases not satisfactory without considering metal complexation with humic substances. HSs are extremely complex, polydisperse mixtures of natural organic compounds formed during the microbial decomposition of plant and animal debris in terrestrial and aquatic environments. Depending on the source materials, degree of decomposition and environmental conditions, HSs of different overall composition are found.[1] For example, aquatic humic and fulvic acids (FAs) are often more hydrophilic and lower in molecular weight than humic and fulvic acids extracted from soil samples. Hydrophilic, low molecular weight humic compounds are more water soluble and therefore expected to be more mobile in soils and groundwater aquifers. On the other hand, humic compounds with high molecular weights or high aromaticity may be more strongly adsorbed to mineral surfaces and therefore are expected to be less mobile.

The development of models describing ion binding to HSs has been complicated by the fact that HSs are extremely heterogeneous, resulting in wide distributions of binding affinities for each ion. One promising model for ion binding to HSs is the "consistent non-ideal competitive adsorption" (NICA) isotherm equation, which can be combined with the Donnan gel model to account for electrostatic effects at different ionic strengths.[2,3] In the NICA model, the overall affinity distribution is described by two pools of binding sites, each with a quasi-Gaussian affinity distribution. The two pools of sites are sometimes conceptually rationalized as carboxylic and phenolic binding sites, respectively.

Even if proton and metal cation binding to a given FA or HA can be fitted with the NICA-Donnan model, the influence of chemical differences between HSs on proton and metal binding are still unclear. In the past, strenuous efforts have been made to characterize the chemical nature of HSs of varying sources, but relatively little is known about the influence of the chemical composition of natural organic matter on proton and trace metal binding.

In this paper, we present a study of the effects of chemical heterogeneity on the binding of protons, Cu(II) and Pb(II) to soil fulvic and humic acids. The approach is to fractionate the soil HSs into fulvic acid, humic acid and four size fractions of the humic acid, respectively. Then, these humic fractions are investigated with a variety of methods to obtain their overall chemical composition. Finally, the humic fractions are used in proton and metal binding experiments to elucidate the effects of chemical differences between the fractions on ion binding. The results are quantitatively described using the NICA-Donnan model in which the concentration of "phenolic-type" binding sites was related to the phenolic carbon content obtained from ^{13}C NMR spectroscopy.

2 EXPERIMENTAL METHODS

2.1 Extraction of Fulvic and Humic Acids

The experiments presented here were conducted with FAs and HAs extracted with IHSS standard procedures from a well-humified organic horizon (H) of a humic Gleysol in northern Switzerland. The fulvic acid was purified using an XAD-8 resin column technique and converted to the protonated form by passing it through a proton-saturated cation exchange resin. The humic acid was treated three times with 0.1 M HCl-0.3 M HF to remove mineral impurities and then dialyzed. The fulvic and humic acids were stored as concentrated stock solutions in the dark at 4°C. Subsamples of all fractions were freeze-dried for chemical analysis. Additional details of the methods used are reported elsewhere.[4]

2.2 Size Fractionation of Humic Acid

Four size fractions of the HA were prepared using a cross-flow hollow fiber ultrafiltration technique described in detail elsewhere.[4] The nominal molecular weight cutoffs of the filter cartridges were 300, 100, 30, and 10 kDa (UFP-E-3A, A/G Technology Corporation). Thus, we obtained HA size fractions which will be referred to as $HA_{>300}$, $HA_{100\text{-}300}$, $HA_{30\text{-}100}$ and $HA_{10\text{-}30}$, respectively. During the entire fractionation procedure, the HA solution (in 1 mM $NaHCO_3$, pH 8.2) was kept under N_2-gas atmosphere to prevent oxidation. Filtrates were immediately passed through a column packed with a proton-saturated cation exchange resin (Amberlite IR-120, Fluka) to remove sodium and carbonate from the HA solutions.

2.3 Characterization of Fulvic and Humic Acids

All FAs and HAs fractions were characterized by size exclusion chromatography (SEC) using a SigmaChrom GFC-1300 column (Supelco). The 30 cm × 7.5 mm column packed with cross-linked polysaccharides was run at a flow rate of 0.5 mL per min. The solutions injected were adjusted to pH 7.5 with a 50 mM Tris-HCl buffer solution containing 100 mM KCl. Elution of HSs was measured by UV-absorption at 280 nm. A molecular weight calibration ranging from 7 to 1000 kDa using globular proteins was performed as recommended by the manufacturer. The fulvic and humic acid fractions were analyzed for elemental composition (C, H, N, S and O) and ash content. In addition, chemical differences between the fractions were studied by UV-VIS absorption, fluorescence, FTIR (transmission mode, KBr pellets) and CP-MAS ^{13}C NMR spectroscopies. Additional details of the characterization methods used are reported elsewhere.[4]

2.4 Acid-Base Titrations

Proton binding by the FA, total HA, and the four humic acid size fractions was studied by acid-base titrations at different ionic strengths. The titration experiments were conducted at 25 ± 1°C using a computer-controlled titration system consisting of four burettes (Dosimat 605, Metrohm) a pH electrode (6.0123.100, Metrohm), and an Ag/AgCl reference electrode (6.0733.100, Metrohm) connected to a personal computer by a Microlink MF18 interface (Biodata, Manchester). The four burettes were filled with CO_2-free deionized water, 0.05 M HNO_3 (Titrisol, Merck), ~0.05 M NaOH and 2 M $NaNO_3$ (Merck, p.a.). All solutions were prepared CO_2 free and were connected with the atmosphere only through a glass tube filled with NaOH on granulated activated carbon (Merck, p.a.) to exclude CO_2 during titrations. The 350-mL teflon titration vessel was continuously flushed with water-saturated, CO_2-free nitrogen gas. After each addition of acid or base, the titrated solution was stirred for two min and electrode readings were recorded when the potential drift had dropped below 0.02 mV min^{-1} or after a maximum equilibration time of 30 min. Typically, experiments were performed by titrating with base (forward titration) followed by backward titration with acid. During each titration cycle, the ionic strength was kept constant within 1% by adding either water or salt solution (2 M $NaNO_3$) to correct for changes in ionic strength due to the acid or base additions. After each cycle, the ionic strength was increased to the next higher level by adding $NaNO_3$ solution. Several forward and backward titrations at different ionic strengths were obtained within a single experiment. Forward and backward titrations gave identical results, except for a small hysteresis (<0.12 mol kg^{-1}) near pH 7. Here, we report only the backward titration results.

2.5 Metal Binding Experiments

The binding of Cu and Pb to fulvic acid, total humic acid, and two humic acid size fractions was investigated by metal titration experiments at pH 4, 6 and 8 using pH and ion selective electrodes (ISE). The four burettes were filled with 0.05 M HNO_3, ~0.05 M NaOH, 1 mM $Cu(NO_3)_2$ (Merck, p.a.) containing 0.1 M $NaNO_3$, and 0.1 M $Cu(NO_3)_2$ for Cu titrations. For Pb binding experiments, $Pb(NO_3)_2$ (Merck, p.a.) solutions were used instead of $Cu(NO_3)_2$ solutions. Free metal cation activities in solution were measured with an ISE for Cu^{2+} (Orion 9429) and Pb^{2+} (Orion 9482), respectively. Metal titrations were carried out in a 200 mL glass vessel with initially 100 mL of solution containing ~1 g/L fulvic or humic acid and 0.1 M $NaNO_3$. To ensure complete dissolution of the humic acid, the pH was first increased to pH 10 by NaOH addition and then lowered to the desired pH value by HNO_3 addition. Before starting metal titrations, the pH was kept constant within a tolerance range of ±0.004 pH units (0.2 mV) for 12 h. This procedure was used to fully equilibrate the humic acid to pH and ionic strength conditions before metal addition was started. During metal titrations, the humic solutions were stirred for four min after each addition of titrant. After metal addition, the pH was automatically readjusted to the desired pH value and then kept constant for 20 min within a tolerance range of ±0.004 pH units (0.2 mV). Electrode readings were recorded when the potential drift was below 0.05 mV min^{-1} for the pH electrode and 0.1 mV min^{-1} for the ISE, or after a maximum equilibration time of 20 min. On average, readjustment of the pH value after metal addition took 60 to 75 min.

The Cu^{2+} and Pb^{2+} ion selective electrodes were calibrated by titrating 50 mL of a 0.15 mM $Cu(NO_3)_2$ or $Pb(NO_3)_2$ solution with a 1.787 mM ethylenediamine (p.a., Fluka) solution. The ethylenediamine solution was prepared under N_2-gas atmosphere using CO_2-free deionized water. To maintain a constant ionic strength throughout the calibration, metal salt and ethylenediamine solutions were prepared in a 0.1 M $NaNO_3$ background. Prior to recording electrode readings, the metal solution was stirred for 2 min after each titrant addition. Electrode readings were recorded when the potential drifts had dropped below 0.1 mV min^{-1} or after a maximum equilibration time of 30 min. Solution speciation was calculated for each recorded data point using the chemical speciation program ECOSAT.[5] Hydrolysis constants of aqueous metal species were taken from Baes and Mesmer.[6] Critical stability constants of ethylenediamine complexes as well as solubility constants of metal oxide and hydroxide phases were taken from Smith and Martell.[7,8] The calculated free metal ion activities were linearly related to the measured ISE voltage readings down to pCu 17 and pPb 9.5, where pMe is the negative logarithm of the free metal ion (Me^{2+}) activity in solution. The performance of both ISEs used was checked before and after each fulvic and humic acid titration.

3 RESULTS AND DISCUSSION

3.1 Chemical Differences between Fulvic Acid and Humic Acid Size Fractions

Size exclusion chromatography (SEC) elution curves for fulvic acid, total humic acid, and four humic acid size fractions obtained by hollow-fiber ultrafiltration are shown in Figure 1a. Generally, longer peak elution times in SEC experiments correspond to lower molecular sizes. Absolute values for the molecular sizes or molecular weights of HSs are difficult to determine by SEC due to the lack of suitable calibration standards with comparable macromolecular properties. The calibration of the SEC column with globular proteins can be used only for quality control and to monitor changes in column performance, but the molecular weights of humic substances may be overestimated based on this calibration. Thus, the values reported in Table 1 are referred to as "apparent" molecular weights based on the SEC method used in this study. Regardless of the absolute molecular weights of the fractions, the results show that the hollow fiber ultrafiltration method was successful in separating the humic acid into four fractions of different apparent molecular weight. The results also suggest that the overall molecular weight of the humic substances is large. For example, the exclusion peak at 4.2 mL observed for the purified humic acid and the largest humic acid size fraction is above the calibration range of the SEC column, which is given as 1000 kDa by the manufacturer. When comparing the apparent molecular weights of the fulvic and humic acids with those from other studies, it has to be kept in mind that we have used soil HSs, which are often of higher molecular weight than aquatic fulvic and humic acids. Another interesting observation is that the fulvic acid fraction had higher apparent molecular weight than the smallest humic acid fraction prepared with a 10-30 kDa nominal filter cutoff. However, it should be kept in mind that the high apparent molecular weight observed by SEC could result from self-association of smaller fulvic and humic acid molecules.[9]

The next questions are how do the chemical compositions of the humic acid fractions vary with apparent molecular weight and what are the chemical differences between the fulvic and humic acid fractions? To answer these questions, let us first examine the elemental composition of the fulvic and humic acid fractions presented in Table 1. Fulvic acid had lower contents of C, H, N and S when compared to humic acid and correspondingly higher O contents. The chemical composition of humic acid size fractions also shows a clear trend: C, H, N and S contents decrease and O content increases with decreasing apparent molecular weight of the humic acid. Note that the smallest humic acid fraction contains even less C, H, N and S and more O than the fulvic acid.

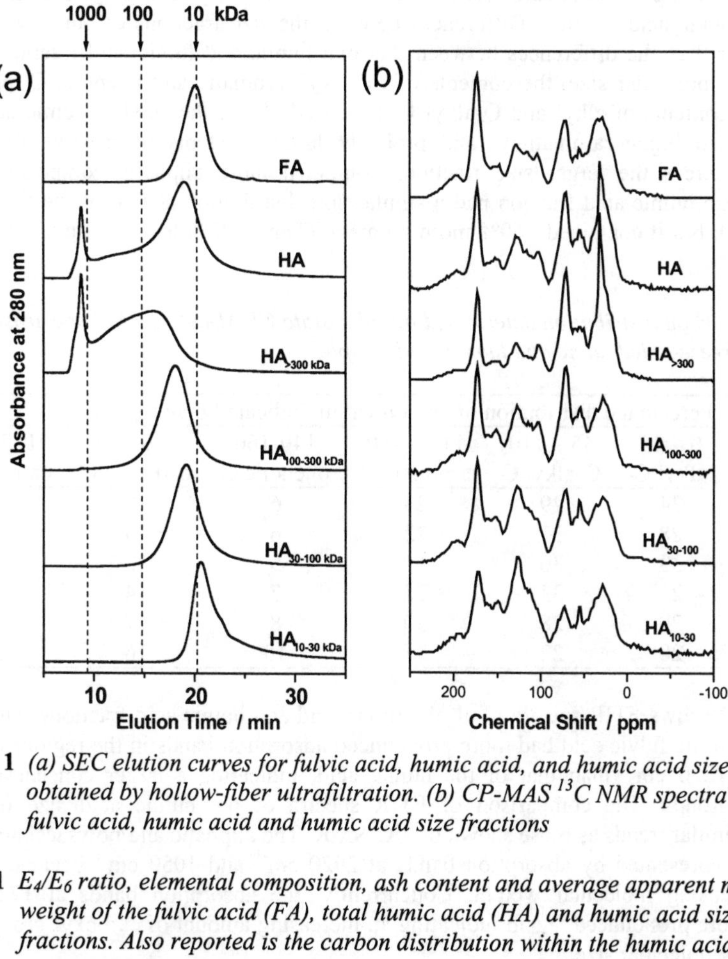

Figure 1 *(a) SEC elution curves for fulvic acid, humic acid, and humic acid size fractions obtained by hollow-fiber ultrafiltration. (b) CP-MAS ^{13}C NMR spectra of the fulvic acid, humic acid and humic acid size fractions*

Table 1 *E_4/E_6 ratio, elemental composition, ash content and average apparent molecular weight of the fulvic acid (FA), total humic acid (HA) and humic acid size fractions. Also reported is the carbon distribution within the humic acid fraction*

Sample	*E_4/E_6*	*Elemental composition /g kg^{-1}*					*% Ash*	$\overline{M}_w$	*% C*
		C	H	N	S	O			
FA	14.3	526	45	17	3	409	0.3	10.5	-
HA	6.6	552	53	33	4	358	< 0.2	> 16.8	100
$HA_{>300}$	6.1	584	58	40	6	312	n. d.	> 63.0	52 ± 6
$HA_{100-300}$	6.2	570	50	30	6	344	n. d.	24.8	7 ± 1
HA_{30-100}	7.5	566	46	24	4	360	n. d.	15.7	34 ± 7
HA_{10-30}	8.6	485	42	16	2	455	n. d.	8.5	3 ± 1

The results of solid-state CP-MAS ^{13}C NMR spectroscopy of freeze dried samples of the fulvic and humic acid fractions are shown in Figure 1b and the quantitative C distribution obtained from peak integration is summarized in Table 2. The fulvic acid contained about 20% more carboxyl C and 10% less alkyl C plus O-alkyl C compared to the total humic acid fraction. Differences between the size fractions of humic acid were even larger than the differences between the total humic and fulvic acid fractions. With decreasing molecular size, the contents of carboxyl, aromatic and phenolic C increased while the contents of alkyl and O-alkyl C decreased. Thus, the smaller humic acid size fractions had higher aromaticity and probably larger contents of reactive functional groups, whereas the larger size fractions contained more aliphatic components. The smallest size humic acid fraction had a similar size distribution as that of the fulvic acid (Figure 1a), but it contained ~60% more aromatic C and ~22% less alkyl plus O-alkyl C (Table 2).

Table 2 *Carbon distribution determined by solid-state CP MAS ^{13}C NMR spectroscopy of freeze-dried fulvic and humic acid samples*

	Percentage distribution of carbon within indicated ppm regions					
Sample	0 – 45 alkyl C	45 – 110 O-alkyl C	110 – 160 aromatic C	140-160 phenolic C	160-185 carboxyl C	185-220 carbonyl C
FA	24	29	23	6	18	6
HA	28	31	22	6	15	5
$HA_{>300}$	33	36	17	5	12	3
$HA_{100-300}$	21	33	25	7	14	5
HA_{30-100}	20	28	30	8	17	6
HA_{10-30}	20	22	37	10	16	5

Figure 2 shows FTIR spectra of all the fulvic acid and humic acid fractions. The FTIR spectrum of the fulvic acid had more pronounced adsorption bands in the regions of 1720 cm^{-1} and 1250 cm^{-1} than that of the humic acid, indicating a larger concentration of carboxyl groups. The comparison of FTIR spectra of the humic acid size fractions revealed similar trends as those shown by ^{13}C NMR. The aliphatic and polysaccharide-like character represented by absorption bands at 2920 cm^{-1} and 1050 cm^{-1} became weaker with decreasing molecular weight. Concurrently, the absorption bands at 1720 cm^{-1} became more pronounced, again indicating an increasing amount of carboxyl groups with decreasing molecular size.

3.2 Proton binding to fulvic and humic acid size fractions

Acid-base titration experiments at different ionic strengths ($NaNO_3$ electrolyte) were conducted to study the charging behavior and proton binding by the fulvic and humic acid fractions. Figure 3a shows a comparison of titration curves for fulvic acid and the total humic acid sample. As expected, the fulvic acid contained more acidic functional groups and exhibited higher charge per unit mass than the total humic acid fraction. The differences in proton binding are even greater than might be expected based on the ^{13}C NMR and FTIR results. However, this can be explained by the fact that carboxylic and phenolic functional groups contribute to the overall proton binding of humic substances and that a significant percentage of the ^{13}C NMR peak intensity in the chemical shift range 160 to 185 ppm is due to esters. A comparison of acid-base titration and ^{13}C NMR data of humic and fulvic acids reported in the literature suggests that the intensity of the carboxylic peak in ^{13}C NMR spectra is primarily due to carboxylic groups for fulvic acids, while esters contribute to at least one third of the intensity in ^{13}C NMR spectra of humic acids.[1]

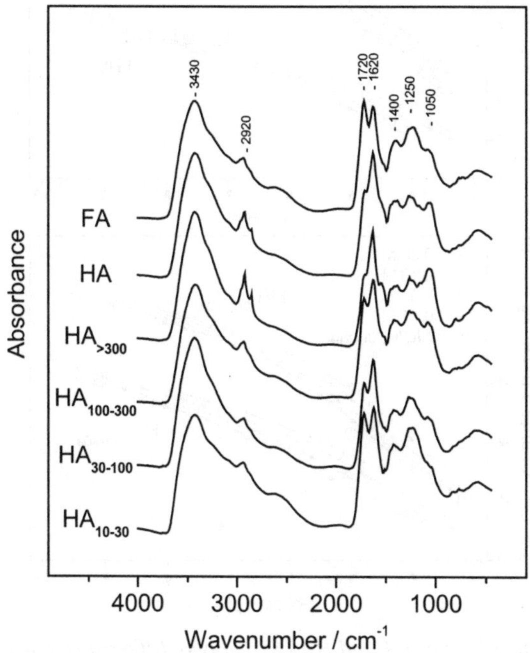

Figure 2 *Transmission FTIR spectra of the fulvic acid, total humic acid and four humic acid size fractions*

Figure 3b shows a comparison of titration curves for the largest and the smallest size fractions of humic acid. Both fractions exhibited similar ionic strength dependence of proton binding and a similar shape of the titration curves that reflects the affinity distribution. The charge per unit mass consistently increased with decreasing molecular size, which again is in qualitative agreement with the ^{13}C NMR and FTIR results. A direct comparison of titration curves for all fulvic and humic acid fractions in 0.1 M $NaNO_3$ is presented in Figure 4. The solid line represents the sum of titration curves for the four humic acid size fractions weighted by their mass fraction of the total humic acid (Table 1). This line should ideally coincide with the titration data for the total humic acid. The slight deviation may be explained by a small loss of low molecular weight humic compounds during the fractionation procedure.[4] The comparison of the fulvic acid with the humic acid titration curves again suggests that the fulvic acid has a much larger content of chargeable functional groups than the humic acid, even when compared only with the smallest humic acid size fraction. This difference is likely due to a much greater content in carboxylic groups and a larger contribution of esters to the 160 to 185 ppm ^{13}C NMR peak in humic acids.

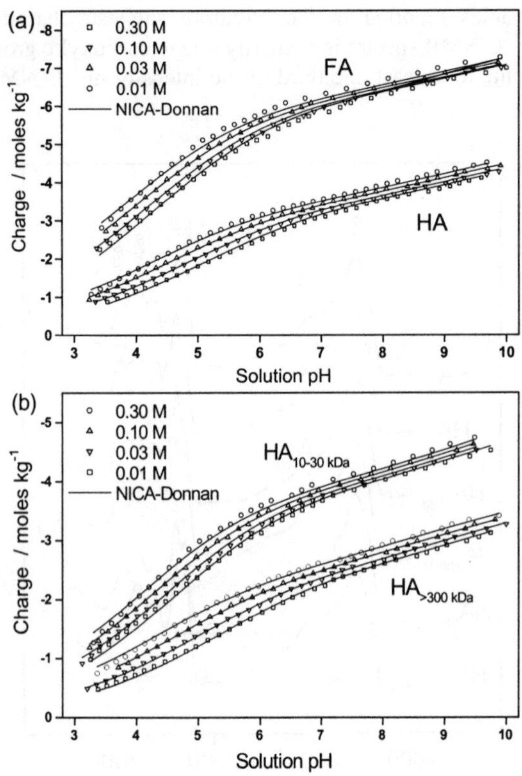

Figure 3 *Acid-base titrations of FA and HA (a) and two different size fractions of the HA (b) at four ionic strengths ($NaNO_3$ electrolyte). The lines represent best-fit descriptions with the NICA-Donnan model*

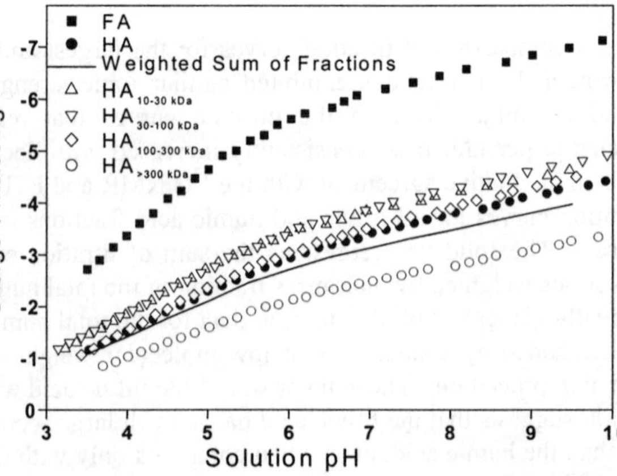

Figure 4 *Acid-base titrations of FA and HA size fractions in 0.1 M $NaNO_3$ electrolyte. The line represents the weighted sum of all HA size fractions and compares well with the titration curve for the total HA*

3.3 Cu(II) and Pb(II) Binding to Fulvic and Humic Acid Size Fractions

Isotherms for Cu(II) and Pb(II) binding to fulvic and humic acid fractions obtained by metal titration experiments with ion selective electrodes (ISE) are presented in Figure 5. Figure 5a shows Cu binding to the fulvic acid and total humic acid fraction at pH 4, 6, and 8 as a function of free Cu concentration in solution. At pH 6 and 8, the humic acid fraction exhibited higher Cu binding than the fulvic acid fraction, despite the larger content of chargeable functional groups in fulvic acid (Figure 4). However, at pH 4 the Cu binding isotherms were very similar for fulvic and humic acid.

Figure 5b shows a comparison of Cu binding isotherms for two different size fractions of the humic acid. At low Cu concentration, both humic acid fractions exhibited similar Cu binding per unit mass. At high Cu concentrations the smaller size fraction had a higher Cu binding capacity than the larger size fraction, which is likely due to a larger number of carboxylic and phenolic functional groups per unit mass.

Pb binding isotherms for the two size fractions of the humic acid are presented in Figure 5c. Again, almost no differences were observed between the two humic acid size fractions, except at high free Pb concentrations in solution, where the binding isotherms started to level off. Obviously, the humic acid size fractions exhibit very similar metal binding behavior despite clear differences in chemical composition. The observed differences in metal binding to fulvic and humic acid fractions were larger despite the apparent similarity of both fractions based on ^{13}C NMR and FTIR.

3.4 Using Chemical Information in NICA-Donnan Modeling

The proton and metal binding data of all fractions were quantitatively described by the NICA-Donnan model developed by Kinniburgh and coworkers.[2,3] The model results are shown as lines in Figures 3 and 5, respectively. The NICA isotherm equation is based on two pools of binding sites, each with a quasi-Gaussian affinity distribution for protons and metal cations, where the overall heterogeneity is conceptually divided into intrinsic heterogeneity parameters for each pool of binding sites and ion-specific heterogeneity parameters for each ion. Thus, the description of acid-base titration data for fulvic or humic acids at different ionic strengths (Figure 3) with the NICA-Donnan model requires 7 adjustable parameters. In the NICA equation, each pool of binding sites is described by a binding site concentration, an average binding constant and a heterogeneity parameter related to the width of the affinity distribution. In addition, the Donnan model has a parameter related to the Donnan volume. To reduce the number of free fitting parameters, we attempted to relate the site concentration of the "phenolic" binding sites in the NICA equation to the concentration of phenolic carbon determined by ^{13}C NMR spectroscopy. In addition, the Donnan model was related to the SEC results for the different humic acid fractions. As shown in Figures 3 and 5, this constrained NICA-Donnan model provided excellent fits to the experimental data and the remaining parameters became more stable and chemically plausible. Details of NICA-Donnan modeling and the resulting parameters are discussed elsewhere.[10,11] In the future, it would be appropriate to also have an independent estimate of the "carboxylic" pool of binding sites in the NICA model from ^{13}C NMR spectroscopy or other methods. However, as pointed out earlier, the "carboxylic C" peak (160-185 ppm) in ^{13}C NMR spectra includes varying contributions from esters.[1] Therefore, the ^{13}C NMR results cannot be directly used to estimate the binding site concentration in the "carboxylic" pool of the NICA model.

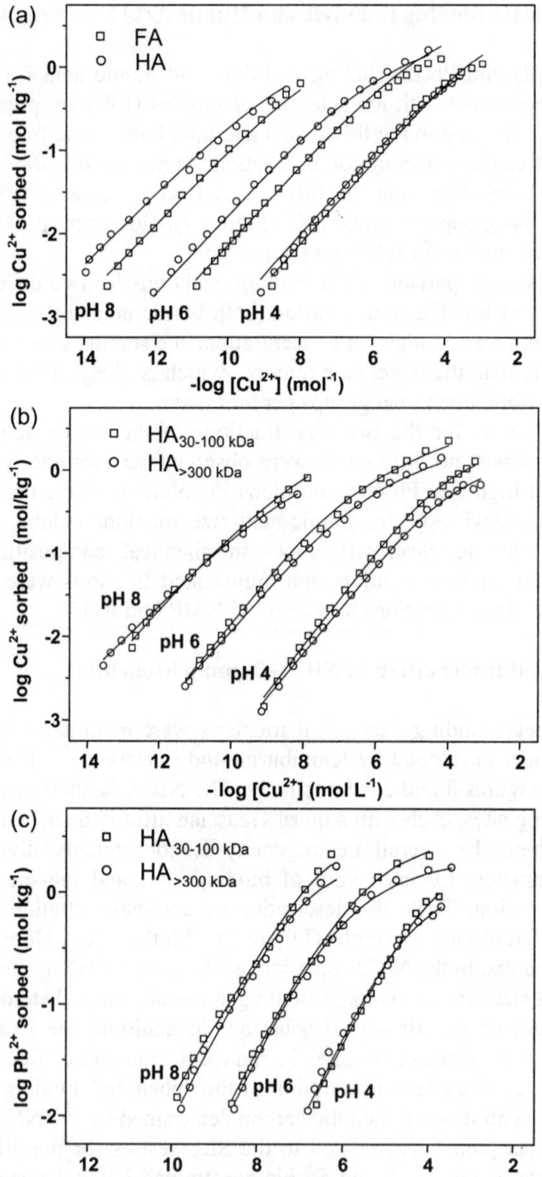

Figure 5 *(a) Cu binding isotherms for FA and HA; (b) Cu binding isotherms for two size fractions of the HA; (c) Pb binding isotherms for two size fractions of HA. Lines represent best-fit descriptions with the NICA-Donnan model*

4 CONCLUSIONS

The results of this study show that humic acid size fractions obtained by hollow fiber ultrafiltration differ in overall chemical composition. Based on ^{13}C NMR spectroscopy, FTIR spectroscopy and elemental (CHNSO) analysis the differences between the humic acid size fractions were greater than the differences between the fulvic acid and total humic acid fractions. The chemical differences between the fulvic acid, humic acid and size fractions of humic acid resulted in large differences in proton binding, which was mostly due to different concentrations of acidic functional groups. However, the differences between humic acid size fractions appeared less pronounced in Cu and Pb binding data. Humic acid had a higher affinity for Cu binding than the fulvic acid except at pH 4. Additional research is necessary to explain these observations at the molecular level.

ACKNOWLEDGEMENTS

We are thankful to H. Knicker and I. Kögel-Knabner for collecting the ^{13}C NMR spectra, C. Milne and D. Kinniburgh for support with NICA-Donnan modeling, and K. Barmettler for technical help in the laboratory. Funding of this research by the Swiss National Science Foundation (NF-21-46957.96) is also gratefully acknowledged.

References

1. R.L. Malcolm, in *Humic Substances in Soil and Crop Sciences: Selected Readings*, eds. P. MacCarthy, C.E. Clapp, R.L. Malcolm and and P.R. Bloom, ASA/SSSA, Madison, WI, 1990.
2. D.G. Kinniburgh, C.J. Milne, M.F. Benedetti, J.P. Pinheiro, J. Filius, L.K. Koopal and W.H. van Riemsdijk, *Environ. Sci. Technol.*, 1996, **30**, 1687.
3. D.G. Kinniburgh, W.H. van Riemsdijk, L.K. Koopal, M. Borkovec, M.F. Benedetti and A.J. Avena, *Colloids Surfaces A,* 1999, **151**, 146.
4. I. Christl, H. Knicker, I. Koegel-Knabner and R. Kretzschmar, *Eur. J. Soil Sci.*, 2000, **51**, 617.
5. M.G. Keizer and W.H. van Riemsdijk, *ECOSAT*, Dept. Environ. Sci., Wageningen Agricultural Univ., Netherlands, 1998.
6. C.F. Baes and R.E. Mesmer, *The Hydrolysis of Cations*, Wiley, New York, 1976.
7. R.M. Smith and A.E. Martell, *Critical Stability Constants Vol. 2: Amines*, Plenum, New York, 1990.
8. R.M. Smith and A.E. Martell, *Critical Stability Constants Vol. 4: Inorganic Complexes*, Plenum, New York, 1990.
9. P. Conte and A. Piccolo, *Environ. Sci. Technol.*, 1999, **33**, 1682.
10. I. Christl and R. Kretzschmar, *Environ. Sci. Technol.*, 2001, **35**, 2505.
11. I. Christl, C.J. Milne, D.G. Kinniburgh and R. Kretzschmar, *Environ. Sci. Technol.,* 2001, **35**, 2512.

CHARACTERIZATION OF TRACE METALS COMPLEXED TO HUMIC ACIDS DERIVED FROM AGRICULTURAL SOILS, ANNELID COMPOSTS, AND SEDIMENT BY FLOW FIELD-FLOW FRACTIONATION-INDUCTIVELY COUPLED PLASMA-MASS SPECTROMETRY (FLOW FFF-ICP-MS)

Thomas Anderson,[1] Laura Shifley,[1] Dula Amarasiriwardena,[1] Atitaya Siripinyanond,[2] Baoshan Xing[3] and Ramon M. Barnes[2,4]

[1] School of Natural Science, Hampshire College, Amherst, MA 01002, USA
[2] Department of Chemistry, University of Massachusetts, Amherst, MA 01003-9336, USA
[3] Department of Plant and Soil Sciences, University of Massachusetts, Amherst, MA 01003, USA
[4] University Research Institute for Analytical Chemistry, Amherst, MA 01002-1869, USA

1 INTRODUCTION

Humic substances (HSs) are vital to maintaining the physical and chemical properties of soil required for healthy microbe activity and plant growth. HSs are carbon-rich, highly functionalized, relatively high molecular weight compounds formed by slow decay and transformation of plant and animal debris. These versatile compounds interact with metal ions, polar and non-polar organic compounds including contaminants and microorganisms in the soil solution, and they form clay-humic aggregates to hold water and air to create a soil matrix.[1-4] Humic substances are divided into three categories: humic acids (HAs; intermediate molecular weight and insoluble in acids but soluble in alkali), fulvic acids (FAs; low molecular weight counterparts, soluble at all pH), and humin (HU; highest molecular weight, poorer sorbents and metal binders than HAs or FAs, insoluble at all pH).

The versatility of HAs makes them critical to buffering equilibria in our environment but uniquely challenging to study. HAs in the soil and aquatic environments form complexes with trace metals through omnipresent ligand functional sites. The trace metal complexation mechanisms at these sites are either metal-ligand interactions or proton displacements via ion exchange.[1] Humic acids contain aromatic and aliphatic carbon structures with many functional groups. They interact with trace metals, minerals and other soil components to form aggregate macrostructures held together by covalent and electrostatic forces.[1,3] Physical properties of HAs and chemical interactions of HAs with toxic elements such as As, Pb and Cd and nutritional elements like Cu, Zn and Mn are important to characterize.

The complexity of HAs and the various methods used to analyze them are apparent in the lack of reproducible and often inconsistent data for HA molecular size and structure. Like many areas in the study of humics, the data depend largely on the methods used in sample isolation and analysis. High performance size exclusion chromatography (HPSEC)[5,6] and capillary electrophoresis (CE)[7] are used to estimate molecular weights of HAs and to determine trace metals complexed to HA molecular fractions in soils and dissolved organic matter (DOM) in water.[8-13] Macromolecular denaturation and the tendency of HAs to aggregate during fractionation often impede HA molecular weight determinations.[2,14]

Methods that separate the HA mixture into molecular fractions without altering composition have long been sought. Flow field-flow fractionation (Flow FFF) accomplishes this objective. Flow FFF, which is based on specific diffusion properties of colloidal macromolecules,[15,16] is an analytical approach that can estimate hydrodynamic colloidal particle diameters, molecular weights and diffusion coefficients.[15,16] These are important parameters in improving understanding of the environmental behavior of trace elements in soil environments. The use of mild physical forces makes flow FFF a unique procedure and a relatively easy one to apply once the operating parameters have been optimized. Field-flow fractionation minimizes interferences and denaturation of macromolecules that may arise from other separation approaches.[17] Flow FFF uses liquid cross flows to separate macromolecules according to their diffusion coefficients, which are related to their molecular size. The identification of trace metals bound to these HA fractions can be examined using inductively coupled plasma-mass spectrometry (ICP-MS), a very sensitive trace elemental analysis technique. The theoretical basis and practical methodology[15,16-18] of flow-FFF and its applications to HA and FA molecular weight characterization have been discussed in the literature.[19,20]

Characterization of colloidal properties of HA can provide important clues in determining the environmental fate of trace metals bound to humic acid molecules. Unique applications of field-flow fractionation-inductively coupled plasma-mass spectrometry (FFF-ICP-MS) were reported for aqueous colloidal materials, clay minerals and suspended particulate matter in river water,[21,22] soil colloids from surface soil horizons,[23] colloidal matter in natural water,[24,25] soil derived HAs[26] and biological materials.[27] Flow FFF, sedimentation FFF and ICP-MS have compatible flow rates so they can linked directly.[16,21-27] The excellent sensitivity of ICP-MS trace metal detection allows analysis of the tightly bound metals in a fractionated HA sample.

In this study we used flow FFF-ICP-MS to analyze six HA samples of different origin. Humic acid functional groups were characterized with spectroscopic techniques. The difference in origin would be expected to produce different HA structures and therefore different spectroscopic properties. Would this difference be evident in the characteristics of the HA samples and in the tightly bound metals? The purpose of this study is to demonstrate the applicability of Flow FFF-ICP-MS to elucidate the colloidal properties of trace metal bound HA molecular fractions.

2 MATERIALS AND METHODS

2.1 Humic Acid Samples and Spectroscopic Characterization

Six HA samples derived from temperate and sub-tropical climate agricultural soils, from contaminated sediments and from annelid compost piles were examined. One HA sample (HACC) was derived from soil from a corn field in Hadley, MA enriched with cow manure and winter legume cover cropping. The second HA (HABV) came from an organic vegetable plot enriched with substantial amounts of cow manure and vegetable compost over the past twenty years with minimal tillage. Sample HASV was extracted from a mule manure annelid composting facility (6-12 months) in San Vincente, Solis Valley, Mexico. Sample HABS was extracted from Blackstone River Valley sediment in Uxbridge, MA. This site is under USEPA Region I designation as the "Rice City Pond 319 project"[28] since it had been previously contaminated by various industrial and municipal wastewater pollutants. Sample HAMX was extracted from a sub-tropical agricultural soil originating from a corn field in Las Reyes, Mexico. Sample HAVM, the second compost derived HA sample, was acquired from a horse manure compost pile (24 months old) near

Piracicaba, Brazil that had been treated with earthworms (see Table 1). Humic acids were extracted from soil using the sodium pyrophosphate extraction method.[29,30] 100 g soil was used for each extraction.

Table 1 *Origins, elemental compositions, % ash, and C/N and E_4/E_6 ratios of HAs*

HA	*Origin*	*%C*	*%N*	*%H*	*C/N*	*% Ash*	*E_4/E_6*
HACC	Soil from corn-field Hadley, MA, USA	52.44	4.03	4.31	13.0	1.1	4.77 ±0.01
HABF	Soil from organic vegetable farm, Amherst, MA, USA	52.88	3.88	4.39	13.6	0.7	4.88 ±0.03
HAMX	Soil from a corn-field Las Reyes, Solis Valley Mexico	52.50	3.58	4.00	14.7	2.1	4.02 ± 0.01
HABS	Sediment, Blackstone Valley, Uxbridge, MA USA	41.79	2.80	4.76	14.9	29.1	6.10 ± 0.59
HASV	Corn, mule manure annelid compost San Vincente, Solis Valley, Mexico	51.44	3.94	5.09	13.1	0.7	6.91 ± 0.07
HAVM	Horse manure compost Piracicaba, Brazil	51.75	3.87	5.42	13.4	1.9	3.88 ± 0.17

The percent ash content was determined by ashing pre-weighed HA samples in porcelain crucibles at 730°C for 6 hours using a Thermoline 6000 muffle furnace. Humic acid samples (2-4 mg) were dissolved in 10 mL of 0.05 M sodium bicarbonate solution. For all HA samples, the average absorbance reading was taken at 465 nm and 665 nm wavelengths using a Hewlett Packard HP8452A UV-visible spectrophotometer (obtained by averaging 465 nm, 466 nm and 664 and 666 nm wavelengths, respectively).

Each humic acid sample (3 mg) was added to 97 mg of spectroscopic grade potassium bromide (Spectratech) powder and ground with an agate motor and pestle. A 100 mg sample of pure KBr was powdered and used it as a blank. The blank and the mixed powder samples were immediately transferred into stainless steel sample holders and the surface was smoothed with a stainless steel blade or a microscope slide. To obtain diffuse reflectance infrared Fourier transform (DRIFT) spectra, 100 scans were collected at a resolution of 16 cm^{-1} with a Midac Series M-2000 Fourier transform infrared spectrophotometer (FTIR) equipped with a DRIFT accessory (Spectros Instruments). The instrument was flushed extensively with nitrogen between each sample analysis to remove carbon dioxide, moisture and hydrocarbons that will interfere with the absorbance peaks. Peaks were recorded as absorbance and converted to the Kubleka-Munk function using GRAMS/32 spectroscopic analysis software (Galactic Corp).

2.2 Flow Field-Flow Fractionation-Inductively Coupled Plasma Mass Spectrometry

Humic acid samples (2-3 mg) were dissolved in a 10 mL aliquot of 30 mM tris-(hydroxymethyl)aminomethane solution in nitric acid (30 mM TRIS-HNO_3 pH 7.3). All flow FFF measurements used 20-µL sample injections and the 30 mM TRIS/HNO_3 carrier fluid. Poly(styrenesulfonate) (PSS) molecular weight standards (6.2, 39 kDa) were used to calibrate the flow FFF system (Model F-1000-FO, FFFractionation LLC). A 3 kDa molecular weight cut-off (MWCO) polyregenerated cellulose acetate Flow FFF membrane was used for duplicate measurements of HA molecular weights. The FFF channel was

27.7 cm long, 2.0 cm wide and 0.0254 cm thick. An HPLC pump (Model-6010, Hitachi) controlled the channel flow rate at 0.90 mL/min at 170-psi pressure, a rate compatible with the ICP-MS nebulizer flow rate. Another HPLC pump (Model 300, Scientific Systems Instruments) controlled the cross flow rate at 2.0 mL/min, which was determined to be the optimum rate to retard the macromolecules' movement without compressing the molecules or adhering them to the FFF channel.[27] The FFF was equipped with a UV detector (Model L4000, Hitachi) to monitor HA chromophores and scattering absorption at 254 nm. The measurements were recorded every 3 seconds in absorbance units.

The flow FFF system was linked with poly-(tetrafluoroethylene) tubing (PTFE 0.3 mm id) to an ICP-MS system (Perkin Elmer Sciex/Elan 6000). The instrument configuration and analytical details were reported earlier.[26,27] The ICP-MS was optimized with a multi-element solution (10 $ngmL^{-1}$ Mg, Rh, Pb, Ce and Ba). The CeO/Ce^{4+} and Ba^{2+}/Ba^{+} ratios were established at less than 0.03 to ensure that oxides formed were minimized and that only singly charged ions were formed, respectively. The ICP-MS recorded 200 data points per element, taking the average of 4 measurements for each data point. The dwell time was set at 100 ms per reading. Other flow FFF-ICP-MS operating parameters are summarized in Table 2.

3 RESULTS AND DISCUSSION

3.1 Spectroscopic Characterization of Humic Acids

3.1.1 Elemental Analyses and E_4/E_6 Ratios. The percent (w/w) C, H and N, the C/N ratio and the E_4/E_6 ratios determined by UV-visible spectroscopy are given in Table 1. The observed C, H and N composition in soil HAs are very similar, and they are in agreement with the ranges reported in the literature for HAs[1] except for sediment derived HA (HABS). The percent H content of compost humic acids (HAVM, HASV) was significantly higher and the percent C content is lower than the other HAs, indicating a higher aliphatic content. The HA derived from semi-tropical soil (HAMX) and the sediment derived HA (HABS) had the highest C/N ratios of 14.7 and 14.9, respectively. Two other soil derived HA samples (HACC, HABF) and compost HA samples (HASV and HAVM) had C/N ratio ranging from 13.0 to 13.6, indicating they are richer in nitrogen functional groups than HAMX or HABS. This probably is due to nitrogen-rich manure amendments in these samples. The percent ash content of all HAs was less than 2.1 except for the sediment derived HA (29.1%), the latter despite HF/HCl washing during the purification procedure.

E_4/E_6 ratios traditionally have been used as a crude assessment of HS maturity; a high ratio indicates a low degree of aromatic condensation and shows the presence of large amounts of aliphatic components.[1] E_4/E_6 ratios of our HAs ranged from 3.88 to 6.91. The E_4/E_6 ratios of the agricultural soil derived HAs are close and typical of mature HAs reported in the literature. Interestingly, HASV, HABS have higher E_4/E_6 ratios than for most mature humic acids.[1] The HASV sample was extracted from 6-12 month old worm compost as opposed to the soil derived HA samples, which have had a longer residence time in soil. In contrast, the horse manure annelid compost sample (HAVM) had the lowest ratio E_4/E_6 = 3.88. This E_4/E_6 ratio could be also related to particle or molecular size[31] but we are unable to find any direct cause for these HA samples.

3.1.2 DRIFT Spectra and O/R ratios. Figure 1 illustrates the DRIFT spectra obtained for soil derived HAs (Figure 1a) and compost and sediment derived HAs (Figure 1b). The peaks are typical of HA spectra reported in literature.[1,12,29,31-34] A large band in the 3500-

2800 cm^{-1} region, which is due to OH stretch, $-CH_2$ symmetric and asymmetric stretches, and amide-NH_2 is followed by a shoulder near 2700-2600 cm^{-1} (COO-H hydrogen bonded). The functional group region has two distinct peaks (C=O of -COOH or C=O stretch of ketonic C=O groups) at 1720 cm^{-1}, and a sharp band at 1260-1230 cm^{-1} followed by the mostly noisy mineral peaks at low wavenumbers. Further examination of the functional group region of soil derived HA spectra (Figure 1a) shows a strong peak at 1720 cm^{-1} (i.e. HAMX), 1670-1630 cm^{-1} bands of aromatic C=C and amide I bend, weak bands around 1450 cm^{-1} (CH_3 asymmetric stretch and CH bend), and strong bands from C=O stretch, -COOH, OH deformation, or COC and OH functional groups around 1260-1220 cm^{-1}. A particularly sharp peak at 1260 cm^{-1} was observed for HACC and HAMX. Weak to moderate peaks at 1080-1030 cm^{-1} due to aliphatic COC and polysaccharides can be seen in all the soil HA spectra. In contrast, the most distinct difference in Figure 1b is the presence of the aliphatic $-CH_2$ asymmetric stretch at 2924 cm^{-1} in all còmpost[34] and sediment derived HA samples. This band is most prominent in HABS, the sediment derived HA. Furthermore, a sharp band at 1080-1030 cm^{-1} can be seen in compost HA samples (HASV and HAVM) indicating the presence of polysaccharides. Oxygen to recalcitrant (O/R) ratios (R_1)[34] for all six HA samples are shown in Table 3.

Table 2 *Fractionation conditions and ICP-MS instrument operating parameters for flow FFF-ICP-MS*

flow FFF conditions:	*Normal Mode (FFFractionation LLC)*
FFF channel dimensions/cm	27.7 × 2.0 × 0.0254
Carrier liquid	30 mM TRIS-HNO_3, pH 7.3
Cross flow rate/mLmin^{-1}	2.0
Channel flow rate/mLmin^{-1}	0.90
Equilibration time/min	1.5
UV wavelength/nm	254
Membrane	3-kDa MW cut-off polyregenerated cellulose
ICP-MS instrument settings and operating parameters	
RF generator frequency/MHz	40
RF forward power/W	1000
Torch	Sciex, short
Spray chamber	Ryton® Scott-type
Nebulizer	Cross –flow (Perkin Elmer)
Nebulizer, gas flow rate/ L min^{-1}	0.92
Auxiliary gas flow rate/ L min^{-1}	1.00
Outer gas flow rate/ L min^{-1}	12
Resolution	1 ± 0.1 at 10% peak maximum
Measurement per peak	100
Dwell time /ms	1
Isotope monitored (m/z)	^{75}As, ^{114}Cd, ^{63}Cu, ^{57}Fe, ^{60}Ni, ^{55}Mn, ^{208}Pb, ^{92}Pt, ^{235}U & ^{64}Zn

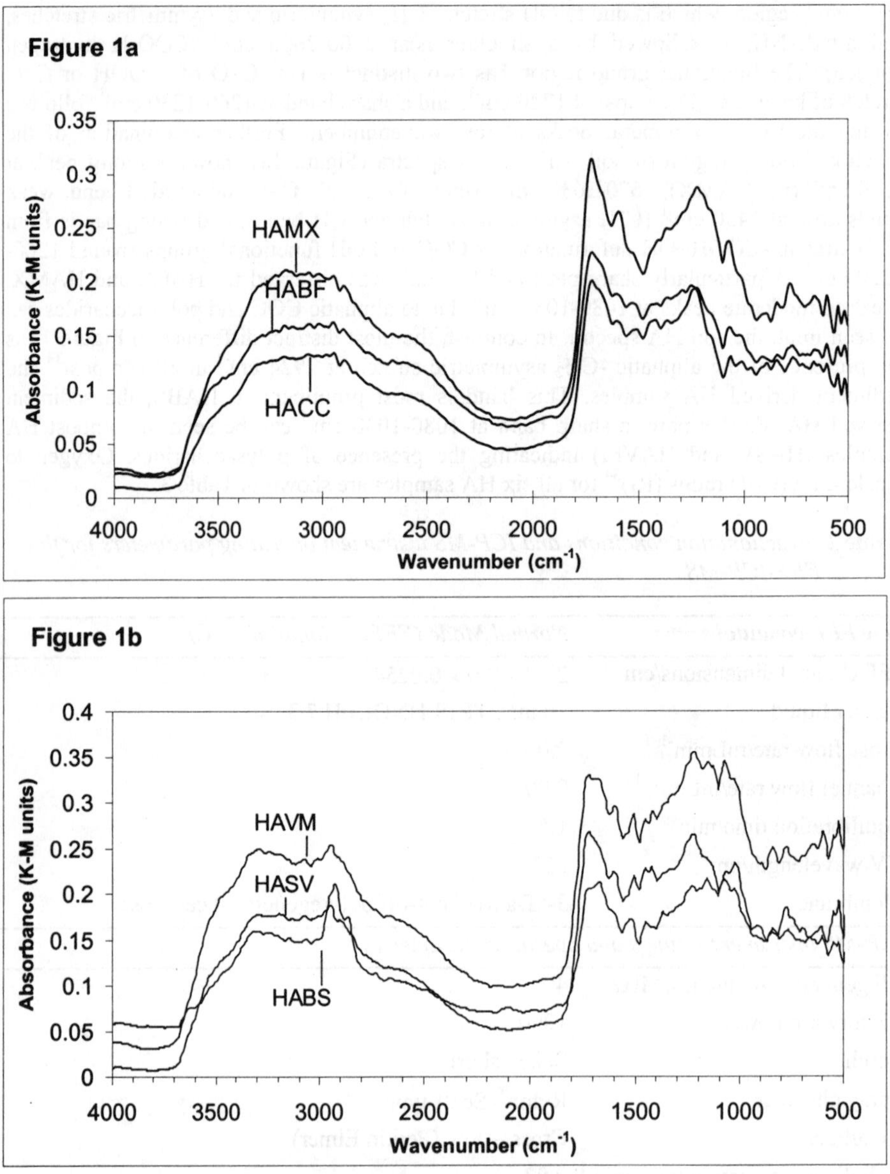

Figure 1 *DRIFT spectra of soil (Fig.1a) and compost and sediment derived humic acids (Fig.1b). See description of the HA sample codes in Table 1. The vertical axes of spectra are represented in Kubelka-Munk units*

R_1 was calculated to illustrate the intensity ratio of ketonic to -COOH centers plus additional functional groups including C=O bend, amide H, alcoholic OH and COC aliphatic ether chains *vs.* recalcitrant CH_2 symmetric and asymmetric stretches and C=C aromatic rings. HAMX had significantly higher O/R ratio R_1 than the rest of the HA samples, indicating an abundance of oxygen functionalities in the HA structure. Horse manure annelid mediated compost HA sample (HAVM) and sediment derived HA (HABS) had the lowest amount of oxygen functional groups compared to recalcitrant functional groups. Thus, the compost sample followed by the sediment samples are the most immature and rich in aliphatics while the soil derived HA samples are more aromatic and mature.

Table 3 *Oxygen to recalcitrant functional group peak height ratios (O/R ratios)[34] of HA by DRIFT spectroscopy*

HA	$R_I = \frac{1727+1650+1150+1160+1127+1050}{2950+2924+2850+1530+1509+1457+1420+779}$
HACC	0.75
HABF	0.63
HAMX	1.59
HABS	0.58
HASV	0.68
HAVM	0.40

3.2 Flow Field-Flow Fractionation of Humic Acids

Flow FFF was used to fractionate HA samples according to their diffusion coefficients and molecular weights. The diffusion coefficients (D_{max}), apparent molecular weight at peak maximum (M_p), and polydispersity of HAs were calculated using Flow FFF relationships discussed in the literature[19,20,26,27] and shown in the Table 4. Molecular weights were determined from the linear calibration functions obtained with PSS molecular standards using a 3 kDa flow FFF membrane. The diffusion coefficient (D) of a macromolecule is an inverse function of its hydrodynamic diameter (d_h).[16,19,20,26,27] For random coil macromolecules, D is non-linearly related to its molecular weight according to Equn. (1),

$$D = A'M^b \quad (1)$$

where constants b and A' depend on macromolecular conformation in the solution and macromolecular-solvent interactions, respectively. As a result, HA molecules with low molecular weights have rapid diffusion properties and emerge first during the fractionation with an emergence time t_1. Those HA molecular fractions with progressively larger molecular weights diffuse with later emergence times t_2, t_3, … t_n. In this way information about the molecular weight distribution (i.e., polydispersity) of the HA sample under consideration can be obtained. Fractograms of all six HA samples were plotted as UV absorbance (at 254 nm) *vs.* molecular weights (Da) (see Figure 2). The differences in the UV absorbance values are due to slight differences of solubility of HA samples in TRIS-HNO_3 and hence the HA concentration in the injection volume. Each HA sample produced a broad monomodal curve indicating the polydispersity. Duplicate analyses performed for each sample gave the average apparent molecular weight at peak maximum (M_p). The M_p values are very similar for all six samples ~ 5,000 Da, and are within the range of 5,000 to 5,600 Da (Table 4). These M_p values are close to the apparent molecular weights obtained for soil derived HA samples reported in the literature using the flow FFF technique.[19,20,26] Somewhat larger molecular weights found from limited replicates measurements for HAMX and HAVM may be attributed to a greater degree of polymerization and/or condensation of individual HA monomers during the humification process. Obviously, hydrodynamic diameters (d_h) of all HAs followed a similar trend, ranging from 4.65 (HAMX, HAVM) to 4.20 nm (HABS). Diffusion coefficients (D_{max}) of the six HA samples ranged from 1.05 x 10^{-6} to 1.16 x 10^{-6} cm^2sec^{-1} and showed very little variance within the experimental error. The molecular weight at the half peak height points are given as a range that indicates the polydisperse nature of HA samples. Polydispersity gives an idea of the variance in molecular size within each heterogeneous HA sample, whereas single apparent molecular weight values provide only limited information. The annelid compost derived HAs HAVM and HASV had the widest polydispersity (10 kDa) and the next most polydisperse sample was HAMX (9.5 kDa) derived from a sub-tropical agricultural site.

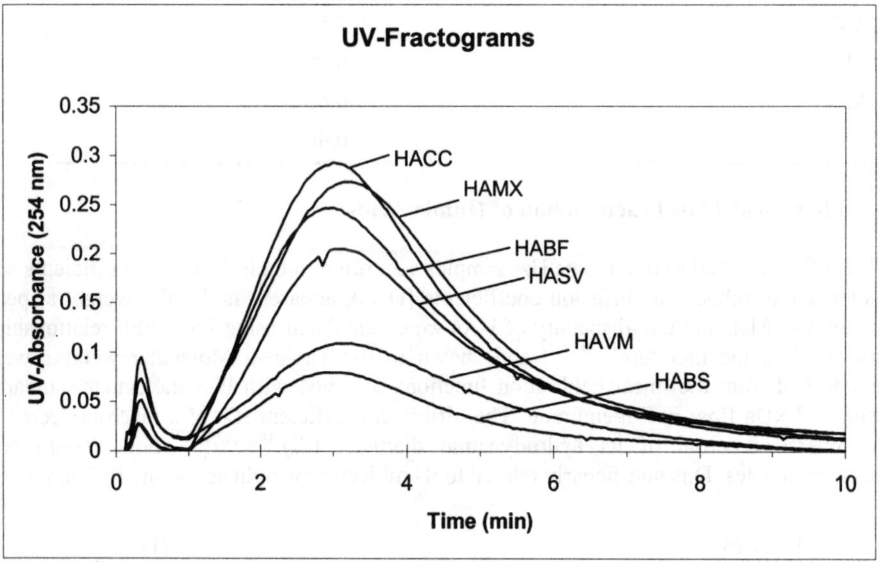

Figure 2 *UV-fractograms of all HA samples. See Table 1 for sample codes and descriptions*

Table 4 *Physical parameters of humic acids obtained from flow FFF*

HA Samples	*Mp n=2 (Da)*	*d_h (nm)*	*10^6 $D_{at\ max}$ (cm^2/sec)*	*Polydispersity (Da)*[a]
HACC	5000	4.35	1.13	9000 ± 200
HABF	5000	4.25	1.15	9200 ± 10
HAMX	5600	4.65	1.05	9500 ± 200
HABS	5000	4.20	1.16	8700 ± 400
HASV	5200	4.35	1.13	10,000 ± 50
HAVM	5500	4.65	1.05	10,000 ± 200

[a] Molecular weight range at half peak height width

On the other hand, the temperate agricultural soil derived HA samples HACC and HABF and the sediment derived HA sample HABS showed the least polydispersity (see Table 4). Overall, temperate soil and sediment derived HAs appeared to be the most uniform in molecular weight distribution while immature compost HAs tend to show greater heterogeneity. The latter could be due to degradation of higher molecular plant and manure materials, and subsequent aggregation and polymerization reactions during the compost maturing process.

3.3 Investigation of Elemental Distributions in Humic Acids by flow FFF-ICP-MS

The trace metals complexed to HA molecular fractions were determined by ICP-MS after fractionation with flow FFF. The analytical details and instrument configuration used here is discussed in earlier reports.[26,27] Ion fractograms (ICP-MS ion signal intensity vs. molecular weight) obtained for ^{63}Cu, ^{60}Ni, ^{208}Pb and ^{64}Zn are shown in Figure 3 for soil, compost and sediment-derived HAs. In addition, ^{75}As, ^{114}Cd, ^{57}Fe, ^{55}Mn, ^{195}Pt and ^{238}U signals were also monitored by ICP-MS. A summary of elements "tightly" bound with the HA samples investigated is shown in Table 5. From the ion fractograms of ^{63}Cu, ^{60}Ni, ^{208}Pb and ^{64}Zn, all HAs are closely matched, with their respective UV-fractograms indicating a high intensity ion signal for the median molecular weight fraction with a diminishing ion signal skewed towards the larger molecular fractions. Relatively elevated signal intensities were observed for ^{208}Pb in HACC, HABF and HASV samples as compared to HAMX or HAVM. Contaminated sediment HA sample (HABS) showed the strongest ion intensity for ^{63}Cu, while soil-derived HA (HACC and HABF) exhibited a low intensity ^{63}Cu signal with identical monomodal elemental distribution.

^{63}Cu and ^{64}Zn have been widely documented in metal binding studies and have been found to have strong HA binding affinities as previously demonstrated by SEC-ICP-MS and flow FFF-ICP-MS analyses.[12,13] A very strong ion intensity signal can be seen for ^{60}Ni except for the HAMX sample, which had an elevated background. The signal to background ratio of ^{75}As ion-fractograms is too small to distinguish statistically, perhaps indicating that this element is "loosely" bound and perhaps stripped from HA during the extraction process (see Table 5). Weak signals of ^{55}Mn appeared to be present in the two soil derived HA samples HACC and HABF. Barely discernable uranium signals were seen in soil (HACC, HABF) and compost (HASV) ion fractograms.

Table 5 *Elements tightly bound to HA median molecular weight fraction as determined by flow FFF-ICP-MS*

HA Sample	*^{63}Cu*	*^{64}Zn*	*^{208}Pb*	*^{57}Fe*	*^{55}Mn*	*^{75}As*	*^{238}U*	*^{60}Ni*
HACC	✓	✓	✓	✓	w	ND	w	✓
HABF	✓	✓	✓	✓	w	ND	w	✓
HAMX	✓	w	✓	NA	✓	ND	✓	✓
HABS	✓	✓	✓	NA	ND	ND	w	w
HASV	✓	✓	✓	✓	ND	ND	w	✓
HAVM	✓	✓	✓	NA	ND	ND	✓	✓

✓- detected, ND - not detected, w- weak signal, NA- not analyzed. In addition, no ^{195}Pt or ^{114}Cd signals could be detected

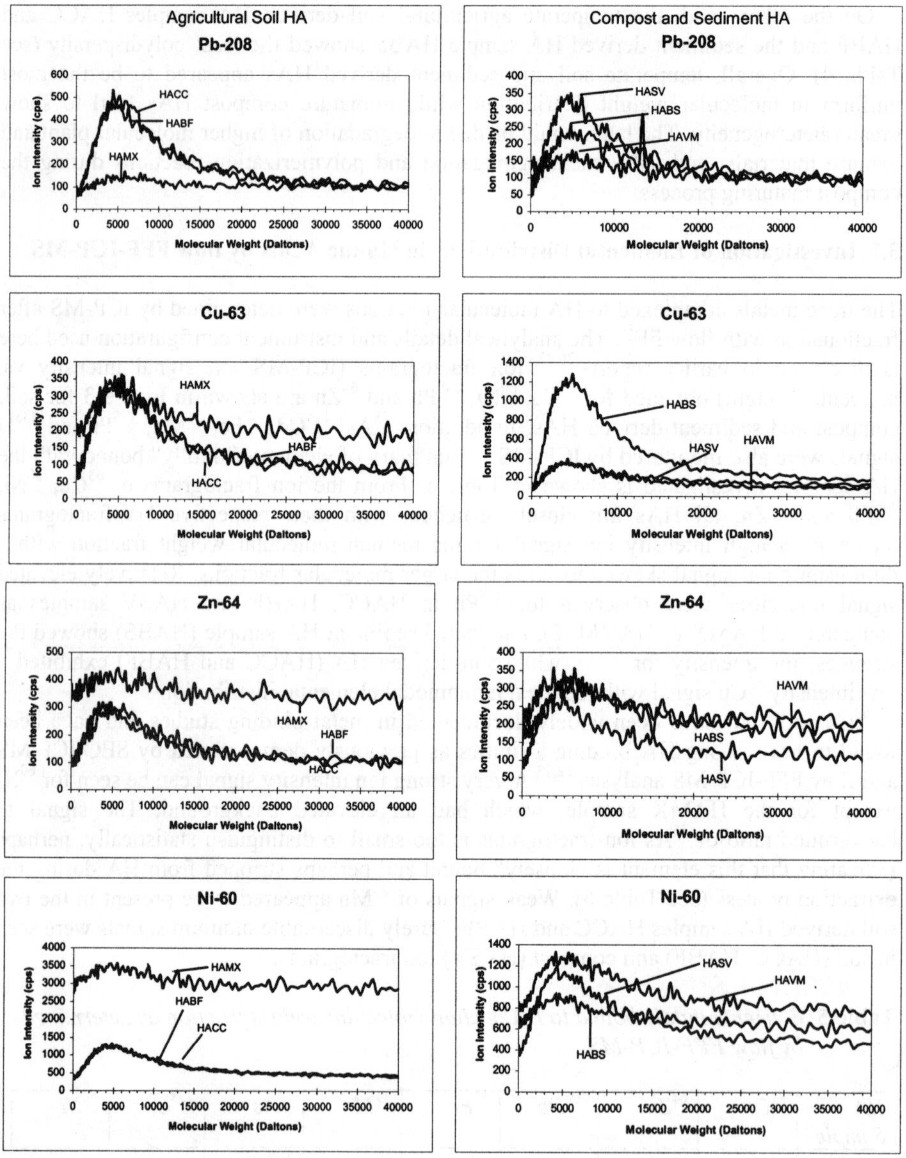

Figure 3 *Elemental fractograms (^{208}Pb, ^{63}Cu, ^{64}Zn and ^{60}Ni) of HA derived from agricultural soils composts and sediments with ICP-MS detection*

'Tightly' binding metals have the propensity to remove metal from the soil environment, while weaker metal HA associations allow trace metal mobility. The observed weak HA binding of As and Mn suggests that these elements are more mobile in soil environments, but the results also may be due to removal of these ions during the extraction process. In contrast, Cu, Zn and Pb ions are still associated with HAs molecules owing to their high complex formation constants. This further demonstrates that HAs play a significant role in mobilization of nutritionally important Cu, Fe and Zn in agricultural soils as well as Pb in contaminated soils.

4 CONCLUSIONS

This study involved the extraction and characterization of six HA samples derived from temperate and sub topical agricultural soils, annelid compost and sediment for functionality using UV-visible and DRIFT spectroscopy. As expected, HAs of different origins showed distinct differences in spectroscopic characteristics. The sub-tropical agricultural sample HAMX is the most mature and appears to be more humified than the other samples. Sediment derived HABS followed by two compost samples (HASV and HAVM) demonstrated abundance of aliphatic moieties, indicating that these are the least mature HAs.

The applicability of flow FFF-ICP-MS for qualitative determination of the distribution of trace metals complexed by HA molecular weight fractions in HAs of different origins was demonstrated. Despite the difference in functional group content, the properties of the HA samples were closely similar. The average apparent molecular weights obtained for all HA samples ranged from 5600-5000 Da, close to those M_p reported in the literature using the same technique.[19,20,27] Annelid compost samples HAVM and HASV had the greatest molecular weight polydispersity, while the sediment-derived HABS sample had the smallest range of molecular weights. Hydrodynamic diameters (d_h) ranged from 4.25 to 4.65 nm for the HAs investigated. The diffusion coefficient, which determines the rate at which molecules can travel though aqueous or soil environments, is an important parameter when examining the mobility of M-HA complexes. The average diffusion coefficients of the six HA samples do not appear to differ significantly and they may have nearly identical mobility ($\sim 1.0 \times 10^{-6}$ cm^2sec^{-1}) in the soil-aqueous interphase.

The transportation of toxic or nutritionally significant elements in aqueous or soil environments is affected, at least in part, by the rate of diffusion of these M-HA associations. Qualitative observation of elemental ion fractograms of HA samples showed strong Cu, Zn, Ni and Pb binding affinities with HA, while As and Mn form weaker metal-HA associations. Contaminated sediment derived humic acid HABS showed elevated levels of Cu tightly bound to its median molecular weight region, along with Pb, Ni and Zn signals. This indicates that a toxic element like Pb could be bound by the HA in the soil organic matter (SOM) layer, making it less likely to migrate through the aqueous phase. However, quantification of HA bound metal fractions by flow FFF-ICP-MS along with total elemental analysis would be necessary to confirm this hypothesis.

The physical parameters (i.e., diffusion coefficients, hydrodynamic diameters) along with information about metal-HA complexes obtained by flow FFF-ICP-MS lend important information about the roles of HAs in agricultural and contaminated soils. With future developments in quantitative analysis of trace metals complexed to HA molecular fractions, flow FFF-ICP-MS could be valuable for studying colloidal and metal complexation properties of soil and aqueous humic substances.

ACKNOWLEDGEMENTS

Dula Amarasiriwardena gratefully acknowledges the financial and instrument support of the National Science Foundation (BIR 951270) and the Kresge Foundation. This research was supported in part by the ICP Information Newsletter, Inc., Hadley, Massachusetts.

Thomas Anderson and Laura Shifley acknowledge the undergraduate research grants awarded by the Howard Hughes Medical Institute (HHMI).

References

1. F.J. Stevenson, *Humus Chemistry: Genesis, Composition, and Reactions*, 2nd Edn., Wiley, New York, 1994.
2. G. Davies, A. Fataftah, A. Cherkasskiy, E.A. Ghabbour, A. Radwan, S.A. Jansen, S. Kolla, M.D. Paciolla, L.T. Sein, Jr., W. Buermann, M. Balasubramanian, J. Budnick and B. Xing, *J. Chem. Soc., Dalton Trans.*, 1997, **21**, 4047.
3. G. Davies and E.A. Ghabbour, *Chemistry and Industry*, June 7, 1999, 426.
4. D.L. Macalady and J.F. Ranville, in *Perspectives in Environmental Chemistry*, ed. D.L. Macalady, Oxford University Press, New York, 1998, Chap. 5, p. 94.
5. Y.P. Chin, G. Alken and E. O'Loughlin, *Environ. Sci. Technol.*, 1994, **28**, 1853.
6. Y.P. Chin and P.M. Gschwend, *Geochim. Cosmochim. Acta.*, 1991, **55**, 1309.
7. M. De Nobili, G. Bragato and A. Mori, in *Understanding Humic Substances: Advanced Methods, Properties and Applications*, eds. E.A. Ghabbour and G. Davies, Royal Society of Chemistry, Cambridge, 1999, p. 101.
8. L. Zernichow and W. Lund, *Anal. Chim. Acta.*, 1995, **300**, 167.
9. L. Rottmann and K.G. Heumann, *Fresenius J. Anal. Chem.*, 1994, **350**, 221.
10. J. Vogl and K.G. Heumann, *Fresenius J. Anal. Chem.*, 1997, **359**, 438.
11. P. Schmitt, A. Kettrup, D. Freitag and A.W. Garrison, *Fresenius J. Anal.Chem.*, 1996, **354**, 915.
12. P. Ruiz-Haas, D. Amarasiriwardena and B. Xing, in *Humic Substances: Structures, Properties and Uses*, eds. G. Davies and E.A. Ghabbour, Royal Society of Chemistry, Cambridge,1998, p. 147.
13. S.A. Bhandari, D. Amarasiriwardena and B. Xing in *Understanding Humic Substances: Advanced Methods, Properties and Applications*, eds. E.A. Ghabbour and G. Davies, Royal Society of Chemistry, Cambridge, 1999, p. 203.
14. A. Piccolo, S. Nardi and G. Concheri, *Chemosphere*, 1996, **33**, 595.
15. J.C. Giddings, *Sep. Sci.*, 1966, **1**, 123.
16. R. Beckett, *At. Spectroscopy*, 1991, **12**, 228.
17. J.C. Giddings, *J. Chromatogr.*, 1989, **470**, 327.
18. J.C. Giddings, M.N. Benincasa, M.K. Liu and P. Li, *J. Chromatog.*, 1992, **15**, 1729.
19. R. Beckett, J. Zhang and J.C. Giddings, *Environ. Sci. Technol.*, 1987, **21**, 289.
20. P.J.M. Dycus, K.D. Healy, G.K. Stearman and M.J.M. Wells, *Sep. Sci. Technol.*, 1995, **30**, 1435.
21. H.E. Taylor, J.R. Garbarino, D.M. Murphy and R. Beckett, *Anal. Chem.*, 1992, **64**, 2036.
22. D.M. Murphy, J.R. Garbarino, H.E. Taylor, B.T. Hart and R. Beckett, *J. Chromatogr.*, 1993, **642**, 459.
23. J.F. Ranville, D.J. Chittleborough, F. Shanks, R.J.S. Morrison, T. Harris, F. Doss, and R. Beckett, *Anal. Chim. Acta*, 1999, **381**, 315.
24. M. Hassellöv, B. Lyvén, C. Haraldsson and W. Sirinawin, *Anal. Chem.*, 1999, **71**, 3497.
25. M. Hassellöv, B. Lyvén and R. Beckett, *Environ. Sci. Tech.*, 1999, **33**, 4528.
26. D. Amarasiriwardena, A. Siripinyanond and R.M. Barnes, in '*Humic Substances: Versatile Components of Plants, Soil and Water*,' eds. E.A. Ghabbour and G.

Davies, Royal Society of Chemistry, Cambridge, 2000, p. 215.
27. A. Siripinyanond and R.M. Barnes, *J. Anal. At. Spectrom.* 1999, **14**, 1527.
28. H. Snook, "Rice Pond 319 Project," Prepared for USEPA Region I, 4th May, 1996.
29. Z. Chen and S. Pawluk, *Geoderma*, 1995, **65**,173.
30. M. Schnitzer, in *Methods of Soil Analysis, Part 2: Chemical and Microbiological Properties*, eds. A.L. Page, R.H. Miller and D.R. Keenly, 2nd ed., Am. Soc. of Agronomy and Soil Sci. Soc. of America, Madison, WI, 1982, Chap. 30, p. 581.
31. M. Schnitzer, in *Soil Organic Matter*, eds. M. Schnitzer and S.U. Khan, Elsevier, Amsterdam, 1978, Chap. 1, p. 1.
32. A.U. Baes and P.R. Bloom, *Soil Sci. Soc. Am. J.*, 1989, **53**, 695.
33. J. Niemeyer, Y. Chen and J.M. Bollag, *Soil Sci. Soc. Am. J.*, 1992, **56**, 135.
34. M.M. Wander and S.J. Traina, *Soil Sci. Soc. Am. J.*, 1996, **60**, 1087.

APPARENT SIZE DISTRIBUTION AND SPECTRAL PROPERTIES OF NATURAL ORGANIC MATTER ISOLATED FROM SIX RIVERS IN SOUTHEASTERN GEORGIA, USA

J.J. Alberts, M. Takács, M. McElvaine and K. Judge

The University of Georgia Marine Institute, Sapelo Island, GA 31327, USA

1 INTRODUCTION

The molecular size of dissolved organic carbon (DOC) has been shown to affect the utilization of these materials by bacteria in a southeastern US river, with the smallest molecular size fractions <1 kDa nominal molecular weight (NMW) being the most efficiently converted into bacterial biomass.[1] In addition, sedimentary microbial assemblages from Georgia salt marshes have been shown to utilize both carbon and nitrogen from humic substance-like materials, which are similar in nature to the DOC brought to the marshes by Georgia rivers.[2] Furthermore, spectral properties of natural organic matter (NOM) are often utilized as a sensitive measure of DOC in natural waters. However, ultraviolet-visible specific absorbance coefficients have been shown to be dependent on molecular size of the NOM[3] and peak positions and relative fluorescence intensities of NOM have been shown to vary with molecular size.[4] Therefore, the chemical nature and distribution of carbon and nitrogen in the DOC size fractions becomes an important consideration not only for chemical transport of organic material and associated pollutants, but also for heterotrophic dynamics in these ecosystems.

Five major rivers draining 89,500 km^2 of forested and agricultural watersheds bring on average 35 km^3/yr of freshwater to approximately 150,000 hectares of salt marsh habitat[5] along a 170 km Georgia coastline.[6,7] These rivers vary in DOC content from approximately 3 to 30 mgC/L.[8] Greater than 50% of the DOC is believed to be humic substances[9,10] and 55-95% of the DOC occurs in relatively large molecular size fractions.[11] In addition, the nitrogen associated with the DOC represents between 50 and 90% of the total dissolved nitrogen in the rivers.[12] These estimates originate in monitoring databases that provide total concentration values over a 20-year period, while the size fraction data reflects summer conditions for two distinct years. While the extant data spans a period of many years, there are few data that systematically compare seasonal and inter-river characteristics of NOM present in southeastern rivers. Thus, it is difficult to predict the dynamics of NOM in these rivers with respect to biological and chemical processes on a seasonal or inter-annual basis.

In this study, six Georgia rivers were sampled in low flow (summer) and high flow (winter) conditions between 1999 and 2001. The DOC was fractionated by ultrafiltration to give insight into seasonal variations of size distribution of DOC and the elemental contents of those materials. In addition, ultraviolet-visible (UV-vis) and fluorescence spectral characteristics of the DOC and its fractions were determined to investigate changes in these parameters, which in the case of UV-vis are often used as surrogate measures of DOC.[13,14] Fluorescence measurements were included to determine if this extremely sensitive spectral measurement that arises from only a very small portion of the overall DOC mixture[15] is evenly distributed across size classes of DOC and whether it could be used as a tracer for specific size classes or the DOC as a whole.

2 MATERIALS AND METHODS

Water samples (approximately 1-2 L) were collected in July 1999, Feb. 2000, June 2000 and Jan. 2001 from the Savannah (Sav), Ogeechee (Ogee), Altamaha (Alt), Satilla (Sat), and St. Mary's rivers (St.M), which drain into the Atlantic Ocean. The Suwannee River (Suw), which drains into the Gulf of Mexico, was sampled in Feb. and June 2000 at Fargo, GA near its origin in the Okefenokee Swamp. The sampling locations of the other rivers have previously been described,[11,16] as has the method of separation by ultrafiltration.[11] Briefly, the samples were transported to the University of Georgia Marine Institute (UGAMI) and stored at 4°C. Samples were filtered through 0.45 μm filters. The resulting DOC was fractionated by means of differential ultrafiltration in a stirred cell (Amicon Corp. Model 401). The ultrafiltration resulted in three fractions <10 ("small"), <50>10 ("medium") and >50 ("large") kDa. The concentration of DOC in the <0.45 μm water samples and in the fractions was determined by high temperature combustion (Shimadzu Corp. Model 500 TOC Analyzer).

2.1 Elemental Analyses

Aliquots of the water were freeze-dried and taken for elemental carbon and nitrogen analyses (Perkin-Elmer Model 2400 CHN Elemental Analyzer). Due to the small sample sizes of some fractions, elemental analyses could not always be conducted. Similarly, the ash content of the freeze-dried material was often not determined. Therefore, the elemental analyses reported here are the atomic C/N ratios, which could be derived from the data without need of the ash contents.

2.2 Ultraviolet-Visible Spectroscopy

Ultraviolet-visible light attenuation of the samples was measured in a Perkin-Elmer Lambda 40 spectrophotometer in a 1-cm-path-length quartz-glass cuvette over the range from 200 to 700 nm.

2.3 Fluorescence Spectroscopy

The total luminescence (TL) spectra were collected on samples diluted to the absorbance at 254 nm of the lowest absorbing sample and buffered to pH 6.9 with 25 mM phosphate buffer solution in order to eliminate, or at least reduce, the effect of changes in pH, ionic strength and the inner filter effect. Total luminescence spectra were acquired with a Perkin-Elmer Model LS50B spectrofluorometer with emission wavelength from 300 nm to 600 nm and excitation wavelength from 220 nm to 415 nm increasing by 5 nm/scan. The excitation and emission slit widths were both 10 nm. The scan speed was 250 nm/min. Instrument validation was conducted every day and spectra of 0.1ppm quinine sulfate in

0.1 M $HClO_4$ solution were determined as the reference. Spectra were accumulated using FLWinlab software (Perkin-Elmer Corp.) and subsequent manipulations were conducted with GRAMS32 (Galactic Industries).

Fluorescence quantum efficiency is a measure of organic matter's ability to transfer absorbed energy to fluorescence emission and gives some insight into the electronic structure of the organic matter. It can be determined by taking the ratio of the integrated fluorescence emission energy and the absorbance at the wavelength of excitation. In this study, the emission curve at 350 nm excitation wavelength was used to calculate relative quantum efficiencies for the NOM samples and size fractions for winter and summer of 2000. The blank was subtracted and the emission peak was integrated over the range 375-575 nm with baseline correction. Values for absorbance were taken at 350 nm from the UV-vis spectra. Quinine sulfate (0.1 ppm in 0.01N H_2SO_4) was used as the reference solution and measured with the same instrument parameters. The formula used to calculate quantum efficiency (QE) is Equn (1)[17]

$$QE_{samp} = QE_{ref}*(OD_{ref}/OD_{samp})*(I_{samp}/I_{ref}) \quad (1)$$

3 RESULTS AND DISCUSSION

3.1 Molecular Size Distribution of DOC

Dissolved organic carbon concentrations in the rivers ranged from about 3 to 40 mgC/L (Figure 1). With the exception of the one data point for the Ogeechee River in summer of 1999, there appears to be a trend of increasing DOC concentration from north to south. The high value in the Ogeechee cannot be readily explained and may be considered an outlier or evidence of a transient event. The Savannah and Altamaha Rivers are Piedmont rivers, with their head waters originating in the Piedmont formations of north Georgia. The Ogeechee River lies between them and has a relatively small watershed. The Satilla, St. Mary's and Suwannee rivers are all Coastal Plains rivers with the head waters in southern Georgia. The St. Mary's and Suwannee rivers arise in the Okefenokee Swamp. Coastal Plains rivers are noted for their highly stained color, high dissolved organic matter content and low particulate loadings. Therefore, the distribution of total DOC among the rivers is consistent with the geology of their watersheds.

There is some indication of increased DOC contents in the rivers during the winter months, which would be consistent with the increased flows in these rivers during that season. However, the trend is not observed in all the rivers, indicating that variations in location and watershed size and usage patterns may play an important part in governing DOC concentrations from year-to-year.

The distributions of DOC in the various size fractions clearly show that during this period the smaller (<10 kDa NMW) size fractions dominate the DOC pool, accounting for 42-86% of the DOC (Figure 1). This distribution is contrary to the distributions observed in these rivers in 1991 and 1993, when this size fraction accounted for 5-26% and 32-45%, respectively, of the DOC.[7] The latter values were summer data and also showed year-to-year differences, with 1993 having a greater DOC pool in the smallest fraction than in 1991. Other than the preponderance of the smallest size fraction in the distribution, there are no readily apparent trends in size fractions with season or river. These observations serve to emphasize the changes that apparently can occur within the DOC pools over time periods of months to years to decades.

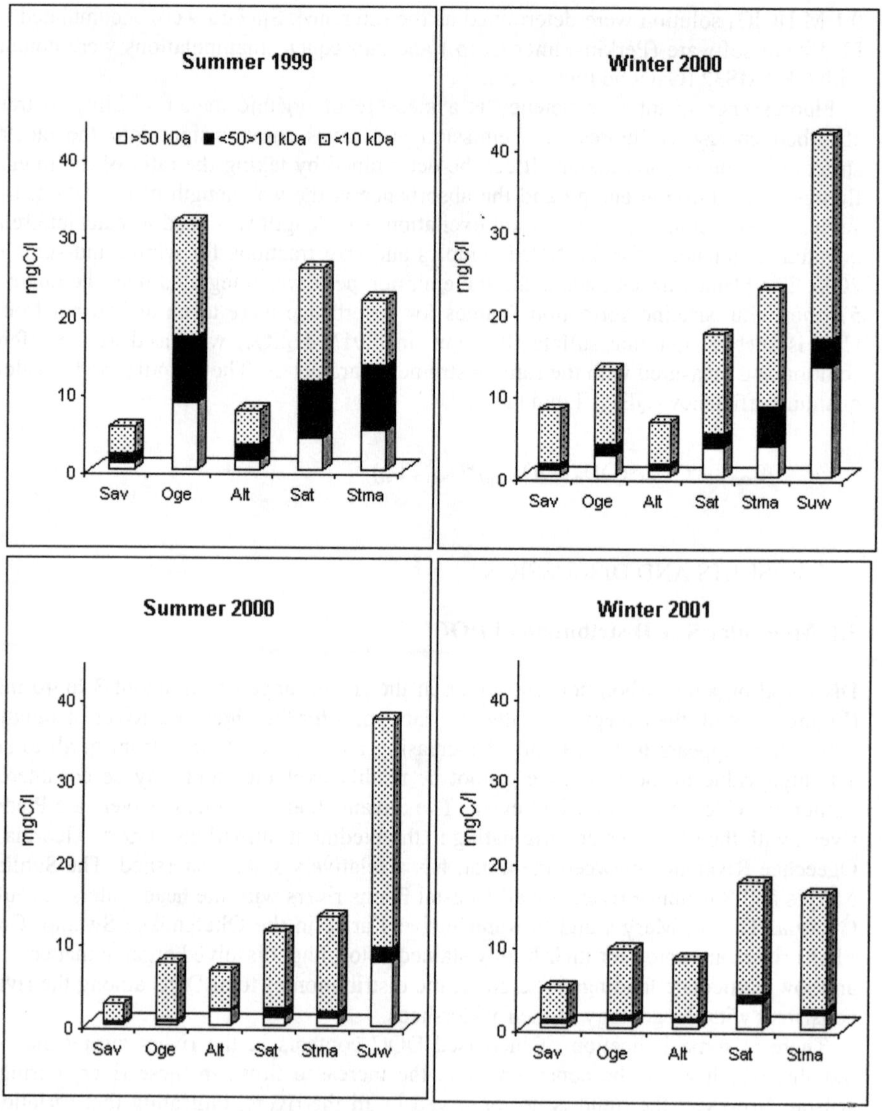

Figure 1 *Dissolved organic carbon concentration of filtered water and ultrafiltered size fractions from six Georgia River, summer 1999 to winter 2001*

Examination of the atomic C/N ratios of the size fractions clearly demonstrates the fact that the middle size fractions (<50>10 kDa) are depleted in nitrogen relative to the DOC in the largest size fractions (Figure 2). This observation is consistent with analyses of smaller sized fulvic acids and larger sized humic acids isolated from the Ogeechee River in 1981,[10] atomic C/N of 73.2 and 51.4, respectively,[17] and multiple samples from the Suwannee River taken between 1977 and 1978, 89.22 ± 12.25 and 53.03 ± 1.86, respectively.[18,19] Furthermore, in both size fractions there appears to be a trend of depletion in nitrogen relative to carbon with year, particularly in the Coastal Plains rivers (Figure 3).

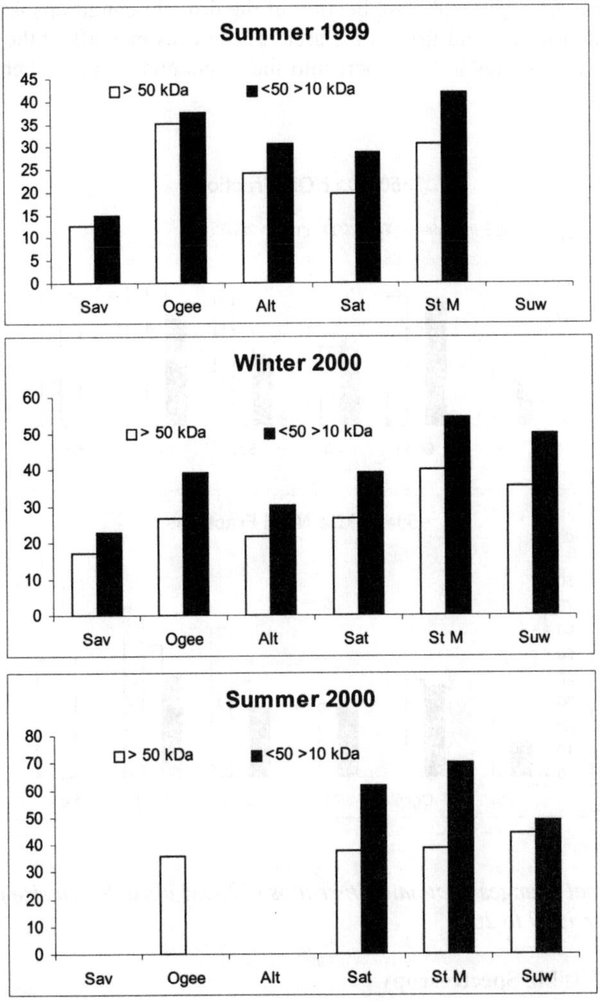

Figure 2 *Atomic C/N ratios of natural organic matter ultrafiltered size fractions from six Georgia rivers, summer 1999 to 2000*

The southeastern United States is in the fourth year of an extended drought. The decreased concentration of the larger sized DOC pools and the depletion of nitrogen may be a result of lower flows and increased biological activity, which utilizes the NOM as a nitrogen source. This explanation would be consistent with the fact that apparent changes are more visible in the Coastal Plains rivers, which are more variable with rainfall than the Piedmont rivers, which have large watersheds and higher volume flows. The Altamaha and Savannah rivers have average monthly flows for the period of 1990 to 1998 of 7,300-39,000 and 9,500-25,000 cubic ft sec^{-1} (cfs), respectively, while the Ogeechee, Satilla, St. Mary's and Suwannee rivers have average monthly flows for the same period of 1,000-6,400, 900-6,200, 300-1,200 and 350-2,700 cfs, resepctively.[8] Thus, many of the changes that are noted in the DOC pools of this data set may arise from biological activity in the rivers and their watersheds, which are being driven by climatic conditions experienced over several years. An additional complication of the drought conditions is the increase in frequency and intensity of wild fires in the area. These fires may affect the amount of the organic matter that is available for export into the rivers and also the chemical nature of that organic matter.

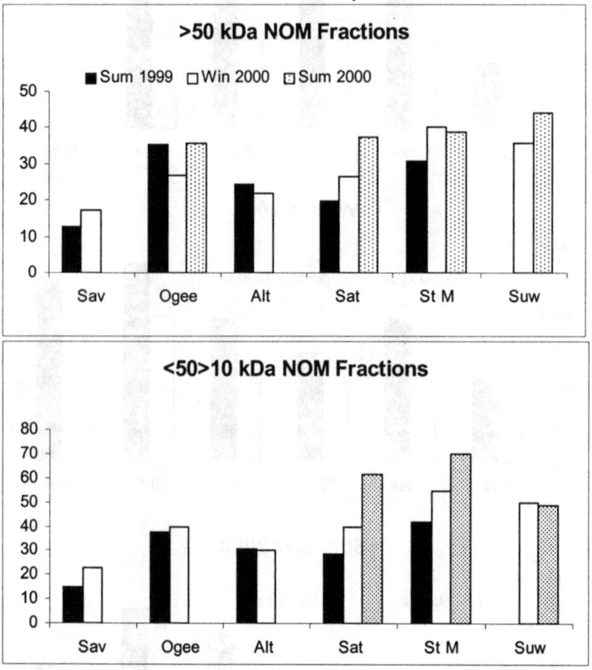

Figure 3 *Temporal changes in atomic C/N ratios of NOM ultrafiltered size fractions, summer 1999 to 2000*

3.2 Ultraviolet-Visible Spectroscopy

The UV-vis spectra of the <0.45μ water samples normalized to carbon content exhibit the usual monotonic featureless decrease in absorbance with increasing wavelength, although there are some differences in intensity of absorbance among the rivers (Figure 4a). The ultrafiltered fractions also show a monotonic decrease in absorbance with increasing wavelength. However, there appear to be significant differences in absorbance as a function of size, particularly in the large versus smaller fractions (Figure 4b).

An apparent specific absorbance of the NOM samples can be calculated by dividing the absorbance at a given wavelength by the concentration of DOC in the sample, thus giving a relative measure of light absorbance by compounds in the various fractions. There is no inherent trend in the apparent specific absorbance as a function of either season or river (Figure 5). However, it would appear that NOM from Coastal Plains rivers may have less variability in absorbance than rivers originating in the Piedmont.

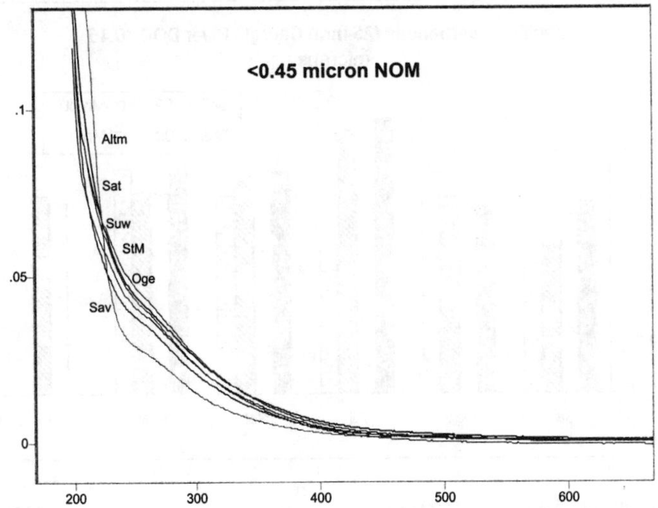

Figure 4a *UV-visible spectra normalized to total organic carbon (TOC) of the <0.45μ NOM from six Georgia Rivers*

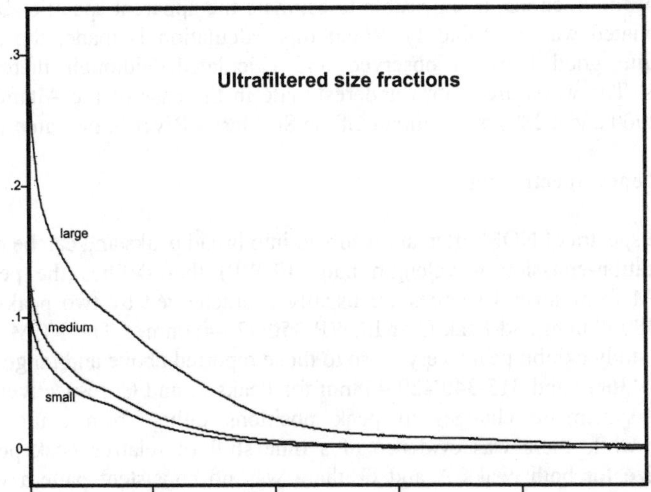

Figure 4b *UV-vis spectra normalized to total organic carbon of the ultrafiltered size fractions of the Satilla River*

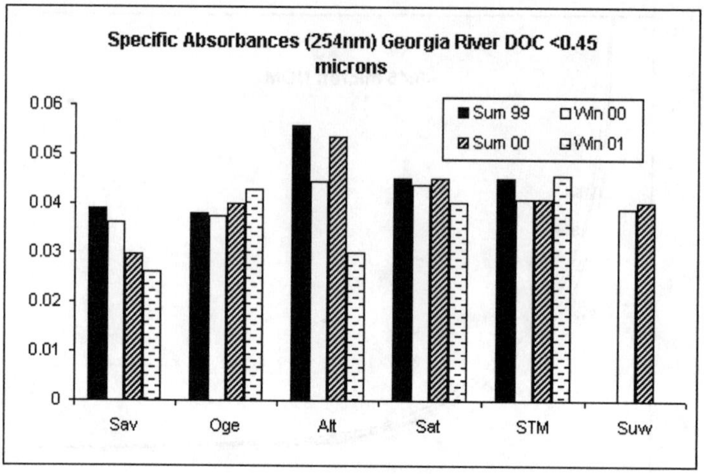

Figure 5 *Specific absorbance of NOM from six Georgia rivers, summer 1999 to winter 2001*

There is a general trend of decreasing apparent specific absorbance with decreasing size of the NOM fractions (Table 1), which agrees with the concept that larger molecules have larger and more complex chromophore systems and hence absorb more light per unit mass. Using the percent recovery of carbon in each fraction and the apparent specific absorbance of the fractions, it is possible to estimate the apparent specific absorbances of the unfractionated waters (Table 1). When this calculation is made, the agreement is generally quite good between observed and calculated, although there are a few discrepancies. The worst are a 35% underestimate in the case of the Altamaha River in Summer of 2000 and a 28% overestimate of the St. Mary's River in the same year.

3.3 Fluorescence Spectroscopy

Fluorescence spectra of NOM often are resolved into broad peaks and can be characterized by the excitation-emission wavelength pair (EEWP) that defines the peak position. Aquatic NOM from several sources are usually characterized by two peaks: Peak A at EEWP 260/380-480nm and Peak C at EEWP 350/420-480nm.[20] The NOM from the six rivers in this study exhibit peaks very close to those reported above and range from EEWP 225-240/401-436nm and 315-340/429-448nm for Peaks A and C, respectively, with little evidence of systematic changes in peak positions either from source or season. Furthermore, while there was evidence of a blue shift of relative peak positions with decreasing size for both peaks A and C, there was no consistent pattern with river or season.

Similarly, the relative fluorescence intensities (RFI) of Peak C did not consistently change with river or season for any size fraction. However, the medium and small size fractions of NOM consistently had a greater RFI values than the larger NOM fractions for all rivers and both seasons (Figure 6). This observation is consistent with the hypothesis that larger molecules, with more extensive conjugated electronic systems, will have greater self absorbance of light energy (inner filter effect) or be capable of dispersing absorbed energy through the system without emitting light (intramolecular conversions). This hypothesis is further supported by the quantum efficiencies of the size fractions, which increase significantly with decreasing molecular size (QE = 0.0032 ± 0.0023, 0.0129 ± 0.0024 and 0.0225 ± 0.0013 for >50, <50 >10 and <10 kDa, respectively, during winter 2000). A similar trend was found in the summer 2000 data (QE = 0.0068 ± 0.0062, 0.0119 ± 0.0025 and 0.0161 ± 0.0038, respectively). There also appears to be a decrease in QE for the size fractions as a function of season, with summer values being smaller than winter values. However, this trend is not entirely supported by the QE data for the <0.45μ NOM samples and requires further verification.

Table 1 *Specific absorbance (A254nm/TOC) of NOM ultrafiltered size fractions from six Georgia rivers as a function of season*[a]

	Savannah	*Ogeechee*	*Altamaha*	*Satilla*	*St. Marys*	*Suwannee*
Sum 99						
<0.45μ	0.0391	0.0381	0.0559	0.0451	0.0453	
(<0.45μ)	(0.0359)	(0.0362)	(0.0516)	(0.0436)	(0.0439)	
>50 kDa	0.0601	0.0557	0.0993	0.0557	0.0624	
<50>10 kDa	0.0380	0.0473	0.0451	0.0479	0.0503	
<10 kDa	0.0231	0.0265	0.0296	0.0322	0.0334	
Win 00						
<0.45μ	0.0361	0.0373	0.0447	0.0440	0.0411	0.0391
(<0.45μ)	(0.0346)	(0.0357)	(0.0410)	(0.0403)	(0.0403)	(0.0349)
>50 kDa	0.0991	0.0519	0.1225	0.0534	0.0579	0.0416
<50>10 kDa	0.0380	0.0430	0.0403	0.0531	0.0558	0.0418
<10 kDa	0.0245	0.0316	0.0297	0.0389	0.0386	0.0287
Sum 00						
<0.45μ	0.0297	0.0401	0.0535	0.0452	0.0408	0.0404
(<0.45μ)	(0.0227)	(0.0279)	(0.0355)	(0.0453)	(0.0524)	(0.0397)
>50 kDa	0.0269	0.0372	0.0556	0.1093	0.0661	0.0631
<50>10 kDa	0.0267	0.0361	0.0275	0.0454	0.0353	0.0491
<10 kDa	0.0230	0.0302	0.0305	0.0359	0.0536	0.0351
Win 01						
<0.45μ	0.0262	0.0429	0.0299	0.0403	0.0457	
(<0.45μ)	(0.0247)	(0.0379)	(0.0274)	(0.0375)	(0.0361)	
>50 kDa	0.0369	0.0625	0.0489	0.0650	0.0635	
<50>10 kDa	0.0197	0.0352	0.0263	0.0508	0.0357	
<10 kDa	0.0234	0.0316	0.0245	0.0339	0.0349	

[a] Values in parentheses are calculated apparent specific absorbances

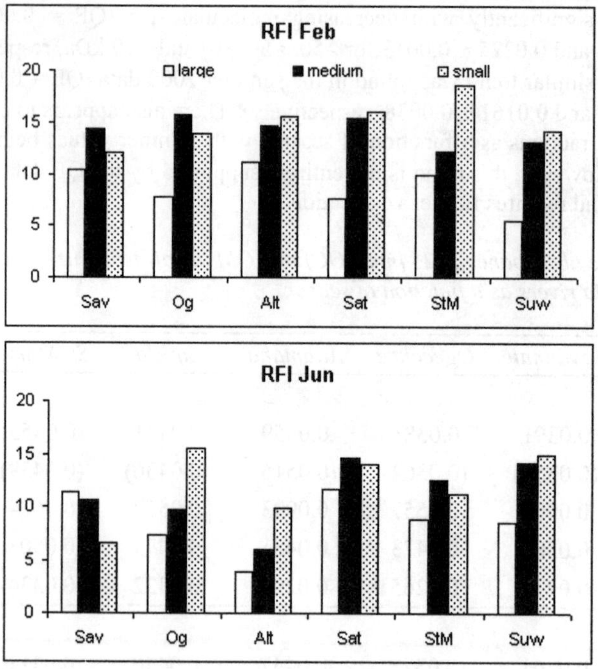

Figure 6 *Relative fluorescence intensity (RFI) normalized to carbon content (pH 6.9) of NOM ultrafiltered size fractions from Georgia rivers, summer and winter 2000*

Finally, the ratios of the RFI of peak C to peak A indicate that peak A, which is the peak excited in the ultraviolet region of the spectrum (λ_{ex} = 225-240 nm), is enhanced relative to peak C in the <0.45μ and all NOM size fractions in the summer relative to the winter (Figure 7). The significance of this observation is unclear, but it does emphasize that seasonal changes are occurring in the fluorescence characteristics of the NOM, possibly as a result of processes in the watersheds.

4 CONCLUSIONS

The DOC concentrations of the rivers during the period of summer 1999 to winter 2001 ranged from 3 to 40 mgC/L and reflect the geology of their watersheds. The DOC during this period is dominated by smaller molecular size material, i.e., 42-86% of the DOC pool in the <10 kDa fraction. There also appears to be a depletion in nitrogen relative to carbon in the medium sized fractions (<50>10 kDa) relative to the larger sized fractions (>50 kDa), which is consistent with previous data from fulvic and humic acids isolated from these rivers.

The UV-vis spectra normalized to carbon content of the rivers for the filtered water samples and the ultrafiltered size fractions show a monotonic decrease with increasing wavelength. The apparent specific absorbances of the size fractions also decrease with decreasing molecular size of the fractions. However, there is no inherent trend in apparent specific absorbance of NOM as a function of river or season.

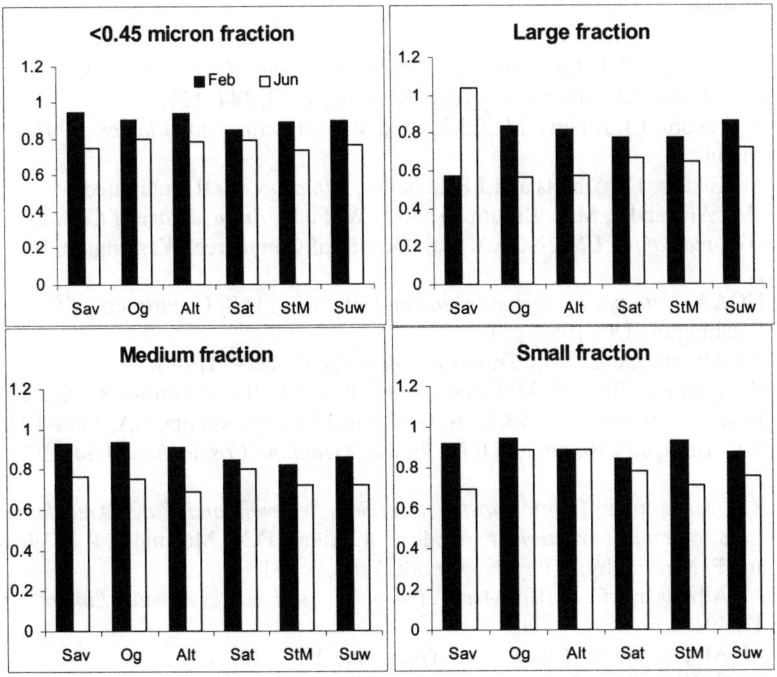

Figure 7 *Ratio of RFI of peak C to RFI of peak A from total luminescence spectra of ultrafiltered size fractions of six Georgia rivers, winter and summer 2000*

Total luminescence fluorescence spectra of the NOM samples primarily show two characteristic peaks with EEWP at 225-240/401-436nm and 315-340/429-448nm, respectively. Although there is some variation in peak positions with decreasing molecular size resulting in a blue shift, there is no observed trend with respect to peak position shifts as a function of river or season. While relative fluorescence intensities showed a similar lack of consistent change with river or season, an increase of RFI with decreasing size is again noted. Quantum efficiencies of the NOM fractions also increased significantly with decreasing molecular size and the ratios of RFI for peaks A and C changed consistently with season.

This study shows that there are changes occurring in the NOM of rivers in the southeastern US that are coupled to processes inherent in the watersheds and that those changes can be affected by climate and watershed usage changes. In addition, some important parameters that might be proposed for use as surrogates to measurement of dissolved organic carbon may be changing relative to environmental conditions and distribution of carbon and nitrogen within the molecular size distribution of the NOM.

ACKNOWLEDGEMENT

The University of Georgia Marine Institute Student Intern Program supported M. McElvaine and K. Judge. This is contribution number 880 from the UGA Marine Institute.

References

1. J.L. Meyers, R.T. Edwards and R. Risley, *Microbial Ecol.*, 1987, **13**, 13.
2. Z. Filip and J.J. Alberts, *Sci. Total Environ.*, 1994, **144**, 121.
3. Y. Yacobi, J.J. Alberts, M. Takács and M. McElvaine, *Aquatic Sci.*, 2001, submitted.
4. M. Takács, J.J. Alberts and J.F. Schalles, *Estuaries*, 2001, submitted.
5. C.E. Alexander, M.A. Broutman and D.W. Field, *An Inventory of Coastal Wetlands of the USA*, NOAA, U.S. Depart. of Commerce, Washington, DC, 1986, p. 1.
6. NOAA, *National Estuarine Inventory Data Atlas*, U.S. Department of Commerce, Washington, DC, 1985, p. 1.
7. J.J. Alberts and Z. Filip, *Trends in Chem. Geol.*, 1994, **1**, 143.
8. W.R. Stokes, III, R.D. McFarlane and G.R. Buell, *Water Resources Data Georgia Water Years 1974 to 1998*, U. S. Geological Survey, Atlanta, GA, 1974-1999.
9. K.C. Beck, J.H. Reuter and E.M. Perdue, *Geochim. Cosmochim. Acta*, 1974, **38**, 361.
10. R.L. Malcolm, in *Humic Substances in Soil, Sediment and Water; Geochemistry, Isolation and Characterization*, eds. G.R. Aiken, D.M. McKnight, R.L. Wershaw and P. MacCarthy, Wiley, New York, 1985, p. 181.
11. J.J. Alberts and C. Griffin, *Arch. Hydrobiol. Spec. Issues Advanc. Limnol.*, 1996, **47**, 401.
12. J.J. Alberts and M. Takács, *Org. Geochem.*, 1999, **30**, 385.
13. J.T.O. Kirk, *Light & Photosynthesis in Aquatic Ecosystems*, Cambridge University Press, New York, 1994, p. 509.
14. K.L. Carder, R.G. Steward, G.R. Harvey and P.B. Ortner, *Limnol. Oceanogr.*, 1989, **34**, 68.
15. P. MacCarthy and J.A. Rice, in *Humic Substances in Soil, Sediment and Water; Geochemistry, Isolation and Characterization*, eds. G.R. Aiken, D.M. McKnight, R.L. Wershaw and P. MacCarthy, Wiley, New York, 1985, p. 527.
16. J.J. Alberts, J.P. Giesy and D.W. Evans, *Environ. Geol. Water Sci.*, 1984, **6**, 91.
17. J.R. Lakowicz, *Principles of Fluorescence Spectroscopy*, 2nd Edn., Kluwer/Plenum, New York, 1999.
18. E.M. Thurman and R.L. Malcolm, in *Humic Substances in the Suwannee River, Georgia: Interactions, Properties and Proposed Structures*, eds. R.C. Averett, J.A. Leenheer, D.M. McKnight and K.A. Thorn, U.S. Geological Survey Open-File Report 87-557, 1989, p. 103.
19. M.M. Reddy, J.A. Leenheer and R.L. Malcolm, in *Humic Substances in the Suwannee River, Georgia: Interactions, Properties and Proposed Structures*, eds. R.C. Averett, J.A. Leenheer, D.M. McKnight and K.A. Thorn, U.S. Geological Survey Open-File Report 87-557, 1989, p. 147.
20. P.G. Coble, *Mar. Chem.*, 1996, **51**, 325.

Models and Theories

MOLECULAR MODELING OF HUMIC STRUCTURES*

A.G. Bruccoleri, B.T. Sorenson and C.H. Langford

Department of Chemistry, University of Calgary, Calgary, AB, Canada T2N 1N4

1 INTRODUCTION

Molecular modeling studies involve three stages: the computational method is selected to describe the intra- and intermolecular interactions; the type of calculation, be it energy minimisation, Monte Carlo conformer search or something else of interest is chosen; and lastly the results are analyzed and compared with experimental data.[1,2] Modeling of complex systems, such as humic mixtures, introduces three specific challenges. Firstly, the size of the molecules makes quantum mechanical methods, where the electrons are treated individually, time intensive and costly. Commonly, larger systems are modelled with molecular mechanics methods; however, by simply treating the atoms in a ball and spring fashion, the bonding and non-bonding interactions are only roughly simulated based on steric and van der Waals interaction parameters validated by a set of simple model molecules. Secondly, humic substances (HSs) exhibit a near 'continuum' of exemplars of the typical organic functional groups.[3,4] Finally, humics are highly dynamic and conformational flexibility and self-assembly of components of the mixtures are important aspects to consider. When a computational study is complete, no X-ray diffraction data and only limited quantitative NMR data are available to validate the significance of the model. Earlier computational studies have exploited 'building-block' structures,[5] most notably the Steelink and TNB (Temple-Northeastern-Birmingham) structures and variations based upon these (*vide infra*) that are representative of experimental functional group distributions. However, care must be taken when using these structures in the modeling of humics, keeping in mind that they are at best a snap-shot of a representative portion of a diverse system. It is noteworthy that previous computational studies of humics have excluded solvation effects or indicated that they were insignificant. Since humics interact with and retain water, solvent effects are expected to have complex and far reaching consequences for the system.

This paper analyzes some of the pitfalls that may occur if the modeler does not carefully analyze the approximations and limitations of a selected computational methodology. Finally, it will be proposed that what we have learned so far from computational modeling of humics can be summarized by four qualitative points.

1. Conformational flexibility is an important feature of humic components and attention is drawn to the role conformational change may play in the functions of humics.

* Some of the figures in this chapter can be seen in colour on the RSC's web site at www.rsc.org/is/books/humicmod.htm

2. Quantum mechanical calculations that explicitly treat electron density can offer approaches to identifying the interactions among components of humics and between humic components and water or xenobiotic organic compounds.
3. The literature of self-assembly may offer significant clues for humic models.
4. Solvation is important, especially because of conformational flexibility.

2 MOLECULAR MECHANICS METHODS

The first published work on computational modeling of humic materials involved the use of molecular mechanics methods.[6-10] "Typical" building-block structures of HSs that have been constructed on the basis of several different lines of experimental evidence that account for elemental analysis and the known distribution of organic functional groups were the subject of computational analysis. These moieties are capable of one or usually more of the following interactions: hydrogen bonding, aromatic stacking interactions, electron donor-acceptor complexation and hydrophobic bonding interactions. Humic structures may be capable of forming a large variety of self-assembly complexes through "non-bonding interactions." Molecular mechanics methods require the choice of a 'correct' force field in order to obtain meaningful results, but are useful because many of the problems may be too large to be considered by quantum mechanical methods. Quantum mechanics deals with the electrons in a system, so that even if some of the electrons are ignored (as in the semi-empirical schemes) a large number of particles must still be considered and the calculations become costly. Force field methods ignore the electronic motions and calculate the energy of the system as a function of the nuclear positions using only empirical functions for the dependence of energy on bond length, angles and steric factors. Molecular mechanics is thus invariably used to perform calculations on systems containing significant numbers of atoms. However, two points should be kept in mind when modeling humic structures:

1. A force field calculation will be precise only for types of structures for which it has been parameterised.
2. Molecular mechanics cannot evaluate properties that depend upon the electronic distribution.

As yet, experimental uncertainties remain as to whether structures needed for HSs have been adequately represented in the construction of force fields. Moreover, various interesting electronic donor-acceptor interactions (charge transfer interactions) and hydrogen bonding may lead the modeler to seek information regarding the electronic distributions.[11]

Molecular mechanics is based on a rather simple model of a system using empirical classical mechanical potential energy functions. In cannot be over-emphasized that since simple functions such as Hooke's law are used, the force field can perform well only for systems that are close analogues of those for which it has been parameterised. Transferability of parameters is a major question for structures as complex as humics. Can parameters developed from data on small 'known-structure' non-polar organic molecules, for example, be used to study much larger complex heterogeneous polymers for which experimental structural data are hardly available?

Most recently an influential conformational study of building block stuctures of a humic acid, derived from the original Steelink and TNB structures, has been reported,[9] see Figures 1 and 2. A comparison of the results of geometric optimization calculations on the

TNB-like model using the same force field on two different computer systems is illustrated in Figure 3.

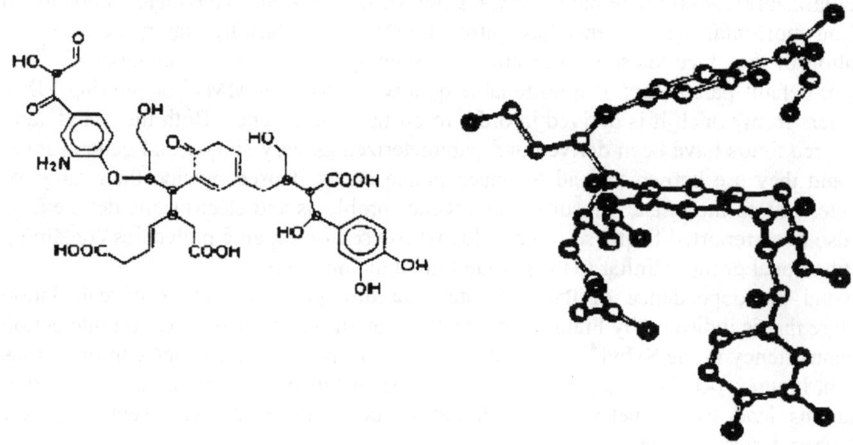

Figure 1 *Molecular and lowest energy conformational structure of the proposed Steelink-like building-block for humics*

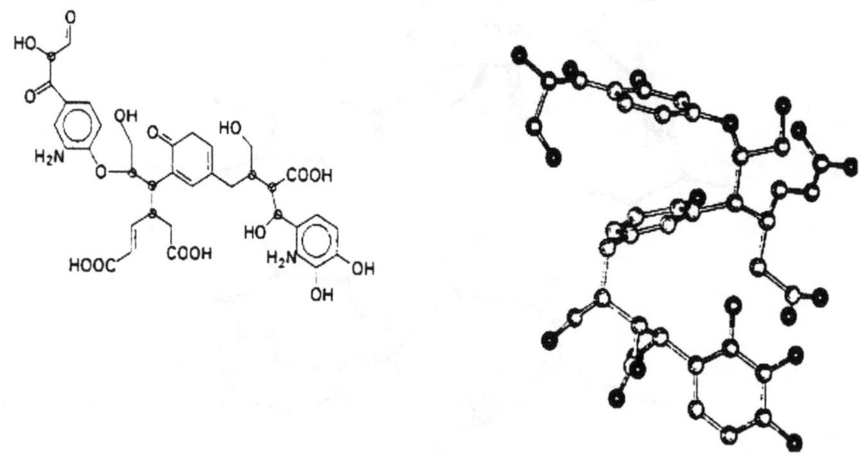

Figure 2 *Molecular and lowest energy conformational structure of the proposed TNB-like building-block for humics*

The conclusions in the Sein, Varnum, and Jansen study included the following points.[9] 1. The most important region of 'interaction' appeared to be the aromatic ring 'stacking' area. 2. A helical structure was found when building a hexamer polymer from the building block unit and geometrically optimizing the structure with the Sybyl® force field (Tripos). 3. Solvation had little effect on the structures.

The force fields used in the study[9] are for the most part only meant for use in pre-optimization calculations. They are good as general-purpose force fields. However, they do not perform conformational calculations well for organic molecules containing polar and aromatic functional groups. Problems with the MM+ and Sybyl® force fields have been documented.[12-18] The MM+ force field of HyperChem is derived from public domain code of the MM2 force field. This force field is based on gas phase structures of small (non-polarizable) monofunctional organics. Both MM2 and MM+ have major problems in modeling molecular systems that have aromatic-aromatic 'stacking' interactions. Overestimation of the 'face-to-face' interactions of such systems is well documented.[19-21] In addition, default parameters of questionable quality are used in MM+ for missing MM2 parameters from which it is derived in order to ensure convergence. Both the Sybyl® and MM+ force fields have been derived and parameterized as very simple and general force fields and they are known to lead to unacceptable errors if used outside their range of parameterization. Inconsistent results, convergence problems and electrostatic deficiencies have also been reported for these force fields when used on organic molecules containing polar functional groups similar to those found in humic material.

Beyond the dependence of the computed conformations on details of calculation procedure that is indicated by Figure 3, overestimation of the aromatic stacking interaction and inconsistency in the Sybyl® force field calculations may well influence conformations shown in Figures 1-3. The results of these molecular mechanics geometrical optimization calculations lead to geometry that is found to be inconsistent with semi-empirical calculations described later.

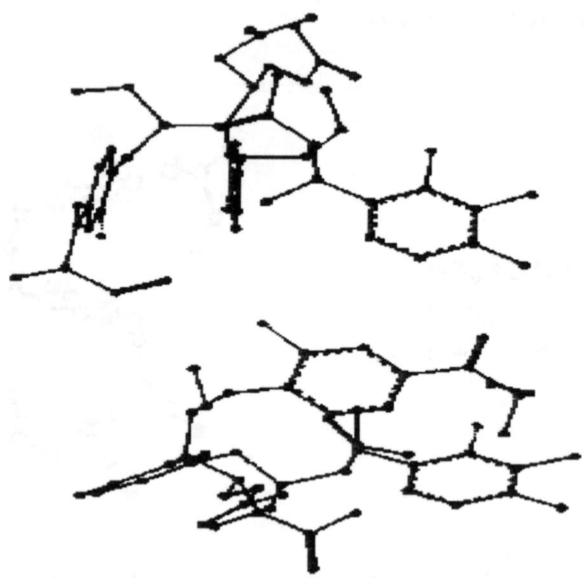

Figure 3 *Comparison of geometrical optimization of the TNB-like model using the Sybyl® force field optimized using the Sybyl 6.1® (Tripos) force field (sketch utility), Top, Fletcher-Powell conjugate-gradient minimization, random conformational search (Sybyl)®. Bottom, geometrical optimization calculations were performed using the IBM RS 6000 and Spartan 4.1.1 (Sybyl®-Tripos) and the Newton-Raphson minimization procedure*

Any errors are due to the limitations of the force field. Such errors increase non-linearly as calculations are extended to polymers of the structure. The inability of these force fields to deal with hydrogen bonding interactions is also evident from the lack of evidence of conformational change when a polar solvent is placed around a molecule containing hydroxyl groups.

An important part of computer modeling studies is the choice among the unavoidable assumptions, approximations and simplifications of the molecular model and computational procedure to minimize the overall inaccuracy. Many of the molecular modeling force fields in use today, including the Sybyl® and MM+ force fields, can be interpreted in terms of a relatively simple four-component picture of the intra- and intermolecular forces within the system. Energetic penalties are associated with the deviation of bond lengths and angles from their 'reference' or 'equilibrium' values, a function describes how the energy changes as bonds are rotated, and finally the force field contains terms that describe the interaction between non-bonded parts of the system. More sophisticated force fields than the Sybyl® and MM+ force fields have additional terms but invariably contain these four components.

A force field must be considered as a single entity. It is not strictly correct to divide the energy into individual components, let alone to take some parameters from one force field and mix them with parameters from another force field. If non-bonding term parameters or the non-bonding functional form of a force field are shown to be invalid for a given system, the force field should be abandoned. The functional terms in a force field are not sufficiently independent of the others to try and attempt to adjust only certain parameters of one of the functional terms. The entire force field would have to be re-parameterized. A force field is generally designed to predict certain properties for a particular molecule and will be parameterized accordingly. While it can be useful in certain cases to try to predict other quantities that have not been included in the parameterization process, great care should be taken in the case of humic structures because of the types of functional groups involved and lack of experimental structural data to either validate or invalidate the model.

Force fields are *empirical*. There is no 'correct' form for a force field. Of course, if one functional form is shown to perform better than another it is likely that form will be favored. However, the problem remains that no statements can be made regarding electronic interactions in hydrogen bonding or non-bonding interactions from the functional non-bonding term in the force field. Most of the force fields in common use do have a very similar form and it is tempting to assume that this must therefore be the 'optimal' functional form. It should always be borne in mind that there may be better forms, particularly if one is developing a force field for a 'new' class of molecules, for example, humic structures. The functional forms employed in molecular mechanics force fields are often a compromise between accuracy and computational efficiency. The most accurate functional form for humic structures may be unsatisfactory for efficient computation if complex hydrogen bonding and non-bonding interactions terms are to be added in order to properly describe humic structure interactions.

Note again that the optimized structures in Figure 3 differ based upon the details of the computational approach, indicating that the force fields used do not function entirely independently of the subjective preferences of the modeler. Assignment of preferred conformation is not simple because the number of possible conformers can be very large. As an example of this consider a simple systematic search technique, limited to single bonds (Y) and excluding flexible rings, of a structure with three single bonds. Each bond can be rotated through a step of 360°/X. As such, if the step is set at 120° (X=3) the number of conformers is 27. By increasing the number of rotatable single bonds to five the number of conformers increases to 243. The increase in possible conformers is governed by the equation X^Y. In some cases steps smaller than 120° are required to avoid missing conformers; hence, large flexible molecules, such as humics, lead to considerable computational complexity. The situation worsens when flexible rings are included. It is apparent that another method is needed to explore the conformational space available to the building block models. A random conformational search (Monte Carlo method) is useful in the study of humic model structures[22-24] The Monte Carlo approach essentially heats the molecule to an enormously high temperature for a short time period, causing it to fluctuate, allows the molecule to cool and then optimises the structure. By varying the upper temperature the Boltzmann distribution is altered and, depending on the need of the modeler, 'local' or 'global' conformational space may be explored. The lesser the temperature increase the more locally space is explored. By using Monte Carlo calculations a large number of low energy conformers can be calculated in a 'relatively' short period of time. Although Monte Carlo searches are possible within quantum mechanical calculations, only molecular mechanics methods are practical for large humic structures. Even then the calculations can be time consuming.

Two different modes of conformer search are available, "single" or "multiple". The former attempts to locate the lowest-energy (global) conformer, while the latter provides a small set of low-energy conformers (including the lowest-energy conformer).

In general, calculated energies have not been reported when the system is modeled in a vacuum. However, even though such energy values are not experimentally significant, the reporting of minimum energy would allow for better comparison of the computational methods used and is suggested for future publications.

3 SEMI-EMPIRICAL QUANTUM MECHANICAL METHODS

Quantum mechanics (QM) explicitly represents the electrons in a calculation and so it is possible to derive properties that depend upon electron distributions. Electron donor-acceptor complexes (charge transfer complexes) and hydrogen bonded complexes are probably best modeled by treating the electrons of the system explicitly. Practical QM methods are also semi-empirical routines because of time and computational power requirements.[25] Molecular mechanics depends on the concept of atom types and parameters associated with these atom types. Since the number of atom types is very large for the 'universe' of possible humic molecules, parameters probably will be missing for humic building-block structures. The number of QM parameters needed for all possible molecules is much smaller. Semi-empirical calculations use parameters associated with specific atoms. These calculations are semi-empirical because their quantum mechanical formulation uses empirical parameters obtained from comparison with experiment to evaluate terms (integrals) in the equations. If parameters are available for the atoms of a given molecule, the semi-empirical calculation has an *a priori* aspect compared to molecular mechanics calculations, allowing exploration of 'new' molecular systems such as humic structures. Semi-empirical molecular orbital methods provide approximate solutions of the Schrödinger equation for the ground state wave function and certain low energy excited states. Attractive features for modeling humic structures include the following points.

1. The semi-empirical methods imply no prior information about the location or geometry of bonds in a molecular system.
2. Parameters for elements (usually derived from empirical data or *ab initio* calculations) are independent of the chemical environment.
3. With simple precautions, semi-empirical methods can describe bond breaking and electron transfer processes.
4. Electrostatic potential energy surfaces may be generated in order to predict self-assembling behavior.

5. Electronic delocalisation and other factors that determine humic geometry and properties may be studied.
6. "Resonance structures", etc. of molecules with intra- and/or intermolecular hydrogen bonding have been suggested to be interesting examples having so-called 'fuzzy' shape.[26-30] It has been suggested that they may be treated more accurately at the quantum mechanical level. A semi-empirical approach to hydrogen bonding may be of value to study properly the extensive hydrogen bonding possibilities in wet humics.

Hydrogen bonding is important to humic structures. An interesting problem was recognised in molecular modeling studies of hydrogen-bonded systems that share similarities to humic building blocks.[27-30] The superposition principle of quantum mechanics gives rise to what some have called "strange" states. The hydrogen atom in a hydrogen bond can sit at different 'fixed' positions. Consideration of appropriate superposition of the two wave functions leaves the hydrogen with no 'fixed' position. The nuclei of a molecule with no apparent fixed position are termed "strange" from the point of view of traditional chemistry. The work of Gilli et al.[27-30] on several hydrogen bonded systems (diketones, carboxylic acids, alcohols, ketoesters, amides, etc.) suggests that these complexes may be best described by *fuzzy* 'non-classical' structures. Humic substances may have 'fuzzy' molecular shape and therefore may best be treated at the quantum mechanical level. The results of the first semi-empirical QM geometric optimization calculations of humic model structures based upon the Steelink and TNB models are presented below (Figures 4 - 6).

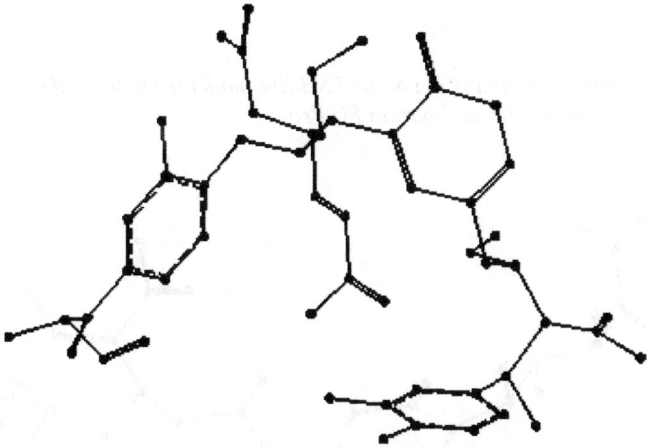

Figure 4 *An example of geometrical optimization of the Steelink-like model using the PM3 semi-emprical method in Spartan PC Pro (Wavefunction Inc.)*

In these cases the modeling work was performed with the PM3 semi-empirical method.[25] The PM3 semi-empirical method is based on MNDO (Modified Neglect of Diatomic Overlap) and the PM3 name derives from the fact that it is the third parameterisation of MNDO, AM1 (the Austin Model) being considered the second. The PM3 Hamiltonian contains essentially the same elements as AM1 but the parameters for the PM3 model were derived using an automated parameterisation procedure devised by J. J. P. Stewart. By contrast, many of the parameters in AM1 were obtained by applying chemical knowledge and 'intuition'. As a consequence, some of the parameters have significantly different values in AM1 and PM3 even though both methods use the same functional form and they both predict thermodynamic and structural properties to approximately the same level of accuracy. PM3 has been found to successfully predict H-bonding.[31] Calculated binding energies of various hydrogen bonded dimers correctly predict experimentally determined association constants.[32,33]

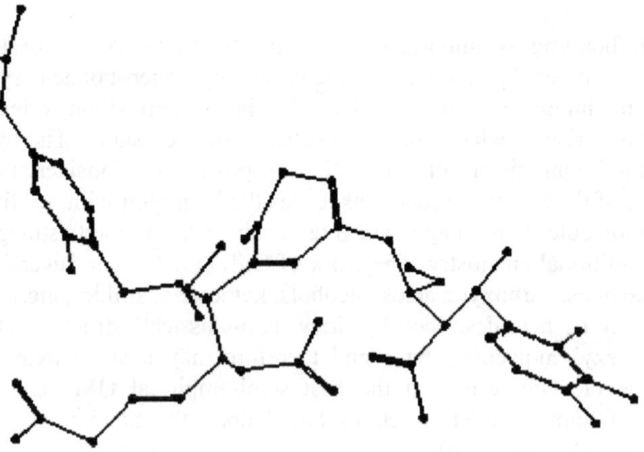

Figure 5 *Geometrical optimization of the TNB-like model using the PM3 semi-empirical method (Geometrical, Spartan PC Pro)*

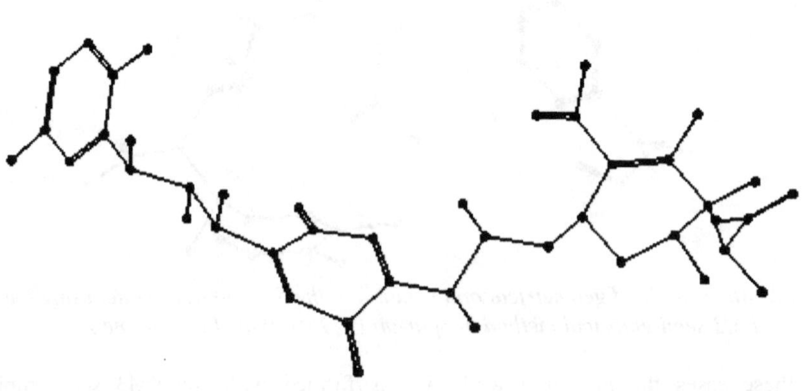

Figure 6 *Geometrical optimization of the Laurentian Fulvic Acid, "LFA" monomer using the PM3 semi-empirical method (Spartan PC Pro)*

There is an interesting similarity in structure of the above three models after semi-empirical optimization. These structures also share a remarkable similarity to some of the optimized models of chemical building-block components in the literature of supramolecular "self-assembly" chemistry called "half tennis ball" structures or "S" and "U" structures.[34-37] One of the self-assembling dimer structures formed by mating the two halves of the tennis ball is illustrated below in Figure 7. Computed dimer structures of the TNB-like and Steelink-like monomers are shown in Figures 8 and 9.

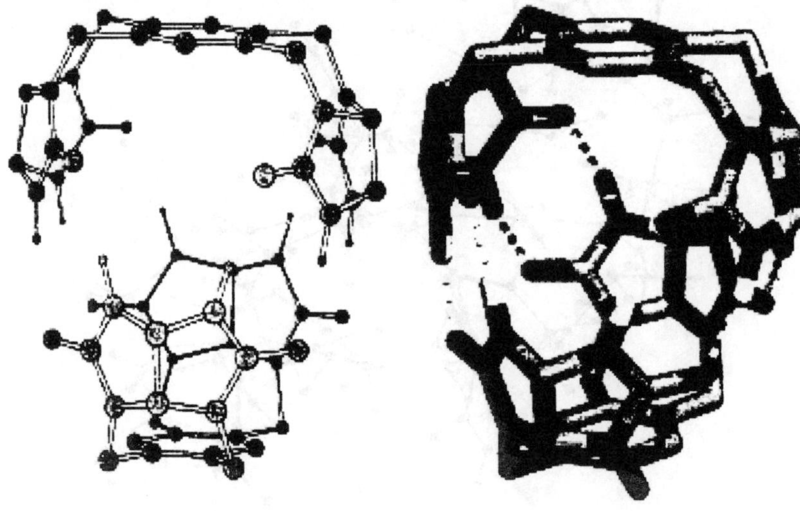

Figure 7 *Self-assembly structure. Pseudospherical ('tennis-ball')*

A molecular modeling study was also performed of binding of a xenobiotic to a LFA dimer. The guest-host complex between the LFA dimer and fluorobenzoic acid after semi-empirical calculations (PM3) is shown in Figure 10. Interestingly, the results of this semi-empirical molecular modeling calculation show the xenobiotic within the humic dimer in a low energy conformation. This phenomenon is remarkably similar to the guest-host interactions that have been well documented in self-assembling capsule studies of hydrogen bonded systems in the literature.[38,39] This leads to two major propositions arising from this study as a contribution to humic chemistry.

1. Semi-empirical quantum mechanical calculations can enrich our perceptions of possible humic structural features.
2. The extensive self-assembly literature may offer valuable guidelines for development of models of aggregation of humic structures. The interactions involve structures closer to those recognized in humics than do biopolymers like proteins and nucleic acids.

The electrostatic potential map for the TNB-like model is illustrated in Figure 11. This is a major addition from QM calculations. Monomers with conformational flexibility that have 'hot' and 'cold', relatively electron rich and electron deficient regions picketed along the edge of the electrostatic potential surface of the molecule can form pseudo-spheres and larger complexes in solution. A simple example can be seen by comparing the electrostatic potential surface of water and ammonia to molecules like methane. 'Hot' and 'cold' regions in water and ammonia predict H-bonding behavior while the electrostatic potential map for methane appears with an even electron distribution and has none of the hydrogen bonding or comparable complex forming potential apparent in water and ammonia.

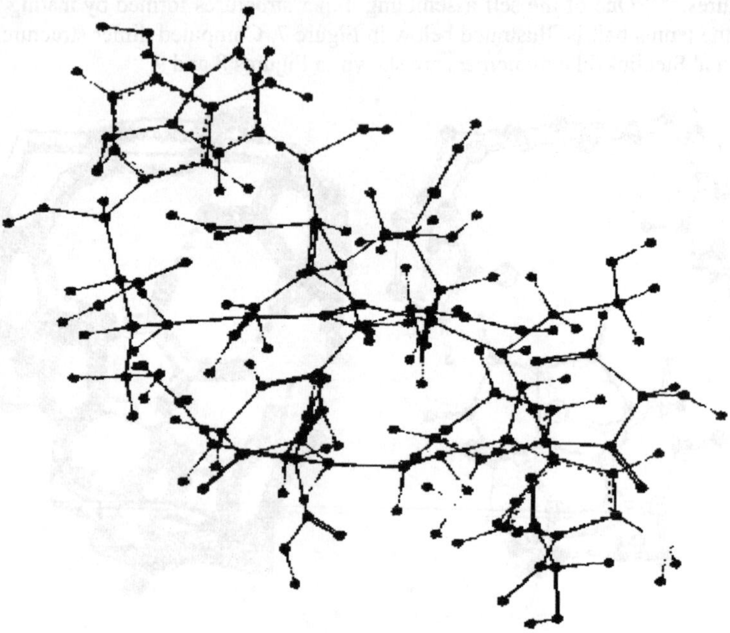

Figure 8 *Geometrical optimization of the TNB-like dimer using the PM3 semi-empirical method (Spartan 4.1.1)*

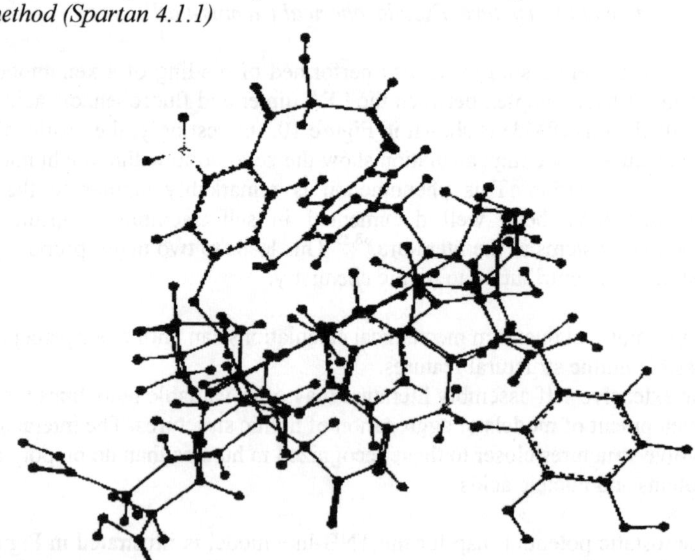

Figure 9 *Geometrical optimization of the Steelink-like dimer using the PM3 semi-empirical method (Spartan 4.1.1)*

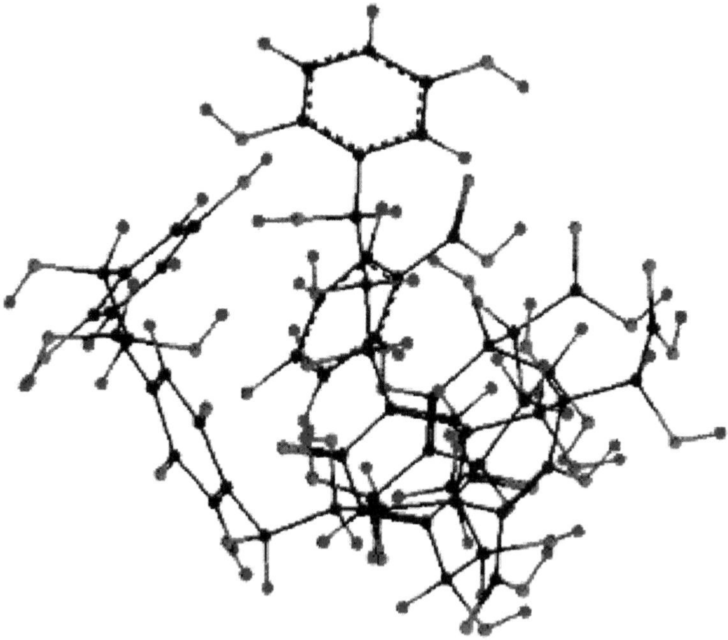

Figure 10 *Guest-host encapsulation of fluorobenzoic acid seen in the center of a pseudosphere created by a LFA dimer model species (Spartan 4.1.1)*

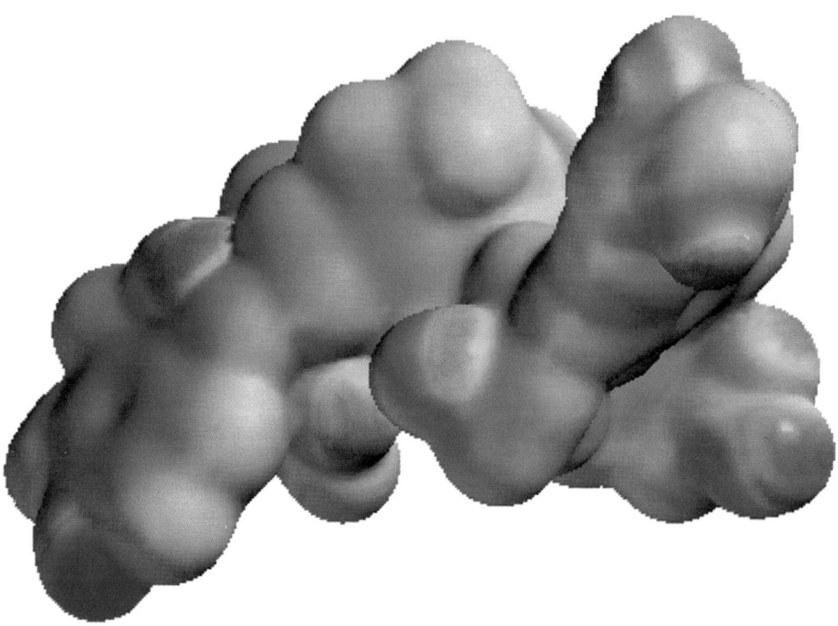

Figure 11 *The electrostatic potential has been mapped onto the electron density surface showing 'hot' and 'cold' regions 'picketed' along the edge of the TNB-like (PM3) optimized structure. The 'hot' and 'cold' regions along the edge of the molecule are a typical trait of self-assembly building-block monomers. (Spartan PC Pro)*

4 MONTE CARLO SEARCH AND SOLVATION EFFECTS

Further work was done to investigate the effects of semi-empirical PM3 calculations (with and without solvent effects) on building block conformers pre-optimized using the Sybyl® force field. Using a Monte Carlo search, 8281 total interactions (heating, cooling and optimization periods) were investigated to obtain the 60 lowest energy conformers. This provides a much more complete exploration of conformational space. Each of these conformers was then optimized using PM3 methods and solvation energies were calculated using the Cramer/Truhlar SM5.4/P model.[40]

Cramer and Truhlar provided the following methodology for the SMx models: 'In the SMx solvation models (with the exception of SM5.0R) solvation effects are included via two terms. The first accounts for electrostatic polarization of the solvent by the generalized Born approximation based on a distributed monopole representation of the solute charges with dielectric screening. Solute polarization is also included self-consistently. The second term is proportional to the solvent-accessible surface area, with a set of proportionality constants (surface tensions) which depend on the local nature of the solute for each atom's or group's interface with the solvent. The SM5.4 model is based on a set of geometry-based functional forms for parameterizing effective coulomb radii, and atomic surface tensions for solutes encompassing a wide variety of functional groups in water, where the functional forms of microscopic surface tensions depend on solute atomic numbers and geometries. In general, the notation SM5.4/AM1 and SM5.4/PM3 without designation of a parameter set implies the use of the special parameters for water, chloroform, benzene and toluene and the general organic parameters for any other solvent'.[41]

Because the molecule is not simply encapsulated and the surface area is directly related to the solvation, it was expected that conformer shape (open or tight) would impact the solvation energies (Figure 12). It was found that there was only weak correlation between the pre-optimized molecular mechanics energy minima and the PM3 calculations. In fact the lowest energy conformer resulted from the PM3 calculation on the 40th lowest energy molecular mechanics conformer found using the Monte Carlo search.

The results of force field calculations have often been presented in terms of the minimum energy conformation. The random search procedure illustrated by the results in Figure 12 clearly shows that even a simple model of a humic structure can have great conformational flexibility. A number of quite different structures can lie within the energy range populated at ordinary temperatures. Of the seven lowest energy structures, three are shown here to highlight the variety that exists (Figures 13-15). Even though the global minimum energy conformation has the lowest energy it may not even be the most highly populated because of the contribution of the vibrational energy levels of low frequency vibrations to the statistical weight of each structure. Also, the global minimum energy conformation may not be the active (i.e., functional) structure. For example, the conformation that binds a contaminant such as a pesticide may not correspond to any minimum on the energy surface of the isolated molecule. At this point the next general proposition this study offers about humic structures emerges.

3. Structures involved in humic mixtures can be expected to exhibit conformational flexibility. No single conformation should be presumed to adequately describe behavior and key processes involving weak interactions may be expected to depend upon conformational rearrangements or selection from among the low energy conformations.

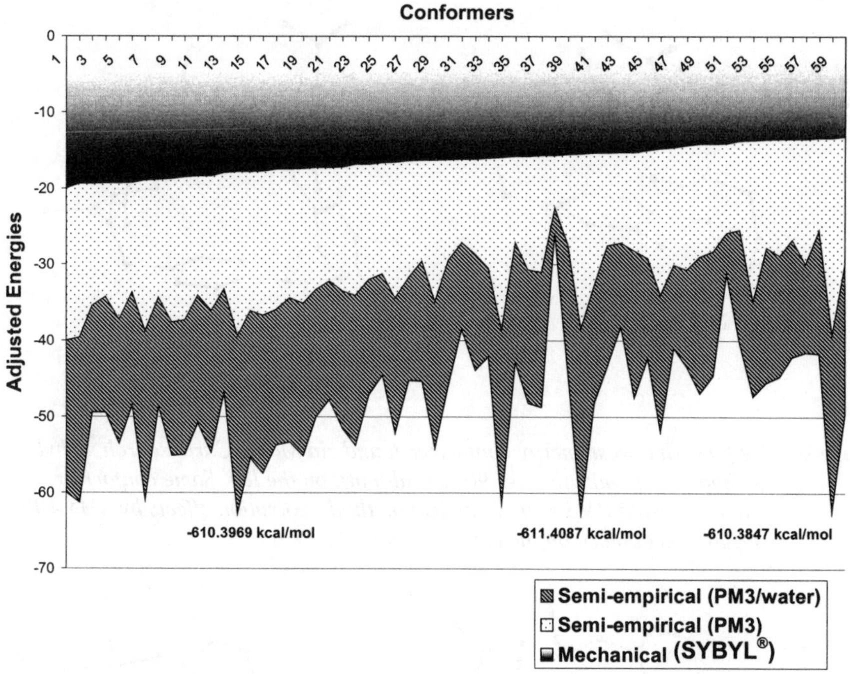

Figure 12 *Energy surface of the lowest sixty building block conformers. Optimized using Spartan PC Pro - Monte Carlo (8281 iterations) using Sybyl® force field followed by geometry optimization using the PM3 force field (with and without solvation effects)*

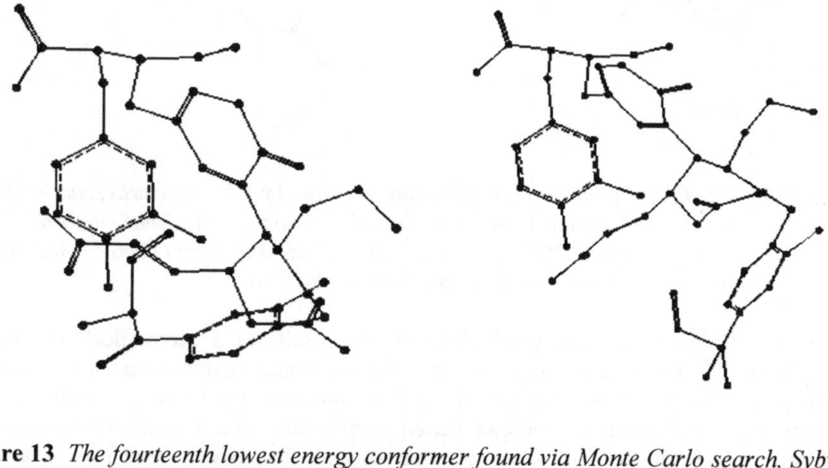

Figure 13 *The fourteenth lowest energy conformer found via Monte Carlo search, Sybyl® mechanical optimization (-6.2603 kcal/mol) on the left. Same conformer optimized using PM3 semi-empirical methods, solvation effects by SM5.4/P (-610.3969 kcal/mol), on the right (Spartan PC Pro)*

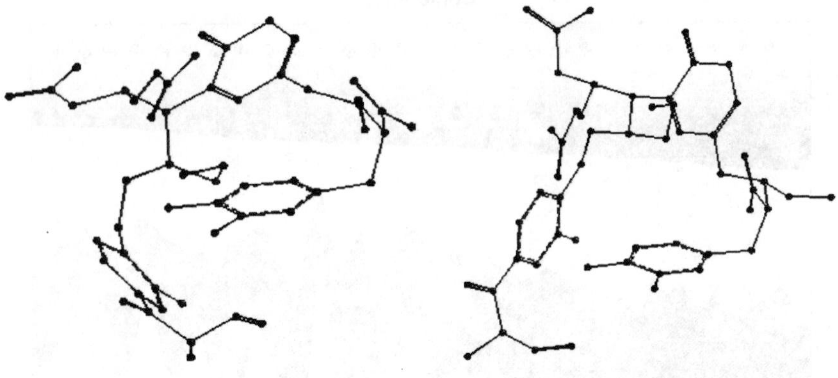

Figure 14 *The fortieth lowest energy conformer found via Monte Carlo search, Sybyl® mechanical optimization (-3.9011 kcal/mol), on the left. Same conformer optimized using PM3 semi-empirical methods, solvation effects by SM5.4/P(-611.4087 kcal/mol), on the right (Spartan PC Pro)*

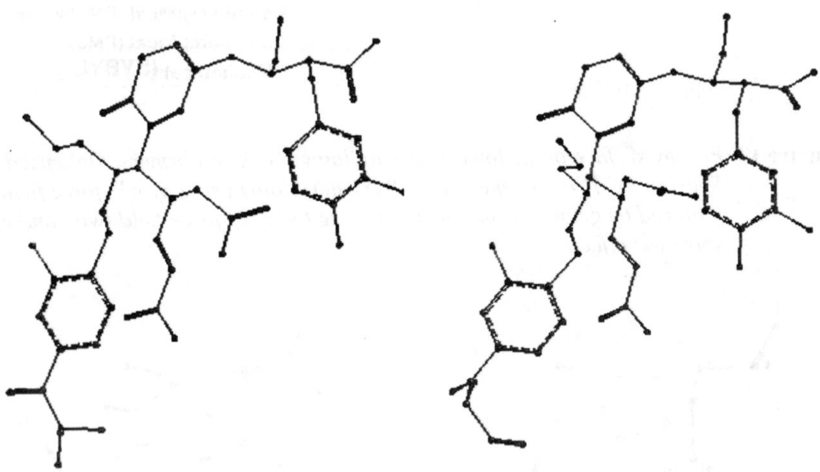

Figure 15 *The fifty-ninth lowest energy conformer found via Monte Carlo search, Sybyl® mechanical optimization (-1.7726 kcal/mol), on the left. Same conformer optimized using PM3 semi-empirical methods, solvation effects by SM5.4/P (-61.3847 kcal/mol), on the right (Spartan PC Pro)*

Solvation can have a profound effect on the results of a geometrical optimization calculation.[42,43] This is especially true when the solute and solvent are polar or when they participate in hydrogen bonding as is the case for humic-water systems. In a current study it was found that solvation energies varied significantly over the different conformers, ranging from 37.80 to 52.90 kcal/mol. The solvent effect is expressed in several ways, including having a strong influence on the energies of different solute conformations or configurations of atoms. Special note should be made of the fact that the nature of solute-solute (humic-humic) and solute-solvent (humic-water) interactions in humic modeling work is dependent on the solvent environment. Water influences the hydrogen-bonding pattern, the humic model structure surface area and hydrophilic-hydrophobic group exposures. The final point to emerge from this study is that:

4. Solvation is important, especially in the light of conformational flexibility.

5 FUTURE WORK

It would be negligent to overlook the rapid progress made elsewhere in the areas of molecular modeling and supramolecular chemistry research with regard to molecular modeling research of humic structures. Molecular modeling is a rapidly developing discipline that has benefited from the dramatic improvements in computer hardware and software over recent years. Calculations that were major undertakings only a few years ago can now be performed using personal computing facilities. Semi-empirical quantum chemical methods are useful when working with humic models, which may have carbohydrate and aromatic containing assemblies and components that may undergo charge-transfer type interactions. The introduction of more powerful computer systems and software will allow for more rigorous quantum mechanical and DFT (Density Functional Theory) based calculations to be performed on humic models. With more powerful resources solvent effects, which will no doubt have complex and far reaching consequences, can be investigated more thoroughly with more sophisticated models of water. Despite the challenge of the lack of unambiguous experimental reference points for the models, computational chemistry can be expected to make a continuing contribution to our understanding of humic systems.

References

1. K.B. Lipkowitz and D.B. Boyd, *Reviews in Computational Chemistry*, Vols. 1-7, New York, VCH, 1990-1996.
2. A. Hinchliffe, *Modeling Molecular Structures*, Chichester, Wiley, 1995.
3. F.J. Stevenson, *Humus Chemistry, Genesis, Composition, Reactions*, 2nd Edn., Chap. 1, Wiley, Toronto, 1992.
4. H. Jenny, *Factors of Soil Formation*, Dover Pub. Inc., New York, 1994, p. 1.
5. C. Steelink, in *Understanding Humic Substances: Advanced Methods, Properties and Applications*, eds. E.A. Ghabbour and G. Davies, Royal Society of Chemistry, Cambridge, 1999, p. 1.
6. H.-R. Schulten and M. Schnitzer, *Naturwissenschaften*, 1995, **82**, 487.
7. H.-R. Schulten and P. Leinweber, *J. Anal. Appl. Pyr.*, 1996, **36**, 1.
8. H.-R. Schulten and M. Schnitzer, *Soil Science*, 1997, **162**, 115.
9. L.T. Sein, Jr., J. Varnum and S.A. Jansen, *Envir. Sci. Technol.*, 1999, **33**, 546.
10. H.-R. Schulten, *Envir. Tox. and Chem.*, 1999, **18**, 1643.
11. A. Bruccoleri, PhD Thesis, University of Calgary, 2000.
12. M. Clark, R. Cramer and N.Opdenbosch, *J. Comp. Chem.*, 1989, **10**, 982.
13. K. Gundertofte, J. Palm, I. Pettersson and A. Stamvik, *J. Comp. Chem.*, 1991, **12**, 200.
14. A. Ghose, E. Jaeger, P. Kowalczyk, M. Peterson and A. Treasurywala, *J. Comp. Chem.*, 1993, **14**, 1050.
15. A. Treasurywala, E. Jaeger and M. Peterson, *J. Comp. Chem.*, 1996, **17**, 1171.

16. *Spartan User Manual* (Wavefunction Inc. 1999), Edn. 4.1, Chap. 3.
17. I. Pettersson, T. Liljefors, *Rev.Comp. Chem.*, 1996, **9**, 167.
18. T. Liljefors, K. Gundertofte, I. Pettersson and P. Norrby, *J. Comp. Chem.*, 1996, **17**, 429.
19. J. Lii, S. Gallion, C. Bender, H. Wikstrom, N. Allinger, K. Flurchick and M. Teeter, *J. Comp. Chem.*, 1989, **10**, 503.
20. J. Lii and N. Allinger, *J. Comp. Chem.*, 1987, **8**, 1146.
21. I. Pettersson and T. Liljefors, *J. Comp. Chem.*, 1987, **8**, 1139.
22. R. Bruccoleri and M. Karplus, *Biopolymers*, 1987, **26**, 137.
23. G. Chang, W. Guida and W. Still, *J. Am. Chem. Soc.*, 1989, **111**, 4379.
24. M. Saunders, K. Houk, Y. Wu, W. Still, M. Lipton, G. Chang and W. Guida, *J. Am. Chem. Soc.*, 1990, **112**, 1419.
25. M. Zerner, *Rev. Comp. Chem.*, 1990, **2**, 313.
26. A. Amann, in *Fundamental Principles of Molecular Modeling*, ed. W. Gans, Plennum Press, N.Y., 1996, p. 55.
27. P. Gilli, V. Bertolasi, V. Ferretti and G. Gilli, *J. Am. Chem. Soc.*, 1994, **116**, 909.
28. G. Gilli, F. Bellucci, V. Ferretti and V. Bertolasi, *J. Am. Chem. Soc.*, 1989, **111**, 1023.
29. V. Bertolasi, P. Gilli, V. Ferretti and G. Gilli, *J. Am. Chem. Soc.*, 1991, **113**, 4917.
30. P. Gilli, V. Ferretti, V. Bertolasi and G. Gilli, in *Advances in Molecular Structure Research*, Vol.2, eds. M. Hargittai and I. Hargittai, JAI Press Inc., Greenwich, CT, 1995.
31. M. Jurema and G. Shields, *J. Comp. Chem.*, 1993, **14**, 89.
32. J. Pranata, S. Wierschke and W. Jorgensen, *J. Am. Chem. Soc.*, 1991, **113**, 2810.
33. T. Murry and S. Zimmerman, *J. Am. Chem. Soc.*, 1992, **114**, 4010.
34. J. Rebek, *Acc. Chem. Res.*, 1999, **32**, 278.
35. C. Seto, J. Mathias and G. Whitesides, *J. Am. Chem. Soc.*, 1993, **115**, 1321.
36. J. Kang, G. Hilmersson, J. Santamaria and J. Rebek, *J. Am. Chem. Soc.*, 1998, **120**, 3650.
37. Y. Tokunaga, D. Rudkevich and J. Rebek, *Angew. Chem.*, 1997, **36**, 2656.
38. J. Rebek, *Chem. Rev.*, 1997, **97**, 1647.
39. M. Sabio and S. Topiol, *J. Mol. Recog.*, 1997, **10**, 59.
40. C. Chambers, G. Hawkins, C. Cramer and D. Truhlar, *J. Phys. Chem.*, 1996, **100**, 16385.
41. Cramer Group Web Site, http://comp.chem.umn.edu/amsol/man.txt
42. M. Connolly, *Science*, 1983, **221**, 709.
43. T. Richmond, *J. Mol. Biol.*, 1984, **178**, 63.

MODELING OF MOLECULAR INTERACTIONS OF SOIL COMPONENTS WITH ORGANIC COMPOUNDS*

G. Haberhauer,[1] A.J.A. Aquino,[1] D. Tunega,[1,2], M.H. Gerzabek[1] and H. Lischka[2]

[1] Department of Environmental Research, Austrian Research Centers, A-2444 Seibersdorf, Austria
[2] Institute for Theoretical Chemistry and Structural Biology, University of Vienna, A-1090 Vienna, Austria

1 INTRODUCTION

Understanding soil processes at a molecular level is crucial for the interpretation of the vast manifold of information available from experiments. The molecular processes are very complex and much effort is being spent to understand them in detail using physical and chemical methods. Molecular modeling is a powerful means of obtaining direct insight into molecular processes by means of computer simulations.[1] This technique has been used in chemistry and molecular biology for a long time, and there have been a number of reported applications in environmental chemistry.[2-17]

Adsorption and desorption processes in soil can occur on organic and inorganic soil material (Figure 1). Distribution coefficients are commonly measured and can be regarded as a standard output of many sorption experiments. Many different processes contribute to the distribution coefficients and these parameters alone do not provide specific information on the contribution of processes like diffusion, chemisorption or physisorption.[18]

More insight may be obtained using computational chemistry tools and might lead to a qualitative understanding of the role and possible contributions of different processes to the overall sorption behavior. For example, the strength of specific sorption interactions could be estimated and compared.

First, a model has to be selected. The model should be a precise description of the process or at least part of the process of interest.[1] The model size is limited by capacity of the computers,[7] which requires reduction of the number of interactions and models to as small as possible. Structural input is commonly obtained from spectroscopic data. The results of the calculations are compared with experimental thermodynamic data to validate the model. Computational chemistry offers alternative approaches of different accuracy. Force fields or so-called empirical derived potential functions have been developed for specific types of compounds (e.g., DNA or proteins). Force fields enable us to handle very large systems and to give precise descriptions of specific compounds. Since force fields are designed for specific groups of molecules, they have to be used with care for untested systems until the application is validated.[1,19] Quantum mechanics (QM) (e.g., the density function theory method (DFT)) gives accurate results and does not need experimental input. But because of their computational demand QM uses are restricted to small systems.

* Some of the figures in this chapter can be seen in colour on the RSC's website at www.rsc.org/is/books/humicmod.htm

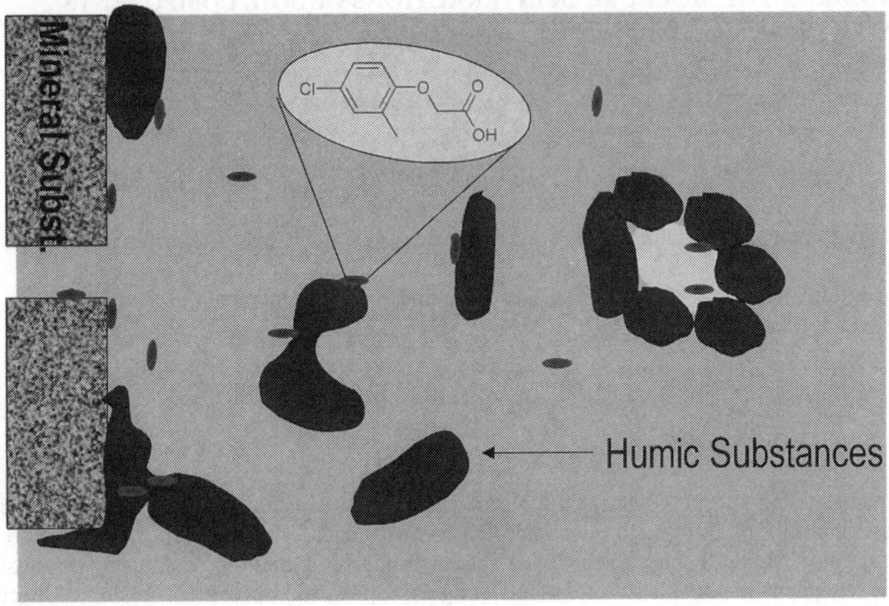

Figure 1 *Schematic representation of possible interactions/distribution of a herbicide (4-chloro-2-methylphenoxyacetic acid (MCPA)) in soil*

The objective of this work was to examine the possibility and investigate the limitations of studying the interactions of model substances with soil constituents by means of molecular modeling techniques. Modeling the whole soil system in full complexity on a molecular level is practically impossible. Thus, one has to introduce several idealizations and simplifications for the construction of appropriate models. We have focused on several particular problems relevant to soil and water processes.

2 COMPUTATIONAL METHODS

The calculations were carried out at the DFT level using the Gaussian98[20] and Turbomole[21,22] programs. We used the BLYP (Becke-Lee-Young-Parr) funçtional and the SVP (split-valence plus polarization) basis set[23] augmented with one set of s and p functions on the oxygen and carbon atoms. The polarized continuum model (PCM) method was used for the calculation of solvent effects.[24-26] In this model the solvent cavity is defined by a set of overlapping spheres and a self-consistent reaction field technique is employed. For the relative dielectric constant ε the value of 78.54 of water at T = 298.15 K was chosen. The United Atom Topological Model[27] was used for defining atomic radii. The standard scaling factor with value 1.2 was used for the definition of the solvent accessible surface. Harmonic vibrational frequencies and thermodynamic quantities (enthalpies and Gibbs free energies) were computed for gas phase structures within the standard harmonic oscillator/rigid rotator/ideal gas approximation. A detailed description of the optimization of the calculation and the theoretical parameters is published.[19,28,29] The cluster model of the single kaolinite layer was derived from the structure of the mineral dickite.[30] The model consists of 78 atoms and contains one ditrigonal tetrahedral ring and one octahedral ring. All dangling bonds were saturated with hydrogen atoms. The resulting chemical formula is $Si_6Al_6O_{36}H_{30}$. The structure of this layer fragment is displayed in Figure 2.

3 RESULTS AND DISCUSSION

Three different examples pertinent to soil reactions are summarized in this paper. The first addresses the interaction of aluminum and organic acids, the second is about clay–mineral interactions and the third deals with possible approaches to estimate the interaction of organic compounds with humic substance moieties.

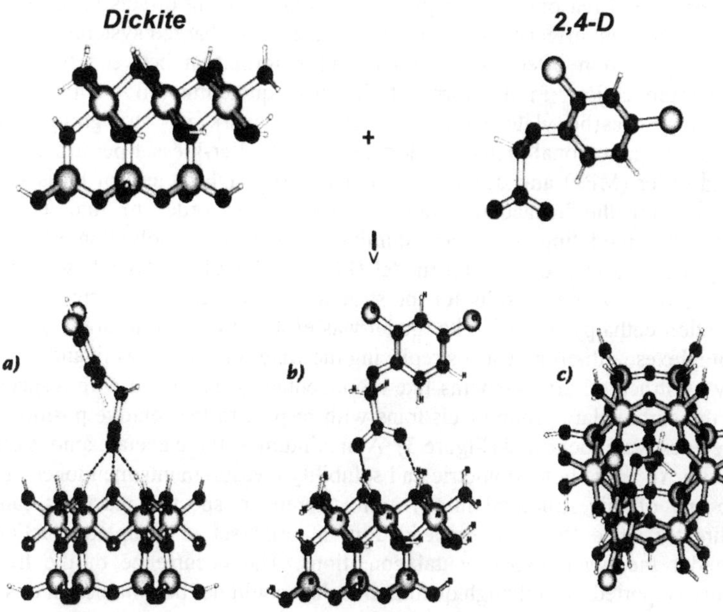

Figure 2 *Modeling of interaction of a model herbicide (2,4-D) on a dickite surface. Top(c) and two side (a,b) views*

Interactions of small organic acids and minerals are of great significance in many geochemical processes. Small acids like citric, oxalic and acetic acids are commonly found at high concentrations in soil solutions and are thought to have an important contribution to complexation and mobilization of certain elements.[27] Complex formation of these acids with aluminum represents a central issue of the way they interact within this system. Aluminum is a major constituent of the soil that becomes mobilized and is phytotoxic at pH < 4.5. The formation of aluminum complexes with organic ligands reduces aluminum toxicity significantly because the penetration of membranes seems to be inhibited for those complexes as compared to "free" aluminum. The aluminum-triggered release of root-derived organic acids such as citric acid is one of the key mechanisms for relieving aluminum toxicity in the rhizosphere.[32,33]

Information on the stability and formation of complexes of Al^{3+} with organic acids such as acetic, oxalic or citric acid and their respective anions can provide fundamental information on the aqueous chemistry and geochemistry of aluminum. In the pH regions of interest (below 6.0) organic acids occur in ionized forms, e.g., as acetate (Ac^-) or oxalate (Ox^{2-} and HOx^-) anions. Complexes of Al^{3+} with up to three ligands have been observed. Several experimental studies (e.g. potentiometric, NMR, IR absorption) have been performed on the complexation of aluminum by organic acids.[34-38] The quantitative analysis of the solubility and potentiometric measurements is performed with complicated thermodynamic models consisting of many coupled dissociation and association equilibria. These studies give global information on the thermodynamic stability of different constituents. However, they do not reveal a detailed picture of individual structural alternatives and their relative stabilities.

Quantum chemical calculations on complexes between Al^{3+} and organic acids are very challenging for several reasons. The major reason is that solvent effects have to be taken into account.[15,39] Solvent effects play an important role since charged systems are involved and charge-compensation effects occur on complex formation between the positively charged aluminum cation (in the form of the hexaaquoaluminum complex) and the negatively charged carboxylate anions. In previous work of our group extensive comparisons of computational methods (Hartree-Fock, Møller-Plesset perturbation theory to the second order (MP2) and density-functional theory (DFT)) and of basis sets have been performed for the hexaaquoaluminum complex[19] in order to find an efficient procedure for the modeling of larger aluminum complexes. Solvation effects were computed by the polarized continuum model (PCM).[24] Finally, a DFT-based procedure was found to give very good results for the structure of the hexaaquoaluminum complex and the solvation enthalpy of Al^{3+}. The model was efficient enough to allow calculations on larger complexes with organic acids replacing the water molecules as ligands.

A variety of structural arrangements like monodentate or bidentate with respect to the bonding of the carboxylate group or cis/trans with respect to the relative position of two acetate molecules were identified (Figure 3). A maximum of three acetate anions can form a complex with Al^{3+}. The potentiometric and solubility measurements mentioned above do not provide any of this structural information. Moreover, specific reactions cannot be observed directly since they are embedded in a large set of coupled reactions and equilibria under the given experimental conditions. The occurrence of the tri-acetate complex is also reported even though quantitative data could not be obtained in this case.

These experiments give no information as to whether these complexes occur as monodentate or bidentate structures. In our calculations the aluminum-hexaaquo complex (Figure 3) represents the basic reactant for the formation of acetate complexes. Large differences between ΔH and ΔG are observed for the formation of bidentate species whereas much smaller differences occur in the case of the monodentate structures (Table 1). This lowering of ΔG is caused by the large positive reaction entropies when bidentate complexes are formed. It is observed for all mono- di- and triacetate bidentate complexes.[19,28,29] The change of entropy is connected with the replacement of two water molecules per acetate during the formation of bidentate species as opposed to monodentate species, where only one water molecule per acetate anion is released. More negative ΔG values favor the formation of bidentate complexes. In fact, the most stable complex is the bidentate triacetate complex (Figure 3).[19] The entropy factor becomes even more dominant at higher temperature. This is in good agreement with the work of Benezeth et al.[40] where the formation of this complex was reported.

In addition to aluminum-acetate, the procedures allowed us to calculate structural and thermodynamic data of aluminum-oxalate[28] and aluminum–citrate[29] complexes. Aluminum–citrate complexes in water solution were found to be the most stable in a calculation comparing the hexaaquoaluminum complex and various aluminum-acetate, aluminum-oxalate and aluminum-citrate complexes. In all calculations the release of water seems to be the driving force for the formation of multiligand complexes.

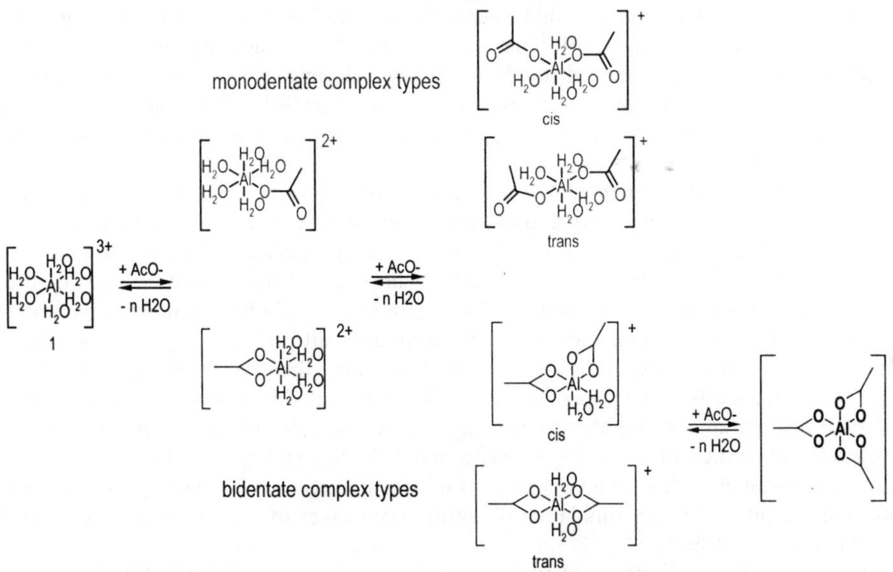

Figure 3 *Reaction scheme of acetate and the hexaqua Al^{3+} complex*

Table 1 *Reaction energies ΔE, reaction enthalpies ΔH, reaction entropic terms TΔS and reaction Gibbs free energies ΔG of the most stable acetate-, oxalate- and citrate-aluminum complexes in the liquid phase*[a,b]

Compound	*ΔE*	*ΔH*	*TΔS*	*ΔG*	*Reference*
	kcalmol^{-1}				
m-$[Al(H_2O)_5Ac]^{2+}$	5.4	5.2	1.7	3.5	19
b-$[Al(H_2O)_2Ac]^{2+}$	19.4	15.9	7.8	8.1	19
m-$[Al(H_2O)_4Ac_2]^{+}$	−13.9	−15.3	-3.4	−11.9	19
b-$[Al(H_2O)_2Ac_2]^{+}$	9.7	6.8	18.5	−11.7	19
b-$[AlAc_3]^0$	0.5	−4.9	23.3	−28.2	19
b-$[Al(H_2O)_4Ox]^{+}$	6.6	5.6	11.9	−6.3	28
b-$[Al(H_2O)_2Ox_2]^{-}$	−7.2	−9.4	22.3	−31.7	28
b-$[AlOx_3]^{3-}$	−13.3	−20.1	27.3	−47.4	28
$[Al(H_2O)_3(HCit_C)]^0$	0.6	---	---	---	29
$[Al(H_2O)_3(HCit_H)]^0$	−26.0	−31.4	25.5	−56.9	29
$[Al(H_2O)_3Cit]^{-}$	-25.2	−36.2	23.6	−59.8	29
$[Al\ (HCit_H)_2]^{3-}$	−63.2	−72.0	45.6	−117.6	29
$[Al\ (Cit)_2]^{5-}$	−61.9	−72.0	51.2	−123.2	29

[a] *T* = 298.15K; [b] m: monodentate, b: bidentate.

Based on this experience, interactions of a model pesticide with regular 001 surfaces of the kaolinite group of clay minerals were investigated. The adsorption processes can be studied experimentally, for example by the measurement of adsorption isotherms. However, it is practically impossible to describe in detail at an atomistic level interaction sites and bonding of the molecular systems to the surface by such macroscopic experiments. It is also very difficult to distinguish experimentally the energetically different adsorption sites on mineral surfaces (e.g., regular 001 surfaces and edge surfaces with broken bonds of clay minerals) or to extract other effects accompanying adsorption (e.g., intercalation).

Computer simulation methods are very useful for giving detailed descriptions of such adsorption processes. Conventional molecular dynamic or Monte Carlo methods have been used in the study of the interaction of water with clay mineral surfaces or inside of the interlayer space.[2-14] In these approaches, empirical potential functions have been used for the energetic description of the system. The determination of a balanced set of parameters for these empirical potentials poses a particularly difficult problem, especially as regards the parameters describing the interaction of aluminum, silicon and oxygen atoms of the clay with the adsorbed molecules. Therefore, Teppen et al.[16] have performed a refinement of force field parameters for atoms occurring in clay minerals that have been applied in the molecular dynamics of kaolinite, gibbsite and beidelite structures. Moreover, a set of empirical parameters describing interactions of organic species with clay minerals has been adapted and applied in the simulation of sorption processes of several organic compounds on clay mineral surfaces.[16,17]

Minerals of the kaolinite group (typical representatives are kaolinite and dickite) have a 1:1 dioctahedral structure. They have a common chemical formula $Al_2Si_2O_5(OH)_4$ and they differ only in the layer stacking. The unit cell of dickite consists of two kaolinite layers and is twice as large as the unit cell of kaolinite. An individual layer consists of two connected sheets – a tetrahedral sheet formed from SiO_4 tetrahedra sharing corners and an octahedral sheet consisting of AlO_6 octahedra sharing edges. Both sheets share a common plane of apical oxygen atoms. One third of all possible octahedral central positions are empty, resulting in octahedral cavities. This causes deformations of the other octahedra occupied by aluminum atoms. Hydroxyl groups participating in hydrogen bonds with basal oxygen atoms of an adjacent layer cover the outer surface of the octahedral sheet. These hydrogen bonds are the main source of the cohesive energy between layers. The tetrahedral side of a layer is characterized by ditrigonal cavities. This surface is built from the plane of basal oxygen atoms each shared by two silicon atoms. These two planes (tetrahedral basal oxygen atoms and octahedral surface hydroxyl groups) are parallel to the crystallographic 001 surface and can be considered as electronically saturated. The interlayer space in kaolinite and dickite is empty, the layers are neutral and no compensating cations are required. Figure 2 shows two different views of the structure of a single dickite layer.

A cluster model approach was used to study the adsorption sites on the 001 surface of kaolinite.[41] The molecules water, acetic acid and acetate are selected as model compounds in order to characterize the structural and energetic aspects of adsorption on the two sides of a single kaolinite layer. These model compounds should be the basis for future investigations of interactions of more complicated molecules such as pesticides with mineral surfaces (Figure 2).

In general, the water molecule and the acetate anion each form several relatively strong hydrogen bonds with the surface hydroxyls on the octahedral side, with interaction energies of –8 and –70 $kcal mol^{-1}$, respectively. These values should be typical for individual interactions with neutral and ionized carboxylic acids. On the other hand, the water molecule and the acetic acid molecule are only weakly bound to the basal oxygen atoms on the tetrahedral side, with interaction energies of about –3 to –4 $kcal mol^{-1}$.[41] From these interaction energies and from the optimized structures it is clear that the two 001 surfaces (octahedral and tetrahedral) of the kaolinite layer differ significantly. The octahedral side offers more possibilities to form hydrogen bonds with the adsorbate than the tetrahedral side. Thus, the octahedral side is more attractive for polar species and adsorption energies will be higher than for the tetrahedral side.

The inclusion of a model for humic substances adds another level of complexity. Due to the highly polymeric and complex structure of HSs,[42] their molecular structures are still unknown. So in contrast to the examples presented above, no structural information on possible sorption surfaces of humic substances is available. Due to the complexity of HSs, molecular analysis is limited to average sum parameters. Structural properties of HSs are best described as distribution probabilities of certain functional groups. The deduction of structural models based on the available functional moiety information is restricted to hypotheses.[43-45] Thus, modeling of interactions of organic compounds with humic substances is restricted to moieties only.[46] We focus on the specific interaction of a model pesticide with functional groups that are known to be present in humic substances (Figure 4).[42]

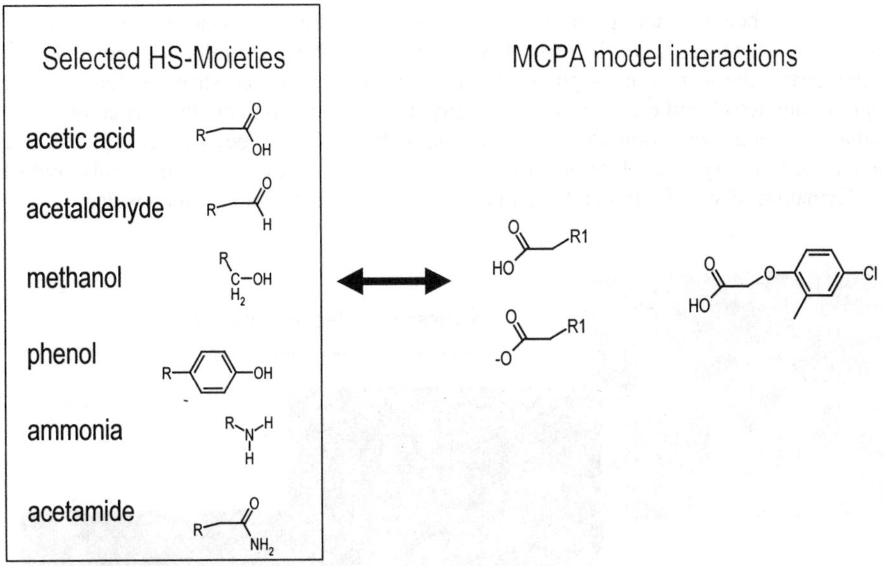

Figure 4 *Selected polar and charged interactions of HS-moieties with a model herbicide*

This is certainly a crude simplification but should at least be a reasonable starting point based on experimental evidence. However, such models can only be used qualitatively to compare different sorption sites. So, in contrast to the previous models of aluminum and dickite interactions that enable a quantitative comparison to experimental data, such simplified models of interactions of organic compounds only allow a relative comparison.

Possible intermolecular interactions of a model pesticide with organic matter moieties are selected. An example of polar and charged single interactions of the model herbicide 4-chloro-2-methylphenoxyacetic acid (MCPA) and 'humic moieties' are displayed in Figure 4. First the interaction energies are calculated in a vacuum. Generally, charged interactions are more stable than polar ones. Calculations of such interactions in a vacuum are a poor approximation for soil solutions.

In real situations in soil solution, these systems will be surrounded by different kinds of environments, ranging from polar to non-polar. Therefore, our interest was directed at investigations of the dependence of interaction energies on the polarity of the environment (as described, for example, by the dielectric constant ε) into which these clusters are embedded (Figure 5). The calculations are done for three different environments. Heptane, on the one hand, represents a very lipophilic environment. The second is dimethylsulfoxide (DMSO), which is more polar and the third is water to simulate a polar environment.

Two general trends are observed for the charged and polar interactions of MCPA (Figure 6). First, with increasing polarity of the environment the charged interactions are no longer more stable than the polar. Second, the relative differences between different types of interactions diminish. We also calculated the free interaction energies from vibrational analysis and quite similar trends are obtained (data not shown).

Our calculation demonstrates that polar and charged interactions depend on the polarity of the environment. So, the more polar the environment the less important seems to be the type of interaction formed. Entropic effects (number of available sorption sites, release of water) could become more pronounced in that case. On the other hand, if a polar herbicide like MCPA manages to diffuse into HS material, which is generally less polar than the water phase, the formation of polar and charged interactions can strongly depend on the type of interaction and can be of high stability. In many experiments the formation of very stable and long lived complexes of pesticides within HSs has been observed. Thus, one factor contributing to the high stability of such complexes of polar compounds could be the formation of specific interactions in more lipophilic environments than water.

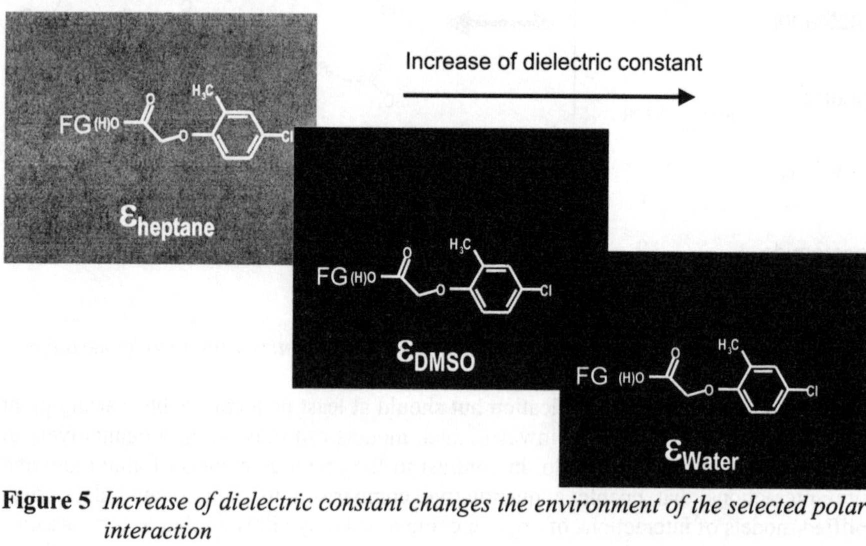

Figure 5 *Increase of dielectric constant changes the environment of the selected polar interaction*

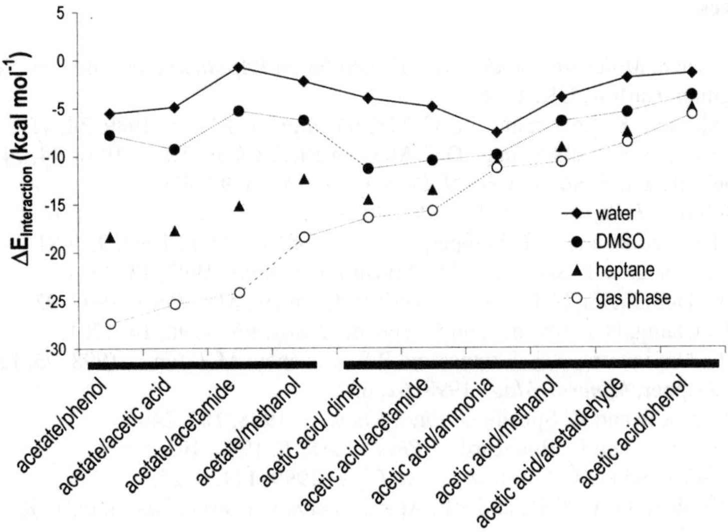

Figure 6 *Calculated interaction energies of selected polar and charged interactions with increasing dielectric constant*

4 CONCLUSIONS

The application of computational chemistry to simulate certain isolated processes in soil requires the identification of suitable models and possible interactions. After proper inclusion of solvation effects and optimization of the calculation parameters the stability of complexes of organic acids with aluminum can be calculated. In that case the release of water seems to be the driving force for the formation of polydentate complexes. The agreement between the experimental and theoretical data is good but not entirely consistent. Such differences still can be due to existing shortcomings of the computational procedures and depend on the models used for analysis of the thermodynamic data.

The lack of structural models of humic substances prevents direct modeling of interactions of organic compounds with humic material. However, interactions between specific functional groups that are known to be present in humic materials and model pesticides can be calculated. Such simplified models can only supply limited information. Since only specific interactions are considered, no information on diffusion, non-specific interactions or partitioning can be obtained. Thus, only relative trends can be estimated. In future studies such trends could be utilized to compare structure-dependent behavior of certain organic compounds with respect to their interaction with selected humic functional groups.

ACKNOWLEDGEMENTS

This work was supported by the Austrian Science Fund, project no. P12969-CHE. We are grateful for technical support and computer time at the DEC Alpha server of the Computer Center of the University of Vienna.

References

1. A.R. Leach, *Molecular Modelling – Principles and Applications*, Addison Wesley Longman, Harlow, UK, 1996.
2. N.T. Skipper, K. Refson and J.D.C. McConnell, *Clay Miner.*, 1989, **24**, 411.
3. N.T. Skipper, K. Refson and J.D.C. McConnell, *J. Chem. Phys.*, 1991, **94**, 7434.
4. A. Delville and S. Sokolowski, *J. Phys. Chem.*, 1993, **97**, 6261.
5. A. Delville, *J. Phys. Chem.*, 1995, **99**, 2033.
6. C.H. Bridgeman and N.T. Skipper, *J. Phys. Condens.-Mat.*, 1997, **9**, 4081.
7. F.R.C. Chang, N.T. Skipper and G. Sposito, *Langmuir*, 1997, **13**, 2074.
8. A.V.C. DeSiqueira, N.T. Skipper and P.V. Coveney, *Mol. Phys.*, 1997, **92**, 1.
9. F.R.C. Chang, N.T. Skipper and G. Sposito, *Langmuir*, 1998, **14**, 1201.
10. A.V.C. DeSiqueira, N.T. Skipper and P.V. Coveney, *Mol. Phys.*, 1998, **95**, 123.
11. N.T. Skipper, *Mineral. Mag.*, 1998, **62**, 657.
12. J. Greathouse and G. Sposito, *J. Phys. Chem. B*, 1998, **102**, 2406.
13. K.S. Smirnov and D. Bougeard, *J. Phys. Chem. B*, 1999, **103**, 5266.
14. R.M. Shroll and D.E. Smith, *J. Chem. Phys.*, 1999, **111**, 9025.
15. J.D. Kubicki, G.A. Blake and S.E. Apitz, *Geochim. Cosmochim. Acta*, 1996, **60**, 4897.
16. B.J. Teppen, K. Rasmussen, P.M. Bertsch, D.M. Miller and L. Schäfer, *J. Phys. Chem B*, 1997, **101**, 1579.
17. B.J. Teppen, C.-H. Yu, D.M. Miller and L. Schäfer, *J. Comp. Chem.*, 1998, **19**, 144.
18. G. Haberhauer, L. Pfeiffer and M.H. Gerzabek, *J. Agric. Food Chem.*, 2000, **48**, 3722.
19. D. Tunega, G. Haberhauer, M. Gerzabek and H. Lischka, *J. Phys. Chem. A*, 2000, **104**, 6824.
20. Gaussian98, Revision A.7: M.J. Frisch, G.W. Trucks, H.B. Schlegel, G.E. Scuseria, M.A. Robb, J.R. Cheeseman, V.G. Zakrzewski, J.A.Momtegomery, Jr., R.E. Stratmann, J.C. Burant, S. Dapprich, V. Millam, A.D. Daniels, K.N. Kudin, M.C. Strain, O. Farkas, J. Tomasi, V. Barone, M. Cossi, R. Cammi, C. Mennucci, C. Pomelli, S. Adamo, J. Clifford, G.A. Ochterski, P.Y. Petersson, B. Ayala, Q. Cui, K. Morokuma, D.K. Malick, A.D. Rabuck, K. Raghavachari, J.B. Foresman, J. Cioslowski, J.V. Ortiz, B.B. Stefanov, G. Liu, A. Liashenko, P. Piskorz, I. Komaroni, R. Gomperts, R.L. Martin, D.J. Fox, T. Keith, M.A. Al-Laham, C.Y. Peng, A. Nanayakkara, C. Gonzalez, M. Challacombe, P.M.W. Gill, B. Johnson, W. Chen, M.W. Wong, J.L. Andres, M. Head-Gordon, E.S. Reploge and J.A. Pople, Gaussian, Inc., Pittsburg, PA, 1998.
21. Turbomole: R. Ahlrichs, M. Bär, M. Häser, H. Horn and C. Kölmel, *Chem. Phys. Lett.*, 1989, **162**, 165.
22. M.v. Arnim and R. Ahlrichs, *J. Comp. Chem.*, 1998, **19**, 1746.
23. A. Schäfer, H. Horn and R. Ahlrichs, *J. Chem. Phys.*, 1992, **97**, 2571.
24. M. Cossi, V. Barone, R. Camini and J. Tomasi, *Chem. Phys. Lett.*, 1996, **255**, 327.
25. S. Miertus and J. Tomasi, *Chem. Phys.*, 1982, **65**, 239.
26. S. Miertus, E. Scrocco and J. Tomasi, *Chem. Phys.*, 1981, **55**, 117.
27. V. Barone, M. Cossi, B. Mennucci and J. Tomasi, *J. Chem. Phys.*, 1997, **107**, 3210.
28. A.J.A. Aquino, D. Tunega, G. Haberhauer, M. Gerzabek and H. Lischka, *Phys. Chem. Chem. Phys.*, 2000, **2**, 2845.

29. A.J.A. Aquino, D. Tunega, G. Haberhauer, M.H. Gerzabek and H. Lischka, *Phys. Chem. Chem. Phys.*, 2001, in press.
30. W. Joswig and V.A. Drits, *N. Jb. Miner. Mh.*, 1986, 19.
31. P.A.W. van Hees, U.S. Lundström and R. Giesler, *Geoderma*, 2000, **94**, 173.
32. D.L. Jones and D.S. Brassington, *Euro. J. Soil Sci.*, 1998, **49**, 447.
33. D.L. Jones and L.V. Kochian, *Plant & Soil*, 1996, **182**, 221.
34. T.L. Feng, P.L. Gurian, M.D. Healy and A.R. Barron, *Inorg. Chem.*, 1990, **29**, 408.
35. J.E. Gregor and H.K.J. Powell, *Aust. J. Chem.*, 1986, **39**, 1851.
36. J.G.R. Hedwig, J.R. Liddle and R.D. Reeves, *Aust. J. Chem.*, 1980, **33**, 1685.
37. L.-O. Öhman and R.B. Martin, *Clin. Chem.*, 1994, **40**, 598.
38. C. Wu, C. Dobrogowska, X. Zhang and L.G. Hepler, *Can. J. Chem.*, 1997, **75**, 1110.
39. D. Kubicki, D. Sykes and S.E. Apitz, *J. Phys. Chem.*, 1999, **103**, 903.
40. P. Benezeth, S. Castet, J.-L. Dandurand, R. Gout and J. Schott, *Geochim. Cosmochim. Acta*, 1994, **58**, 4561.
41. D. Tunega, G. Haberhauer, M.H. Gerzabek and H. Lischka, submitted.
42. E.A. Ghabbour and G. Davies, eds., *Humic Substances – Versatile components of plants, soils and water*, Royal Society of Chemistry, Cambridge, 2000.
43. L.T. Sein, Jr., J.M. Varnum and S.A. Jansen, *Environ. Sci. Technol.*, 1999, **33**, 546.
44. H.-R. Schulten and M. Schnitzer, *Soil Science*, 1997, **162**, 115.
45. H.-R. Schulten and P. Leinweber, *Biol. Fertil. Soils*, 2000, **30**, 399.
46. M.H. Gerzabek, A.J.A. Aquino, G. Haberhauer, D. Tunega and H. Lischka, *Bodenkultur*, 2001, **52**, 137.

BINDING OF HYDROPHOBIC ORGANIC COMPOUNDS TO DISSOLVED HUMIC SUBSTANCES: A PREDICTIVE APPROACH BASED ON COMPUTER ASSISTED STRUCTURE ELUCIDATION, ATOMISTIC SIMULATIONS AND FLORY-HUGGINS SOLUTION THEORY*

Mamadou S. Diallo,[1,2] Jean-Loup Faulon,[3] William A. Goddard III[1] and James H. Johnson, Jr[2]

[1] Materials and Process Simulation Center, Beckman Institute, California Institute of Technology, Pasadena, CA 91125, USA
[2] Department of Civil Engineering, Howard University, Washington, DC 20015, USA
[3] Computational Biology and Materials Technology Department, Sandia National Laboratories, Albuquerque, NM 8785-0710, USA

1 INTRODUCTION

Natural organic matter (NOM) in the environment can be broadly divided into two classes of compounds: non-humic substances (for example, polysaccharides and amino acids) and humic substances.[1,2] Humic substances (HSs) are formed by the slow decay of animals and plant remains and are ubiquitous in soils, sediments and aquatic systems.[1,2] HSs constitute 60 to 70% of soil organic matter (SOM) and 30 to 50% of dissolved organic matter (DOM).[1] They are broadly divided into fulvic acids (FAs), humic acids (HAs) and humin. FA is operationally defined as soluble in water at all pH; HA is defined as the HS fraction that is insoluble in water at pH <2 and soluble at higher pH, whereas humin is insoluble in water at all pH values.

The interactions of hydrophobic organic compounds (HOCs) with dissolved FAs determine to a large extent their mobility, bioavailability and toxicity in aquatic systems.[1-3] The uptake of HOCs by dissolved FAs and HAs has been the subject of many experimental and theoretical investigations.[4-12] Carter and Suffet[4] used equilibrium dialysis to measure the uptake of 2,2-bis(4-chlorophenyl)-1,1,1-trichloroethane (p,p'-DDT) by dissolved Pakim Pond HA, Boonton Reservoir Sediment HA and Aldrich HA. They found that more than 75% of the total DDT in solution at pH = 8.3 was bound to the dissolved HA in all cases. Carter and Suffet[4] also found the extent of binding of p,p'-DDT to depend on solution pH, ionic strength, Ca^{2+}, the origin of the HA samples and their concentration.

Chiou et al.[5] carried out direct measurements of the water solubility of p,p'-DDT, 2,4,5,2',5'-PCB, 2,4,4'-PCB, 1,2,3-trichlorobenzene and lindane in low concentration aqueous solutions ($\leq$ 100 mg/L) of dissolved Sanhedron soil FA/HA and Suwanee River FA/HA. They reported a significant enhancement of water solubility and found the extent of this enhancement to depend on solution pH, solute size and polarity and dissolved FA/HA concentration and origin.

Chin and Weber[6] used equilibrium dialysis to measure the binding constants of p-dichlorobenzene, 1,2,4-trichlorobenzene, *cis*-chlordane and 2,5,2'-PCB in aqueous solutions of Aldrich HA. They found that "the relatively high degree of correlation supports a general hypothesis that the binding constant is strongly dependent upon the hydrophobicity of the compound." Schlautman and Morgan[7] used fluorescence quenching

---

* Some of the figures in this chapter can be seen in colour on the RSC's website at www.rsc.org/is/books/humicmod.htm

to examine the effects of solution chemistry on the uptake of polycyclic aromatic hydrocarbons (PAHs: anthracene, pyrene and perilene) by Suwanee River HA in aqueous solutions. They reported that in NaCl solutions PAH binding to the humic compound "generally decreased with increases in pH (constant ionic strength) and generally decreased with increasing ionic strength (fixed pH) NaCl solutions." More recently, Poerschmann and Kopinke[12] employed solid phase microextraction to assay the binding of alkanes (n-heptane, n-octane, n-nonane, n-decane, n-undecane, n-dodecane and n-tridecane), PCBs (PCB-1, PCB-15, PCB-52, PCB-118 and PCB-153) and 5 PAHs (naphthalene, fluorene, phenanthrene, pyrene and chrysene) to dissolved Roth HA. They also reported a "strong K_{DOM}-K_{OW} correlation over the entire range of analyte hydrophobicities."

Consistent with the widely held view that HSs consist of 3-D polymeric matrices, the Flory-Huggins solution theory has emerged as the preferred thermodynamic framework for describing HOC binding to dissolved HSs during the last 15 years. Chin and Weber[7] have evaluated the ability of the Flory-Huggins solution theory to describe HOC binding to dissolved Aldrich HA. They found that the model overestimated the measured binding constants in most cases. Schlautman and Morgan[7] also reported the "failure of a Flory-Huggins partitioning (i.e., dissolution) model to consistently characterize the environment of the humic substances." One major problem with these assessments of the predictive capability of the Flory-Huggins solution theory has been the lack of reliable thermodynamic data for HSs. This is the primary reason why Chin and Weber[6] selected methylsalicylate as an HA surrogate and used its thermodynamic properties (for example, its 2-D solubility parameters) to "calibrate" their Flory-Huggins model of HOC binding to dissolved Aldrich HA. This also is the reason why Schlautman and Morgan[7] evaluated their input thermodynamic data (for example, solubility parameters) by fitting measured solute binding data to the Flory-Huggins model. Fitting measured HOC-DOM/SOM binding constants to a Flory-Huggins model has in fact become the "standard" method for determining the solubility parameters of HSs[6-7,12-14] because direct measurements of their bulk thermodynamic properties are not feasible in most cases. Although Poerschmann and Kopinke[12] have recently employed HSs solubility parameters estimated by fitting binding constant data to show that "literature based partition coefficients of DOM fit very well" their "new solubility parameter concept" of HOC sorption/partitioning to SOM/DOM, the predictive capability of this "universal one-parameter" concept remains to be established.

In this paper we describe a hierarchical approach for predicting the binding of organic solutes to dissolved HSs *without using any empirically derived thermodynamic parameter.* This novel approach combines computer assisted structure elucidation (CASE) with atomistic simulations and Flory-Huggins solution theory to estimate the constants of binding organic solutes to dissolved FAs and HAs (Figure 1). To illustrate this approach we use quantitative and qualitative structural data as input to the CASE program SIGNATURE to generate a sample of 16 3-D structural models of Chelsea HA. We then use these models as starting structures to carry out constant pressure and constant temperature (NPT) molecular dynamics simulations followed by energy minimization to estimate the strain energies, bulk densities and Hildebrand solubility parameters of the computer generated structural models. The estimated bulk densities and solubility parameters are compared with literature values to select a sample of 6 Chelsea HA model isomers as representative structural models that "best match" the input analytical data. We then show how this representative sample of 3-D models can be used to calculate reliable estimates of the thermodynamic input parameters needed to calculate the constants of binding HOC to dissolved Chelsea HA *without using any empirically derived thermodynamic parameter*.

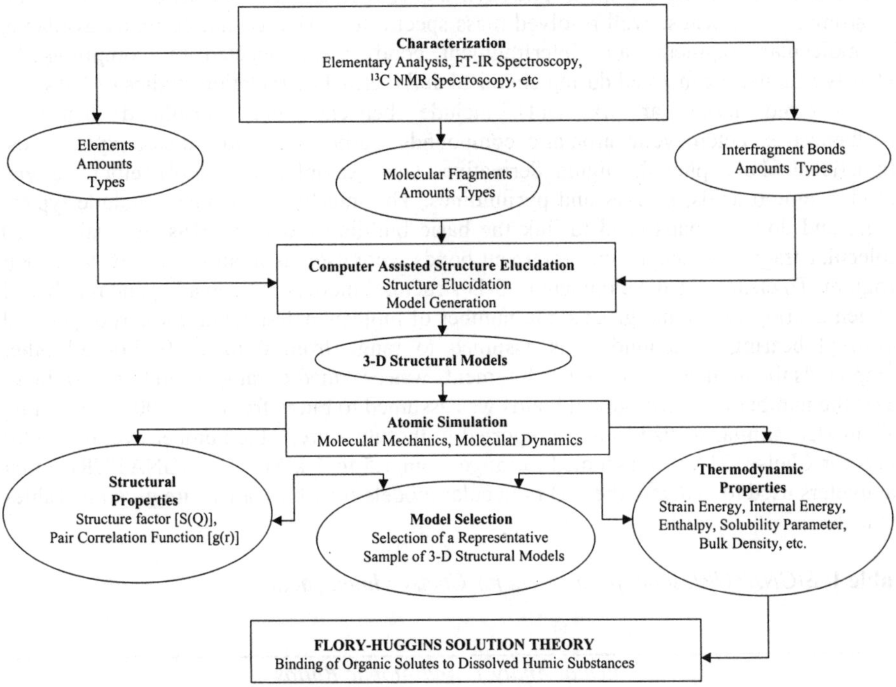

Figure 1 *Hierarchical approach for predicting HOC binding to dissolved HS*s

2 COMPUTATIONAL METHODS AND PROCEDURES

2.1 Computer Assisted Structure Elucidation of Humic Substances

Computational chemistry is increasingly being used to probe the molecular processes that control contaminant uptake by organic and inorganic geosobents.[15,16] The starting point of any molecular level investigation of the physicochemical behavior of a given compound is the bond topology, that is, a list of connections between all its atoms. Because FAs and HAs are operationally defined as "solubility" classes of compounds, the development of 2-D and 3-D structural models for these compounds has been a major challenge to environmental and soil chemists. In this paper, we used the stochastic generator of chemical structure (SIGNATURE) developed by Faulon[17] to generate 3-D structural models for Chelsea HA, a model soil HA that has been evaluated by Huang and Weber.[18] Faulon[17] has shown that by using a stochastic approach it is possible to generate a sample of structural models that statistically represents the entire population of all the possible models that can be built from a given set of analytical data. The required input data for the SIGNATURE program include: (i) atomic ratios, (ii) functional group ratios, (iii) structural groups centered on carbon atoms and hydrogen atoms, (iv) the nature and amount of molecular fragments and interfragment bonds and (v) the number average molecular weight. The elemental analysis and ^{13}C NMR data for Chelsea HA were taken from Huang and Weber.[18] Consistent with the ^{13}C NMR data, oxygen was assumed to consist of 1/3 sp^2 O and 2/3 sp^3 O. Nitrogen was also assumed to consist of 2/3 sp^3 N and 1/3 aromatic N. Because well resolved mass spectra for CHA are not currently available, the molecular fragments and interfragment bonds were selected by compiling the extensive database generated during chemical and thermal degradation studies of HAs.[19-24] The selected molecular fragments include benzene and substituted benzenes, alkylbenzenes, heterocyclic aromatic compounds, carboxylic acids, dicarboxylic acids, aliphatic alcohols, phenols, lignin derivatives (for example, resorcinol), ethers, esters, sulfides, amino acids, purines and pyrimidines. The interfragment bonds include typical single and double bonds used to link the basic building blocks of HSs. A total of 110 molecular fragments and 13 interfragment bonds were used as input to the SIGNATURE program. To ensure the development of 3-D structural models with an adequate number of oxygen bearing functional groups, the number of molecular fragments for carboxylic and hydroxyl bearing compounds was assumed to range from 0 to 1000. For all other compounds the number of molecular fragments was assumed to range from 0 to 100. In all cases the number of interfragment bonds was assumed to range from 0 to 100. To generate 3-D models similar in size to those reported in the literature,[25] the number average molar mass for Chelsea HA was assumed to range from 1.5 to 5 kDa. The SIGNATURE input parameters used to generate the 3-D molecular models for Chelsea HA are given in Tables 1 and 2.

Table 1 *SIGNATURE input parameters for Chelsea humic acid*

Elemental Analysis and Atomic Ratios				
Element	*% Weight*[a]	*Atomic Ratios*[b]	*SIGNATURE Input Parameter*[c]	*SIGNATURE Input Parameter*[d]
C	51.3			
H	4.00	93.40	74.7	112.1
O	39.7	58.00	46.4	69.6
S	0.90	0.66	0.52	0.79
N	4.12	6.90	5.51	8.26
		^{13}C NMR Data[e]		
	Carbon	*% Integrated Area*	*SIGNATURE Input Parameter*[g]	*SIGNATURE Input Parameter*[h]
	Aliphatic	14	11.2	16.8
	Aromatic	18	14.4	21.6
	C-O[f]	32	25.6	38.4
	Carboxylic	18	14.4	21.6

[a] Organic normalized elemental analysis data taken from ref. 18; [b] Atomic ratios normalized per 100 C atoms; [c] Minimum number of atoms per 100 C atoms, assumes a standard error of 20% for the elemental analysis data; [d] Maximum number of atoms per molecule assuming a standard error of 20% for the elemental analysis data;[e] ^{13}C NMR data taken from ref. 18; [f] Methoxyl and other oxygen substituted aliphatic carbon; [g] Minimum percentage of carbon atoms assuming a standard error of 20% for the integrated area; [h] Maximum percentage of carbon atoms assuming a standard error of 20% for the integrated area.

Table 2 *SIGNATURE input parameters for Chelsea humic acid: molecular fragments*[a] *and interfragment bonds*[b]

Molecular Fragments	*Molecular Fragments*	*Molecular Fragments*	*Molecular Fragments*	*Molecular Fragments*	*Interfragment Bonds*
Benzene	COOH-CH_2-COOH	Ethylphenol	Phenylalanine	Benzoic Acid	Csp^3_ Csp^3
Benzofuran	COOH-$CH_2)_2$-COOH	Propylphenol	Serine	Salicylic Acid	Csp^3_Csp^2
Benzothiophene	COOH-$CH_2)_3$-COOH	Buthylphenol	Threonine	Palmitic Acid	Csp^3_C_R
Methoxybenzene	COOH-$CH_2)_4$-COOH	Pentylphenol	Tryptophan	Stearic Acid	Csp^3_H
1,2,3,4-Tetramethylbenzene	COOH-$CH_2)_5$-COOH	1,2,3-Trihydroxylbenzene	Tyrosine		Csp^3_N
1,3,6-Trimethylbenzene	COOH-$CH_2)_6$-COOH	2,3,4,5-Tetracarboxyphenol	Valine	Dimethyl sulfide	Csp^3_O
1,2,6-Trimethylbenzene	COOH-$CH_2)_7$-COOH	2,3,4-Trihydroxybenzoic acid	Aspartic acid	Dimethyl ether	Csp^3_N
1,4,5-Trimethylbenzene	COOH-$CH_2)_8$-COOH	2,3,6-Tricarboxyphenol	Glutamic acid	Pyridine	Csp^3_S
m-Xylene	Cyclohexane	2,4,5-Trihydroxytoluene	5-Methylcysteine	Pyrole	C_R_C_R
o-Xylene	Methylcyclohexane	2,4-Dicarboxyphenol	Adenine	Phenol	C_R_Csp^2
p-Xylene	Methyl alcohol	2,4-Dihydroxybenzoic acid	Arginine	Methylphenol	C_R_H
Ethylbenzene	Ethyl alcohol	3,4,5-Trihydroxybenzoic acid	Cytosine	Furan	C_R_N
Propylbenzene	Propyl alcohol	3,4-Dihydroxybenzoic acid	Guanine	Indane	C_R_O
Butylbenzene	Butyl alcohol	3,5-Dihydroxybenzoic acid	Histidine	Indene	
Pentylbenzene	Pentyl alcohol	3-Hydroxybenzoic acid	Lysine	Quinoline	
1,2-Dihydroxybenzene	Dibutylether	4-Hydroxybenzoic acid	Uracil		
1,3-Dihydroxybenzene	Diethylether	m-Creosol	Alanine		
Indole	Dipropylether	Methylphloroglucinol	Asparagine		
Hydroquinone	Ethylbutylether	o-Cresol	Cysteine		
Methylnaphthalene	Ethylmethylether	p-Cresol	Glutamine		
Maleic acid	Ethylpropylether	Phloroglucinol	Glycine		
2,4-Dihydroxytoluene	Methylbutylether	Resorcinol	Isoleucine		
Methyl	Methylpropylether	Ester	Leucine		

[a] To ensure the generation of 3-D structural models with the adequate amount of oxygen bearing functional groups, the number of molecular fragments for selected carboxylic and hydroxyl bearing compounds was assumed to range from 0 to 1000. For all other compounds, the number of molecular fragments was assumed to range from 0 to 100; [b] The number of interfragment bonds was assumed to range from 0 to 100.

2.2 Estimation of Bulk Phase Thermodynamic Properties of Humic Substances from Molecular Dynamics Simulations

Molecular dynamics (MD) simulations have emerged as one of the most widely used theoretical methods of investigation of condensed phase systems.[26] In an MD simulation, a finite number of molecules are allowed to interact via a prescribed force field FF in a finite domain. The motions induced by the FF are deterministic. Thus, the positions, velocities and energies of the molecules can be determined by solving the corresponding Newton's equations of motion. Current MD simulation codes implement a step-by-step strategy to solve these equations of motion for the N-body system describing the molecular system of interest. MD simulation methodology was initially developed to model condensed phase systems in the microcanonical (NVE) ensemble. In the earlier 1980s, extended forms of MD were developed for simulating condensed systems in other ensembles of practical importance such as the isobaric and isothermal (NPT) ensemble and the isochoric and isothermal (NVT) ensemble.[26]

Equilibrium statistical mechanics provides the link between MD simulations and classical thermodynamics.[26] First-order bulk thermodymamic properties such as internal energy, pressure and density are calculated by time averaging of the corresponding microscopic quantities.[26] Second-order bulk thermodynamic properties such as specific heat, isothermal compressibility and thermal expansion coefficient can also be calculated from MD simulations.[26] Critical structural properties of amorphous systems (for example, FAs and HAs) such as the pair correlation function (g(r)) can also be estimated from MD simulations.[26]

In this paper we use molecular mechanics (that is, energy minimization) and NPT MD simulations to calculate the strain energy and bulk thermodynamic properties (for example, molar volume, solubility parameter and bulk density) of the SIGNATURE generated 3-D structural models for Chelsea HA. Each model was first energy minimized (rms force = 0.1 Kcal/molA^{-1}). and subjected to 15 ps of simulated annealing ($T_{initial}$ = 300 K, T_{final} = 600 K and $T_{increment}$ = 50 K) using the Cerius[2] molecular modeling software.[27] Each annealed model was then placed in a 3-D cell with periodic boundary conditions and packed to a bulk density of 1.0 g/cm^3 using the Amorphous Builder of Cerius.[2] The models were then subjected to energy minimization (rms force of 0.1 Kcal mol^{-1}A^{-1}) to remove the packing-induced bad contacts. Each minimized 3-D periodic model was subsequently subjected to 25 ps of constant pressure and constant temperature (NPT) molecular dynamics (MD) simulations at T = 300 K followed by energy minimization until its bulk density remained constant. Only two cycles of MD simulations followed by energy minimization were needed in most cases to achieve this goal. The Universal Force Field (UFF)[28] was used in all the MD simulations and energy minimization. The charge equilibration (Qeq) procedure[29] was used to evaluate all partial atomic charges. Ewald summation[30] was employed to calculate the long range electrostatic and van der Waals interactions for all the periodic systems. Conversely, these interactions were treated directly with a cut-off radius of 30 Å for the non-periodic systems. The Berendsen thermal coupling method (time constant 0.1 ps)[31] and the Andersen pressure control method (cell mass prefactor 0.04)[32] were employed in all NPT MD simulations. After completion of the MD simulations and energy minimization runs, the cell volume (V_p), condensed phase strain energy (E_p) and gas phase strain energy (E_{np}) were calculated for each model. The molar volume (V_m), bulk density ρ and cohesive energy (E_c) for each 3-D Chelsea HA model were expressed as Equns. (1) to (3), respectively,

$$V_m = N_a V_p \tag{1}$$

$$\rho = \frac{V_m}{M_n} \tag{2}$$

$$E_c = -(E_p - E_{np}) \tag{3}$$

where N_a is Avogadro's number and M_n is the number average molecular weight of the SIGNATURE generated Chelsea HA model. Following Barton,[33] the Hildebrand solubility parameter (δ) is expressed as in Equn. (4).

$$\delta = \sqrt{\frac{E_c}{V_m}} \tag{4}$$

Equns. (1) through (4) provide a rigorous thermodynamic framework for calculating the bulk thermodynamic properties of HSs from a given source once reliable 3-D structural models have been generated.

2.3 Flory-Huggins Model for Binding of Hydrophobic Organic Compounds to Dissolved Humic Substances

The uptake of an organic solute i by dissolved HSs is readily expressed as an equilibrium binding constant K_{bi}, Equn. (5),

$$K_{bi} = \frac{C_{iHS}}{C_{ia}} \tag{5}$$

where C_{iHS} and C_{ia} are, respectively, the amount of solute bound per unit mass of dissolved HSs and the concentration of "free" solute in solution (mass/L). Following Chin and Weber,[6] K_{bi} can be expressed in terms of the solute activity coefficients γ_i^{HS} and γ_i^a in the aqueous phase and dissolved HS pseudophase as shown in Equn. (6),

$$\mathrm{Ln}\,(K_{bi}) = \mathrm{Ln}\,(\gamma_i^{HS}) - \mathrm{Ln}\,(\gamma_i^{a}) + \mathrm{Ln}\,(\frac{V_w}{V_{HS}}) \tag{6}$$

where V_w and V_{HS} are the molar volumes of water and the HS host, respectively. Assuming that HSs consist of 3-D polymeric matrices, Chin and Weber[6] used the Flory-Huggins solution theory to express $\mathrm{Ln}\,\gamma_i^{HS}$ as in Equn. (7),

$$\mathrm{Ln}\,(\gamma_i^{HS}) = \mathrm{Ln}\,(V_i) - \mathrm{Ln}\,(V_{HS}) + 1 + \chi_i^{HS} \tag{7}$$

where V_i and V_{HS} are the molar volumes of the solute and the HS host, respectively, and χ_i^{HS} is their Flory-Huggins interaction parameter. Combination and rearrangement of Equns. 6 and 7 gives Equn. (8),[6]

$$\log(K_{bi}) = \log(\gamma_i^a) + \log(\frac{V_w}{V_i}) - \log\rho_{HS} - \frac{(1+\chi_i^{HS})}{2.303} \quad (8)$$

where ρ_{HS} is the bulk density of the HS host. The Flory-Huggins interaction parameter χ_i^{HS}, which a measure of the "compatibility" between solute and HS host, usually has both enthalpic (χ_{ih}^{HS}) and entropic (χ_{is}^{HS}) components, Equn. (9).

$$\chi_i^{HS} = \chi_{ih}^{HS} + \chi_{sh}^{HS} \quad (9)$$

For nonpolar solutes (for example, HOCs) that primarily interact with dissolved HS through dispersion interactions, χ_{ih}^{HS} is given by the Hildebrand-Scatchard Equn. (10),[33]

$$\chi_{ih}^{HS} = \frac{V_i}{RT}(\delta_i^2 - \delta_{HS}^2) \quad (10)$$

where δ_i and δ_{HS} are the Hildebrand solubility parameters of the solute and HS host, respectively, R is the ideal gas constant and T is the solution temperature. The entropic interaction parameter χ_{ci}^{HS} accounts for "deviations from the ideal entropy of mixing" caused by differences in molecular size (that is, molar volume) between a solute and its HS host. Its magnitude is determined primarily by the solute-HS critical interaction parameter χ_{ci}^{HS}, Equn. (11).[6]

$$\chi_{ci}^{HS} = 0.5(1 + \frac{1}{(V_{HS}/V_i)^{0.5}})^2 \quad (11)$$

When χ_i^{HS} is greater than χ_{ci}^{HS}, the solute and its HS host are "highly dissimilar in size." In this case, χ_{si}^{HS} is equal to 0.5, the limiting value for χ_{ci}^{HS} as $V_{HS}/V_i \rightarrow \infty$. Conversely, when χ_i^{HS} is less than χ_{ci}^{HS}, the solute and HS host have "approximately equal molar volume". In this case, χ_{si}^{HS} is equal to 0. Equations 8 through 11 provide a rigorous thermodynamic framework for estimating HOC uptake by dissolved HS if the pertinent thermodynamic data for the solute and HS host are available.

3 RESULTS AND DISCUSSION

3.1 Structure Elucidation

Generation of the 3-D molecular models for Chesea HA by SIGNATURE was carried out in two steps: (i) determination of the set of molecular fragments and interfragment bonds that best match the input data from Tables 1 and 2 and (ii) generation of a sample of 20 structural models from the molecular fragments and interfragment bonds found in step (i). The first step was carried out by running the "elucidation" mode of the SIGNATURE program.[17] In this mode the program solves the "signature" equations to determine the amounts and types of molecular fragments and interfragment bonds that best match the input data. A simulated annealing search[17] of 10 cycles ($T_{initial}$ = 100 K, T_{final} = 1000 K and $T_{increment}$ = 100 K) was used to solve these equations. The SIGNATURE output list of molecular fragments that best match the input data is given in Table 3. It consists of 8 carboxyl group-bearing compounds, 2 aliphatic alcohols, 2 hydroxyl group-bearing

Table 3 *SIGNATURE output parameters for Chelsea humic acid: list of molecular fragments and interfragment bonds that best match the input data*

Molecular Fragments	*Number of Fragments*	*Interfragment Bonds*	*Number of Interfragment Bonds*
Salicylic acid	1	$Csp^3\_C_R$	5
$COOH-CH_2-COOH$	2	Csp^3_O	6
$COOH-(CH_2)_2-COOH$	1	$C_R\_C_R$	7
$COOH-(CH_2)_5-COOH$	1	C_R_O	3
Ethyl alcohol	1		
Propyl alcohol	1		
2,3,4,5-Tetracarboxyphenol	1		
2,4,5-Trihydroxytoluene	2		
2,4-Dicarboxyphenol	1		
2,4-Dihydroxytoluene	2		
3,4-Dihydroxybenzoic acid	1		
3-Hydroxybenzoic acid	2		
Cysteine	1		
Glycine	1		
Isoleucine	1		
Serine	1		
Glutamine	1		
Thymine	1		

Table 4 *SIGNATURE output parameters for Chelsea humic acid model predictions for atomic ratio versus experimental analytical input data*[a]

Element	*Atomic Ratios*		
	Experiment	*Model Prediction*	*Error (%)*
C			
H	93.40	92.5	0.96
O	58.00	58.3	-0.52
S	0.66	0.8	26.00
N	6.90	5.8	16.00
Total			9.57

[a] Atomic ratios normalized per 100 C atoms

aromatic compounds, 5 amino acids and 1 pyrimidine. The SIGNATURE output list of interfragment bonds is also given in Table 3. It consists of 5 $C_{sp3}\_C_R$ bonds, 6 C_{sp3}_O bonds, 7 $C_R\_C_R$ bonds and 3 C_R_O bonds. The combination of these molecular fragments and interfragment bonds yields structural models of Chelsea HA with the molecular formula $C_{120}H_{111}O_{70}N_7S_1$. The number average molar mass of these models is equal to 2801 Da. This value is consistent with our assumption of a number average molar mass for Chelsea HA ranging from 1.5 to 5 kDa and it also compares well with literature values.[25] However, as shown in Table 4 the match between the experimental and the SIGNATURE estimated atomic ratios is not perfect. The overall deviation between input and model atomic ratios is approximately 10%. Although the errors in the estimated atomic ratios for H and O are low (less than 3%), the errors in the estimated atomic ratios for N (16%) and S (26%) are relatively large. We can reduce these errors to less than 5% if additional experimental data (for example, 1-D and 2-D $^1H/^{13}C$ NMR, MS, and so on) are available.

3.2 Model Generation

The second step of the model building process for Chelsea HA was carried out by running the "prediction" mode of SIGNATURE.[17] In this mode, SIGNATURE used a *simple random sampling without replacement* approach to generate 3-D models by randomly connecting the molecular fragments and interfragments found in step 1. The bonding sites were assigned to ensure that the SIGNATURE generated structural models for Chelsea HA will incorporate basic building blocks of HA such as the substituted aromatics, carboxylic acids, amino acids and so on described in section 2. The molecular fragments used to generate the 3-D structural models for Chelsea HA are given in Figure 2 with all bonding sites shown in green.

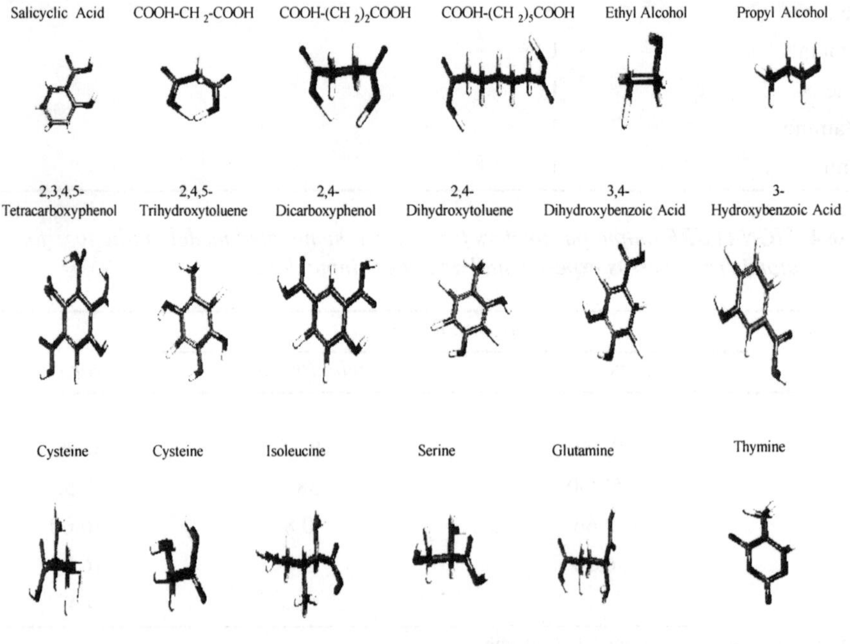

Figure 2 *SIGNATURE generated building blocks for Chelsea humic acid with bonding sites shown in green*

A total of 18 molecular fragments with 42 bonding sites and 21 interfragment bonds were randomly connected to generate a sample of 20 3-D models for CHA. Four of the generated structural models were discarded because they contained unrealistic bonding schemes (for example, SIGNATURE forced the alkyl chain of a molecular fragment to go through an aromatic ring in order to bond with another fragment). The remaining 16 Chelsea HA structures were subjected to energy minimization followed by NPT MD simulations using the procedures described in section 2.2. After completion of these simulations, the strain energy (E_{strain}), molar volume (V_m), bulk density (ρ), condensed phase strain energy (E_p) and gas phase strain energy (E_{np}) were calculated for each model. The estimated strain energies, bulk densities and solubility parameters for the 16 3-D Chelsea HA models are shown in Figure 3. The bulk densities of HSs have been estimated to range from 1.2 to 1.4 g/cm^3.[6,7,12-14] Several investigators have reported estimates of solubility parameters of HA ranging from 10.0 to 13.5 cal$^{1/2}$ /cm$^{3/2}$.[6-8,12-14] Based on these literature values we selected a sample of 6 structural models (CHA Model # 1, CHA model # 4, CHA model # 7, CHA model # 17 and CHA model # 18) as a representative sample of Chelsea HA 3-D structural models (Figure 4). Each of these models has a molecular formula of $C_{120}H_{111}O_{70}N_7S_1$ with a molar mass of 2801 Da. The average sample density (ρ_{CHA}) is equal to 1.28 ± 0.04 g/cm^3. The average sample molar volume (V_{mCHA}) is 2188.28 cm^3/mole. The average sample solubility parameter (δ_{CHA}) is 10.8 ± 1.2 cal$^{1/2}$/cm$^{3/2}$, which is equivalent to 22.0 ± 2.4 J$^{1/2}$/cm$^{3/2}$ and comparable with an average of 25.5 ± 1.5 J$^{1/2}$/cm$^{3/2}$ obtained by Poerschmann and co-workers[14,34] from a good many sorption measurements with Roth HA.

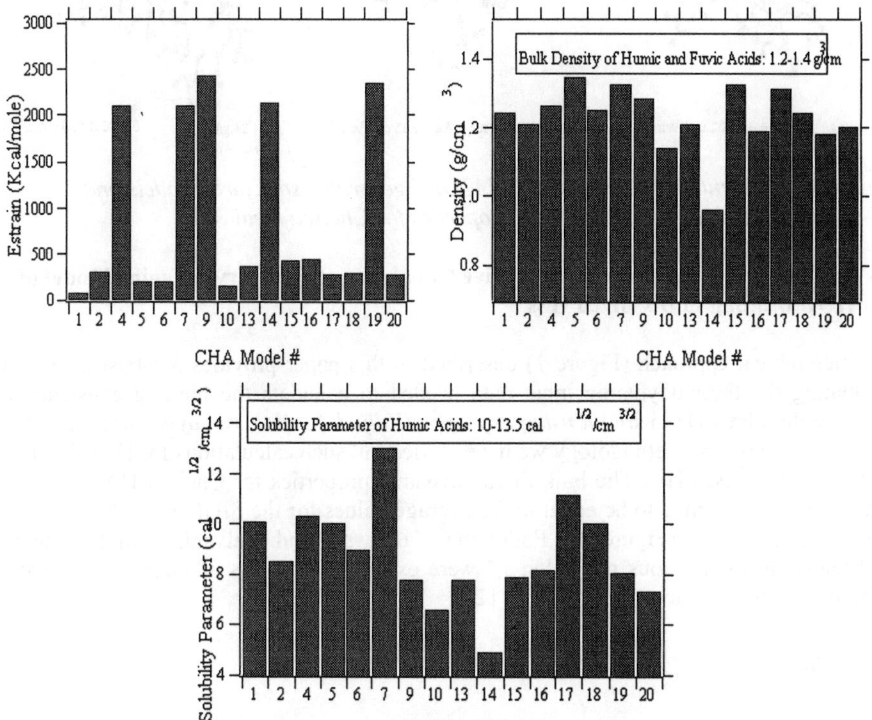

Figure 3 *Estimated strain energies, bulk densities and solubility parameters for Chelsea humic acid model isomers from molecular simulations*

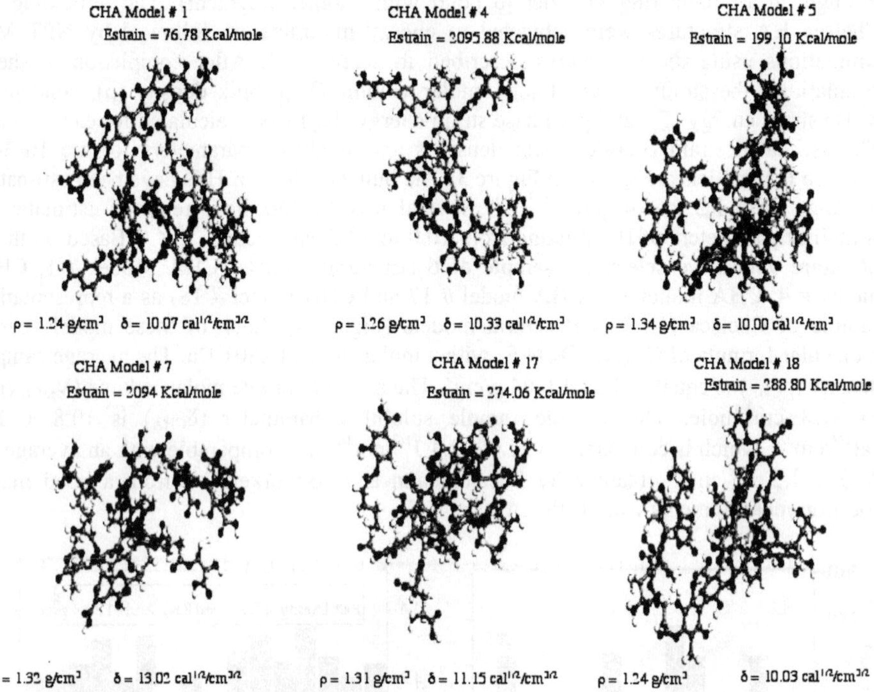

Figure 4 *Representative sample of SIGNATURE generated structural models and selected bulk thermodynamic properties for Chelsea humic acid*

3.3 Preliminary Assessment of Predictive Capability of the Flory-Huggins Model of HOC Binding to Dissolved HSs

The hierarchical approach (Figure 1) described in this paper provides a robust means of calculating the thermodynamic input data needed to calculate the binding constants of HOCs to dissolved HSs *without using any empirically derived thermodynamic parameter.* To illustrate this new methodology we have carried out such calculations for HOC binding to dissolved Chelsea HA. The bulk thermodynamic properties for Chelsea HA (ρ_{HA}, δ_{HA} and V_{HA}) were assumed to be equal to the average values for the SIGNATURE generated Chelsea HA sample (Figure 4). Following Yalkowski and Valvani,[35] solute activity coefficients in the aqueous phase log γ_i^w were expressed in terms of their octanol-water partition coefficients (log K^i_{OW}), Equn. (12).

$$\log \gamma_i^w = \log K^i_{OW} + 0.82 \tag{12}$$

The solute-HS enthalpic interaction parameter (χ_{ih}^{HS}) was estimated using Equation 10. Because Chelsea HA and the organic solutes have widely different molar volumes (see the values of their molar volume ratios and critical interaction parameters (χ_{ci}^{CHA}) in Table 4), the solute-HS entropic interaction parameter (χ_{ci}^{CHA}) was assumed to be equal to 0.5. All the data used to calculate the binding constants of the organic solutes to dissolved Chelsea HA are given in Table 4. *No adjustable parameter was used in these calculations.*

Table 4 *Physicochemical data used to estimate the binding constants of selected hydrophobic organic compounds to dissolved Chelsea humic acid*

	Compound	δ_i cal$^{1/2}$/cm$^{3/2}$ [a]	V_i cm^3/mole [b]	log (V_w/V_i) [c]	log K^i_{OW} [d]	log γ_i^{w} [e]	V_{CHA}/V_i [f]	χ_{ci}^{CHA} [g]
1	Tetrachloroethylene	9.2	90.2	-0.700	2.53	3.35	24.3	0.72
2	Toluene	8.9	106.8	-0.770	2.69	3.51	20.5	0.74
3	Naphthalene	9.9	111.5	-0.790	3.38	4.2	19.6	0.75
4	Phenanthrene	9.8	158	-0.940	4.46	5.28	13.9	0.80
5	Anthracene	9.9	150	-0.920	4.54	5.36	14.6	0.79
6	Fluorene	9.7	138	-0.885	4.18	5	15.9	0.78
7	Biphenyl	8.3	177	-0.990	3.95	4.77	12.4	0.82
8	p,p'-DDT	8.8	222	-1.09	5.98	6.8	9.9	0.87
9	1,4-Dichlorobenzene	9.7	112.5	-0.800	3.36	4.18	19.5	0.75
10	1,2,4-Trichlorobenzene	9.3	159	-0.950	3.98	4.8	13.7	0.80
11	n-Octane	7.6	164	-0.959	5.05	5.87	13.3	0.81
12	n-Nonane	7.7	179	-0.997	5.45	6.27	12.2	0.83
13	n-Decane	7.8	195	-1.034	5.98	6.8	11.2	0.84
14	n-Undecane	7.8	211	-1.068	6.51	7.33	10.4	0.86
15	n-Dodecane	7.8	227	-1.100	7.04	7.86	9.6	0.87
16	n-Tridecane	7.9	243	-1.129	7.57	8.39	9.0	0.89
17	Pyrene	10.1	179	-0.997	5.19	6.01	12.2	0.83
18	Chrysene	10.1	196	-1.036	5.5	6.32	11.2	0.84

[a] δ_i (cal$^{1/2}$/cm$^{3/2}$): Solute Hildebrand solubility parameter. Data are from refs. 6 and 12; [b] V_i (cm^3/mole): Solute molar volume. Data are from refs. 6 and 12; [c] log (V_w/V_i): Estimated logarithm of water-solute molar volume ratio. The molar volume of water V_w (18.02 cm^3/mole) was from ref. 6; [d] log K^i_{OW}: logarithm of solute octanol-water partition coefficient. Data are from refs. 6 and 12; [e] log γ_i^{w}: logarithm of solute activity coefficient in the water phase. Estimated using Equn. (12); [f] V_{CHA}/V_i: Estimated solute-Chelsea HA molar volume. Solute molar volumes were from refs. 6 and 12. The estimated molar volume of Chelsea HA was from Figure 4; [g] χ_{ci}^{CHA}: solute-Chelsea HA critical interaction parameter. Estimated using Equn. (11).

As a preliminary assessment of the predictive capability of our new approach (Figure 1), we compare the magnitudes of the calculated organic carbon normalized binding constants (log K_{bi}, OC) of HOC to dissolved Chelsea HA to measured HOC log K_{bi}, OC for Aldrich HA[6] and Roth[12] HA. As shown in Figure 5 and Table 5, the magnitudes of the calculated log K_{bi}, OC of HOC to dissolved Chelsea HA compare well with the corresponding measured binding constants of HOC to dissolved Aldrich HA and Roth HA.

We are currently carrying out a more extensive assessment of our approach to describe the linear and nonlinear sorption of HOCs to soil HA *without using any adjustable parameter.*

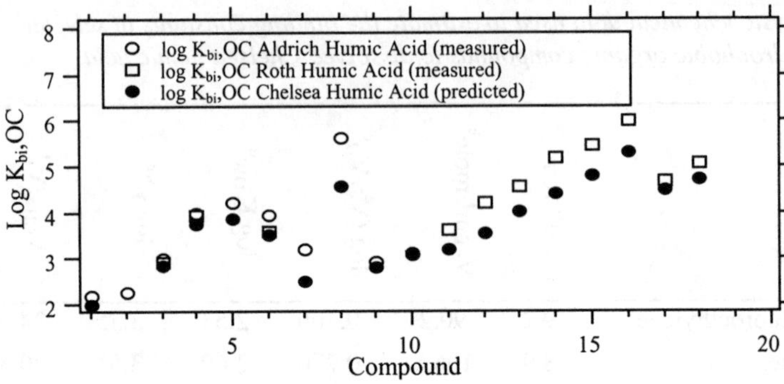

Figure 5 *Predicted log K_{bi}, OC and measured log K_{bi}, OC for HOC binding to dissolved Aldrich HA and Roth HA*

Table 5 *Predicted log K_{bi}, OC for HOC binding to dissolved Chelsea HA vs. Measured log K_{bi}, OC for HOC binding to dissolved Aldrich and Roth HAs*

Compound		*measured log K_{bi}, OC*		*predicted log K_{bi}, OC, Chelsea HA*[c]
		Aldrich HA[a]	*Roth HA*[b]	
1	Tetrachloroethylene	2.20	NA	2.01
2	Toluene	2.27	NA	1.99
3	Naphthalene	3.02	2.93	2.87
4	Phenanthrene	4.00	3.93	3.75
5	Anthracene	4.21	NA	3.88
6	Fluorene	3.95	3.58	3.52
7	Biphenyl	3.22	NA	2.50
8	p,p'-DDT	5.61	NA	4.59
9	1,4-Dichlorobenzene	2.92	NA	2.81
10	1,2,4-Trichlorobenzene	3.11	NA	3.12
11	n-octane	NA	3.62	3.21
12	n-nonane	NA	4.24	3.54
13	n-decane	NA	4.57	4.00
14	n-undecane	NA	5.20	4.40
15	n-dodecane	NA	5.48	4.79
16	n-tridecane	NA	6.0	5.29
17	pyrene	NA	4.70	4.48
18	chrysene	NA	5.06	4.74

[a] Measured log K_{bi}, OC for HOC binding to dissolved Aldrich humic acid (ref. 6); [b] Measured log K_{bi}, OC for HOC binding to dissolved Roth humic acid (ref. 12); [c] Predicted log K_{bi}, OC for HOC binding to dissolved Chelsea humic acid.

4 CONCLUSIONS

This paper describes a hierarchical approach for predicting the binding of hydrophobic organic compounds (HOCs) to dissolved HS *without using any empirically derived thermodynamic parameter.* This novel approach combines computer assisted structure elucidation (CASE) with atomistic simulations and Flory-Huggins solution theory to estimate the constants of binding HOCs to dissolved FAs and HAs. To illustrate this new approach we used quantitative and qualitative structural data as input to the CASE program SIGNATURE to generate a sample of 16 3-D structural models of Chelsea HA. We then used these models as starting 3-D structures to carry out constant pressure and constant temperature (NPT) molecular dynamics simulations followed by energy minimization to estimate the strain energies, bulk densities and Hildebrand solubility parameters of the computer generated structural models. The estimated bulk densities and solubility parameters are compared with literature values to select a sample of 6 Chelsea HA model isomers as representative structural models that "best match" the input analytical data. We then showed how this representative sample of 3-D models can be used to calculate reliable estimates of the thermodynamic input parameters needed to calculate the binding constants of HOC to dissolved Chelsea HA *without using any empirically derived thermodynamic parameter for the HS host.* As a preliminary assessment of the predictive capability of this new approach, we compare the magnitudes of the calculated organic carbon normalized binding constants (log K_{bi},OC) of HOC to dissolved Chelsea HA with measured HOC log K_{bi},OC for Aldrich HA[6] and Roth[12] HA, two commercial HAs with similar properties. We find that the magnitudes of the calculated log K_{bi},OC of HOC to dissolved Chelsea HA compare very well with the corresponding measured constants of HOC binding to dissolved Aldrich HA and Roth HA. We are currently carrying out a more extensive assessment of our approach to describe the linear and nonlinear sorption of HOCs to soil HA *without using any adjustable parameter.*

ACKNOWLEDGEMENTS

This research was conducted in the Department of Civil Engineering, Howard University and at the Materials and Process Simulation Center of the Beckman Institute, California Institute of Technology. Funding for this research was provided by The University of Michigan, Michigan State University and Howard University through the Great Lakes and Mid-Atlantic Hazardous Substance Research Center under Grant R-825540 from the Office of Research and Development, U.S. Environmental Protection Agency. Partial funding for this research was also provided to Howard University and Caltech by the Department of Commerce under Cooperative Agreement 7NANB8HO102. The content of this publication does not necessarily represent the views of any of the funding agencies. We also acknowledge the Institute of Multimedia Applications at Howard University for providing additional computer resources for this project. MSD wishes to thank Dr. Weilin Huang (Drexel University) and Dr. Walter Weber (University of Michigan) for providing samples and analytical data for Chelsea humic acid.

References

1. M.N. Jones and N.D. Bryan, *Adv. Coll. Int. Sci.*, 1998, **78,** 1.
2. G. Davies and E.A. Ghabbour, eds., *Humic Substances: Structures, Properties and*

Uses, Royal Society of Chemistry, Cambridge, 1998.
3. F.J. Stevenson, *Humus Chemistry: Genesis, Composition, Reactions*, 2nd Edn., Wiley, New York, 1994.
4. C.W. Carter and I.H. Suffet, *Environ. Sci. Technol.*, 1982, **16**, 735.
5. C.T. Chiou, R.L. Malcolm, T.I. Britton and D.E. Kile, *Environ. Sci, Technol*, 1986, **20**, 502.
6. Y. Chin and W.J. Weber, Jr., *Environ. Sci. Technol.*, 1989, **27**, 976.
7. M.A. Schlautman and J.J. Morgan, *Environ. Sci. Technol.*, 1993, **27**, 961.
8. J.F. McCarthy and B.D. Jimenez, *Environ. Sci. Technol.*, 1985, **19**, 1072.
9. D.T. Gauthier, E.C. Shane, W.F. Guerin, W.R. Seitz and C.L. Grant, *Environ. Sci. Technol.*, 1986, **20**, 1162.
10. P.J. Hasset and E. Millicic, *Environ. Sci. Technol.*, 1985, **19**, 628.
11. A.D. Sita, *J. Phys. Chem. Ref. Data*, 2001, **30**, 187.
12. J. Poerschmann and F.E. Kopinke, *Environ. Sci. Technol.*, 2001, **35**, 1142.
13. C.T. Chiou, P.E. Porter and D.W. Schmedding, *Environ. Sci. Technol.*, 1983, **17**, 227.
14. F.E. Kopinke, J. Poerschmann and U. Stottmeister, *Environ. Sci. Technol.*, 1995, **29**, 941.
15. L.G. Akim, G.W. Bailey and S.M.A. Shevchenko, in *Humic Substances: Structures, Properties and Uses*, eds. G. Davies and E.A. Ghabbour, Royal Society of Chemistry, Cambridge, 1998, p. 133.
16. S.M.A. Shevchenko and G.W. Bailey, *J. Mol. Struct. Theochem.*, 1998, **422**, 259.
17. J.L. Faulon, *J. Chem. Comput. Sci.*, 1994, **34**, 1204.
18. W. Huang and W.J. Weber, Jr., *Env. Sci. Technol.*, 1997, **31**, 2562.
19. S.M. Griffith and M. Schnitzer, in *Humic Substances II: In Search of Structures*, eds. M.H.B. Hayes, P. MacCarthy, R.L. Malcolm and R.S. Swift, Wiley, New York, 1989, p. 69.
20. J.W. Parsons, in *Humic Substances II: In Search of Structures*, eds. M.H.B. Hayes, P. MacCarthy, R.L. Malcolm and R.S. Swift, Wiley, New York, 1989, p. 98.
21. F.J. Stevenson, in *Humic Substances II: In Search of Structures*, eds. M.H.B. Hayes, P. MacCarthy, R.L. Malcolm and R.S. Swift, Wiley, New York, 1989, p. 121.
22. P.G. Hatcher and D.J. Clifford, *Org. Geochem.*, 1994, **21**, 1081.
23. J. Del Rio and P.G. Hatcher, in *Humic and Fulvic Acids: Isolation, Structure, and Environmental Role*, eds. J.S. Gaffney, N.A. Marley and S.B. Clark, ACS Symposium Series 651, 1996, p. 79.
24. R. Kuckuck, W. Hill, P. Burba and A.N. Davies, *Fresenius J. Anal. Chem.*, 1994, **350**, 528.
25. R. Beckett, Z. Jue and J.C. Giddings, *Environ. Sci. Technol.*, 1997, **21**, 289.
26. A.R. Leach, *Molecular Modeling: Principles and Applications*, Addison Wesley-Longman, New York, 1996.
27. *Cerius² User Guide*, Molecular Simulations, San Diego, 1997.
28. K. Rappé and W.A. Goddard III, *J. Phys. Chem.*, 1991, **95**, 8340.
29. A.K. Rappé, C.J. Casewit, K.S. Colwell, W.A. Goddard III and W.M. Skiff, *J. Am. Chem. Soc.*, 1992, **114**, 10024.
30. N. Karasawa and W.A. Goddard III, *J. Phys. Chem.*, 1989, **93**, 7320.
31. H.J.C. Bereendsen, J.P.M. Postma, W.F. van Gunsteren, A. DiNola and J.R. Haak, *J. Phys. Chem.*, 1984, **81**, 3684.
32. H.C. Andersen, *J. Chem. Phys.*, 1980, **72**, 2384.
33. A.F.M. Barton, *CRC Handbook of Solubility Parameters and Other Cohesion*

Parameters, CRC Press, Boca Raton, FL, 1991.

34. J. Poerschmann, in *Humic Substances: Versatile Components of Plants, Soil and Water*, eds. E.A. Ghabbour and G. Davies, Royal Society of Chemistry, Cambridge, 2000, p. 165.
35. S.H. Yalkowski and S.C. Valvani, *J. Pharm. Sci.*, 1980, **69**, 912.

Images, Oxidation and Humification

ATOMIC FORCE MICROSCOPY (AFM) STUDY OF THE ADSORPTION OF SOIL HA AND SOIL FA AT THE MICA-WATER INTERFACE

A.-M. Tugulea,[1] D.R. Oliver,[2] D.J. Thomson[2] and F.C. Hawthorne[1]

[1]Department of Geological Sciences, University of Manitoba, Winnipeg MB R3T 2N2, Canada
[2] Department of Electrical and Computer Engineering, University of Manitoba Winnipeg, MB R3T 5V6, Canada

1 INTRODUCTION

1.1 The Significance of Clay-Humic Complex Morphology

Adsorption of humic molecules on clay minerals improves their geochemical stability and provides an active surface for cation exchange and sorption of organic compounds.[1-3] The importance of clay-humic complexes in contaminant binding and transport has been demonstrated.[4-10] Experimental data suggest that the interaction between humic substances and clay particles influences the conformation of the humic molecules and consequently the sorptive properties of the clay-humic complex.[11,12] Structural information about clay-humic complexes is needed to design appropriate models for clay-humic complex formation and stability under diverse environmental conditions and to understand the sorption of ions and molecules on clay-humic complexes.[13,14] Very little direct information is available concerning the mechanism of humic-fraction/surface interactions with clays and the structural characteristics of the reaction products. Relevant information about clay-humic complexes includes the topography of clay-humic surface (degree of coverage, thickness of the organic layer) and the influence of environmental parameters (humic fraction concentration, pH, ionic strength and presence of complexing cations).

1.2 Mica (Muscovite) as a Model for the Surface of Clay Minerals

Mica (muscovite) can be used as a model for clay mineral surfaces in atomic force microscopy (AFM) studies.[15,16] Muscovite plates are commercially available. They are inexpensive and easy to cleave to give a clean surface. Freshly cleaved mica provides an atomically smooth surface (flatness better than 0.3 nm over areas of tens of square microns)[16] that is similar to some of the surfaces of phyllosilicate minerals frequently found in clay fractions of soils and sediments. The muscovite surface is simple and negatively charged at intermediate pH values, with uniform characteristics on a relatively large scale and little intrinsic topographic variation (Figure 1).

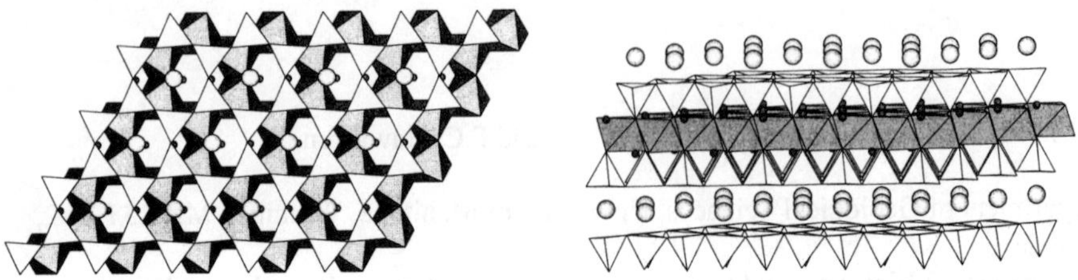

Figure 1 *Muscovite structure*

1.3 AFM in the Study of Humic and Fulvic Acid Morphology

AFM is a relatively new imaging technique used in the study of humic acid (HAs) and fulvic acid (FAs) morphology.[15-18] Compared with electron microscopy, AFM provides more flexibility in sample preparation, three-dimensional images and, in most cases, better image resolution.[15] A variety of humic materials have been studied by AFM using different materials as atomically-flat solid supports: polished graphite,[18] glass[17] and mica.[15,16] The three-dimensional images can be used to derive information about humic particle shape, dimensions, degree of aggregation and surface coverage under different adsorption conditions. Contact-mode AFM[16-18] as well as tapping-mode AFM[15] have been used to study humic acid morphology. Both methods have advantages and disadvantages when used on fragile humic samples. In an attempt to provide relevant environmental information for samples in their normal hydrated state, liquid humic-mica samples were studied.[15]

1.4 Objectives of this Work

In the present work, IHSS Standard Soil HA and FA[19] and freshly cleaved muscovite were used to model adsorption of a natural HA and FA to clay minerals and describe the morphology of the resulting surface complex.

It is well documented both experimentally and theoretically that in clay-humic mixtures both precipitation of the clay and humic colloids and adsorption of humic aggregates on the clay surface occur.[20] In order to avoid deposition of humic aggregates on the mica surface by sedimentation and study only the humic material that truly adsorbs to the mica surface, adsorption was done with the mica plate immersed vertically in the humic solution. The contact time was 24 h. Adsorption experiments were done at different humic material concentrations, pH and ionic strength. Contact-mode AFM was used for imaging the dried mica surface after adsorption.

2 MATERIALS AND METHODS

2.1 Solutions of Standard Soil HA/FA (IHSS)

IHSS Standard Soil HA and IHSS Standard Soil FA from the International Humic Substances Society collection were used.[19] Stock solutions with a concentration of 400 mg/L in double distilled water were prepared and kept under continuous stirring. The stock solutions were stirred at least 5 days before using them for the preparation of the working solutions. The concentration of the HA or FA in the working solutions was 0.4, 4 and 40 mg/L. The pH of the working solutions was adjusted to pH 1.5 with a minimum volume of HCl solution and to pH 7 and pH 10.5 with a minimum volume of NaOH solution. Some of the working solutions were adjusted to 0.01 M NaCl concentration with solid NaCl.

2.2 Mica (Muscovite) Plates

Mica (muscovite) was purchased from Asheville-Schoonmaker Mica Company, Newport News, VA. A fresh mica surface was prepared by removing the old mica surface using clear adhesive tape and the sample was immediately immersed in the working solution.

2.3 Adsorption Experiments

The mica plates were suspended vertically in 5 mL glass beakers containing the working solution. In some adsorption experiments, a second mica plate with one surface covered with adhesive tape was placed horizontally in the beaker. The plates were removed from the solution after 24 hours and the excess solution drained as well as possible by capilliarity. The samples were air dried for 2-4 h.

2.4 AFM Imaging

Images were obtained using a contact-mode, constant-force scanning microscope (SPM). A Park Scientific Instruments SPM operating at ambient room temperature and in atmospheric conditions was used throughout these investigations. The instrument was calibrated using a standard calibration grating (Ultrasharp/Silicon-MTD Ltd. TGZ02), which has a period of 3 μm (xy) and a step height of 104 nm (z). NT-MTD SCS12/W2C cantilevers were used in all experiments.

During SPM operation, the probe is brought into contact with the surface and this contact is maintained at a constant force while the probe is scanned in a raster pattern across the sample. The probe is mounted on a rectangular cantilever that bends as the force is applied. During topography measurements, a constant probe-sample contact force was maintained by detecting variations in the position of a laser beam reflected from the upper surface of the cantilever. The SPM controller[21,22] adjusts the z position of the sample to compensate for these variations. This technique and instrument are capable of achieving topographic resolution of less than 1 nm. A general description of this and similar techniques may be found in reference 23. The images obtained via the data-acquisition boards[21] and the software interface[22] were analyzed and formatted for presentation using Image SXM software.[24] On each sample between 6 and 12 different locations were imaged in order to account for spatial variability.

3 RESULTS AND DISCUSSION

3.1 Mica Blanks

A freshly cleaved mica surface was scanned; the variability in the surface relief was under 0.2 nm (Figure 2a). The mica surface (an aluminosilicate surface) is expected to be reactive under extreme pH conditions and the purity of double-distilled water and inorganic reagents may be insufficient at the sensitivity level of the analytical method used. In order to account for any changes in surface morphology not related to the interaction with humic colloids, blanks were prepared by immersing freshly cleaved mica plates for 24 h in the electrolyte solution with no humic material added. The variability of the mica surfaces that were treated in this way was found to be an order of magnitude higher than the variability of the freshly cleaved surface (Figures 2b, 2c and 2d). Similar data have been reported for mica immersed in 0.1 M NaCl at pH values from 3 to 8.[15] There seems to be no significant difference in variability between blank samples with respect to pH, but this may be due to insufficient sensitivity of the method.

3.2 Soil HA and FA Sorption at the Mica Surface as a Function of Concentration

Adsorption experiments were conducted with different humic acid concentrations (0.4, 4, 40 and 100 mg/L). At higher concentrations of the HA, a thick organic layer was formed on the mica surface and prevented proper imaging of the surface, at least by contact-mode AFM. At low concentrations (0.4 mg/L) there was little detectable sorption of humic acid on the mica surface, especially in experiments using 0.01 M NaCl as background electrolyte (Figure 3). This effect was more noticeable for FA solutions at concentrations of 40 mg/L and higher (Figure 4). In view of these results, a humic acid solution with a concentration of 4 mg/L and a fulvic acid solution with a concentration of 0.4 mg/L were used in the subsequent experiments.

3.3 Mica Plate Position in Solution

Under different pH and ionic strength conditions in solution, both sedimentation and adsorption of humic aggregates on the mineral surface may occur. Positioning the mica surface vertically in the humic solution should reduce deposition of humic aggregates by sedimentation. A number of samples obtained by placing mica plates vertically and horizontally in the same humic solution were imaged.

The differences between the vertical samples and the horizontal samples were more marked for the experiments in which 0.01 M NaCl was used as the background electrolyte (Figure 5), suggesting that precipitation and sedimentation of the humic colloids predominate under higher ionic strength conditions. The differences between the vertical and the horizontal samples were insignificant for experiments in which FA at pH 10.5 was used because under these conditions the mica surface seems to remain uncovered. The subsequent experiments were done with the mica plate immersed vertically in the humic solution in order to avoid deposition of humic aggregates on the mica surface by sedimentation and to study only the humic material that interacts with the mica surface.

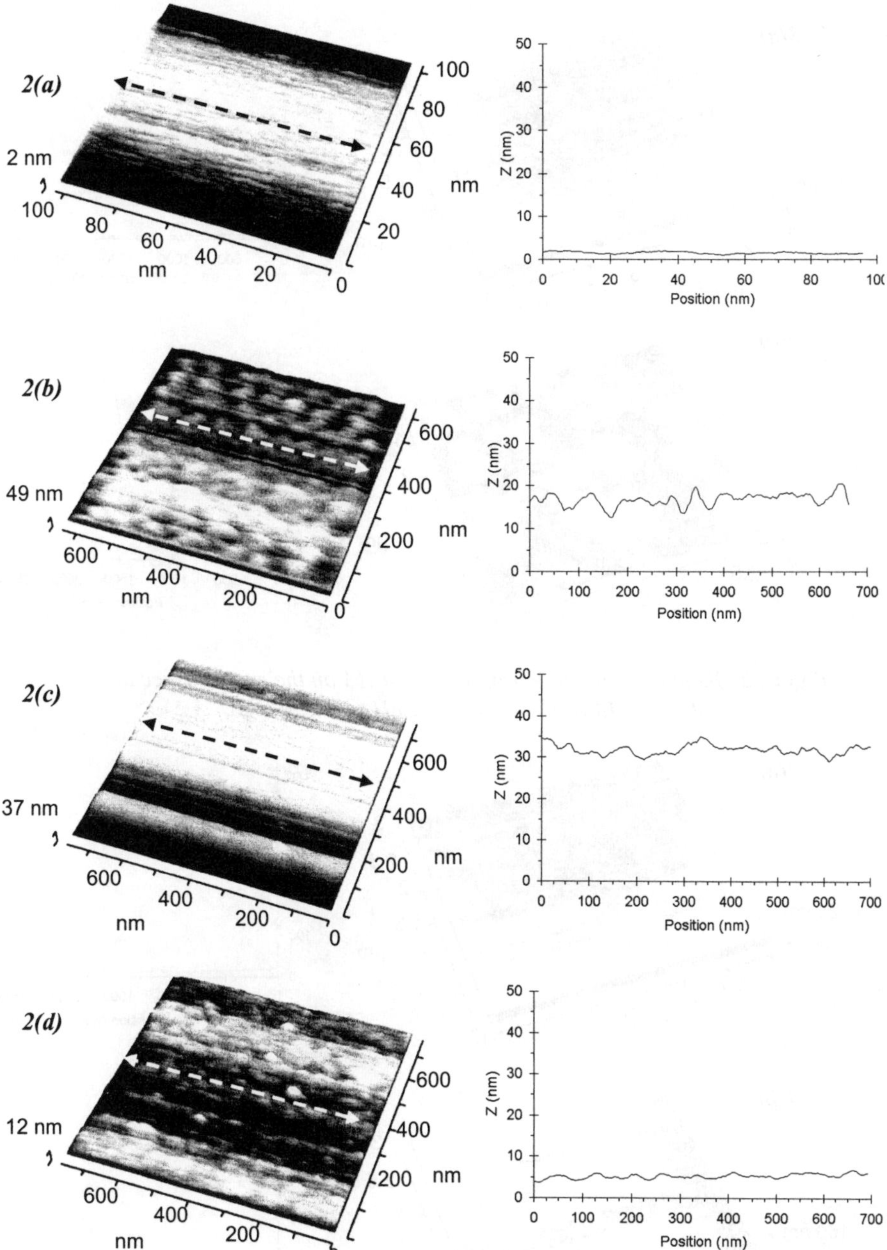

Figure 2 *3D images and line profiles of the mica surface. (a) Freshly cleaved, (b) After 24 h immersion in water at pH 1.5, (c) After 24 h immersion in water at pH 7, (d) After 24 h immersion in water at pH 10.5*

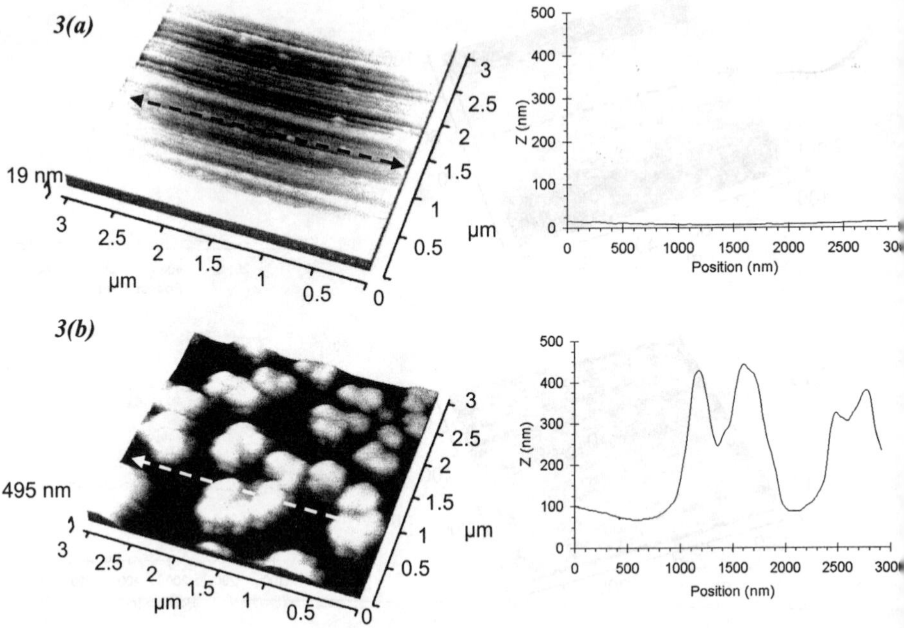

Figure 3 *3D images and line profiles of soil HA on the mica surface at pH 6.1, 0.01 M NaCl. (a) HA conc. 0.4 mg/L, (b) HA conc. 4 mg/L*

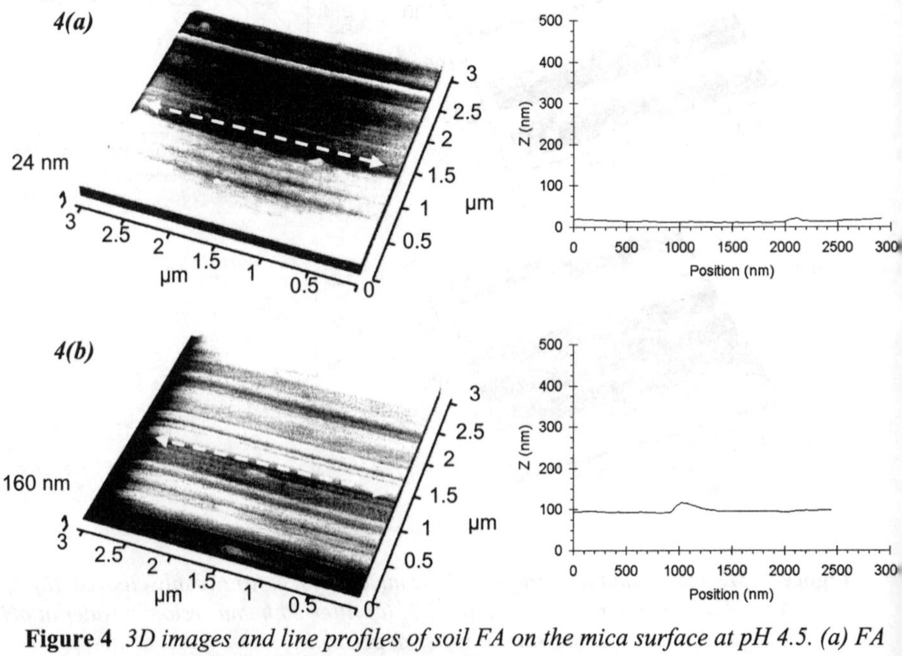

Figure 4 *3D images and line profiles of soil FA on the mica surface at pH 4.5. (a) FA conc. 4 mg/L, (b) FA conc. 40 mg/L*

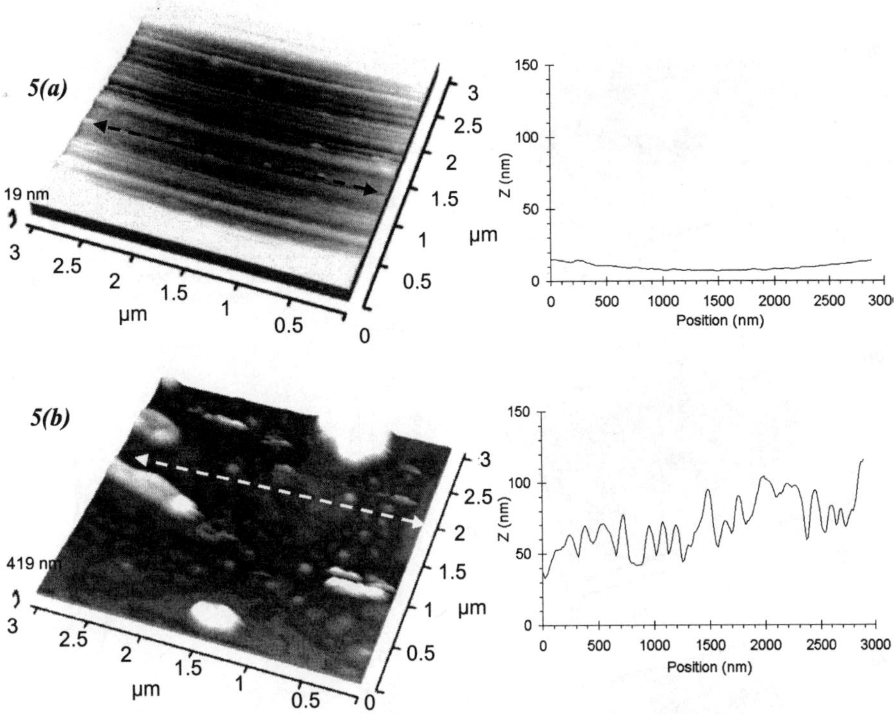

Figure 5 *3D images and line profiles of soil HA on the mica surface at pH 6, NaCl 0.01 M, HA conc. 0.4 mg/L. (a) Mica plate placed vertically in solution, (b) Mica plate placed horizontally in the solution*

3.4 Soil FA Sorption at the Mica Surface as a Function of pH

Results of the study of soil FA (0.4 mg/L) sorption on mica as a function of pH are presented in Figure 6. The dimensions of the fulvic-acid aggregates decrease with increasing pH (80-100 x 250 x 250 nm aggregates seem to be typical at pH 1.5 and 40-60 x 150 x 150 nm aggregates at pH 7). At pH 10.5 there is almost no organic layer on the mineral surface, as predicted by theory.

3.5 Soil HA Sorption at the Mica Surface as a Function of pH

Figure 7 summarizes the results of the study of soil HA (4 mg/L) adsorbed on mica. As predicted by theory, the amount of humic acid covering the mica surface decreases with increasing pH. At pH 1.5, large disc-shaped humic acid aggregates (120 x 250 x 250 nm) cover a significant part of the mineral surface; at pH 7, the humic acid aggregates have a similar disc shape but smaller dimensions (25-40 x 60 x 60 nm), whereas at pH 10.5 just a few large aggregates (150 x 250 x 300 nm) occur. The presence of 0.01M NaCl as the background electrolyte dramatically increases the amount of humic material on the mica surface: very large composite aggregates (200-500 x 500 x 500 nm) cover most of the mineral surface. The humic acid aggregates appear to be larger than the fulvic acid aggregates.

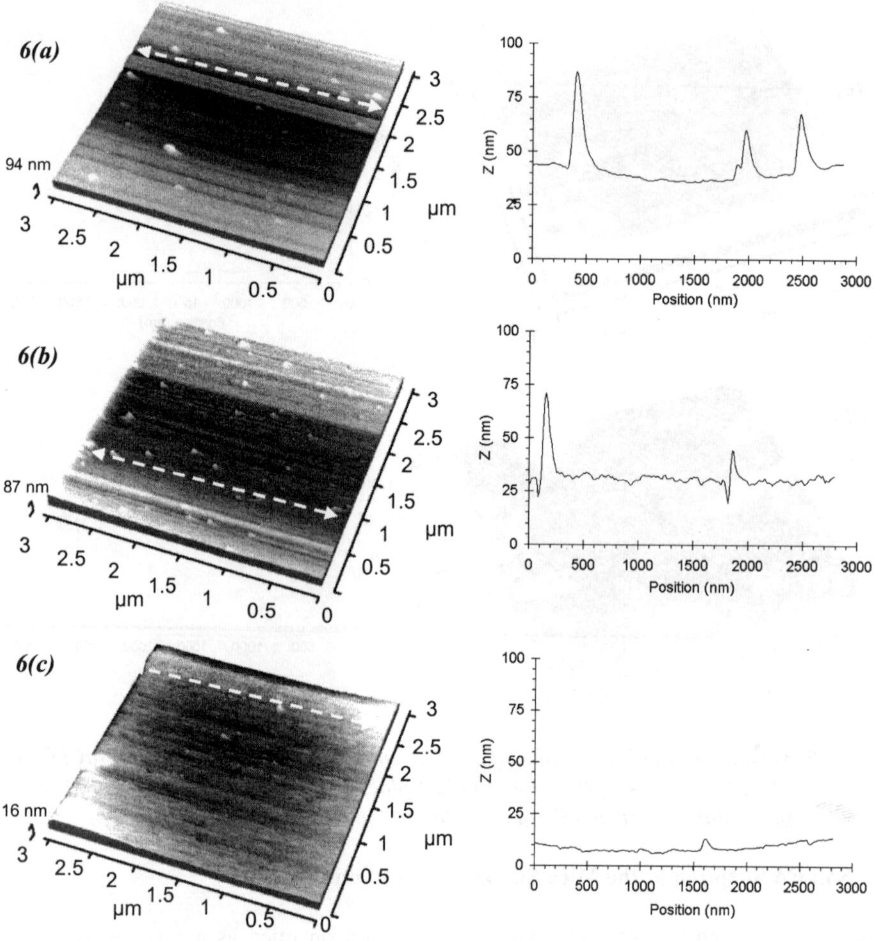

Figure 6 *3D images and line profiles of soil FA on the mica surface, FA conc. 0.4 mg/L/. (a) at pH 1.5, (b) at pH 7, (c) at pH 10.5*

4 CONCLUSIONS

The variability of mica surfaces immersed in electrolyte solutions for 24 h is an order of magnitude larger than the variability of the freshly cleaved mica surface, which makes it difficult to distinguish small aggregates (height smaller than 4 nm) and to estimate the degree of coverage of the surface. In order to distinguish between adsorbed humic colloids and background, each sample has to be compared to the appropriate blank.

At higher concentrations of the HA in solution, a thick organic layer formed on the mica surface, making proper imaging of the surface difficult and limiting the amount of information that can be obtained from the images (there was very little topographic variation). This effect was more noticeable for FA solutions at concentrations of 40 mg/L and higher.

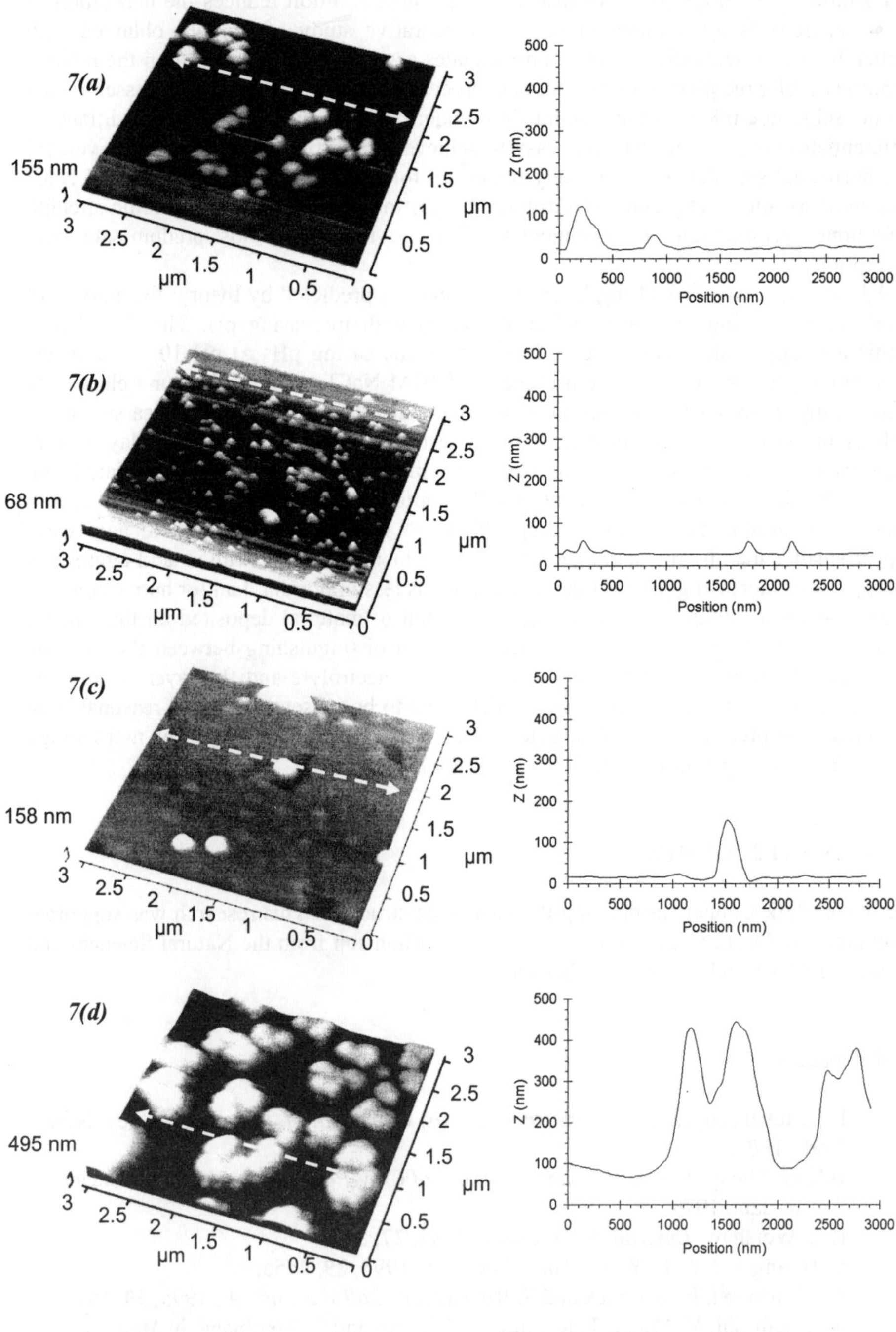

Figure 7 *3D images and line profiles of soil HA on the mica surface, HA conc. 4 mg/L. (a) at pH 1.5, (b) at pH 7, (c) at pH 10.5, (d) at pH 6.1, NaCl 0.01 M*

Positioning the mica plates vertically in the humic solution reduces the deposition of humic aggregates by sedimentation. The comparative study of samples obtained with vertically and horizontally positioned mica plates can provide information on the relative importance of precipitation-sedimentation processes versus adsorption processes in the humic substance-mica surface interaction under various environmental conditions, a differentiation that is not otherwise easy to achieve. The differences between the vertical and horizontal samples was especially important for experiments in which 0.01 M NaCl was used as the background electrolyte, suggesting that under higher ionic-strength conditions precipitation and sedimentation of the humic colloids predominate over adsorption.

Adsorption of soil HA (4 mg/L) on mica occurs as predicted by theory: the amount of humic acid covering the mica surface decreases with increasing pH. The disc-shaped humic acid aggregates also decrease in size with increasing pH. At pH 10.5, just a few large aggregates are formed. The presence of 0.01M NaCl as the background electrolyte dramatically increases the amount of humic material accumulated on the mica surface as well as the size of the aggregates. This suggests that ionic strength may play a more significant role than pH in determining both the amount of humic material associated with the clay fraction of soils and sediments and the conformation of the organic interface, with important consequences for the hydrophobicity of this surface. Fulvic acid aggregates appear smaller than humic acid aggregates. The dimensions of the fulvic acid aggregates decrease with increasing pH but the variation seems less important than for humic acid.

Even semi-quantitative estimation of the amount of material deposited on the mineral surface is a challenge, mainly due to difficulties in distinguishing between the intrinsic variability of the mica surface after treatment with electrolyte and flat layers of organic macromolecules (less than 4 nm high), which seem to be present and could reasonably be expected to be present on the mica surface. This is an important aspect of the morphology of clay-humic complexes and requires further work.

ACKNOWLEDGEMENTS

We thank Mark Cooper for drawing the muscovite structure. This research was supported from grants to Dr. F. C. Hawthorne and Dr. D. J. Thomson from the Natural Sciences and Engineering Research Council of Canada.

References

1. F.J. Stevenson, *Humus Chemistry. Genesis, Composition, Reactions*, Wiley, New York, 1982.
2. B.K.G. Theng, *Formation and Properties of Clay-Polymer Complexes*, Elsevier, Amsterdam, 1979.
3. R.L. Wershaw, *Environ. Sci Technol.*, 1993, **27**, 814.
4. C. Huang and Y.-L. Yang, *Water Research*, 1995, **29**, 2455.
5. A.J. Fairhurst, P. Warwick and S. Richardson, *Colloids Surf. A.*, 1995, **99**, 187.
6. Y. Takahashi, Y. Minai, T. Kumura, Y. Meguro and T. Tominaga, in *Material Research Society Symposium Proceedings*, 1995, p. 189.
7. R. Celis, L. Cox, M.C. Hermosin and J. Cornejo, *J. Environ. Qual.*, 1997, **26**, 472.
8. E. Barriuso, M. Schiavon, F. Andreux and J.M. Portal, *Chemosphere*, 1991, **22**,

1131.
9. J.J. Pignatello, F.J. Ferrandino and L.Q. Huang, *Environ. Sci. Technol.*, 1993, **27**, 1563.
10. A. Pusino, W. Liu and C. Gessa, *J. Agricult. Food Chem.*,1994, **42**, 1026.
11. E.M. Murphy, J.M. Zachara and S.C. Smith, *Environ. Sci. Technol.*, 1990, **24**, 1507.
12. E.M. Murphy, J.M. Zachara, S.C. Smith, J.L. Phillips and T.W. Wietsma, *Environ. Sci. Technol.*, 1994, **28**, 1291.
13. M. Rebhun, R. Kalabo, L. Grossman, J. Manka and C. Rav-Acha, *Water Research*, 1992, **26**, 79.
14. J.A. Davis and D.B. Kent, *Rev. Mineral.*, 1990, **23**, 178.
15. M. Plaschke, J. Romer, R. Klenze and J.I. Kim, *Coll. Surf. A*, 1999, **160**, 269.
16. A. Liu, R.C. Wu, E. Eschenazi and K. Papadopoulos, *Coll. Surf. A*, 2000, **174**, 245.
17. M. Shevchenko, Y.S. Yu, L.G. Akim, G.W. Bailey, *Holzforschung*, 1998, **52**, 149.
18. A. Liu and P.M. Huang, in *Understanding Humic Substances: Advanced Methods, Properties and Applications*, eds. E.A. Ghabbour and G. Davies, Royal Society of Chemistry, Cambridge, 1999, p. 87.
19. P. MacCarthy, R.L. Malcolm, M.H.B. Hayes, R.S. Swift, M. Schnitzer and W.L. Campbell, in *Transactions of the International Congress of Soil Science*, 13-th, Hamburg, West Germany, 1986, p. 378.
20. E. Tombacz, M. Gilde, I. Abraham and F. Szanto, *Appl. Clay Sci.*, 1990, **5**, 101.
21. A. Lemus, MSc Thesis, University of Manitoba, 1996.
22. D. Shimizu, MSc Thesis, University of Manitoba, 1998.
23. R. Wiesendanger, in *Scanning Probe Microscopy and Spectroscopy-Methods and Applications*, Cambridge University Press, New York, 1994, Chap. 2, p. 210.
24. *Image SXM is a version of NIH Image* (http://rsb.info.nih.gov/nih-image/) specifically adapted for scanning microscope images and can be obtained from http://reg.ssci.liv.ac.uk/.

THE INFLUENCE OF CATECHOL HUMIFICATION ON SURFACE PROPERTIES OF METAL OXIDES

C. Liu and P.M. Huang

Department of Soil Science, University of Saskatchewan, Saskatoon SK S7N 5A8, Canada

1 INTRODUCTION

Organic and inorganic components are closely associated in soil.[1] Products resulting from these interactions play an important role in affecting the physical, chemical and biological properties of soils and thus exert significant effects on soil and environmental quality. The binding of metal ions such as cadmium by mechanical mixtures of humic substances and soil minerals has been reported by Vermeer et al.[2] Their results indicate that the sum of the amounts of Cd ions bound by humic acid (HA) and by hematite is larger than the amount of Cd ions bound by their mechanical mixture due to the interaction between the negatively charged HA and the positively charged iron oxide. Furthermore, metal oxides often demonstrate the power to catalyze the transformation of natural organic compounds.[3] Metal oxides, especially manganese(IV) and iron(III) oxides, are very reactive in promoting the oxidative polymerization of phenolic compounds and the subsequent formation of humic substances.[3-6] However, the surface chemistry of the resultant metal oxide-humic complexes remains obscure.

The surface of metal oxide-humic complexes in soil is the region of their interactions with nutrients, heavy metals and organic pollutants. The surfaces of soil particles act as efficient sorbents for anions, cations and uncharged molecules. Therefore, the surface characteristics of the complexes can greatly affect the transformation and transport of nutrients and inorganic and organic pollutants in soils and associated environments. It has been reported that surface properties such as surface charge, surface area, surface morphology and surface geometry of Fe oxides formed under the influence of citrate significantly modified the rates and mechanisms of their binding of anions such as phosphate and cations such as lead.[7]

In this study, the influence of catechol humification on the surface properties, including the specific surface, surface charge, point of zero salt effect (PZSE) and fine-scale surface features of synthetic Al, Fe and Mn oxides and the residence time effect of the interaction on the surface properties were investigated.

2 MATERIALS AND METHODS

2.1 Preparation of Metal Oxides

The chemicals used in this study were reagent grade. The deionized distilled water was autoclaved. Aluminum oxide was synthesized by titrating 800 mL of 6.25 x 10^{-3} M $AlCl_3$ solution with 0.2 M NaOH solution under vigorous stirring until the pH was 8.2. The rate of titration was 60 mL h^{-1}. The solutions were then made up to 1000 mL so that the final Al concentration was 5 x 10^{-3} M. The suspensions were aged for 5 d at 23.5 ± 0.5°C in a polypropylene container. The precipitate was dialyzed against deionized distilled water until Cl^- free and then freeze-dried. The product was a mixture of gibbsite and bayerite (see Results and Discussion).

Iron(III) oxide was synthesized by titrating 0.15 M $Fe(NO_3)_3$ solution with 13% aqueous ammonia to pH 5.5. The suspensions were aged for 2 d at 23.5 ± 0.5°C, dialyzed against deionized distilled water and then freeze-dried. The iron(III) oxide product could not be identified because it was amorphous (see Results and Discussion).

Manganese oxide was synthesized by the method of McKenzie.[8] After preparation, the precipitates were dialyzed against deionized distilled water and then freeze-dried. The synthesized Mn oxide was birnessite (see Results and Discussion).

2.2 Preparation of the Reaction Systems

Five hundred milligrams of freeze-dried Al, Fe or Mn oxide was added to 85 mL autoclaved deionized distilled water. The pH of the suspensions was adjusted to 6.0 with 0.1 M NaOH. The volume of the suspensions was brought to 90 mL with autoclaved deionized distilled water. Ten milliliters of 1 M catechol stock solution (pH 6.0) were mixed with the metal oxide suspension. Twenty milligrams of thimerosal (sodium ethylmercurithiosalicylate) was immediately added to the suspensions to inhibit the growth of microorganisms. The final thimerosal concentration was 0.02% (w/v). The final concentrations of catechol and metal oxide were 0.1 M and 5 g L^{-1}, respectively. Catechol solution alone as the control was also prepared by the same procedure except that no metal oxides were added. The suspensions were, respectively, shaken for 1, 5, 10 and 20 d at 25.5°C and then centrifuged for 25 min at 20000 *g*. The solid reaction products were dialyzed against deionized distilled water and then freeze-dried.

2.3 Characterization of the Reaction Products

The supernatants of the reaction systems were analyzed by UV-visible spectroscopy on a Beckman Model DU 650 spectrophotometer and Fourier transform infrared spectroscopy (FTIR) on a Bio-Rad Model 3240 infrared absorption spectrophotometer using a liquid cell. The pure metal oxides and the solid reaction products were also examined by FTIR by the KBr pellet technique. One milligram of the sample was mixed with 200 mg of KBr for the FTIR analysis. Twenty milligram samples of the pure metal oxides and the solid reaction products were examined by X-ray powder diffraction (XRD) analysis on a Rigaku diffractometer with Fe-Kα radiation filtered by a graphite monochromator at 40 kV and 130 mA. The X-ray diffractograms were recorded from 4° to 60°2θ with 0.005°2θ steps at a speed of 0.5°2θ per min.

The C contents of the pure metal oxides and the solid reaction products were determined by mass spectrometry with a Europa Scientific 20/20 mass spectrometer coupled to a gas/solid/liquid preparation module. The contents of Al, Fe and Mn in the pure metal oxides and the solid reaction products were determined by atomic absorption spectroscopy (AAS) after the sample was digested with a HNO_3-$HClO_4$-H_2SO_4 mixture. The ash content of the solid product was determined by heating the sample in a muffle furnace at 600°C for 4 h.[9]

2.4 Surface Properties of the Reaction Products

For atomic force microscopy (AFM) analysis, 5 mg of the pure metal oxides or the solid reaction products were dispersed in 15 mL deionized distilled water by ultrasonification (Sonifier Model 350) at 150 watts for 2 min in an ice bath. One drop of the suspension was deposited on a watch glass and air-dried overnight at room temperature (23.5 ± 0.5°C). The watch glass was then fastened to a magnetized stainless steel disk (diameter 12 mm) with double-sided tape. The 3-dimensional AFM images were obtained under ambient conditions with a NanoScopeTMIII atomic force microscope (Digital Instruments). The imaging areas were 2 µm x 2 µm. The scanner type was 1881E and the scanner size was 15 µm. A silicon nitride cantilever with a spring constant of 0.12 N/m was used in the contact mode. The scanning rate was 22 Hz. The AFM cantilever was changed frequently to avoid experimental artifacts. Furthermore, the scanning area and scanning angle were often changed by entering different area and angle parameters to detect artifacts caused by adhesion of sample particles to AFM tips.

The specific surface area of the solid samples was measured from a 5 point BET N_2 adsorption isotherm[10] obtained with a Quantachrome Autosorb-1 apparatus. Prior to N_2 adsorption, the 100 mg samples were outgassed for 24 h at 10 mTorr and room temperature (23.5 ± 0.5°C). During N_2 adsorption the solids were thermostated in liquid N_2 (-195 to -196°C). The point of zero salt effect (PZSE) and the net surface charge of the solid samples were determined in 0.01, 0.1 and 1 M NaCl solutions by the potentiometric method of Parks and de Bruyn[11] as modified by Atkinson et al.[12] The automatic titration as described by Sakurai et al.[13] was conducted with a Metrohm Model 682 titroprocessor.

3 RESULTS AND DISCUSSION

Compared with catechol solution alone, the darkening of the supernatant of the Al, Fe and Mn oxide-catechol systems was enhanced to various extents. The supernatant absorption spectra in the wavelength range 350 to 600 nm of the metal oxide-catechol systems at the end of different reaction periods (Figure 1) indicate that the formation of solution products of catechol humification increased steadily with increasing reaction period in all three metal oxide-catechol systems. The formation of highly colored solution products in the Mn oxide-catechol system was much faster than with the Fe oxide-catechol system, which was in turn faster than the Al oxide-catechol system (Figure 1). The color of the suspension of the Mn oxide-catechol system changed to dark brown immediately after mixing the Mn oxide with catechol. A dark solution was observed in the Fe oxide-catechol system after reaction for 5 d. However, the catalytic power of Al oxides was much less. After reaction for 5 d, the color of Al oxide particles changed to dark green but the supernatant did not significantly change. Darkening of the supernatant of the Al oxide-catechol system was observed after reaction for 10 d and was greatly enhanced after reaction for 20 d.

The difference FTIR spectra between the catechol solution and the supernatant of the metal oxide-catechol systems (Figure 2) revealed oxidation and ring cleavage of catechol catalyzed by metal oxides as indicated by the absorption bands at 1631-1652 cm^{-1}. These absorption bands are due to vibration of carboxylate groups $-COO^-$ and aromatic C=C

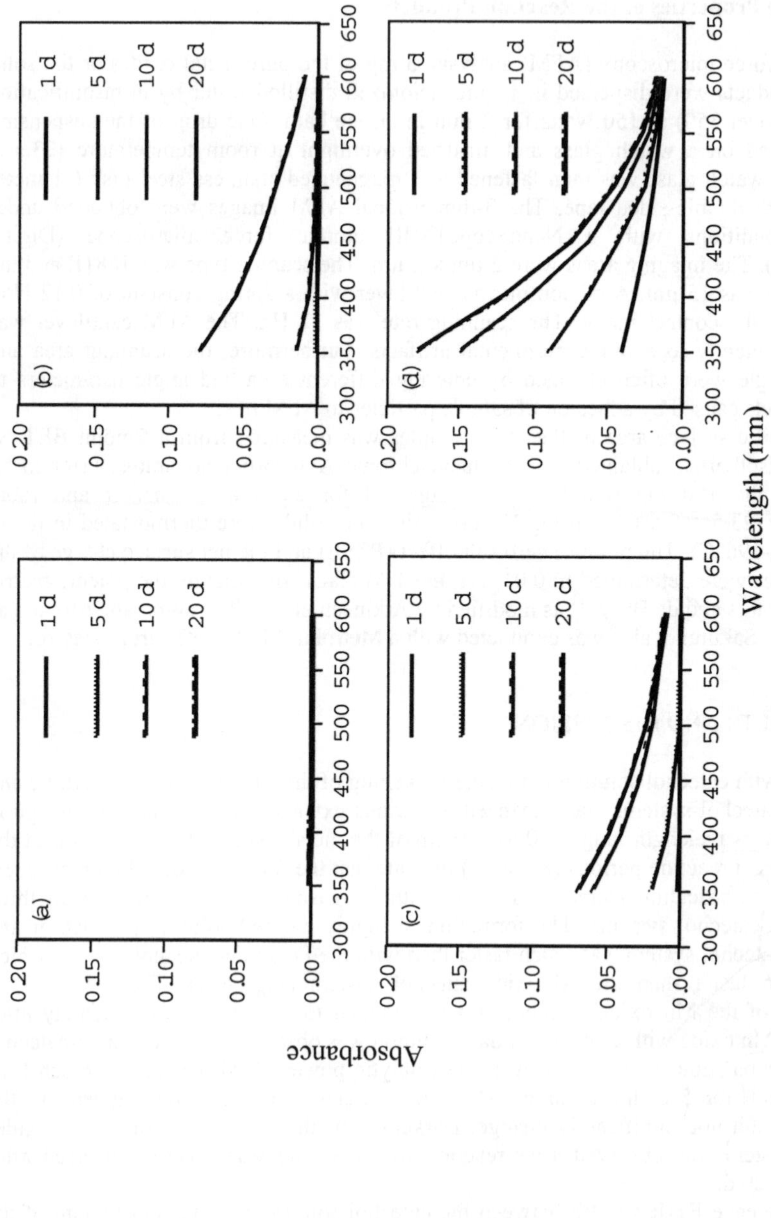

Figure 1 *UV-visible spectra of (a) catechol solution and the supernatants of (b) Al oxide-, (c) Fe oxide- and (d) Mn oxide-catechol systems*

double bonds conjugated with C=O groups.[14] The difference FTIR data also show that the Mn oxide was much more powerful in catalyzing the ring cleavage and oxidation of catechol than Fe and Al oxides. This is consistent with previous reports.[3,4,15] A very recent paper by Matocha et al.[16] also reported that the rate of reductive dissolution of birnessite by catechol is very rapid. The evolution of CO_2 in bubbles observed when catechol was mixed with the Mn oxide also provided evidence for the oxidation of catechol catalyzed by Mn oxide. The metal concentration in the supernatant of the metal oxide-catechol systems in the present study (Table 1) shows that much more Mn oxide was dissolved than for Fe and Al oxides, resulting in more Mn ions released into the solution. Furthermore, more Fe ions from the Fe oxide were released into solution than Al ions from the Al oxide.

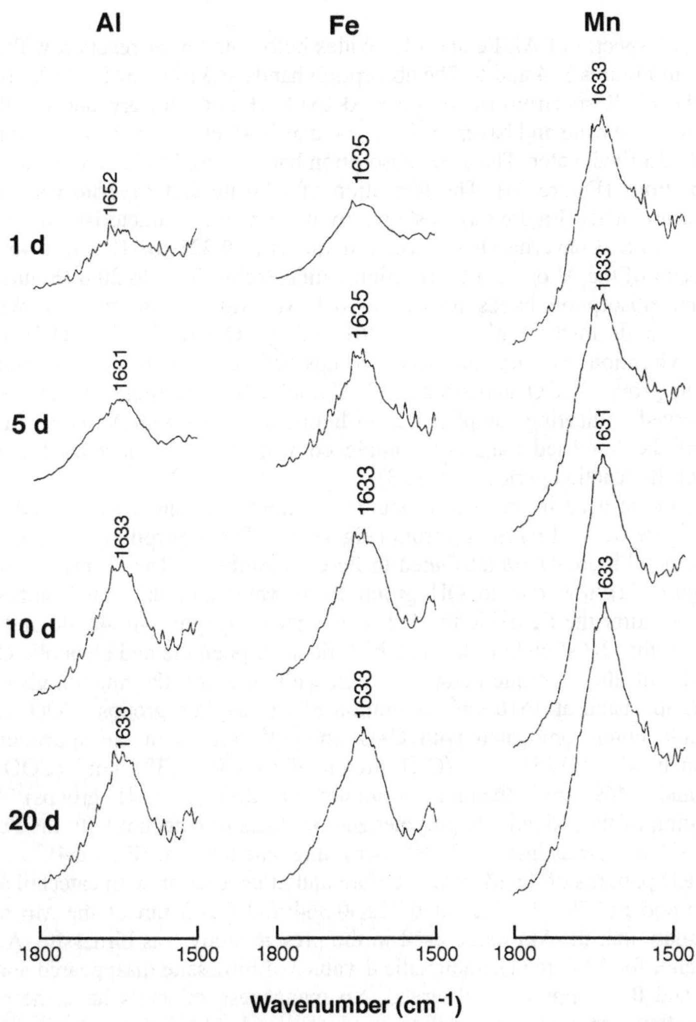

Figure 2 *Difference FTIR spectra between catechol solution and the supernatants of the catechol-metal oxide systems after different reaction periods*

Table 1 *Elemental contents of the supernatant after reaction of metal oxides with catechol for 20 days*

	Al	*Fe*	*Mn*
		mg L-1	
Al oxide	7	nd[a]	nd
Fe oxide	nd	49	nd
Mn oxide	nd	nd	159

[a] Not detectable

The FTIR spectra of Al, Fe and Mn oxides before and after reaction with catechol are presented in Figures 3, 4 and 5. The absorption bands at 3460, 3521, 3559, 3619 and 3664 cm^{-1} in the FTIR spectrum of the pure Al oxide (Figure 3a) are due to OH vibrations possibly from gibbsite and bayerite.[17] The band at 1646 cm^{-1} was from the vibration of OH groups of adsorbed water. The other absorption bands at 527 to 1026 cm^{-1} are attributed to Al-O vibrations (Figure 3a). The formation of gibbsite and bayerite was confirmed by XRD measurements (Figure 6a) as shown by the gibbsite characteristic d-values at 0.436 and 0.483 nm and bayerite characteristic d-values at 0.221, 0.318 and 0.469 nm. In the FTIR spectra of the Al oxide after reaction with catechol for 1 to 20 d (Figure 3b and 3e), besides the absorption bands present in the Al oxide before reaction with catechol, absorption bands in the region from 1261 cm^{-1} (C-O stretch of COOH) to 1495 cm^{-1} (bending vibration of aliphatic C-H groups)[14,18] and 1610 cm^{-1} (vibration of the carboxylate groups $-COO^-$ and aromatic C=C double bond conjugated with C=O groups)[14] were observed, indicating complexation of humic materials with Al oxide. In general, the amount of the humified materials complexed with Al oxide increased gradually with increase of the reaction period (Figure 3).

The Fe oxide used in the present study was noncrystalline, as indicated by the XRD pattern (Figure 6c) and FTIR spectrum (Figure 4a). The absorption bands at 444-452 and 573-580 cm^{-1} (Figure 4) are attributed to Fe-O vibrations.[17] The absorption band at 3294 cm^{-1} (Figure 4a) was due to OH group from water and this band shifted to higher wavenumber after the Fe oxide reacted with catechol (Figure 4b, 4c, 4d and 4e). This is attributed to the OH stretching from carboxylic acid, phenolic and alcoholic OH groups.[14] Compared with the Fe oxide before reaction with catechol, the much higher intensity of the absorption band at 1616 cm^{-1} (vibration of carboxylate groups $-COO^-$ and aromatic C=C double bond conjugated with C=O groups) along with the appearance of other absorption bands at 1253 cm^{-1} (C-O stretch of COOH), 1381 cm^{-1} (COO^- symmetric stretch) and 1469 cm^{-1} (bending vibration of aliphatic C-H groups)[14,18] indicates complexation of the oxidatively polymerized products of catechol with the Fe oxide. The intensity of those bands increased with increasing reaction time (Figure 4).

The XRD patterns of the Mn oxide before and after reaction with catechol are shown in Figure 6e and 6f. The d-values at 0.722, 0.364 and 0.243 nm of the Mn oxide before reaction show that the Mn oxide used in the present study was birnessite. After reaction with catechol for 20 d, the characteristic d-values of birnessite disappeared and new peaks at 0.665 and 0.283 nm were observed. No manganese minerals have these d-values,[19] indicating that some new crystalline products, probably organic compounds such as carbohydrates, were formed after reaction for 20 d. The d-values at 0.67 and 0.28 nm have been observed for starch and other carbohydrates.[20] The FTIR spectra of birnessite before and after reaction with catechol were dramatically different (Figure 5). The absorption bands at 512 and 527 cm^{-1} are due to Mn-O vibrations.[17] The bands at 3440-3447 cm^{-1}

resulted from vibration of OH of Mn oxide[17] and the OH stretching of carboxylic acid, phenolic and alcoholic OH groups.[14] Very strong absorption bands in the range from 1261 cm^{-1} (C-O stretch of COOH) to 1480 cm^{-1} (vibration of aliphatic C-H groups) and 1566 cm^{-1} (asymmetric stretch of COO- groups) of the Mn oxides after reaction with catechol were observed even after a 1-d reaction time. The intensity of the bands increased with increasing reaction time. An absorption band at 860 cm^{-1}, which was not observed in the Fe or Al oxide-catechol system, appeared in the Mn oxide-catechol system. It may be attributed to carbohydrates,[21] which is in accord with the XRD data (Figure 6f). Its intensity increased with increasing reaction time. The appearance of this band in the Mn oxide-catechol system along with the very strong absorption bands at 1261 to 1566 cm-1 indicate much stronger catalytic ability of Mn oxide than Fe and Al oxides.

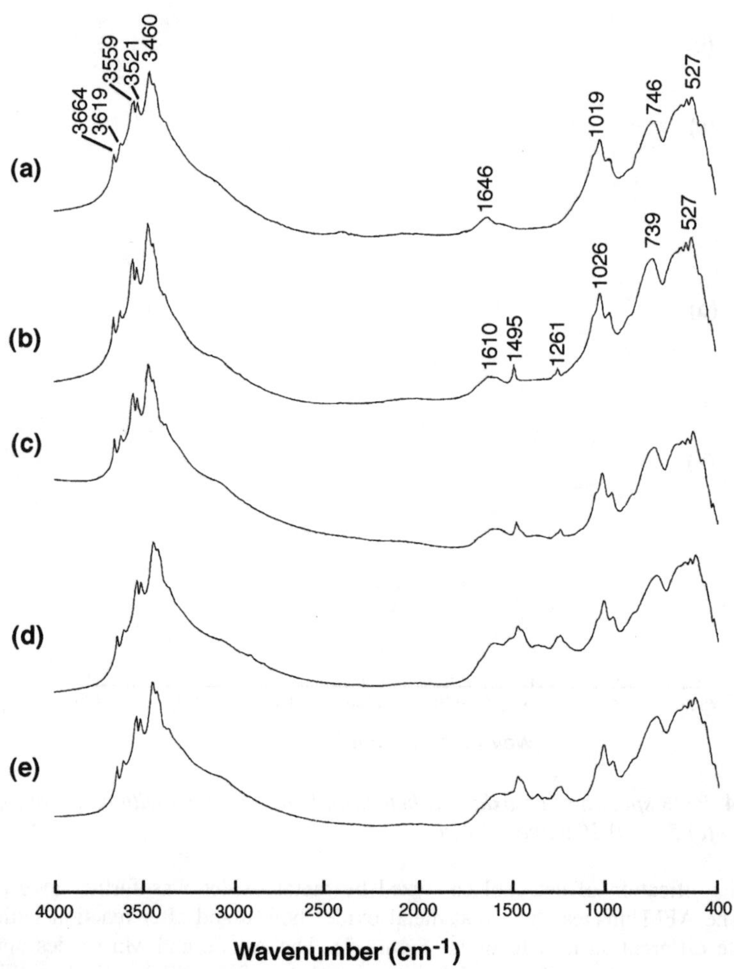

Figure 3 *FTIR spectra of Al oxide (a) before and after reaction with catechol for (b) 1 d, (c) 5 d, (d) 10 d and (e) 20d*

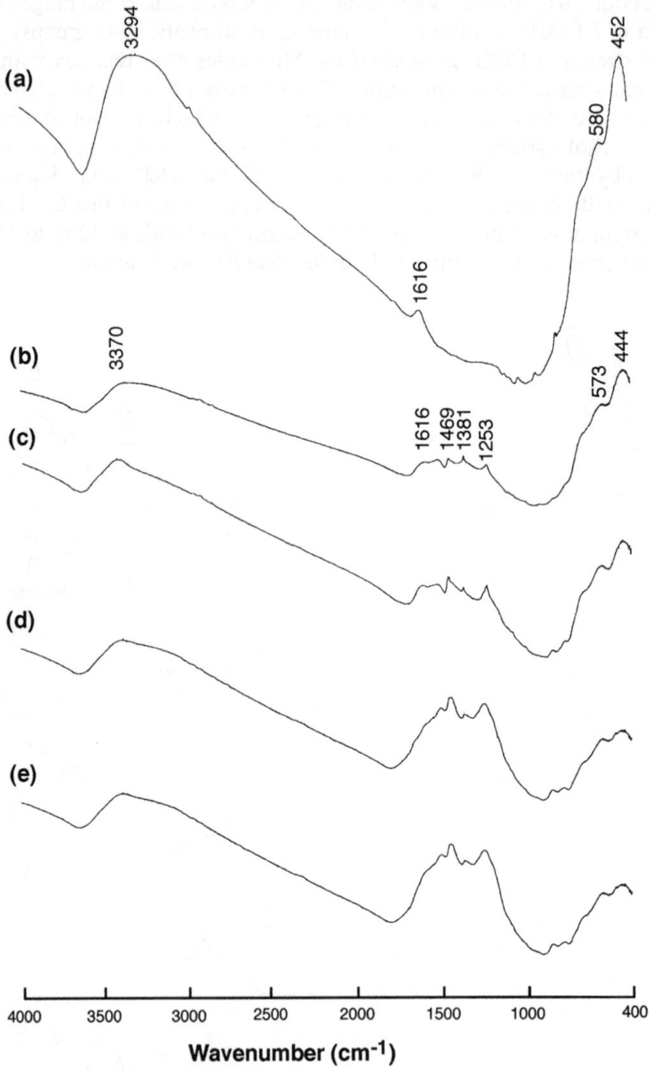

Figure 4 *FTIR spectra of Fe oxide (a) before and after reaction with catechol for (b) 1 d, (c) 5 d, (d) 10 d and (e) 20d*

The humification of catechol catalyzed by metal oxides was further investigated by AFM. The AFM images show that metal oxides before and after reaction with catechol had quite different surface features (Figure 7). The Al, Fe and Mn oxides appeared as layer or irregular shape (Figure 7a), irregular shape (Figure 7c) and plate (Figure 7e), respectively. Spheroidal particles were observed in all three of the metal oxides after reaction with catechol for 20 d (Figure 7b, 7d and 7f). These spheroidal particles were humic materials formed from catechol and/or metal (oxide)-humic complexes. Our previous research showed that the IHSS standard soil HA has a spheroidal shape.[22,23]

The C contents of the solid reaction products in the metal oxide-catechol systems (Table 2) show that more organic components were complexed with oxides in the Mn oxide-catechol system compared with Fe and Al oxide-catechol systems. This may explain the higher intensity of FTIR absorption bands in the region from 1261 to 1566 cm^{-1} observed in the solid reaction products of the Mn oxide-catechol system compared with Fe and Al oxide-catechol systems (Figures 3, 4 and 5). The ash-free C content of the solid reaction products in the metal oxide-catechol systems ranged from 559 to 613 g C kg^{-1}, which is different from the C content of catechol (654.5 g C kg^{-1}) and is similar to the C content of soil HAs. The C content of soil HAs[24,25] ranges from 537 to 604 g C kg^{-1} (the C content of the IHSS standard soil HA was 581.3 g C kg^{-1}).

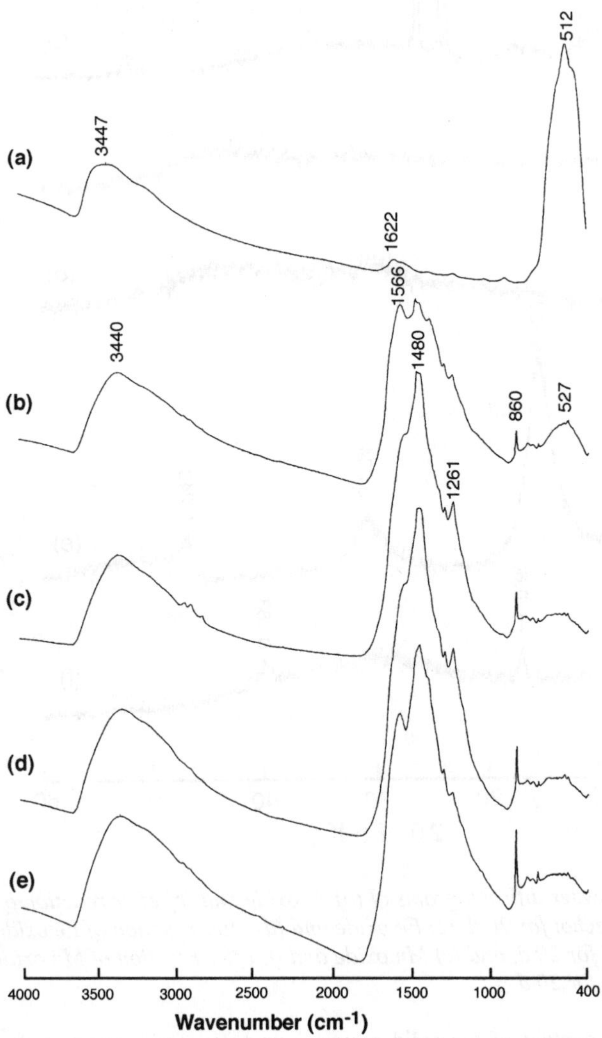

Figure 5 *FTIR spectra of Mn oxide (a) before and after reaction with catechol for (b) 1 d, (c) 5 d, (d) 10 d and (e) 20d*

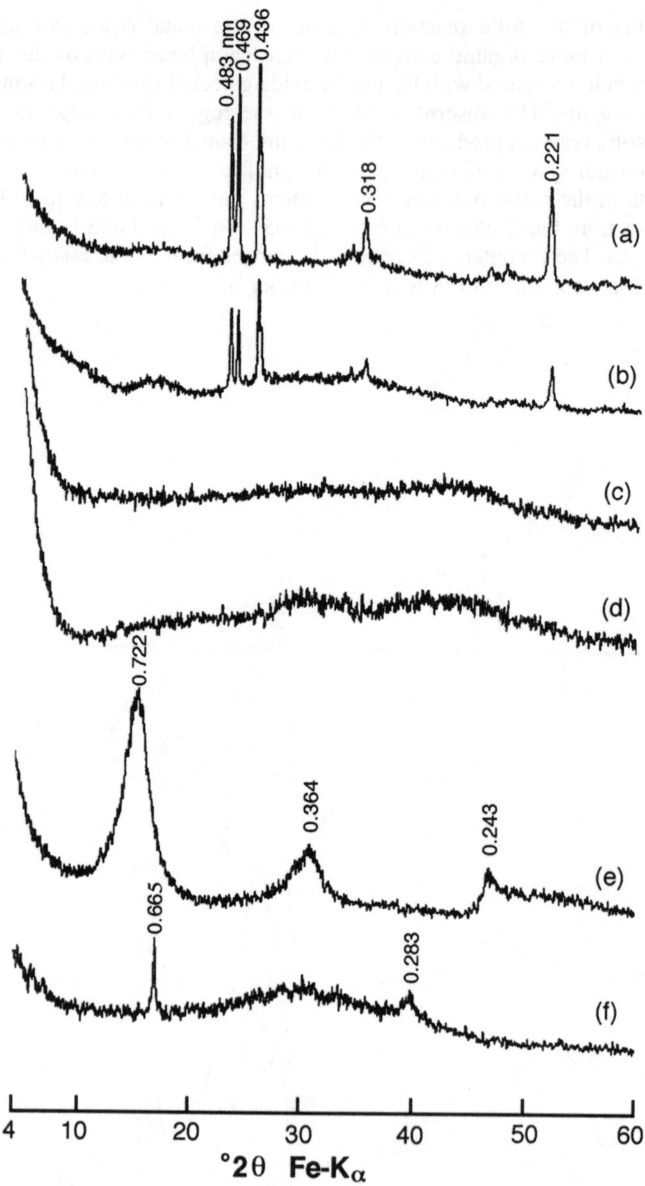

Figure 6 *X-ray powder diffractograms of (a) Al oxide and (b) after reaction of Al oxide with catechol for 20 d, (c) Fe oxide and (d) after reaction of Fe oxide with catechol for 20 d, and (e) Mn oxide and (f) after reaction of Mn oxide with catechol for 20 d*

The ash-free C content of the solid reaction products was highest in the Mn oxide-catechol system, intermediate in the Fe oxide-catechol system and lowest in the Al oxide-catechol system. The lower Al, Fe and Mn contents in the solid reaction products of the metal oxide-catechol systems than those of the metal oxides before reaction with catechol are due to dissolution of some metal oxide particles and the incorporation of humified materials. The ash content of the solid reaction products in the metal oxide-catechol systems was in the order Mn oxide < Fe oxide < Al oxide. These data along with UV-visible spectra, FTIR and AFM evidence reveal that the degree of catechol humification in the metal oxide systems was in the order Mn oxide > Fe oxide > Al oxide.

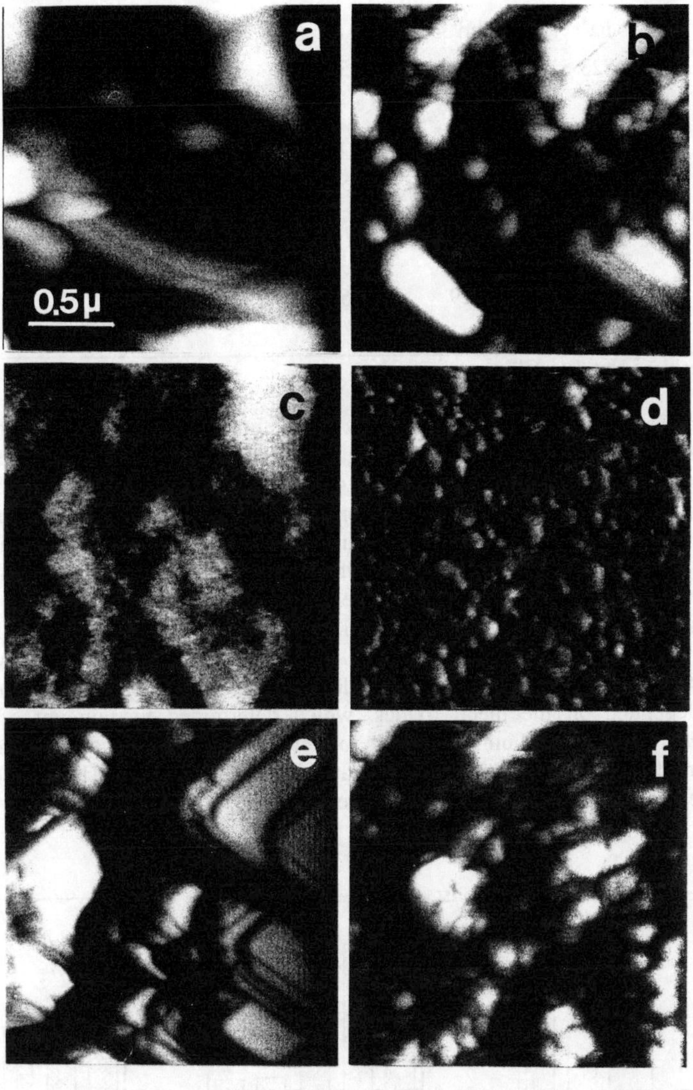

Figure 7 *Atomic force micrographs of (a) the Al oxide and (b) the solid phase product of Al oxide after reaction with catechol for 20 d, (c) the Fe oxide and (d) the solid phase product of Fe oxide after reaction with catechol for 20 d, and (e) the Mn oxide and (f) the solid phase product of Mn oxide after reaction with catechol for 20 d. The image scale is 0.5 μ*

Table 2 *Elemental contents of metal oxide before and after reaction with catechol for 20 days*

	Ash	*C*	*Ash-free C*	*Al*	*Fe*	*Mn*
			g kg^{-1}			
Al oxide						
before	na[a]	nd[b]	na	312	nd	nd
after	823	99	559	278	nd	nd
Fe oxide						
before	na	nd	na	nd	438	nd
after	758	142	589	nd	357	nd
Mn oxide						
before	na	nd	na	nd	nd	282
after	463	329	613	nd	nd	72

[a] Not applicable; [b] Not detectable

The specific surface area of the metal oxides before and after reaction with catechol was estimated with the N_2-BET method (Figure 8). The Fe oxide had the highest specific surface area due to its noncrystalline nature, whereas the Al oxide had the lowest specific surface area because crystalline gibbsite and bayerite were present (Figure 6). Only mesopores with diameter between 2 and 50 nm were observed in the Al and Mn oxides. In addition to mesopores, some micropores with diameter less than 2 nm were observed in the Fe oxide (Figure 8). The specific surface area of the metal oxides steadily decreased with increasing of reaction time. The specific surface area of the Al, Fe and Mn oxide-humic complexes formed at the end of a 20-d reaction period, respectively, were 49%, 93% and 66% of the respective metal oxides. The extent of the specific surface area decrease depended on the metal oxides, which determined the amount of humified organics incorporated. The humic materials formed from catechol appeared to enhance the aggregation of metal oxide particles, as indicated by the AFM images (Figure 7), resulting in decrease of the specific surface area (Figure 8).

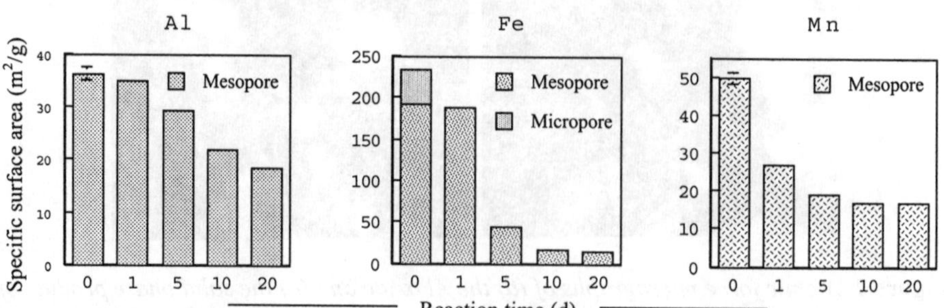

Figure 8 *Specific surface area of metal oxides before and after reaction with catechol. The standard errors, which are too small to be drawn on the bar diagram, are not shown*

It is interesting to note that regardless of the specific surface area of metal oxides before reaction with catechol, the specific surface area of the solid reaction products in the three metal oxide-catechol systems was about 20 m^2 kg^{-1}. Although micropores were present in the Fe oxide before reaction with catechol, they were not observed in the Fe oxide-humic complexes. The N_2-BET surface area of organic matter largely depends on sample pretreatment, i.e., whether it is oven-dried, air-dried or lyophilized.[26] It should also be noted that N_2 is subject to molecular sieving at -196°C due to activated diffusion in micropores < ~1 nm. Therefore, the N_2-BET method may severely underestimate the total specific surface area of organic matter or oxide-organic complexes.

One of the most important properties of Al and Fe oxides is their surface structure and the resultant dependence of surface charge on pH. In the presence of water, the surface of Al and Fe oxides becomes completely hydroxylated and/or hydrated and this particular surface is very reactive.[27,28] The PZSE of the metal oxides before and after reaction with catechol are shown in Figure 9. The PZSE values of the Al, Fe and Mn oxides before reaction with catechol were, respectively, 7.5, 7.8 and 3.2. The PZSE value of metal oxides is directly related to the pK values of M-OH and M-OH_2 surface functional groups. The pK values of Fe-OH_2 and Fe-OH of Fe oxides and those of Al-OH_2 and Al-OH of Al oxides are about 4.3, 9.7, 7.0 and 8.8, respectively.[29,30] The pK values for the Mn-OH and Mn-OH_2 groups of Mn(IV) oxide are not available.[31] The PZSE of a metal oxide generally is half of the sum of pK values of M-OH_2 and M-OH when the metal oxides is present in a simple electrolyte system such as $NaNO_3$.[32] The high PZSE values of Fe and Al oxides revealed that more OH_2 groups (which carry positive charges) than OH groups (which carry negative charges) were present on their surfaces at pH 6, whereas the Mn oxide had predominantly OH groups on its surface at the same pH. Thus, the higher PZSE of Al and Fe oxides than that of Mn oxides indicates that more positive charges were present on Al and Fe oxides than Mn oxides at the same pH. This is shown by the net surface charges of the metal oxides at different pH values in Figure 10.

The PZSE of Al and Fe oxides decreased after they reacted with catechol, whereas the PZSE of Mn oxide increased after reaction with catechol. This is attributed to the complexation of these metal oxides with humic materials formed from catechol. The apparent mechanisms of alteration of surface properties of metal oxides caused by catechol humification are illustrated in Figure 11. The humic macromolecules complex with the surfaces of Fe and Al oxides by replacing OH groups and especially dominant OH_2 groups with their COOH or OH functional groups, thus partially neutralizing the

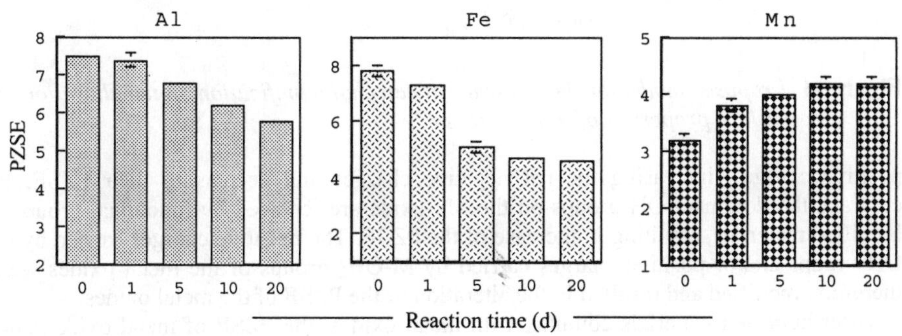

Figure 9 *Point of zero salt effect of metal oxides before and after reaction with catechol. The standard errors, which are too small to be drawn on the bar diagrams, are not shown*

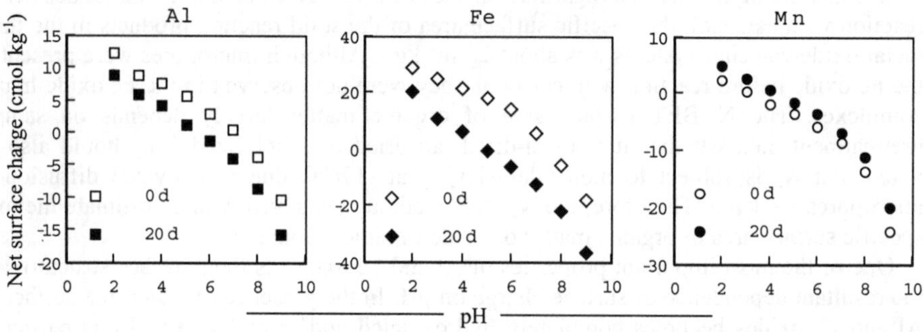

Figure 10 *Net surface charge of metal oxides before and after reaction with catechol. The precision of the determination of net surface charge is within the size of the symbols*

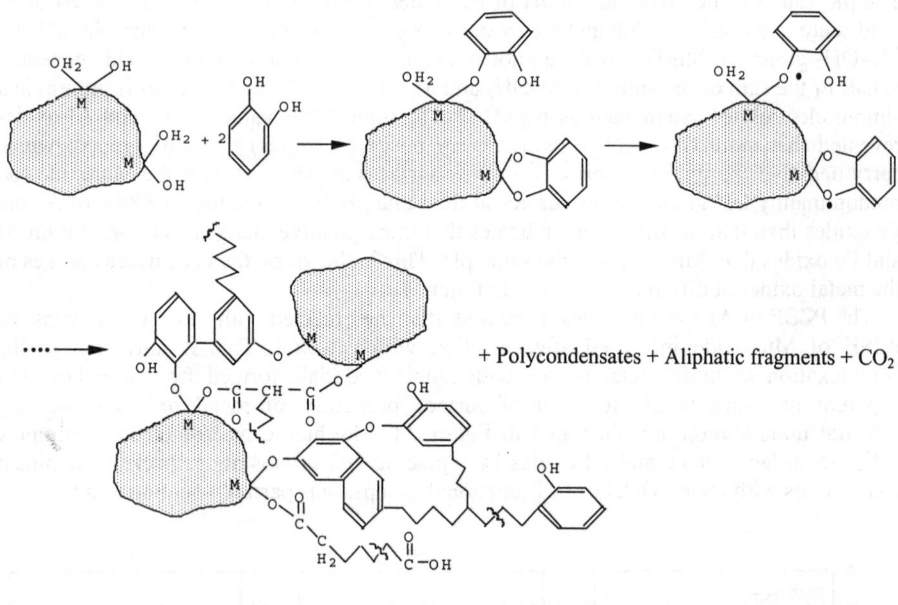

Figure 11 *Proposed model for the influence of catechol humification on the alteration of surface properties of metal oxides*

positive charges, increasing the net negative charges and decreasing their PZSE. By contrast, the dominant OH groups on the Mn oxide are replaced by functional groups of humified materials, resulting in increase of the PZSE. The negative charges carried by M-OH groups and/or positive charges carried by M-OH_2 groups of the metal oxides were, therefore, modified and resulted in the alteration of the PZSE of the metal oxides.

After humified materials complex with metal oxides, the PZSE of metal oxide-humic complexes can be further influenced by the pK values of functional groups such as COOH and phenolic OH of the humic materials. The mean pK values of COOH and phenolic OH groups of proton-humate complexes are about 4.5 and 10, respectively.[33-35] The substitution of a group on the complexing organics may influence the acidity of the donor groups, which is caused by relative electron repulsion or attraction of the group.[36] Therefore, the pK values of COOH and phenolic OH groups of proton-humate complexes may range from 1 to 8 and from 6 to 14, respectively.[35] When the pH value of the system (such as 6) is higher than the pK value of the COOH groups, more than half of their protons would be dissociated, leading to the development of more negative charges on the surfaces, resulting in the alteration of the PZSE. This pK value effect on the PZSE should greatly depend on the nature and amount of humic materials complexed with metal oxides and the amount of original M-OH_2 and M-OH groups left on the metal oxide surfaces. If most of the original M-OH_2 and M-OH groups on the metal surfaces are replaced by the functional groups of humic materials, the PZSE of the metal oxide-humic complexes should be near the pK values of the functional groups of the humic materials. This can explain why the PZSE values of the Fe oxide- and Mn oxide-complexes did not increase with increase of reaction time from 10 d to 20 d. The PZSE values of the Fe oxide-humic complex formed after 10- and 20-d reaction times were 4.7 and 4.6, respectively, while the Mn oxide-humic complex formed after 10- and 20-d reaction times had the identical PZSE value, i.e., 4.2 (Figure 9). These PZSE values are very close to the pK values of COOH groups of humic materials. The proposed model (Figure 11) also explains enhancement of the aggregation of metal oxides by catechol humification after reaction with catechol.

The catalytic effectiveness of a metal ion depends upon its ability to complex with ligands and to shift electron density and molecular conformation in ways favorable for the reaction.[3,37] The Mn oxide had the highest catalytic power of the three metal oxides. This is probably because of the lower electronegativity of Mn. The electronegativity values of Mn, Fe, Al, H and O are, respectively, 1.55, 1.83, 1.61, 2.20 and 3.44.[38] Catechol acts as a hard Lewis base and Al, Fe and Mn are hard Lewis acids. When Al, Fe or Mn replaces H in catechol to form metal-catechol complexes, the electron cloud delocalizes from phenolic oxygen into the π-orbital formed from overlap of the 2p orbitals of the aromatic C atoms, thus accelerating the formation of semiquinone free radicals and their coupling to form polycondensates. The semiquinone free radicals formed appeared to be partially transformed, through ring cleavage, to aliphatic fragments, resulting in the development of carboxyl groups and subsequent decarboxylation and CO_2 release (Figure 11). The electron cloud around the Mn-O bond in the Mn oxide-phenolic complex should be more delocalized than that around the Al-O bond in the Al-catechol complex and especially the Fe-O bond in Fe oxide-phenolic complex due to the lower electronegativity of Mn than those of Al and Fe. Consequently, the accelerating effect of Mn oxide on the humification of catechol was much stronger than Fe and Al oxides, as observed in the present study.

Redox reactions play an important role in many abiotic catalyses.[3] Aluminum oxide is not subject to redox reaction. The standard electrode potential (E^0) values of the overall redox reaction in the Fe(III) oxide-catechol and Mn(IV) oxide-catechol systems are +0.071 V and +0.509 V, respectively, as indicated by the following reactions:[4]

$$Fe^{3+} + e^- = Fe^{2+} \qquad E^0 = 0.770\ V \qquad (1)$$

$$MnO_2 + 4H^+ + 2e^- = Mn^{2+} + 2H_2O \qquad E^0 = 1.208\ V \qquad (2)$$

$$C_6H_4(OH)_2 = C_6H_4O_2 + 2H^+ + 2e^- \qquad E^0 = -0.6992\ V \qquad (3)$$

The positive E^0 values of the overall redox reactions indicate that the reactions are thermodynamically feasible and catechol oxidation can thus be accelerated by Fe oxide and especially Mn oxide. When the Mn oxide and catechol solution were mixed together, the Mn oxide easily oxidized catechol as indicated by the evolution of CO_2. This also explains the stronger catalytic ability of the Mn oxide than Fe and Al oxides in accelerating catechol oxidation.

In addition, the lower PZSE and more negative charges of the Mn oxides than the Fe and Al oxides could also enhance the oxidation of catechol. More negative charges of the Mn oxide may favor the binding of protons released from catechol (Equn. 3) and subsequently increase the catalytic reaction rate. Therefore, the catalytic ability of a metal oxide in catechol transformation depends upon the ability of the metal ions to complex with ligands, to shift electron density and molecular conformation and to favor the binding of protons.

4 CONCLUSIONS

The results obtained in this study show that black solid products were formed in all the metal oxide-catechol systems and that the extent of formation of black products in the metal oxide-catechol systems was in the order Mn oxide > Fe oxide > Al oxide. The surface properties of the metal oxides were substantially modified by catechol humification. The specific surface area of the Al, Fe and Mn oxide-humic complexes formed at the end of a 20-d reaction period, respectively, decreased by 49%, 93% and 66% compared with the metal oxides. This is due to aggregation of the metal oxide-humic complexes. Catechol humification substantially decreased the PZSE of Al and Fe oxides and, by contrast, increased that of Mn oxide apparently through two mechanisms: (1) replacement of some of the bound H_2O from $M\text{-}OH_2$ and/or the bound OH from M-OH groups on the oxide surfaces by functional groups such as COOH and phenolic OH of humic substances; and (2) dissociation of protons from the functional groups of the humic substances and the metal oxides. Compared with the respective metal oxides, the Al and Fe oxide-humic complexes had higher net negative charges and Mn oxide-humic complexes had lower net negative charges at the same pH. The results of this study are of fundamental significance in understanding the influence of humification on the surface properties of naturally occurring metal oxides and the potential impact on the transformation and transport of nutrients and inorganic and organic pollutants in soils and associated environments.

ACKNOWLEDGEMENTS

This research was supported by Research Grant GP2383- and Equipment Grant EQP156628-Huang of the Natural Sciences and Engineering Research Council of Canada.

References

1. M. Schnitzer, in *Environmental Impact of Soil Component Interactions. Vol. 1. Natural and Anthropogenic Organics*, eds. P.M. Huang, J. Berthelin, J.-M. Bollag, W.B. McGill and A.L. Page, CRC/Lewis Publishers, Boca Raton, FL, 1995, p. 3.
2. A.W. Vermeer, J.K. Mcculloch and L.K. Koopal, *Environ. Sci. Technol.,* 1999, **33,** 3892.
3. P.M. Huang, in *Handbook of Soil Science*, ed. M.E. Sumner, CRC Press, Boca Raton, FL, 2000, p. B303.
4. H. Shindo and P.M. Huang, *Soil Sci. Soc. Am. J.,* 1984, **48,** 927.
5. M.C. Wang and P.M. Huang, *Soil Sci.*, 2000, **165**, 737.

6. M.C. Wang and P.M. Huang, *Soil Sci.*, 2000, **165**, 934.
7. C. Liu and P.M. Huang, *Soil Sci. Soc. Am. J.*, 1999, **63**, 65.
8. R.M. McKenzie, *Mineral Mag.*, 1971, **38,** 493.
9. C.M. Preston and R.H. Newman, *Can. J. Soil Sci.,* 1992, **72,** 13.
10. S.J. Gregg and K.S.W. Sing, *Adsorption, Surface Area and Porosity*, 2nd Edn., Academic Press, London, 1982.
11. G.A. Parks and P.L. de Bruyn, *J. Phys. Chem.*, 1962, **66**, 967.
12. R.J. Atkinson, A.M. Posner and J.P. Quirk, *J. Phys. Chem.*, 1967, **71**, 550.
13. K. Sakurai, Y. Ohdate and K. Kyuma, *Soil Sci. Plant Nutr.,* 1989, **35**, 89.
14. P. MacCarthy and J.A. Rice, in *Humic Substances in Soil, Sediment, and Water*, eds. G.R. Aiken, D.M. McKnight, R.L. Wershaw and P. MacCarthy, Wiley, New York, 1985, p. 527.
15. M.C. Wang and P.M. Huang, *Sci. Total Environ.,* 1992, **113,** 147.
16. C.J. Matocha, D.L. Sparks, J.M. Amonette and K. Kukkadapu, *Soil Sci. Soc. Am. J.*, 2001, **65,** 58.
17. H.W. van der Marel and H. Beutelspacher, *Atlas of Infrared Spectroscopy of Clay Minerals and Their Admixtures*, Elsevier, New York, 1976.
18. N.A. Marley, J.S. Gaffney and K.A. Orlandini, in *Humic and Fulvic Acids. Isolation, Structure, and Environmental Role*, eds. J.S. Gaffney, N.A. Marley and S.B. Clark, American Chemical Society, Washington, DC, 1996, p. 96.
19. M.E. Morse, B. Post, S. Weissmann and H.F. McMurdie, *Mineral Powder Diffraction File. Search Manual,* JCPDS International Centre for Diffraction Data, Swarthmore, PA, 1980.
20. V. Plantchot, P. Colonna, A. Buleon and D. Gallant, in *Starch Structure and Functionality*, eds. P.J. Frazier, P. Richmond and A.M. Donald, Royal Society of Chemistry. Cambridge, 1997, p. 141.
21. N.B. Colthup, L.H. Daly and S.E. Wiberley, *Introduction to Infrared and Raman Spectroscopy*, Academic Press, Boston, 1990.
22. C. Liu and P.M. Huang, in *Understanding Humic Substances: Advanced Methods, Properties, and Applications*, eds. E.A. Ghabbour and G. Davies, Royal Society of Chemistry, Cambridge, 1999, p. 87.
23. C. Liu and P.M. Huang, in *Humic Substances: Versatile Components of Plants, Soil and Water*, eds. E.A. Ghabbour and G. Davies, Royal Society of Chemistry, Cambridge, 2000, p. 37.
24. M. Schnitzer and S.U. Khan, *Humic Substances in the Environment*, Dekker, New York, 1972.
25. F.J. Stevenson, *Humus Chemistry*, 2nd Edn., Wiley, New York, 1994.
26. G. Guggenberger and K.M. Haider, in *Interactions between Soil Particles and Microorganisms and the Impact on the Terrestrial Ecosystem*, eds. P.M. Huang, J.-M. Bollag and N. Senesi, IUPAC Book Series on Analytical and Physical Chemistry of Environmental Systems Vol. 8, Wiley, Chichester, 2001 (in press).
27. U. Schwertmann and R.M. Taylor, in *Minerals in Soil Environments*, 2nd Edn., eds. J.B. Dixon and S.B. Weed, Book Ser. 1. SSSA, Madison, WI, 1989, p. 379.
28. P.M. Huang and A. Violante, in *Interactions of Soil Minerals with Organics and Microbes*, eds. P.M. Huang and M. Schnitzer, SSSA Spec. Pub. no.17, SSSA, Madison, WI, 1986, p. 159.
29. C. Liu and P.M. Huang, *Geoderma*, 2001 (in press).
30. C.V. Toner IV and D.L. Sparks, *Soil Sci. Soc. Am. J.,* 1995, **59,** 395.
31. F.L. Wang, P.M. Huang and J.R. Bettany, in *Proceeding of the 11th International Clay Conference*, ed. H. Kodama, 1997, International Association of Study Clays, Ottawa, Canada, 1999, p. 505.

32. D. Dzombak and F.M.M. Morel, *Surface Complexation Modeling: Hydrous Ferric Oxide*, Wiley, New York, 1990.
33. A.T. Stone and A. Torrents, in ref. 1, p. 275.
34. B. Manunza, S. Deiana, V. Maddau, C. Gessa and R. Seeber, *Soil Sci. Soc. Am. J.*, 1995, **59,** 1570.
35. E.M. Perdue, in *Humic Substances in Soil, Sediment, and Water*, eds. G.R. Aiken, D.M. McKnight, R.L. Wershaw and P. MacCarthy, Wiley, New York, 1985, p. 493.
36. A.E. Martell and M. Calvin, *Chemistry of Metal Chelate Compounds*, Prentice-Hall, New York, 1952.
37. M.R. Hoffmann, *Environ. Sci. Technol.*, 1980, **14,** 1061.
38. W.W. Porterfield, *Inorganic Chemistry. A Unified Approach*, Harper International SI Edition. London, 1983.

SPECTROSCOPIC EVALUATION OF HUMIN CHANGES IN RESPONSE TO SOIL MANAGEMENTS

Guangwei Ding,[1] Jingdong Mao,[1] Stephen Herbert,[1] Dula Amarasiriwardena[2] and Baoshan Xing[1]

[1] Department of Plant and Soil Sciences, Stockbridge Hall, University of Massachusetts, Amherst, MA 01003, USA
[2] School of Natural Science, Hampshire College, MA 01002, USA

1 INTRODUCTION

Humic substances (HSs) constitute the bulk of soil organic matter (SOM).[1] They may be described as a series of yellow-to-black acidic polyelectrolytes with highly variable molecular weights.[2] HSs can be defined in various ways, such as the one based on the solubility in aqueous acids and bases.[3] Following this scheme, HSs include humic acids (HAs), fulvic acids (FAs), and humin (HU).[4] HA is the fraction extracted from soils with alkaline solutions and is precipitated on acidification, whereas FA is the fraction that remains in solution. Regardless of pH, the fraction that is not solubilized by base solutions is referred to as the HU, a high proportion of SOM being in this fraction.[5-9]

HSs are heterogeneous mixtures.[1,9-12] Investigations indicate that there are distinct compositional differences between HSs from different sources, soil types, soil managements and climates.[12-15] For example, Wander and Traina[15] reported that the percentages of carbon and nitrogen in the HU fraction were higher in conventional rotation soils than that in rotation with cover crops. Zalba and Quiroga[16] observed that the FA fraction is sensitive to agronomic and environmental factors. They investigated 114 surface samples of Hapludolls, Hapulustolls and Entisols ranging from sandy to silty loam from a wide region in Argentina and concluded that FA carbon was more influenced by the farming system than total carbon.

To our knowledge, there have been relatively few studies on HU structural and compositional changes because of its low carbon and high magnetic particle contents.[9] Some related studies indicate that the low solubility of HU may be due to several factors including a higher proportion of poorly decomposed plant material, higher molecular weight of the humified material and strongly associated organo-mineral complexes.[9,17-19]

The structures of HU are poorly understood. A better understanding of HU structures and compositions would help to determine their genesis and roles in soil. Spectroscopic techniques allow a rapid quantitative investigation of organic compounds without any chemical pretreatment.[1,13] In our previous paper[20] we demonstrated that the HAs from rye alone cover were more aromatic and less aliphatic than those from a vetch/rye cover system. In this investigation, solid-state cross-polarization magic-angle-spinning total sidebands suppression (CPMAS-TOSS) ^{13}C nuclear magnetic resonance (NMR) and diffuse reflectance fourier transform infrared spectroscopy (DRIFT) techniques were used to characterize the compositional and structural changes of HU fractions under different cover crop systems.

2 MATERIALS AND METHODS

2.1 Site Description and Sampling

A field study was conducted in 1990 at the University of Massachusetts Agronomy Research Farm in South Deerfield, Massachusetts. The soil at the experiment site is a Hadley fine sandy loam (coarse, mixed, mesic Fluventic Dystrudept) and low in SOM (~2%). It is a typical soil in the intensively cropped Connecticut River Valley in Massachusetts. Cover crop treatments consisted of hairy vetch + rye, rye alone, and no cover crop seeded at the rate of:

(i) control (no cover crop) (C1, no crop, no fertilizer; C4, no crop, with fertilizer)
(ii) rye (125 kg/ha) (R1, rye alone, no fertilizer; R4, rye alone, with fertilizer)
(iii) hairy vetch + rye (46 + 65 kg/ha) (VR1, vetch/rye, no fertilizer; VR4, vetch/rye, with fertilizer)

Nitrogen fertilizer rates were 0, 67, 135 and 202 kg N/ha using NH_4NO_3 for the different treatments. Soil samples were collected (1998) from the top (0-25 cm) soil using a 10 by 5 cm core sampler (with 0 and 202 kg N/ha treatments). Samples were immediately brought into the laboratory, stored at 4 °C while in transit and maintained field moist until processed. Detailed soil sample information was reported elsewhere.[20]

2.2 Preparation of HU Fractions

The HU fractions were prepared using methods similar to those described previously with some modifications.[5,9] Air-dried soil samples (< 2 mm) were first stirred in water to remove a poorly decomposed light fraction which floated in water. They were then extracted with 0.1 M $Na_4P_2O_7$ (three times) under nitrogen for 24 hours, shaking at room temperature. After removal of the solution, the insoluble residue (designated as "crude humin") was washed with deionized water until the pH was neutral, dried at 70°C and then ground. 200 mL of 1.0 M HF was added to 20 g of crude humin and a magnetic stir bar in a plastic centrifuge bottle. The bottle was shaken for 4 hours and the stir bar was taken out to remove any magnetic particles (e.g., Fe) clinging to it. The bottles were centrifuged at 3000 rpm, the HF was replaced with fresh solution and the stir bar was replaced. The HF treatment was done a total of 6 times, but the stir bar was used only for the first 3-4 treatments as the yield of magnetic particles declined sharply with each treatment. After the last HF wash the de-ashed humin was rinsed into a shallow dish with deionized water, left to air-dry and ground.

2.3 Diffuse Reflectance Fourier Transform Infrared (DRIFT) Analysis

All HU samples were analyzed with an infrared spectrophotometer (Midac Series M 2010) equipped with a DRIFT accessory (Spectros Instruments). HU fractions were powdered with an agate mortor and pestle and stored over P_2O_5 in a dry box. Three-mg solid HU samples were then mixed with 97 mg of KBr and reground to powder consistency. A sample holder was filled with the powder mixture. A microscope glass slide was used to smooth the sample surface. At the beginning of analysis, the diffuse-reflectance cell that contained the samples was flushed with nitrogen for 10 minutes to reduce the interference from carbon dioxide and moisture. The sample compartment was supplemented with anhydrous $Mg(ClO_4)_2$ to further reduce atmospheric moisture.

The DRIFT spectra were acquired with a minimum of 100 scans collected at a resolution of 16 cm^{-1}. The spectrometer was calibrated with the background, which consisted of powdered KBr and scanned under the same environmental conditions as the sample-KBr mixtures. Absorption spectra were converted to a Kubelka-Munk function using the Grams/32 software package (Galactic Corporation). Peak assignments and intensity (by height) ratio calculations were made following the methods of Niemeyer et al.,[21] and Wander and Traina.[22] We used ratios of labile (O-containing) and recalcitrant (C and H and/or N) functional groups to compare HU spectra of products from different treatments because organic oxygen is present in all major SOM functional groups (carboxyl, hydroxyl, phenol, carbonyl and alcohol) and they are associated with SOM binding features. Ratios of reactive and recalcitrant functional groups were generated by measurement of peak heights relative to spectral baselines followed by summation and division of the peaks of interest.

2.4 Solid-State ^{13}C NMR Spectroscopy

For CPMAS-TOSS experiments, approximately 300-400 mg of a HU sample was packed in a 7-mm-diameter zirconia rotor with a Kel-F cap. The spectrum was run at 75 MHz (^{13}C) in a Bruker MSL-300 spectrometer. The spinning speed was 4.5 kHz. A ^{1}H 90° pulse was followed by a contact time (t_{cp}) of 500 μs and then a TOSS sequence was used to remove sidebands.[23,24] The 90° pulse length was 3.4 μs and the 180° pulse was 6.4 μs. The recycle delay was 1 s with the number of scans about 100,000. The details were reported elsewhere.[24]

3 RESULTS AND DISCUSSION

3.1 Solid-State ^{13}C NMR Analysis of HU

The CPMAS-TOSS ^{13}C NMR spectra were used for comparison. After evaluating different solid-state ^{13}C NMR techniques, Xing et al.[24] reported that CPMAS-TOSS could be used to eliminate spinning sidebands from magic angle spinning spectra. They demonstrated that CPMAS-TOSS has two advantages over CP-MAS. The first is that a good TOSS can eliminate all the sidebands so that the spectra show only the true peaks for a HA sample. The second advantage is that implementation of CPMAS-TOSS can avoid baseline distortion arising from dead time. Xing et al. concluded that CPMAS-TOSS was consistently better than CPMAS for both quantification and qualitative comparison. They recommended that CPMAS-TOSS be used instead of CPMAS for measurements using a ≥ 300 MHz spectrometer.

Humin is the least studied humic fraction because it is difficult to separate from the mineral matrix. For our purpose, de-ashed organic-matter-enriched HU fractions were prepared using 1.0 M HF accompanied by magnetic stir bars to remove ferromagnetic iron particles. Six HU ^{13}C NMR spectra are shown in Figure 1. Functional group assignments are shown in Table 1.

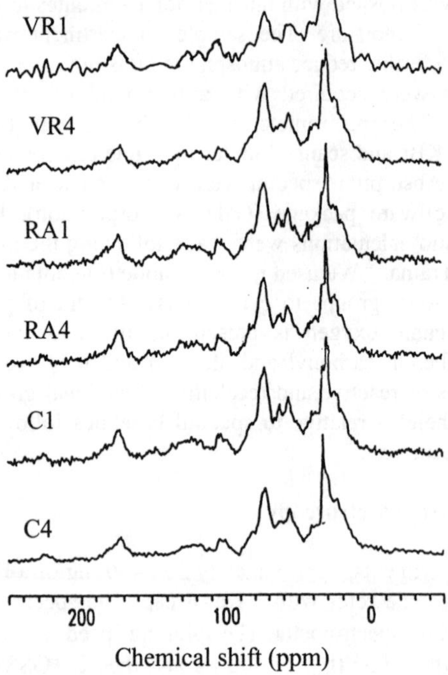

Figure 1 *CPMAS-TOSS ^{13}C NMR spectra of HU isolated from different cover crop systems*

Table 1 *Peak assignments for chemical shifts of NMR spectra*[a]

Ranges /peaks	*Chemical shift, ppm*	*Functional groups*	*Notes*
1	0-50	C, CH, CH_2, CH_3	Unsubstituted aliphatic C (e.g., alkanes, fatty acids)
2	50-60	C, CH_3O-, CH_2-O-, CH, CH-NH	N-alkyl (e.g., amino acid-, peptide, and protein-C) + methoxyl-C
3	60-96	CHOH, CH_2OH, CH_2-O-	Saccharide, alcohol, ether
4	96-108	O-CH-O, CH	Carbohydrates
5	108-145	C, CH	Aromatic C consisting of unsubstituted and alkyl substituted aromatic C, alkenes also resonate in this region
6	145-162	$C-O^-$, C-OH	Phenolic C
7	162-190	COO^-, COOH, C-O-C=O	Carboxyl C (includes the carboxylate ion, COO^-)
8	190-220	C=O, HC=O	Ketone, quinone and aldehyde

[a] These assignments are from Stevenson,[1] Preston,[19] and Schnitzer and Preston[25]

The main features of HU occur at about 30 ppm, which is assigned to be aliphatic carbons, especially those in long -$(CH_2)_n$- chains.[25,26] In the aliphatic region, small peaks can be observed at about 55 ppm that are due to methoxyl carbon. Another major peak is at 75 ppm, which is assigned to oxygen-alkyl carbon including the ring carbons of carbohydrates and the side-chains of lignin. The next peak at about 105 ppm is associated with di-oxygen-alkyl carbon, including the anomeric carbon of carbohydrates. The peak at 174 ppm is assigned to carboxyl, amide and ester carbons.

Integrating the spectra based on the standard chemical shift ranges gives the carbon percentage data in Figure 2. The aliphatic carbon content (0-108 ppm) in vetch/rye plots with or without nitrogen fertilizers was greater than that from the rye alone system. The same situation was observed for HA.[20] This may be associated with loss of the most easily metabolizable carbohydrates in the rye alone system. HU fractions in the rye system may be more stabilized than those from the vetch/rye cover system. The alkyl carbon content of HU (0-50 ppm) (Figure 2A) was much higher than that in the FA fraction. This result suggests that the HU fractions contain more cutin and suberin, which are highly resistant to decomposition.[9] However, Skjemstad et al.[27] reported that an accumulation of alkyl material was considered to be due not to selective preservation, but rather to an increase in cross-linking of the long chain alkyl material occurring during humification.

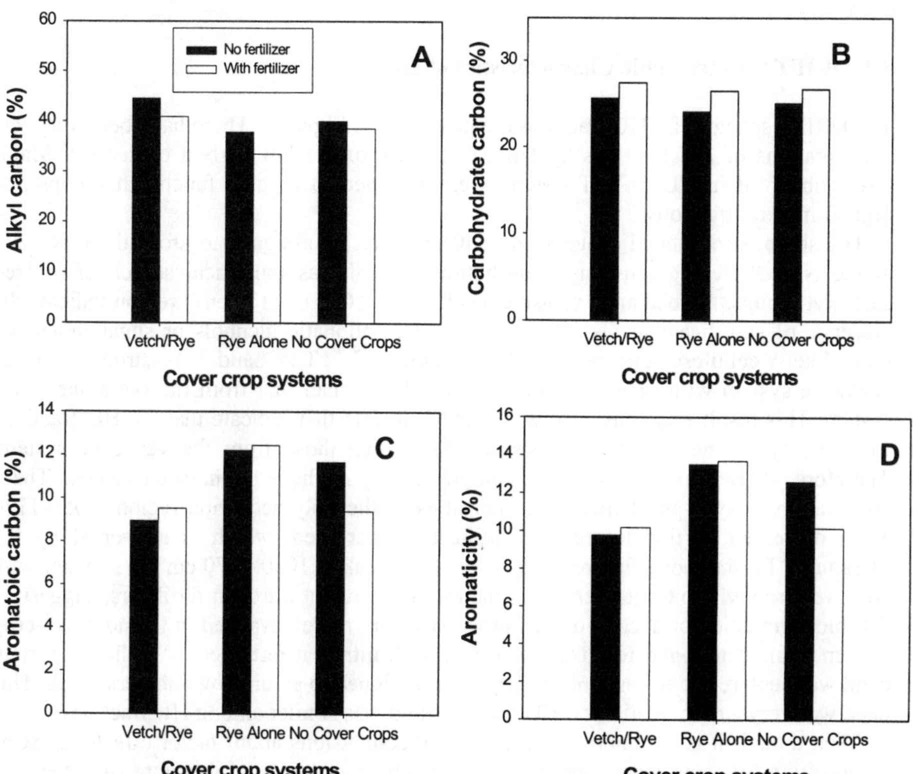

Figure 2 *Solid-state ^{13}C integration data under different cover crop systems: A. Alkyl-C (0-50 ppm): B. Carbohydrate-C (60-96 ppm); C. Aromatic-C (108-162 ppm); D. Aromaticity (108-162 ppm)/(0-162 ppm)*

Another interesting feature (Figure 2B) is that fertilization was associated with a small increase of relative intensity for the carbohydrate-C region (60-96 ppm) and a similar decrease for the alkyl region (0-50 ppm) (Figure 2A). These differences in composition were observed consistently in the three different cover crop systems. A similar result was reported by Preston and Newman.[9] The ratio of alkyl carbon (0-50 ppm)/oxygen-alkyl carbon (60-96 ppm) was higher for the treatments without nitrogen fertilizers of all three cover crop systems than that with nitrogen fertilizer. The di-oxygen-alkyl (96-108 ppm) intensity showed the same trend and the ratio of alkyl carbon/(oxygen- + di-oxygen-alkyl carbon) also decreased for all the three cover crop systems with nitrogen fertilizers (data not shown). These results indicate a lower degree of decomposition of HU from the fertilized plots. A possible explanation is that these HU fractions reflected larger plant inputs with nitrogen fertilizers.

The aromatic carbon (108-162 ppm) intensity was higher in the rye alone cover system with or without nitrogen fertilizers than that under the vetch/rye cover system (Figure 2C). The aromaticity [(108-162 ppm)/(0-162 ppm)] was higher in HU from the rye system than that from the vetch/rye system (Figure 2D). This result was consistent with our previous report[20] that the HA fractions in the rye alone system were more aromatic than those from the vetch/rye cover system. Therefore, the decomposition and humification processes are affected by the plant residue characteristics, including nitrogen content and carbon chemistry.

3.2 DRIFT Spectroscopic Characteristics of HU

The DRIFT spectra of 6 HU fractions are presented in Figure 3. There have been very few investigations of HU fractions by DRIFT because of the low carbon contents.[5,9] Major peaks observed in HU DRIFT spectra were assigned to organic functional groups and organo-mineral fractions.

The sharp, strong band centered at 1640 cm^{-1} can be assigned to aromatic, C=C, and amide N-H stretching plus any contribution from the asymmetrical stretch of ionized carboxyl groups.[28] Broad and extensive bands in the 1050 to 1170 cm^{-1} region indicate the presence of significant levels of polysaccharides (aliphatic alcohols or sugar moieties), most likely cellulose and residual hemicellulose.[29,30] The band was stronger in the vetch/rye system with or without nitrogen fertilizers than that from the rye alone cover system. This result was consistent with our NMR data that indicate that the HU fractions from the rye alone system were less aliphatic than those from the vetch/rye system. Therefore, DRIFT and ^{13}C NMR are complementary methods to characterize HSs. These observations may suggest that peak intensities in the polysaccharide region (1050-1170 cm^{-1}) decreased as the degree of humification increased, which is supported by the literature.[21] Furthermore, Figure 3 reveals that the peak at 1050-1170 cm^{-1} is stronger with no cover crop with nitrogen fertilizers than in that without nitrogen fertilizers, suggesting that biodegradation or decomposition processes were more advanced in the no-cover-crop system without nitrogen fertilizers than that with nitrogen fertilizers. Another important band was centered at around 660 cm^{-1} and was assigned to an unknown mineral peak. This band was very strong in all our 6 HU fractions and was similar among HU fractions.

Even though it is common practice to draw conclusions about factor effects on SOM composition by using single spectra of composite fractions,[1,21,29] this approach does not provide detailed information about cover crop effects on the characteristics of SOM. In this study we used ratios of labile (O containing) and recalcitrant (C and H and/or N) functional groups to compare the spectra. Ratios of peak heights from DRIFT spectra of HU fractions are presented in Table 2.

The O/R ratios of HU isolated from the vetch/rye cover system with or without nitrogen fertilizer treatment had a higher R1 than that from the rye alone system. We expect labile SOM to be relatively depleted in SOM pools isolated from the rye alone system. The relatively high O/R ratios of the vetch/rye soil may imply that SOM in the vetch/rye cover system was more biologically active.

Compared with HU fractions, HAs have low molecular weights and are rich in oxygen-containing groups such as carboxyl, phenol and/or enolic OH, alcoholic OH and C=O of quinones. Thus, HA is more polar and mobile than HU. This investigation is consistent with our previous conclusion[20] that HA O/R ratios under a no cover crop system with or without nitrogen fertilizers (1.0 and 0.85) were higher than for HU (0.64 and 0.73).

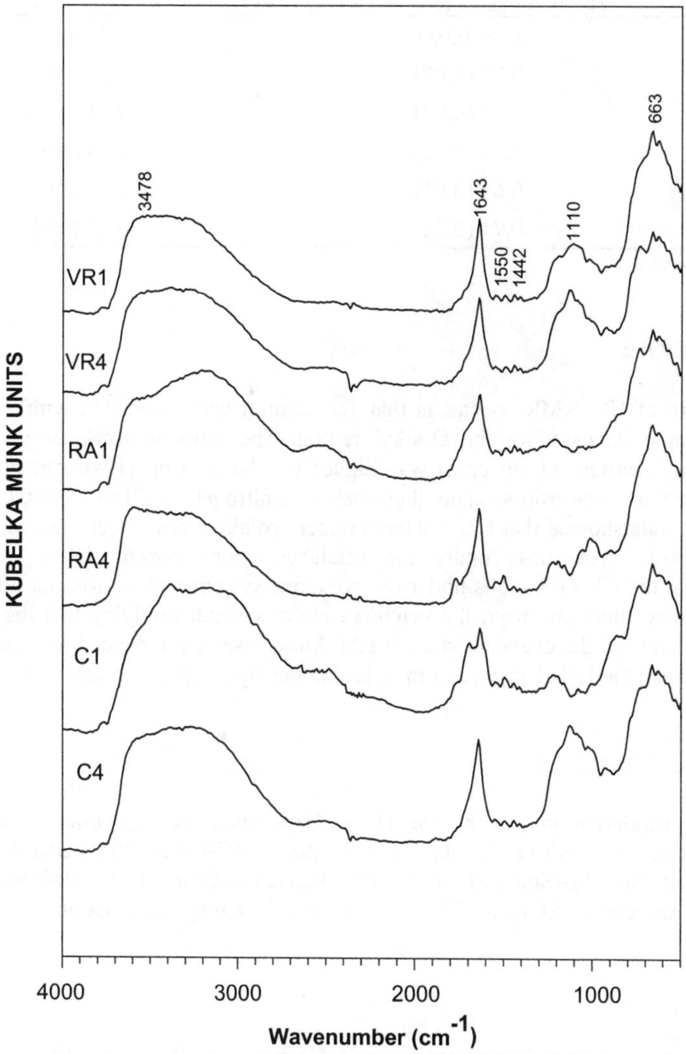

Figure 3 *DRIFT spectra of HU under different cover crop systems*

This study indicates that the O/R ratios of HU provide sensitive indices that reflect the effects of cover crop system management. This result was in general agreement with information obtained from physically isolated fractions. Similar results were reported by Wander and Traina.[15,22] Our results are also consistent with the trends observed by others who had demonstrated that management practices can lead to changes in the HU fractions.[9]

Table 2 *Ratios of selected peak heights from DRIFT spectra of HU*

HU samples	R_1 $\frac{1727+1650+1160+1127+1050}{2950+2924+2850+1530+1509+1457+1420+779}$	R_2 $\frac{1727}{1457+1420+779}$
VR1	0.75 (0.03)[a]	0.26 (0.01)
VR4	0.77 (0.03)	0.27 (0.01)
RA1	0.58 (0.02)	0.24 (0.01)
RA4	0.68 (0.02)	0.26 (0.01)
C1	0.64 (0.03)	0.27 (0.01)
C4	0.73 (0.02)	0.33 (0.01)

[a]standard deviations

4 CONCLUSIONS

The main feature of HU NMR spectra is that fertilization was associated with a slight increase of relative intensity for the O-alkyl region. The ratio of alkyl carbon (0-50 ppm)/oxygen-alkyl carbon (60-96 ppm) was higher for the treatments without nitrogen fertilizer in all three cover crop systems than that with nitrogen fertilizer. The CPMAS-TOSS ^{13}C NMR data showed that HU fractions under rye alone cover were less aliphatic and more aromatic than those under the vetch/rye cover system. DRIFT results demonstrated that the HU O/R ratios under the rye alone system with or without nitrogen fertilizer were lower than HU from the vetch/rye cover system, implying that the cover crop systems influenced the characteristics of HU. More research is needed to clarify the relationship between the HU changes and their functional significance in agricultural soils.

ACKNOWLEDGEMENTS

This work was supported in part by the U. S. Department of Agriculture, National Research Initiative Competitive Grants Program (98-35107-6319), the Federal Hatch Program (Project No. MAS00773), a Faculty Research Grant from University of Massachusetts (Amherst), NSF Grant BIR 9512370 and the Kresge Foundation.

References

1. F.J. Stevenson, *Humus Chemistry: Genesis, Composition, Reactions*, 2nd Edn., Wiley, New York, 1994.
2. M.H.B. Hayes, in *Humic Substances, Peats and Sludges: Health and*

Environmental Aspects, eds. M.H.B. Hayes and W.S. Wilson, Royal Society of Chemistry, Cambridge, 1997, p. 3.
3. G.R. Aiken, D.M. McKnight, R.L. Wershaw and P. MacCarthy, in *Humic Substances in Soil, Sediment, and Water*, eds. G.R. Aiken, P. MacCarthy, R.L. Malcolm and and R.S. Swift, Wiley, New York, 1985, p. 1.
4. M.H.B. Hayes, P. MacCarthy, R.L. Malcolm and and R.S. Swift, in *Humic Substances II: In Search of Structures*, eds. M.H.B. Hayes, P. MacCarthy, R.L. Malcolm and and R.S. Swift, Wiley, Chichester, 1989, p. 3.
5. C.M. Preston, M. Schnitzer and J.A. Ripmeester, *Soil Sci. Soc. Am. J.*, 1989, **53**, 1442.
6. G. Almendros and J. Sanz, *Soil Biology Biochem.*, 1991, **23**, 1147.
7. F.J. Stevenson, *Micronutrients in Agriculture*, 2nd Edn., SSSA, Madison, WI, 1991, p. 145.
8. G. Almendros and J. Sanz, *Geoderma*, 1992, **53**, 79.
9. C.M. Preston and R.H. Newman, *Geoderma*, 1995, **68**, 229.
10. C.M. Preston and B.A. Blackwell, *Soil Sci.*, 1985, **139**, 88.
11. J.F. Domarr, *Can. J. Soil Sci.*, 1979, **59**, 349.
12. C.M. Ciavatta, M. Govi, L. Sitti and C. Gessa, *Commun. Soil Sci. & Plant Anal.*, 1995, **26**, 3305.
13. O. Francioso, S. Sanchez-Cortes, V. Tugnoli, C. Ciavatta, L. Sitti and C. Gessa, *Appl. Spectrosc.*, 1996, **50**, 1165.
14. O. Francioso, C. Ciavatta, S. Sanchez-Cortes, V. Tugnoli, L. Sitti and C. Gessa, *Soil Sci.*, 2000, **165**, 495.
15. M.M. Wander and S.J. Traina, *Soil Sci. Soc. Am. J.*, 1996, **60**, 1081.
16. P. Zalba, and A.R. Quiroga, *Soil Sci.*, 1999, **164**, 57.
17. C. Saiz-Jimenez, B.L. Hawkins and G.E. Maciel, *Organic Geochem.*, 1986, **9**, 277.
18. M. Krosshavn, I. Kogel-Knabner, T. Southon and E. Steinnes, *Soil Sci.*, 1992, **43**, 473.
19. C.M. Preston, *Soil Sci.*, 1996, **161**, 144.
20. G. Ding, D. Amarasiriwardena, S. Herbert, J. Novak and B. Xing, in *Humic Substances: Versatile Components of Plants, Soil and Water*, eds. E.A. Ghabbour and G. Davies, Royal Society of Chemistry, Cambridge, 2000, p. 53.
21. J. Niemeyer, Y. Chen and J.M. Bollag, *Soil Sci. Soc. Am. J.*, 1992, **56**, 135.
22. M.M. Wander and B.R. Traina, *Soil Sci. Soc. Am. J.*, 1996, **60**, 1087.
23. K. Schmidt-Rohr and H.W. Spiess, *Multidimensional Solid-State NMR and Polymers,* Academic Press, London, 1994.
24. B. Xing, J. Mao, W.-G. Hu, K. Schmidt-Rohr, G. Davies and E.A. Ghabbour, in *Understanding Humic Substances: Advanced Methods, Properties and Applications*, eds. E.A. Ghabbour and G. Davies, Royal Society of Chemistry, Cambridge, 1999, p. 49.
25. M. Schnitzer and C.M. Preston, *Soil Sci. Soc. Am. J.*, 1986, **50**, 326.
26. M. Schnitzer, H. Kodama and J.A. Ripmeester, *Soil Sci. Soc. Am. J.*, 1991, **55**, 745.
27. J.O. Skjemstad, P. Clark, A. Golchin and J.M. Oades, in *Driven by Nature. Plant Litter Quality and Decomposition*, eds. G. Cadisch and K. E. Giller, CAB International, Wallingford, 1997, p. 253.
28. J.R. Ertel and J.L. Hedges, *Geochim. Cosmochim. Acta*, 1985, **49**, 2097.
29. Y. Inbar, Y. Chen and Y. Hadar, *Soil Sci. Soc. Am. J.*, 1989, **53**, 1695.
30. P. MacCarthy and J.A. Rice, in *Humic Substances in Soil, Sediment and Water*, eds. G.R. Aiken, P. MacCarthy, R.L. Malcolm and and R.S. Swift, Wiley, New York, 1985, p. 527.

EFFECTS OF CLEAR-CUTTING ON STRUCTURE AND CHEMISTRY OF SOIL HUMIC SUBSTANCES OF THE HUBBARD BROOK EXPERIMENTAL FOREST, NEW HAMPSHIRE, USA

David A. Ussiri and Chris E. Johnson

Department of Civil and Environmental Engineering, Syracuse University, Syracuse, NY 13244, USA

1 INTRODUCTION

Soil organic matter (SOM) is composed of non-living organic material of plant, animal and microbial origin. It is a heterogeneous mixture of compounds with a wide range of structural and functional features. Soil organic matter is often divided into humic and non-humic substances.[1] Humic substances (HSs) are amorphous, highly transformed, and darkly colored materials that contain no recognizable plant, animal or microbial structures and cannot be identified as belonging to an established group of organic compounds. Humic substances comprise up to 80% of SOM.[2]

Soil organic matter exerts a major influence on the biological and physical properties of soil.[3] SOM also influences soil fertility, soil development and various soil chemical properties including acid-base chemistry, buffer capacity, pH, cation exchange capacity (CEC) and metal complexation and transport. Transport of organic matter and associated metals is believed to play a central role in the genesis of Spodosols.[4,5] SOM is also important in the binding of anthropogenic organic chemicals, potentially reducing their toxicity.[6-8]

Clear-cutting alters the organic C cycle through removal of above-ground tree biomass for wood products. This induces changes in the forest ecosystem that extend to a wide range of other biogeochemical processes. For example, inputs of litter to the forest floor decrease for several years following clear-cutting.[9] Forest harvesting also causes changes in the soil environment: removal of the forest canopy decreases interception of atmospheric moisture, reduces transpiration and increases soil temperatures due to increased solar radiation at the soil surface. These conditions will stimulate microbial activity and therefore increase the rate of organic matter decomposition.[10-16] Clear-cutting also removes an important nutrient sink-plant uptake.[17] Accelerated decomposition will therefore cause increased leaching of plant nutrients. Elevated hydrologic losses of NO_3^- and the nutrient cations Ca^{2+}, Mg^{2+} and K^+ have been observed after clear-cutting.[18-22] Changes in decomposition patterns also influence soil properties such as CEC, pH and acidity by altering the amount of SOM and the balance of basic and acidic cations on the exchanger.[23,24]

Changes in decomposition rates and patterns after clear-cutting are also likely to influence the amount, structure and chemistry of HSs. Research in the northern hardwood forest has suggested that changes in soil chemical properties following clear-cutting are the result of changes in both the quality and quantity of soil organic matter.[13,24-26] However, changes in SOM as a result of clear-cutting have generally been inferred from changes in critical ratios such as C/N, C/SOM, N/SOM and CEC/SOM.[13,24,26] In this study we used ^{13}C nuclear magnetic resonance (NMR) spectroscopy and chemical titration methods to examine the chemistry and structure of HSs fractions extracted from a managed watershed at the Hubbard Brook Experimental Forest. In this paper the following research questions will be examined: (i) Did clear-cutting change the concentration and composition of HSs? (ii) Did clear-cutting result in preferential loss of any HSs fractions? (iii) Did clear-cutting result in increased acidic functional groups? (iv) What are the possible implications of the results to soil chemistry?

2 MATERIALS AND METHODS

2.1 Soils

The Hubbard Brook Experimental Forest (HBEF) is located in the White Mountain National Forest in central New Hampshire (45° 56′N and 71° 45′W). Soils are predominantly well-drained, coarse-loamy, mixed, frigid Typic Haplorthods with a 3-15 cm organic layer at the surface.[27] These soils are acidic (pH<4.5) with low base saturation and cation exchange capacity.[23] Average soil depth is approximately 60 cm.[23] Details regarding the climate, geology and hydrology at the HBEF have been published elsewhere.[18,27,28] Vegetation at the HBEF is mostly northern hardwoods, which are dominated by American beech (*Fagus grandifolia*), sugar maple (*Acer saccharum*) and yellow birch (*Betula alleghaniensis*). Pockets of the spruce-fir forest type, consisting of red spruce (*Picea rubens*), balsam fir (*Abies balsamea*) and white birch (*Betula papyrifera*) are found at the highest elevations and on exposed slopes.[29]

This research was conducted on watershed 5 (W5) at HBEF (Figure 1), a 23 ha watershed which was dominated by a 65 year-old second-growth northern hardwood forest prior to clear-cutting. The watershed was clear-cut in a whole-tree harvest in the winter of 1983-1984. All trees with diameter at breast height greater than 5 cm were cut and whole trees (boles and branches) were removed. Soils were intensively sampled on W5 prior to clear-cutting (1983) and in post-harvest years 3, 8 and 15 (1986, 1991 and 1998). Samples were collected in layers and by genetic horizons. Soil pits were excavated at 60 sites at each of the four sampling periods. These soils have been archived. More details of the sampling technique have been published elsewhere.[23,30,31] A subset of genetic horizon samples was analyzed in this study. We used samples collected from 6 pits in each sampling year selected to represent the range of elevation (Figure 1). Each of the selected pits exhibited the full suite of horizons found in HBEF Spodosols.

2.2 Extraction, Fractionation and Purification of Humic Substances

The extraction procedure used in this study is summarized in Figure 2. Humic substances were extracted with 0.5M NaOH at a soil:solution ratio of 1:10 (mass:volume) under an N_2 atmosphere.[3] The extracted humic substances were fractionated into humic acid (HA) and fulvic acid (FA) by the Stevenson method.[3] Polysaccharides were isolated from the fulvic acid using a modified version of the method described by Cheshire et al.[32] The yields of the freeze-dried humic substance fractions are presented in Table 1.

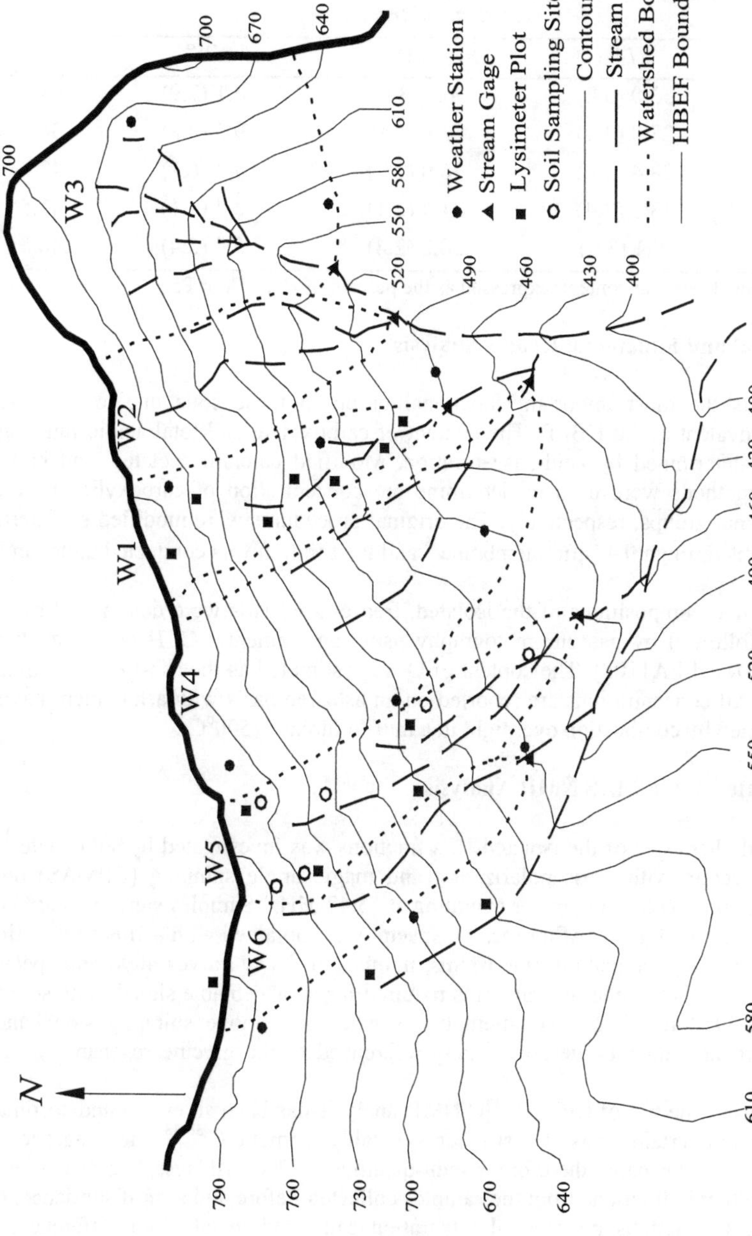

Figure 1 *Map of the Hubbard Brook Experimental Forest, showing the location of experimental watersheds, including sampling locations used in this study*

Table 1 *HSs fractions extracted from each horizon, expressed as a percentage of SOM, determined by loss-on-ignition (500°C). Residue refers to the unextractable organic matter. Values are averages for all years, HA, FA and polysaccharide (PS) yield are expressed on an ash-free basis*

Horizon	*Extracted fractions*			*Residue*
	HA[a]	*FA*[a]	*PS*[a]	
Oa	33.0 (3.2)	6.6 (3.0)	4.1 (7.9)	48.2
E	23.0 (3.3)	9.1 (3.4)	9.3 (4.8)	58.4
Bh	28.4 (3.0)	26.0 (2.7)	3.3 (6.2)	43.4
Bs1	13.3 (3.4)	31.4 (2.1)	2.4 (3.1)	47.5
Bs2	9.1 (3.9)	30.2 (2.4)	5.4 (5.4)	48.5

[a] Values in brackets are ash content, expressed as the percent of HA, FA or PS

2.3 Elemental and Functional Group Analysis

Carboxylate is the most important functional group in these soils due to the acidic conditions prevalent at the HBEF. The content of carboxylate and total acidic functional groups were determined by acid-base titration. Modified calcium acetate[33] and barium hydroxide[34] methods were used to determine the concentration of carboxylic and total acidic functional groups, respectively. The original procedures were modified by filtering the suspensions through 0.45-μm membrane[35] and the titration was conducted under an N_2 atmosphere.

The elemental compositions of the isolated, freeze-dried HSs were determined by dry combustion followed by gas chromatography using an elemental C, H and N analyzer (Carlo Erba Model EA1108). The content of O was estimated as the ash-free mass minus C, H and N. All concentrations are reported on an ash-free basis with ash content having been determined by combustion overnight in a muffle furnace (500°C).

2.4 Solid-state ^{13}C CP/MAS NMR Analysis

The structural chemistry of the isolated HSs fractions was investigated by solid state ^{13}C NMR spectroscopy with cross polarization and magic angle spinning (CPMAS) on a Bruker AMX 300 spectrophotometer operating at 75.47 MHz. Samples were spun at 5kHz in a zirconia rotor within a MAS probe. All spectra were obtained with a 1ms contact time and 1s recycle time, acquisition time 61 ms, number of decay curves 4096 and spectral width 33112 Hz. The number of transients required for an acceptable signal to noise ratio ranged from 3918 to 99224. No attempt was made to remove spinning side bands. Chemical shift assignments were externally referenced to the glycine resonance at 176 ppm.

In some cases, the use of CPMAS ^{13}C NMR analysis has been shown to underestimate the resonances of certain types of C structures, notably aromatic C.[36-38] The integration of spectral regions in this paper therefore is semi-quantitative. Nevertheless, because our goal is to compare humic fractions from the samples collected before and after disturbance, the estimation of C fractions by spectral integration can yield insight into differences in structural properties.

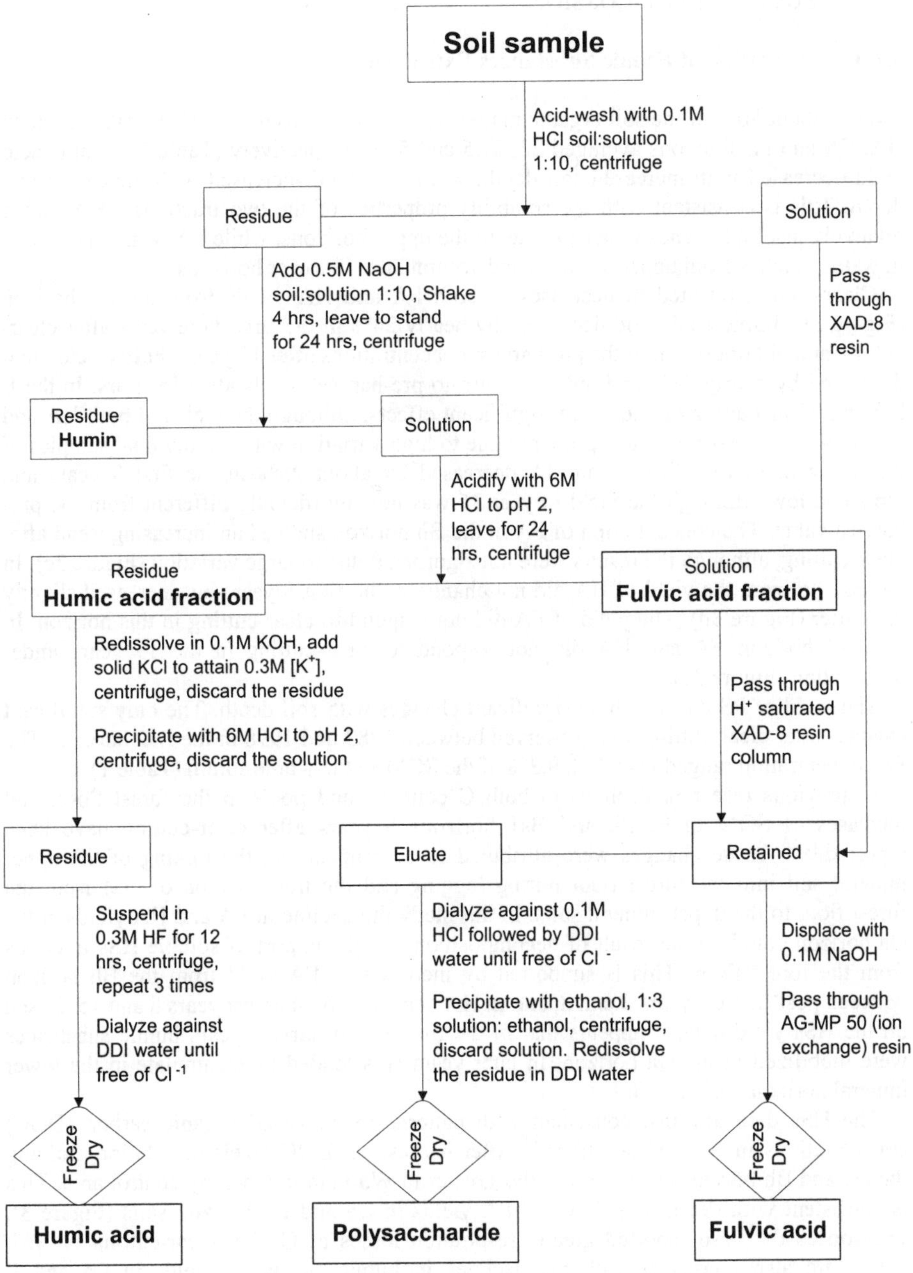

Figure 2 *Procedure for extraction, isolation and purification of humic substances*

3 RESULTS AND DISCUSSION

3.1 Concentration of Humic Substances Extracted

Overall about 50% of the soil organic matter was extracted from the HBEF soils, of which HA, FA and PS fractions averaged 22, 21.5 and 5.5% respectively (Table 1). Humic acid yield decreased with increased soil depth, while FA yeild increased with increasing soil depth. This is consistent with the solubility properties of the two fractions - HA, being relatively insoluble, tends to accumulate in the upper horizons, while FA, which is soluble in water, leaches from upper horizons and accumulates in lower horizons.

Clear-cutting resulted in decreases of both FA and HA yields from the Oa horizon (Figure 3). Humic acid yield decreased by nearly 24% in the first three years after clear-cutting and did not return to the pre-harvest concentrations after 15 years. Fulvic acid yield decreased by nearly 47% and did not return to pre-harvest levels after 15 years. In the E horizon, clear-cutting did not show significant effects, although the yields of both HA and FA showed decreasing trends, probably due to high variation within individual samples of this horizon. In the Bh horizon HA decreased by about 25% in the first 3 years and remained low, although the yield in year 15 was not significantly different from the pre-harvest value. The concentration of FA in the Bh horizon showed an increasing trend after clear-cutting, although the results were not significant due to large variation (Figure 3c). In the Bs1 horizon, the yield of HA did not change in the first 8 years but increased slightly thereafter (Figure 3d). The yield of FA did not respond to clear-cutting in this horizon. In the Bs2 horizon FA and HA did not respond to clear-cutting in the 15 years under observation (Figure 3e).

The yield of PS did not show significant changes with soil depth. The only significant changes after clear-cutting were observed between 1983 and 1998 in the Bs2 horizon. The PS concentration ranged from 2 to 9.3% of the SOM in the whole solum (Table 1).

In previous research, declines in both C contents and pools in the forest floor, and increases of SOM in E, Bh and Bs1 horizons 8 years after clear-cutting have been reported.[13,26] These changes were attributed to decomposition, the mixing of the upper mineral soil into the forest floor during logging and the translocation of OM from the forest floor to the upper mineral soils. In this study the decline in FA and HA yields in the Oa horizon could be the result of decomposition and the export of soluble HSs fractions from the forest floor. This is supported by increases in FA yield from the Bh horizon between post-harvest years 3 and 8, and in HA between post-harvest years 8 and 15. Based on the HSs yield data, it appears that in the first 3 post-harvest years humic substances were mobilized in the Oa horizon. In later years HSs tended to accumulate in the lower mineral horizons (Bh and Bs1).

The HSs data are also consistent with patterns in dissolved organic carbon (DOC) concentrations in W5 soil solutions.[13] After harvesting, DOC levels in solution draining the Oa and Bh horizons were up to 40% greater in W5 than in a nearby control area. This is consistent with declines in HA and FA yields in Oa and Bh horizon soils (Figure 3). Johnson et al.[13] also reported greater respiration losses of C after clear-cutting on W5, which are also consistent with the declines in humic substance yields in Oa and E horizons.

3.2 Elemental Compositions of the Fractions

The elemental compositions of the extracted fractions are presented in Table 2. The carbon content of HAs ranged from 488 to 618 g kg^{-1}, with a coefficient of variation of 4.14%.

This range falls within the range of C content of soil HAs compiled by Rice and MacCarthy.[39] Carbon contents of FAs ranged from 477 to 543 g kg^{-1} with a coefficient of variation of 2.9%. The mean C content of FA in this study is higher than the average C content of soil FA and falls in the higher end of soil FA C content ranges compiled by Rice and MacCarthy.[39] This could be due to the isolation technique used in this study in which polysaccharides, which have lower C content, were isolated from FA. In most other studies, polysaccharides are included as part of FA. Soil depth did not affect the C content of either HA or FA.

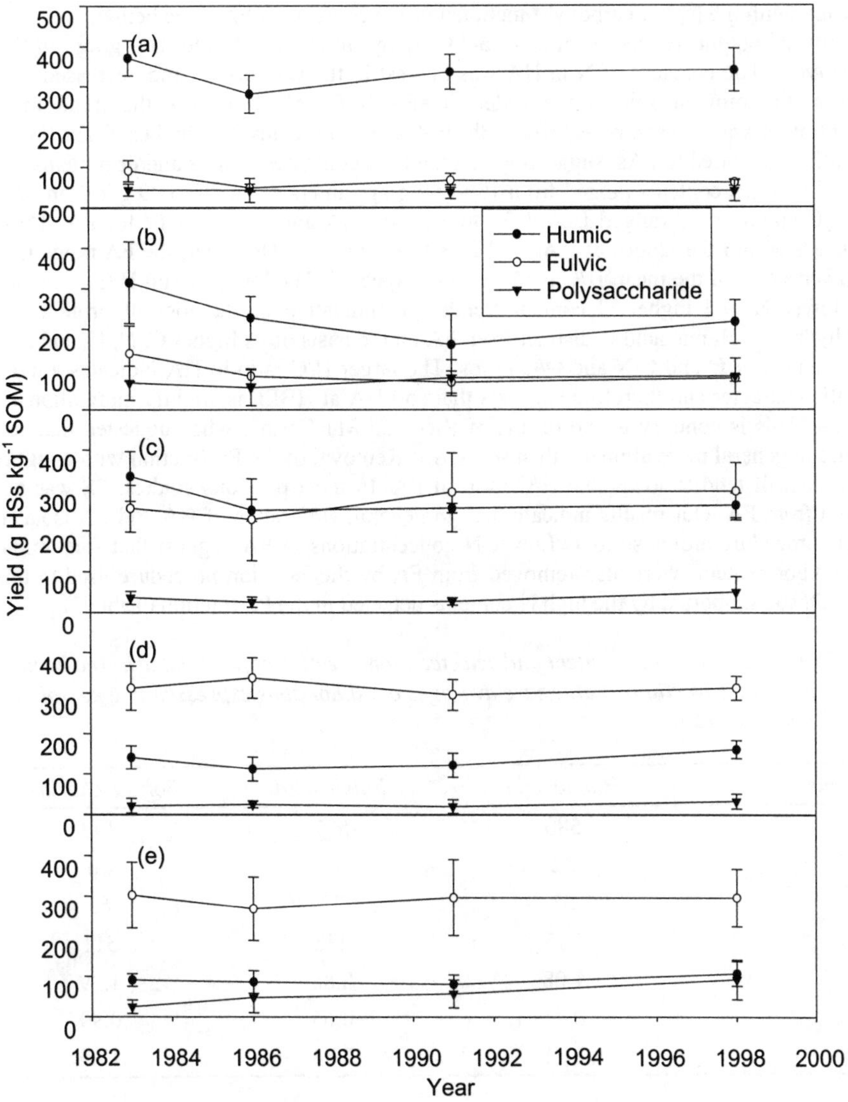

Figure 3 *Yields of humic substances (HSs) extracted from various soil horizons: (a) Oa (b) E (c) Bh (d) Bs1 (e) Bs2. Error bars represent standard deviations*

Estimated O content of humic acid ranged from 293 to 441 g kg^{-1} with coefficient of variation of 8.6%. The range for FA was 403 to 481 g kg^{-1} with a coefficient of variation of 3.5%. These ranges fall within the ranges of O concentrations of soil humic and fulvic acid reported by other workers.[39-41] Oxygen to C atomic ratios ranged from 0.36 to 0.68 for humic acid and 0.56 to 0.96 for fulvic acid, respectively. The O/C ratio has been suggested as a good indicator of humic substances sources.[42] The O/C ratios for humic and fulvic acids in our study are consistent with the suggested ratios for soil humic substances.[39,42] Oxygen contents and O/C ratios were significantly higher in fulvic acid than in humic acid fractions. This indicates that FA is more oxidized than HA and is consistent with the higher carboxyl functional group contents of FAs (see below).

Nitrogen contents ranged from 22.6 to 43.3 g kg^{-1} in HA and 9.5 to 21.9 g kg^{-1} in FA, respectively. The conctent of N in HA was highest in the Oa horizon and decreased with depth in the mineral soil in the order Oa=E>Bh>Bs1>Bs2. In FA the trend in N concentration was reversed: Bs2=Bs1=Bh>E=Oa. Humic acids had higher N and lower C/N ratios compared to FAs, suggesting that the HAs contained lignins and/or proteins.

The hydrogen contents ranged from 19 to 69 g kg^{-1} in HAs and 29 to 49 g kg^{-1} in FAs. The H/C atomic ratio ranged from 0.37 to 1.36 for HA and 0.71 to 1.17 for FA. These results are within the ranges of HAs and FAs from soils.[39,42] However, the FA range is at the lower end, and the mean H/C is lower than expected. The lower H and H/C combined with lower N and higher C content trends are indicative of the loss of protein and carbohydrate.[42] Humic acid is distinct from FA on the basis of its higher C, N, H, and H/C ratios and lower O and C/N and O/C ratios. The larger H/C ratio in HA indicates a more aliphatic character and therefore suggests that soil HA at HBEF is slightly more aliphatic than FA. This is contrary to the results of Rice and MacCarthy, who suggested that soil FAs are in general more aliphatic than soil HA.[39] Removal of the PS fraction with a higher H/C ratio will tend to lower the H/C ratio of FA. In most previous studies, PS was not isolated from FA. Our results indicate that, in general, H/C ratios of HA and FA isolated by this procedure are close to 1. Lower N concentrations in FA suggest that free amino acids and/or proteins were also removed from FA by the isolation procedure used in this study. This is supported by the high N contents detected in the PS fraction (Table 2).

Table 2 *Average elemental content and selected atomic ratios of purified, ash-free humic substance fractions. Values are averages of all horizons expressed in g kg^{-1} of humic substance*

Elements	*Humic acid*	*Fulvic acid*	*Polysaccharides*
C	580	507	399
N	32	15	26
H	53	38	63
O	335	438	511
H/C	1.08	0.88	1.99
O/C	0.44	0.65	0.99
C/N	21.1	39.4	17.9

The elemental C, N, H and O contents and the atomic H/C and O/C ratios in the polysaccharide fraction are consistent with values reported by Cheshire et al.[32] The isolated polysaccharide fraction had an elemental composition close to carbohydrate

(CH_2O). The nitrogen contents of polysaccharide fractions suggests that this fraction also contains considerable amounts of secondary amides and amino acids.[32]

3.3 Acidic Functional Groups

Concentrations of acidic functional groups in humic and fulvic acids are presented in Table 3. In humic acids, total acidity and carboxylic acidity ranged from 474 to 798 $cmol_c$ kg^{-1} and 265 to 383 $cmol_c$ kg^{-1}, respectively. In the fulvic acid samples total acidity and carboxylic acidity ranged from 790 to 1100 $cmol_c$ kg^{-1} and 613 to 834 $cmol_c$ kg^{-1}, respectively. These ranges were consistent with the ranges reported by Schnitzer for cool, acid soils.[40] Fulvic acids contained more acidic functional groups than HAs (Table 3). In both HAs and FAs, carboxylic acid accounted for 50 to 78% of the total acidic functional groups. This is consistent with Stevenson's conclusion that the total acidity of FAs is greater than that of HAs, with the carboxyl group being most important.[43] Comparison of total and carboxylic acid functional groups in different horizons revealed that HAs and FAs from Oa and E horizons had lower acidic functional group concentrations than those from lower mineral horizons (Bh, Bs1 and Bs2). This suggests that the HAs and FAs found in deeper mineral horizons are older and more oxidized than humic substances found in E and Oa horizons.

In general there were few significant changes in acidic functional group contents in humic acids after clear-cutting (Table 3). Total acidic groups were lower in post-harvest year 8 in Bh and Bs1 in humic acids compared to pre-harvest values. There were no significant changes in carboxylic functional groups in the HA fraction. In FAs, carboxylic functional group concentration showed an increasing trend in all horizons and the increase was significant in E and Bs2 horizons (Table 3).

Table 3 *Concentrations of acidic functional groups ($cmol_c$ kg^{-1}) in humic (HA) and fulvic (FA) acids as affected by clear-cutting*

Horizon	*Carboxylic acidity*				*Total acidity*			
	1983	*1986*	*1991*	*1998*	*1983*	*1986*	*1991*	*1998*
HA								
Oa	304	265[a]	303	279	589	556	632	603
E	302	279[a]	306	301	474	615	556	677
Bh	362	368	347	358	688	763	582	786
Bs1	341	364	345	347	779	780	631	713
Bs2	349	322	383	317	537	612	798	643
FA								
Oa	613	663	679	608	790	800	870	802
E	624	619	640	721	862	859	835	845
Bh	748	755	786	762	920	1100[b]	940	1060
Bs1	758	796[b]	797[b]	700	964	1030	991	1010[b]
Bs2	745	720	834	762	911	1080	1010	957

[a] significantly lower than 1983 values ($p = 0.05$), [b] significantly higher than 1983 values ($p = 0.05$)

3.4 CP/MAS ^{13}C NMR Analysis

Examples of solid-state CP/MAS ^{13}C NMR spectra of the isolated fractions are presented in Figure 4. The ^{13}C NMR spectra of PS, FA and HA exhibit distinct peaks in the alkyl (0-50 ppm), O-alkyl (50-110 ppm), aromatic (110-160 ppm) and carbonyl (160-220 ppm) C ranges. In the PS fractions the peaks at 16.7 and 22.7 ppm are most likely due to aliphatic C chains. In the 50-110 ppm region aliphatic C substituted by O and N are usually observed.[44] The peaks at 71.7 and 102.4 ppm correspond to carbohydrate C. The relatively high N content in the PS fraction (Table 2) suggests that the PS fraction may contain amino acids whose C will contribute to the intensity of the peak at 71.7 ppm and therefore overlap with carbohydrate peak. Spectra of the PS fractions suggest that they contain relatively minor amounts of C in the aromatic and carbonyl regions (Figures 4 and 5). In FAs and HAs, peaks in the alkyl C region were broad and centered around 36.1 ppm for FAs and 30 ppm for HAs. These peaks are likely due to C in long alkyl chains $(CH_2)_n$, although other alkyl carbon may also contribute.[44,45] The peaks at 78 and 109 ppm in FAs, and at 56, 73 and 80 ppm in HAs correspond to carbohydrate C. In view of the high N content of HA, most likely there is also a considerable amount of lignin that contributes to the signals in the lower and upper end of this region.

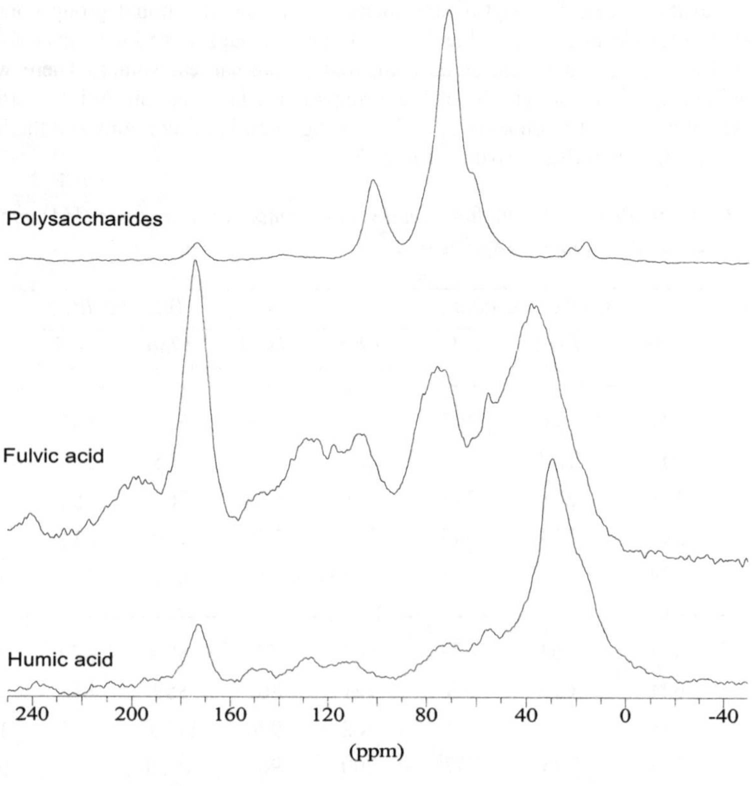

Figure 4 *Solid-state CPMAS ^{13}C NMR spectra for humic fractions isolated from soils at the Hubbard Brook Experimental Forest, New Hampshire*

In the aromatic C range, FAs showed a peak at 130 ppm and HAs had a peak at 127 ppm. The signals at the lower end of the aromatic C range (110-130 ppm) are usually attributed to unsubstituted aromatic rings such as alkylbenzenes.[46,47] The peak at 173 ppm in HAs and 175 ppm in FAs is due to C of carboxyl groups. Amides and esters also could contribute to this peak.

Results from integration of major NMR spectral regions of the extracted HSs fractions are presented in Figure 5. Integration of the major regions of ^{13}C resonance revealed that C in the HAs fraction was mainly alkyl (47 to 61% of total C) and O-alkyl (19 to 24% of total C). Aromatic C (11 to 15%) and carbonyl C (10 to 13%) were less abundant. In contrast, FAs had significantly lower alkyl C contents (31 to 38% of total C) and significantly greater O-alkyl (28 to 31% of total C) and carbonyl (18 to 22% of total C) content compared to HAs. Aromatic C (14.9 to 17.4% of total C) was also a minor component in FAs, and its concentration did not differ significantly from the HAs. The C composition of PS was primarily alkyl and O-alkyl (nearly 75% of total C, Figure 5). Fulvic acids had relatively larger proportions of O-alkyl and carbonyl C compared to HAs. Since carboxyl C contributes most of the area in the carbonyl C region, this result is consistent with our findings from elemental and chemical analysis discussed in the previous sections.

The C distribution in the PS samples revealed that O-alkyl C was the main C component (63 to 86.6% of total C). Combined with the spectra (Figure 5) and the elemental analysis, this provides further evidence that our procedure isolated most of the free carbohydrate from FA fractions. Signals for O-alkyl C in FAs likely represent carbohydrate structures bound in larger humic molecules.

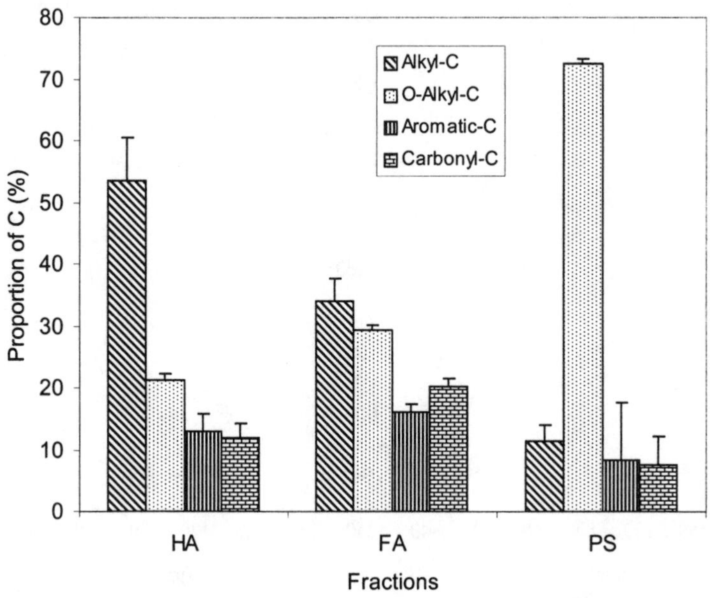

Figure 5 *Average distributions of C in fractions based on ^{13}C NMR analysis. Samples were averaged over all horizons and years. Error bars represent standard deviations*

In general, FAs had significantly higher carbonyl C than HAs, and in both FAs and HAs carbonyl C was lower in the Oa horizon than in the mineral horizons in both pre- and post-harvest (Figures 6, 7 and 8). This is consistent with our chemical data (see Table 3). Clear-cutting caused a slight decrease in carbonyl C of the Oa horizon in both HAs and FAs (Figures 6, 7 and 8).

Signal intensities for alkyl C (0-50 ppm) increased with soil depth in HAs in both pre- and post-harvest samplings (Figure 7). However, this increase with depth was more dramatic in post-harvest samples. In contrast, O-alkyl C decreased with depth in both pre- and post-harvest samples. The decrease was also more pronounced in the post-harvest samples. These patterns could have resulted from the increased biodegradation of OM. Similar results have been reported for whole soils[48] and soil solutions.[49] Preferential biological degradation has been suggested for the generally observed decrease in carbohydrate content with depth.[50,51] Our data indicate that the fraction of carbohydrate C was actually greater in the Oa horizon 15 years after harvesting than it was prior to harvesting (Figure 7). We attribute this to the input of decomposing logging residues ("slash") and roots to the forest floor after cutting. Consistent with this hypothesis, the aromatic C fraction in humic acids of post-harvest soils was greater than before cutting (Figure 7).

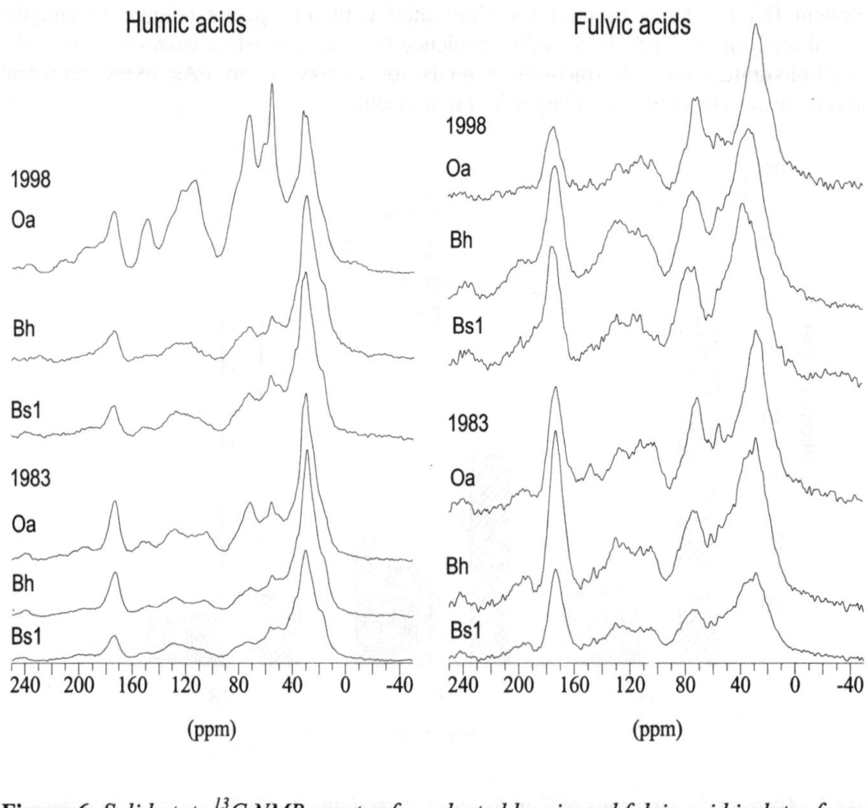

Figure 6 *Solid-state ^{13}C NMR spectra for selected humic and fulvic acid isolates from pre-harvest (1983) and post-harvest (1998) soil samples*

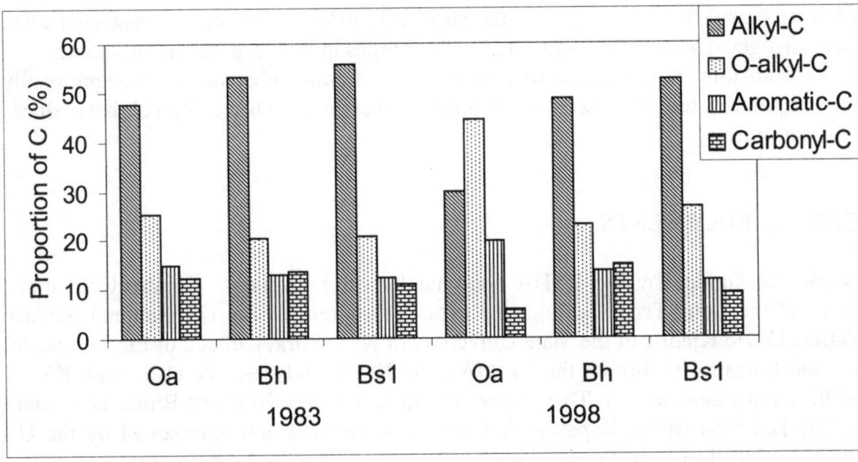

Figure 7 *Distribution of C analyses of HAs based on ^{13}C NMR in pre- and post-harvest soils*

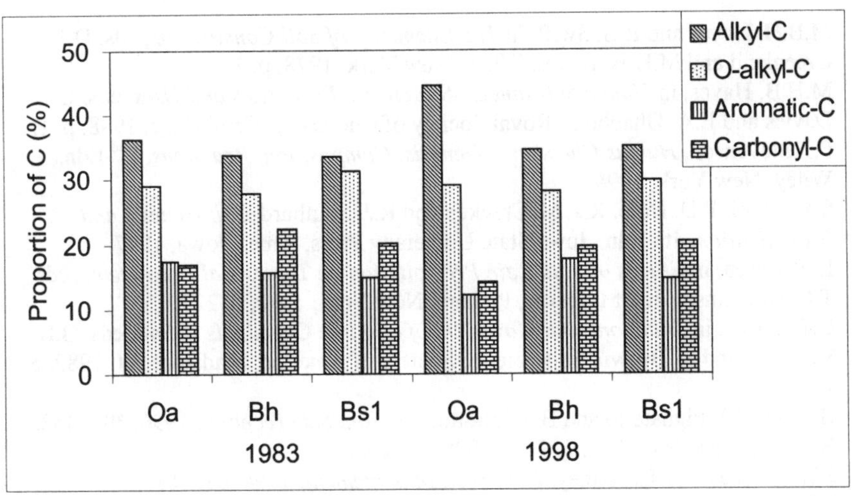

Figure 8 *Distribution of C in FAs based on ^{13}C NMR analysis of pre- and post-harvest soils*

4 CONCLUSIONS

The application of chemical analysis and a spectroscopic technique (^{13}C NMR) were found suitable for investigations of soil organic matter and the possible structural changes induced by clear-cutting at the HBEF. Clear-cutting caused decreases in HA and FA yields in forest floor and upper mineral horizons and increased FA yields in the lower mineral horizons. Also, the acidity of FA in the Bs1 horizon was significantly higher in the post-harvest samples. NMR data confirmed the chemical analyses and further suggested some structural changes that could be associated with changes in the decomposition regime.

Results from this study suggest that even though humic substances are operationally defined compounds, they are useful in identifying changes due to ecological disturbances such as clear-cutting.

ACKNOWLEDGEMENTS

This work was funded by the USDA National Research Initiative Competitive Grants Program and the Long-Term Ecological Research Program of the US National Science Foundation. David Kiemle of the State University of New York, College of Environmental Science and Forestry performed the CP/MAS ^{13}C NMR analyses. We also thank K'o H. Dai for his advice and support. This is a contribution from the Hubbard Brook Ecosystem Study. The Hubbard Brook Experimental Forest is operated and maintained by the US Forest Service Northeastern Research Station, Newtown Square, Pennsylvania.

References

1. M.B.H. Hayes and R.S. Swift, in *The Chemistry of Soil Constituents*, eds. D.J. Greenland and M.H.B. Hayes, Wiley, New York, 1978, p. 179.
2. M.H.B. Hayes, in *Humic Substances: Structures, Properties and Uses*, eds. G. Davies and E.A. Ghabbour, Royal Society of Chemistry, Cambridge, 1998, p. 1.
3. F.J. Stevenson, *Humus Chemistry: Genesis, Composition, Reactions*, 2nd Edn., Wiley, New York, 1994.
4. S.W. Buol, F.D. Hole, R.J. McCracken and R.J. Southard, *Soil Genesis and Classification*, 4th Edn., Iowa State University Press, Ames, Iowa, 1997.
5. L. Petersen, in *Effects of Acid Rain Precipitation on Terrestrial Ecosystems*, eds. T.C. Hutchinson and M. Havas, Plenum, New York, 1980, p. 223.
6. C.T. Chiou, in *Reactions and Movement of Organic Chemicals in Soil*, eds. B.L. Sawhney and K. Brown, Soil Science Society of America, Madison, WI, 1989, p. 1.
7. B. Xing, J.J. Pignatello and B. Gigliotto, *Environ. Sci. Technol.*, 1996, **30**, 2432.
8. B. Xing, *J. Environ. Health, Pt B*, 1998, **33**, 293.
9. J.W. Hughes and T.J. Fahey, *For. Ecol. Management*, 1994, **63**, 181.
10. F.H. Bormann, G.E. Likens, T.G. Siccama, R.S. Pierce and J.S. Eaton, *Ecol. Monogr.*, 1974, **44**, 255.
11. K.G. Mattson and W.T. Swank, *Biol. Fert. Soils*, 1989, **7**, 247.
12. T.R. Moore, *Water Resour. Res.*, 1989, **25**, 1321.
13. C.E. Johnson, C.T. Driscoll, T.J. Fahey, T.G. Siccama and J.W. Hughes, in *Carbon Forms and Functions in Forest Soils*, eds. J.M. Kelly and W.W. McFee, American Society of Agronomy, Madison, WI, 1995, p. 463.
14. J.L. Clayton and D.A. Kennedy, *Soil Sci. Soc. Am. J.*, 1985, **49**, 1041.
15. O.Q. Hendricksen, L. Chatarpaul and D. Burgess, *Can. J. For. Res.*, 1989, **19**, 725.
16. R.A. Dahlgren and C.T. Driscoll, *Plant Soil*, 1994, **158**, 239.
17. R.B. Romanowicz, C.T. Driscoll, T.J. Fahey, C.E. Johnson, G.E. Likens and T.G. Siccama, *Soil Sci. Soc. Am. J.*, 1996, **60**, 1664.
18. F.H. Bormann and G.E. Likens, *Pattern and Process in a Forested Ecosystem*, Springer-Verlag, New York, 1979.

19. G.D. Mroz, M.F. Jurgensen and D.J. Frederick, *Soil Sci. Soc. Am. J.*, 1985, **49**, 1552.
20. L.K. Mann, D.W. Johnson, D.C. West, D.W. Cole, J.W. Hornbeck, C.W. Martin, H. Riekirk, C.T. Smith, W.T. Swank, L.M. Tritton and D.H. van Lear, *For. Sci.*, 1988, **34**, 412.
21. G.E. Likens, C.T. Driscoll, D.C. Buso, T.G. Siccama, C.E. Johnson, G.M. Lovett, D.F. Ryan, T.J. Fahey and W.A. Reiners, *Biogeochem.*, 1994, **25**, 61.
22. G.E. Likens, C.T. Driscoll, D.C. Buso, T.G. Siccama, C.E. Johnson, G.M. Lovett, T.J. Fahey, W.A. Reiners, D.F. Ryan, C.W. Martin and S.W. Bailey, *Biogeochem.*, 1998, **41**, 89.
23. C.E. Johnson, A.H. Johnson and T.G. Siccama, *Soil Sci. Soc. Am. J.*, 1991, **55**, 505.
24. C.E. Johnson, R.B. Romanowicz and T.G. Siccama, *Can. J. For. Res.*, 1997, **27**, 859.
25. K.E. Snyder and R.D. Harter, *Soil Sci. Soc. Am. J.*, 1985, **49**, 223.
26. C.E. Johnson, *Can. J. For. Res.*, 1995, **25**, 1346.
27. G.E. Likens and F.H. Bormann, *Biogeochemistry of a Forested Ecosystem*, 2nd edn, Springer-Verlag, New York, 1995.
28. C.A. Federer, L.D. Flynn, C.W. Martin, J.W. Hornbeck and R.S. Pierce, *USDA Forest Services General Technical Report*, NE, 1990, p 141.
29. R.H. Whittaker, F.H. Bormann, G.E. Likens and T.G. Siccama, *Ecol. Monogr.*, 1974, **44**, 233.
30. T.G. Huntington, D.F. Ryan and S.P. Hamburg, *Soil Sci. Soc. Am. J.*, 1988, **52**, 1162.
31. C.E. Johnson, A.H. Johnson, T.G. Huntington and T.G. Siccama, *Soil Sci. Soc. Am. J.*, 1991, **55**, 497.
32. M.V. Cheshire, J.D. Russell, A.R. Fraser, J.M. Bracewell, G.W. Robertson, L.M. Benzing-Purdie, C.I. Ratcliffe, J.A. Ripmeester and B.A. Goodman, *J. Soil Sci.*, **43**, 1992, 359.
33. J.R. Wright and M. Schnitzer, *Nature (London)*, 1959, **184**, 1462.
34. M. Schnitzer and S.U. Khan, *Humic Substances in the Environment*, Dekker, New York, 1972.
35. E.M. Perdue, in *Humic Substances in Soil, Sediment and Water. Geochemistry, Isolation and Characterization*, eds. G.R. Aiken, D.M. McKnight, R.L. Wershaw and P. MacCarthy, Wiley, New York, 1985, p. 493.
36. R. Frund and H.D. Ludemann, *Sci. Tot. Environ.*, 1989, **81/82**, 157.
37. P. Kinchesh, D.S. Powlson and E.W. Randal, *European J. Soil Sci.*, 1995, **46**, 125.
38. J.-D. Mao, W.-G. Hu, K. Schmidt-Rohr, G. Davies, E.A. Ghabbour and B. Xing, *Soil Sci. Soc. Am. J.*, 2000, **64**, 873.
39. J.A. Rice and P. MacCarthy, *Org. Geochem.*, 1991, **17**, 635.
40. M. Schnitzer, in *Soil Organic Matter Studies*, International Atomic Energy Agency, Braunsweig, 1977, p. 117.
41. Y. Chen, N. Senesi and M. Schnitzer, *Geoderma*, 1978, **19**, 87.
42. C. Steelink, in *Humic Substances in Soil, Sediment and Water: Geochemistry, Isolation and Characterization*, eds. G.R. Aiken, D.M. McKnight, R.L. Wershaw and P. MacCarthy, Wiley, New York, 1985, p. 457.
43. F.J. Stevenson, in *Humic Substances in Soil Sediment and Water*, eds. G.R. Aiken, D.M. McKnight, R.L. Wershaw and P. MacCarthy, Wiley, New York, 1985, p. 13.
44. M. Schnitzer and C.M. Preston, *Plant Soil*, 1983, **75**, 201.
45. L.P. Linderman and J.Q. Adams, *Anal. Chem.*, 1971, **43**, 1245.

46. P.G. Hatcher, R. Rowman and M.A. Mattingly, *Org. Geochem*., 1980, **2**, 77.
47. M. Schnitzer and C.M. Preston, *Soil Sci. Soc. Am. J.*, 1986, **50**, 326.
48. K.H. Dai, C.E. Johnson and C.T. Driscoll, *Biogeochem.*, 2001, in press.
49. K.H. Dai, M.B. David and G.F. Vance, *Biogeochem.*, 1996, **35**, 339.
50. I. Kogel-Knabner, *Forest Soil Organic Matter: Structure and Formation*, University of Bayreuth, Germany, 1992.
51. I. Kogel-Knabner, in *Soil Biochemistry*, eds. J.M. Bollag and G. Stotsky, Dekker, New York, 1993, p. 101.

SIGNIFICANCE OF BURNING VEGETATION IN THE FORMATION OF BLACK HUMIC ACIDS IN JAPANESE VOLCANIC ASH SOILS

H. Shindo[1] and H. Honma[2]

[1] Faculty of Agriculture, Yamaguchi University, Yamaguchi 753-8515, Japan
[2] Forensic Science Laboratory, Yamaguchi Prefectural Police Headquarters, Yamaguchi 753-8504, Japan

1 INTRODUCTION

Japan is a typical volcanic country and volcanic ash soils developed in deposits of materials such as ash are widely distributed. These soils have a thick black or brownish black A horizon with a high humus content. The black color is due to the presence of Type A humic acids (HAs) with a high degree of darkening.[1]

Several hypotheses referred to as the lignin, polyphenol and sugar-amine condensation theories have been proposed for the mechanisms of formation of humic substances (HSs).[2] In the case of Japanese volcanic ash soils it has been assumed that grassland plants are the major source responsible for the abundance of humus and that burning is necessary to maintain a grassland for a long time, since forest is the climax vegetation under the meteorological conditions prevailing in Japan.[3] Furthermore, burning vegetation not only has been often practiced by man (including shifting populations) but also occurred from wild fires.[4] The objective of this study was to evaluate the role of burning vegetation in the formation of black humic acids by investigating the distribution of charred plant fragments in volcanic ash soils in Japan and comparing the physicochemical and spectroscopic properties of the humic acids obtained from the charred plant residues after the prescribed burning of a grassland and from volcanic ash soils.

2 MATERIALS AND METHODS

2.1 Distribution of Charred Plant Fragments in Japanese Volcanic Ash Soils

2.1.1 Soil Samples. Twenty four soil samples were collected at a depth of 15 - 70 cm from 20 profiles of volcanic ash soils in Hokkaido, Honshu and Kyushu Islands, Japan (Table 1). Each soil sample was air-dried and passed through a 2 mm mesh sieve. Then, plant remains such as stems in the samples were carefully removed under a stereomicroscope. The humic acids in all the soils used were found to be of Type A.

2.1.2 Isolation of Charred Plant Fragments. Sodium polytungstate solution (specific gravity (s.g.) 2.8 gcm^{-3}) was purchased from Sigma-Aldrich and diluted with water to prepare separation media of s.g. 1.6 and 2.0 gcm^{-3}.

Table 1 *Sampling sites and organic-C contents of the soils used*

Sample No.	*Soil profile*	*Site*	*Depth, cm*	*Organic-C, g kg^{-1}*	*Note*
1	P1	Hiroshima, Hokkaido	24-35	70.0	Forest
2	P2	Erimo, Ho kkaido	19-35	125	Grassland
3	P2	Erimo, Hokkaido	35-45	121	Grassland
4	P3	Shizukuishi, Iwate	27-41	78.3	Grassland
5	P4	Takizawa, Iwate	32-49	39.4	Grassland
6	P5	Kanagasaki, Iwate	23-35	112	Forest
7	P5	Kanagasaki, Iwate	35-63	115	Forest
8	P6	Imaichi, Tochigi	32-46	127	Forest
9	P7	Kanuma, Tochigi	23-45	56.1	Upland Field
10	P8	Tukuba, Ibaragi	30-48	31.9	Grassland
11	P9	Fujinomiya, Shizuoka	15-27	181	Forest
12	P10	Fujinomiya, Shizuoka	30-45	118	Forest
13	P10	Fujinomiya, Shizuoka	45-65	84.7	Forest
14	P11	Iwata, Shizuoka	39-62	47.4	Forest
15	P12	Hidaka, Hyogo	15-48	47.0	Forest
16	P13	Hidaka, Hyogo	24-51	126	Forest
17	P14	Sekikane, Tottori	24-34	125	Forest
18	P15	Ohota, Shimane	22-47	146	Grassland
19	P16	Miyoshi, Hiroshima	32-42	58.2	Forest
20	P17	Nishigoshi, Kumamoto	30-50	92.6	Grassland
21	P18	Kobayashi, Miyazaki	22-66	78.3	Paddy Field
22	P19	Miyakonojo, Miyazaki	30-37	88.0	Paddy Field
23	P20	Miyakonojo, Miyazaki	31-44	83.8	Grassland
24	P20	Miyakonojo, Miyazaki	44-70	95.6	Grassland

Two grams of the soil sample were placed in a 50 mL glass centrifuge tube and oven-dried at 90°C overnight. To the tube, 10 mL of s.g. 2.0 gcm^{-3} sodium polytungstate solution was added. The tube was closed with a stopper and inverted gently by hand 10 times. The soil particles adhered to the stopper were returned into the tube using 10 mL of the same solution. The tube containing the soil suspension was sonified at 39 kHz and 100 W for 15 min and then centrifuged at 1700 *g* for 20 min. After centrifugation, the upper parts containing floating particles were transferred into a 100 mL Squibb glass separatory funnel. The precipitates remaining in the tube were washed with 10 mL of s.g. 2.0 gcm^{-3} sodium polytungstate solution several times by centrifugation until no floating particles were detected. The upper parts were combined in the separatory funnel and allowed to stand overnight to separate the floating and precipitate fractions. The precipitate fraction was removed. The floating fractions were ultrafiltered using a membrane filter (0.1 μm, Advantec Toyo) and washed thoroughly with water. The residue on the filter was dislodged with a microspatula and placed in a 50 mL glass centrifuge tube. To the tube, 10 mL of s.g. 1.6 gcm^{-3} sodium polytungstate solution was added. The subsequent procedure, except for sonification, was the same as described above. Thus, two fractions with different specific gravity, s.g. 1.6-2.0 gcm^{-3} (1.6-2.0 fraction) and s.g. <1.6 gcm^{-3} (<1.6 fraction), were isolated. These fractions were oven-dried at 90°C and analyzed.

2.1.3 Microscopic Observation. The morphology of the <1.6 fraction and 1.6-2.0 fraction was observed with a stereomicroscope and a scanning electron microscope.

2.1.4 Determination of Organic-C. The organic-C content was determined by the dichromate oxidation method.[5] In this study, an adequate amount of the <1.6 fraction or whole soil was digested with oxidant at about 200°C for 30 min in a 100 mL flask with a condenser. After cooling, the organic-C content was determined titrimetrically.

2.2 Characterization of Humic Acids Obtained from Charred Grassland Plants after Prescribed Burning

2.2.1 Charred Grassland Plants Produced by Burning. Fire was set in a grassland in the Akiyoshi plateau, Mine, Yamaguchi, Japan. The dominant vegetation of this grassland was represented by dense Susuki (*Miscanthus sinensis* Anderss) and Kenezasa (*Pleioblastus pubescens* Nakai) plants. Fire ran quickly on the soil surface and charred plants, together with half-charred plants, were scattered all over the area. Only charred grassland plants were collected and used for the extraction of humic acids.

2.2.2 Isolation of Humic Acids. The amounts of NaOH-extractable materials from the charred grassland plants were extremely small. It is known that dilute HNO_3 treatment can result in a large amount of NaOH-extractable materials from oven-charred Susuki plants.[6,7] Thus, the charred grassland plants were oxidatively degraded with diluted HNO_3 (1:4) as follows. A 1 g aliquot of the charred plants was suspended in 50 mL of HNO_3 (1:4) and the flask (100 mL) containing the suspension was heated at about 200°C for 1 h. After cooling, the HNO_3-treated residue was collected on a suction filter (Advantec Toyo No. 4A), washed with 40 mL of water three times and then oven-dried at 90°C for 24 h.

The charred grassland plants after HNO_3 treatment were shaken with 0.1 M NaOH at 30°C for 48 h using a ratio of solution to material of 150:1.[8] After very dark supernatants were separated from the materials by centrifugation, the alkaline extracts were acidified with diluted HCl (1:1) to pH 2.0 and allowed to stand overnight at room temperature. The precipitate was redissolved in 0.1 M NaOH and then reprecipitated. This purification procedure was repeated once. The precipitate obtained was then dialyzed against water and freeze-dried. For reference, the humic acid of a soil was also isolated by the same method (but with no HNO_3 treatment) from the A horizon of volcanic ash soil of the Forest Experimental Station, Tottori University, Hiruzen, Okayama, Japan.

2.2.3 Degree of Darkening. The degree of darkening of the humic acids obtained from the charred grassland plants after HNO_3 treatment (P-HA) and from the A horizon of the volcanic ash soil (S-HA) was measured by the method of Kumada et al.[9] Humic acid was dissolved in 0.01 M NaOH and then the spectrum of the solution in the region 230 to 700 nm was immediately measured with a JASCO Ubest-50 spectrophotometer. The Δlog K value is the logarithm of the ratio of the absorbance of humic acid at 400 nm to that at 600 nm. The RF value is the absorbance of humic acid at 600 nm multiplied by 1,000 and divided by the number of mL of 0.02 M $KMnO_4$ consumed per 30 mL of humic acid solution.[9,10]

2.2.4 Elemental Composition. Elemental analyses for C, H and N contents were conducted by conventional organic microanalysis.[6] The oxygen content was calculated by subtracting the weights of ash, C, H and N from the total weight. The elemental compositions of the products are expressed on a moisture and ash-free basis.

2.2.5 Infrared Spectra. The IR spectra of the humic acids were measured as described by Kumada and Aizawa[11] using a Shimadzu FTIR-8600PC Fourier transform infrared spectrophotometer.

2.2.6 ^{13}C-NMR Spectra. ^{13}C-NMR spectra were obtained at a ^{13}C resonance frequency of 75.45 MHz with a JEOL JNM-alpha 300 solid NMR system using solid-state cross-polarization magic-angle spinning (CP-MAS) and total suppression of sidebands (TOSS) techniques for eliminating spinning sidebands.[12] The operating conditions were as follows: CP contact time, 5ms; recycle delay time, 10s; scans, 8,000-10,000; temperature, 26-27°C.

2.2.7 X-ray Diffraction. X-ray diffraction was measured by the powder method using a Rigaku RINT 2200V x-ray diffractometer with Cu_{α} radiation and a graphite monochromator. The operating conditions were as follows: scanning angles 5 - 50° (2θ); scanning speed 2° min^{-1}.

3 RESULTS AND DISCUSSION

3.1 Distribution of Charred Plant Fragments in Japanese Volcanic Ash Soils

Figure 1 shows a photograph of the <1.6 fraction observed under a stereomicroscope. Charred plant fragments, which are brown or black particles, were the main component of the < 1.6 fraction. Charred plant fragments in the 1.6-2.0 fraction occurred in association with mineral-humus colloids, which made the analytical data hard to interpret. Therefore, the relationship between the organic-C contents of the charred plant fragments and its whole soil was examined for the < 1.6 fractions. Figure 2 shows the relationship between the organic-C content of a soil (from Table 1) and the contribution to this carbon from the < 1.6 fraction. The latter, with units g C from < 1.6 fraction/kg whole soil, is obtained by multiplying the yield of < 1.6 fraction (column 2 of Table 2) by the carbon-C content of this fraction (column 3 of Table 2). Figure 2 shows that the organic-C content of the soils increases with the contribution from charred plant fragments ($r = 0.49^{**}$). The percentage of the total organic-C that comes from charred plant fragments can be calculated by dividing the data in column 4 of Table 2 by the corresponding data in column 5 of Table 1. The results range from 0.4 to 5.4%, with 9 of the 24 samples having 3% or more of their organic-C originating from charred plant remains (Figure 3).

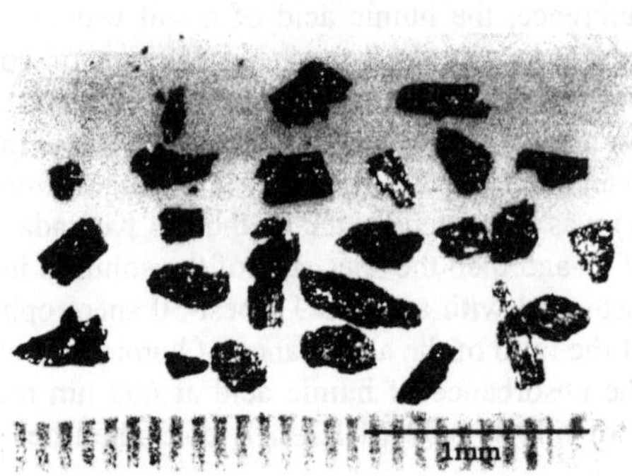

Figure 1 *Charred plant fragments in the < 1.6 fraction*

Table 2 *Yields and organic-C content of the <1.6 fraction*

Sample No.	*Yield g kg^{-1} soil*	*Organic-C content of the fraction, g kg^{-1}*	*Organic-C content due to whole soil, g kg^{-1}*
1	11.4	217	2.47
2	1.68	261	0.438
3	2.87	273	0.784
4	2.63	369	0.970
5	4.61	384	1.77
6	17.1	327	5.59
7	9.17	430	3.94
8	6.84	381	2.61
9	3.42	347	1.19
10	4.00	336	1.34
11	14.2	303	4.30
12	5.28	284	1.50
13	2.34	317	0.742
14	2.73	250	0.683
15	4.51	349	1.57
16	9.68	439	4.25
17	4.88	282	1.38
18	21.4	285	6.10
19	0.77	525	0.40
20	7.26	228	1.66
21	11.4	369	4.21
22	3.14	109	0.342
23	2.11	212	0.447
24	1.81	329	0.595

The results obtained reveal that charred plant fragments are widely distributed in Japanese volcanic ash soils containing Type A humic acids with a high degree of darkening. Plant fragments are the main component in the light fraction of the soils studied. Previous studies have indicated that charred plant fragments were detected in soils of the UK[10], USA,[13] New Zealand,[14] Bolivia[15] and the Philippines.[16] These findings suggest that charred plant fragments are common in various soils.

3.2 Characterization of Humic Acids Obtained from Charred Grassland Plants after Prescribed Burning

Kumada et al.[9] proposed that soil HAs could be classified into Types A, B, P and Rp, based on their optical properties (ΔlogK and RF values), the degree of darkening being in the order Types A>B≈P>Rp. Table 3 shows the ΔlogK and RF values and elemental compositions of P-HA and S-HA.

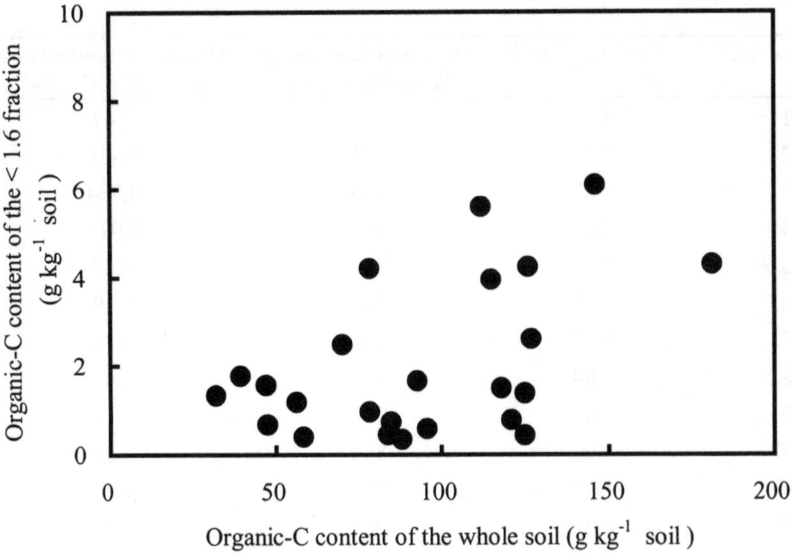

Figure 2 *Relationship between the organic-C contents of the < 1.6 fraction and whole soil*

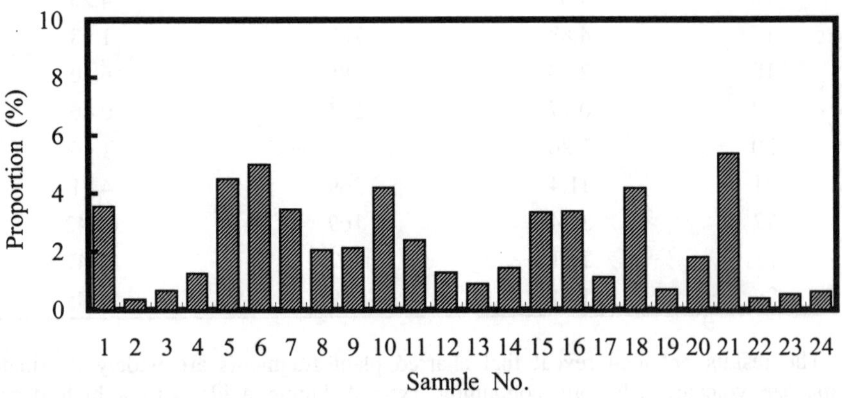

Figure 3 *Proportion of organic-C in the <1.6 fraction*

Table 3 *Degree of darkening and elemental composition of P-HA and S-HA*

Humic acid	*Degree of darkening*		*Elemental composition (g kg^{-1})*			
	ΔlogK	*RF*	*C*	*H*	*N*	*O*
P-HA	0.580	149	606	29.8	42.0	322
S-HA	0.533	141	599	29.5	26.2	346

ΔlogK and RF values for the P-HA and S-HA correspond to all the criteria (ΔlogK<0.7 and RF>80) for Type A humic acids, although the ΔlogK and RF values of the former were somewhat higher than those of the latter. Type A humic acids generally have higher carbon and oxygen contents and lower hydrogen and nitrogen contents than other types of humic acids. The elemental composition of the P-HA indicated fair agreement with that of the S-HA. These analytical data of P-HA, except for the nitrogen content, fall within the ranges of those of Type A humic acids in Japanese soils.[17]

The UV-vis absorption curves of the P-HA and S-HA are shown in Figure 4. The shapes of their spectra are similar to each other and to those of Type A soil humic acids.[9,10]

The IR absorption spectrum of P-HA resembled that of S-HA, with a C=O stretching vibration (1,700 cm^{-1}) and a C=C stretching vibration (1,600 cm^{-1}), although small peaks were more numerous in the spectrum of S-HA (Figure 5).

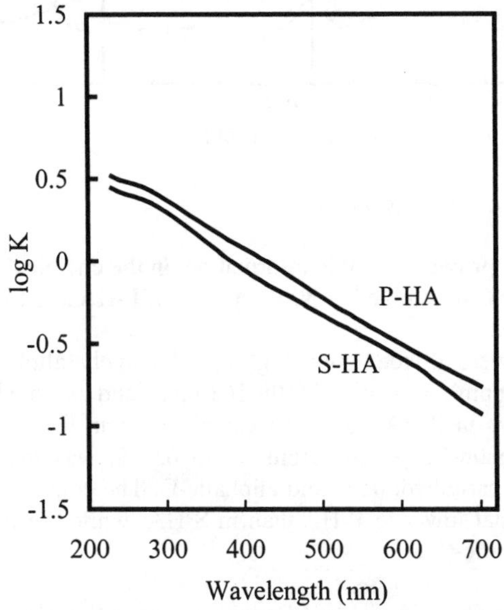

Figure 4 *Absorption curves of P-HA and S-HA*

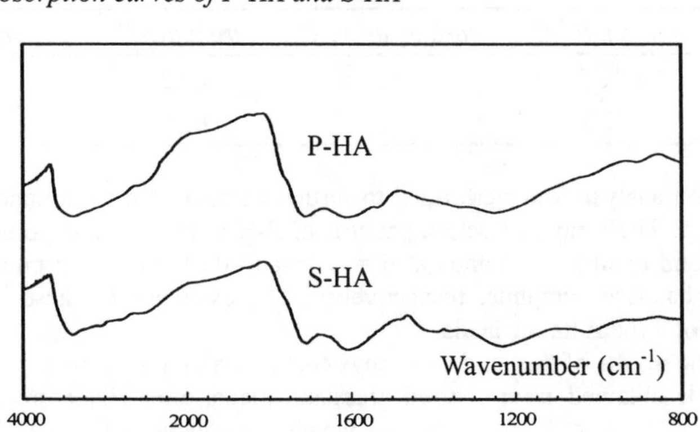

Figure 5 *Infrared spectra of P-HA and S-HA*

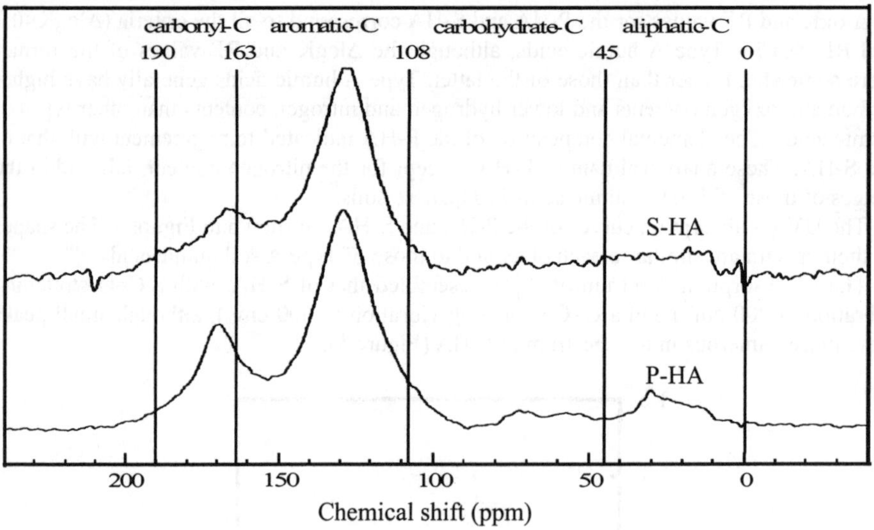

Figure 6 *^{13}C NMR spectra of P-HA and S-HA*

^{13}C NMR analysis can provide valuable information on the chemical structures of humic substances. Figure 6 shows the ^{13}C NMR spectra of P-HA and S-HA. Their spectra were similar to each other.

The carbon species were divided into 4 groups,[18] namely aliphatic-C (0-45 ppm), carbohydrate-C (45-108 ppm), aromatic-C (108-163 ppm) and carbonyl-C (163-190 ppm) and their relative contents in P-HA and S-HA calculated from the ^{13}C NMR spectra are given in Table 4. Aromatic-C, ranging from 60 to 63 %, was the highest, generally followed by carbonyl-C, carbohydrate-C and aliphatic-C. The contents of aromatic-C and carbonyl-C were somewhat lower in P-HA than in S-HA, while the reverse was true for carbohydrate-C and aliphatic-C.

Table 4 *Relative contents (%) of individual C species to total C of the P-HA and S-HA*

Humic acid	*aliphatic-C*	*carbohydrate-C*	*aromatic-C*	*carbonyl-C*
P-HA	12.2	13.7	60.2	13.9
S-HA	9.5	9.5	63.0	17.9

X-ray diffraction analysis was performed to further compare the nature and properties of P-HA and S-HA. The x-ray diffraction patterns of P-HA and S-HA are similar to each other (Figure 7) and exhibit two bands at d = 0.35 and 0.21 nm, corresponding to the (002)- and (10)- bands of graphite, respectively.[19] The existence of these bands is an important feature of Type A humic acids.[19,20]

According to the results of this study, the physicochemical and spectroscopic properties of the humic acids obtained from charred grassland plants after HNO_3 treatment were similar to those of Type A humic acids in Japanese volcanic ash soils. This HNO_3 treatment of the charred grassland plants may involve oxidative degradation such as depolymerization and carboxylation by the fission of saturated rings and polycyclic aromatic structures as well as cleavage of ester linkagès.[21] Although charred plants produced by burning vegetation display a very high resistance to microbial decomposition,[6] it appears that part of the charred plants are subject to oxidative degradation under the influence of oxygen and moisture (terrestrial conditions) during a long period of time after burning and converted to humic acids, as in the case of HNO_3 treatment in our study. Furthermore, charred plant fragments are commonly detected in Japanese volcanic ash soils. We propose, therefore, that burning vegetation by man and wild fires is one of the possible mechanisms for the formation of black humic substances in soil and the associated ecosystems.

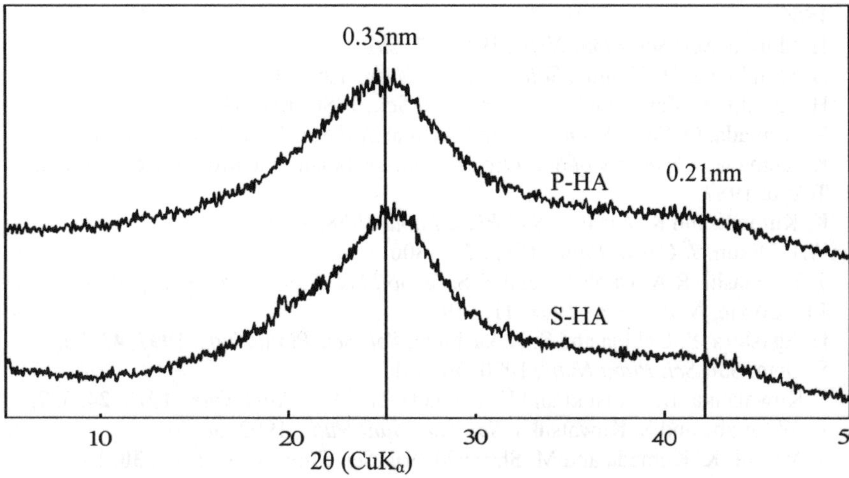

Figure 7 *X-ray diffraction patterns of the P-HA and S-HA*

4 CONCLUSIONS

Charred plant fragments are widely distributed in Japanese volcanic ash soils containing black humic acids. The physicochemical and spectroscopic characteristics of the humic acids obtained from the charred plant residues, which were oxidatively degraded with dilute HNO_3 after prescribed burning, were similar to those of humic acids in volcanic ash soils. The findings indicate that burning vegetation merits close attention as one of the possible mechanisms for the formation of black humic acids in Japanese volcanic ash soils, since it is considered that part of charred plants produced by burning are subjected to oxidative degradation and converted to humic acids in soil ecosystems, as in the case of HNO_3 treatment.

ACKNOWLEDGEMENTS

We thank Dr. S. Hiradate, Ministry of Agriculture, Forestry and Fisheries, Japan, for technical assistance. We also thank members of the Cooperative Research Projects on Ando soils, especially, Dr. M. Honna and Dr. S. Yamamoto, Tottori University, Japan for supplying the volcanic ash soils.

References

1. S. Arai, T. Honna and Y. Oba, in *Ando Soils in Japan*, ed. K. Wada, Kyushu University Press, Fukuoka, 1986.
2. F.J. Stevenson, *Humus chemistry,* Wiley, New York, 1982.
3. Ministry of Agriculture and Forestry, Japanese Government, *Volcanic Ash Soils in Japan,* Sakurai-Koseido, Tokyo, 1964.
4. J.S. Pyne, P.L. Andrews and R.D. Laven, *Introduction to Wildland Fire*, 2nd Edn., Wiley, New York, 1996.
5. M.M. Kononova, *Soil Organic Matter*, 2nd Engl. Edn., Pergamon Press, Oxford, 1966.
6. H. Shindo, *Soil Sci. Plant Nutr.*, 1991, **37**, 651.
7. H. Shindo and H. Honma, *Soil Sci. Plant Nutr.*, 1998, **44**, 675.
8. H. Shindo, Y. Matsui and T. Higashi, *Soil Sci.*, 1986, **141**, 84.
9. K. Kumada, O. Sato, Y. Ohsumi and S.Ohta, *Soil Sci. Plant Nutr.*, 1967, **13**, 151.
10. K. Kumada, *Chemistry of Soil Organic Matter,* Japan Scientific Societies Press, Tokyo, 1987.
11. K. Kumada and K. Aizawa, *Soil Plant Food*, 1958, **3**, 152.
12. W.T. Dixon, *J. Chem. Phys.*, 1982, **77**, 1800.
13. T. Takahashi, R.A. Dahlgren and T. Sase, *Soil Sci. Plant Nutr.*, 1994, **40**, 617.
14. J.D. Cowie, *N. Z. J. Sci.*, 1968, **11**, 459.
15. K. Egashira, S. Uchida and S. Nakashima, *Soil Sci. Plant Nutr.*, 1997, **43**, 25.
16. S. Ohta, *Soil Sci. Plant Nutr.*, 1990, **36**, 561.
17. S. Kuwatsuka, K. Tsutsuki and K. Kumada, *Soil Sci. Plant Nutr.*, 1978, **24**, 337.
18. A. Watanabe and S. Kuwatsuka, *Soil Sci. Plant Nutr.*, 1992, **38**, 31.
19. Y. Matsui, K. Kumada and M. Shiraishi, *Soil Sci. Plant Nutr.*, 1983, **30**, 13.
20. M. Schnitzer, H. Kodama and J.A. Ripmeester, *Soil Sci. Soc. Am. J.*, 1991, **55**, 745.
21. M. Schnitzer and J.R. Wright, *Soil Sci. Soc. Am. Proc.*, 1960, **24**, 273.

Organic Ores and Analysis of Commercial HSs

LEONARDITE AND HUMIFIED ORGANIC MATTER

D.M. Ozdoba,[1] J.C. Blyth,[1] R.F. Engler,[2] H. Dinel[3] and M. Schnitzer[3]

[1] Specialty Products Division, Luscar Ltd., Edmonton, Alberta, Canada T5L 3C1
[2] Luscar Ltd., Edmonton, Alberta, Canada T5K 3J1
[3] Eastern Cereal and Oilseed Research Centre, Agriculture and Agri-Food Canada, Ottawa, Canada K1A 0C6

1 INTRODUCTION

Leonardite and humified organic matter have been mentioned synonymously for many years, yet what do they have in common? Leonardite was named after Dr. A. G. Leonard in recognition of his work and is associated with lignite coal reserves. Leonardite and similar materials from oxidized sub-bituminous coals and carbonaceous shales are good sources of humic substances or organic material. Further work by Schnitzer, Dinel and co-workes[1] has shown that these organic materials can exhibit similar chemical properties to those of humified soil organic matter.

The term soil organic matter is generally used to represent the organic constituents in the soil, including identifiable high molecular weight organic materials, simpler substances (sugars, amino acids) and humic substances, as characterized by Stevenson.[2] Hayes and Himes[3] have further defined humic substances as being "humus that is composed of organic macromolecular substances arising in soils from which they were derived. Humic acids, fulvic acids and humin materials are regarded as humic substances." As a result, the study of humic substances is critical to demonstrating the link between known lignite deposits and soil organic matter.

The objective of this work was to characterize the different geological deposits of leonardite type materials in North America, the key factors associated with them and some common applications of such materials.

2 MATERIALS AND METHODS

2.1 Materials

Six locations were identified in North America to represent the most common reserves of leonardite or similar materials containing humified organic matter as shown in Figure 1. Leonardite is an oxidized form of lignite, which is a soft brown, coal-like substance that is associated with location 2 (South Eastern Saskatchewan) and location 3 (North Dakota). Currently, the North Dakota reserves have several locations from which the leonardite material is mined and processed.

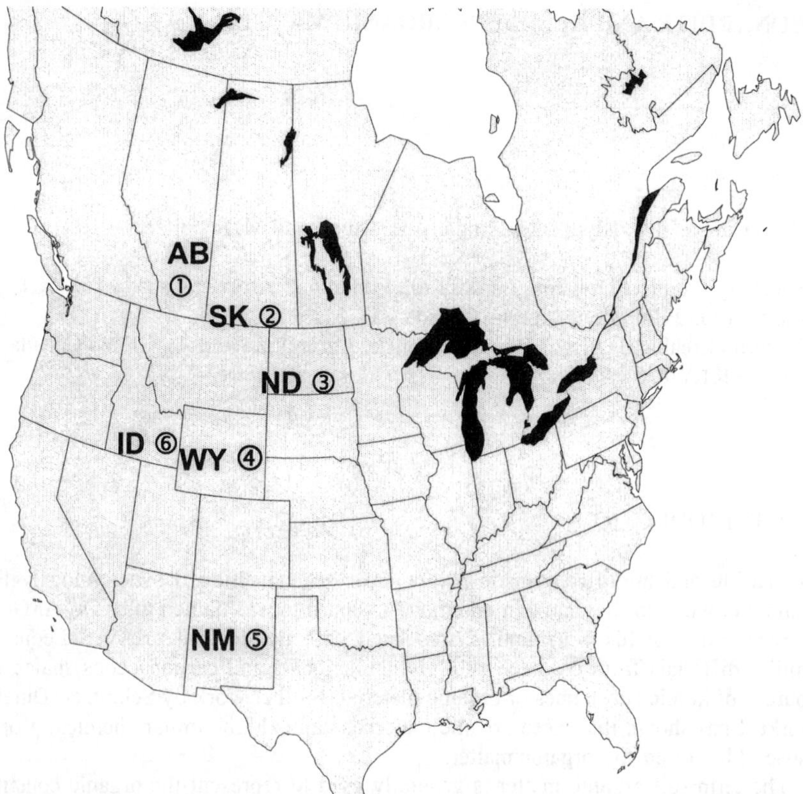

Figure 1 *North American deposits of humic substances*

In addition to leonardite, other sources of rich organic materials such as humalite exist. Humalite is defined by Hoffman et al.[4] as a naturally occurring organic material that is highly oxidized or weathered, brownish-black in color and found adjacent to sub-bituminous coals fields in Alberta. Reserves of humalite exist at location 1.

Location 4 (Wyoming) and location 5 (New Mexico) are also associated with weathered sub-bituminous coal fields and carbonaceous shales. New Mexico contains several locations in which such materials are mined and processed today.

Location 6 (Idaho) is significantly different from the above locations and is more closely associated with carbonaceous shales.

2.2 Methods

All samples were oven-dried and then heated at 700°C for 3h to determine their ash content. Total carbon, hydrogen and nitrogen were analysed by dry combustion and total sulfur by oxygen-flask combustion. Geological factors and characterization data were from previously determined sources.

Humic acid content was measured by two methods; the A&L method, which is a qualitative method using light refraction and the CDFA (California) method, which is more quantitative.

3 RESULTS AND DISCUSSION

The formation of organic matter and coal is best illustrated in Figure 2, which shows how oxygen content and compaction affect the characteristics of organic matter or coal. Organic materials are all formed from plant material. Buckman and Brady[5] describe peat as an unconsolidated soil material consisting largely of undecomposed, or slightly decomposed, organic matter accumulated under conditions of excessive moisture. Based on the formation process shown in Figure 2, humus is formed from peat. This in turn leads to humic substances, which have been defined by MacCarthy et al.[6] as heterogeneous mixtures of naturally occurring organic materials. All of the materials or ores described in the six locations come from being weathered or exposed to the elements over millions of years and thus can be considered as a form of organic matter. Also, the data indicate that humalite is quite different from the other ore deposits in terms of purity. One of the reasons is the origin of the material: humalite originates in poorly drained fresh water swamps and not salt water deposits.

Ash content is extremely important when comparing each of the deposits in terms of purity and humic acid content. The A&L (qualitative) and CDFA (quantitative) are common methods to measure humic acid. However, one of the main problems is in the consistency of reporting, since some methods report humic acid and others humic substance. Furthermore, neither method is conducted on an "ash-free" basis, which often leads to inaccuracies in reporting or communicating humic acid content. Based on the six locations, humalite from Alberta contained the lowest amount of ash (11.1%) compared to the Idaho deposit (84.7%), which was extremely high in ash and silica (carbonaceous shale origin). Leonardite from the North Dakota and Saskatchewan deposits are somewhat higher at 20 –22% ash as they are "essentially salts of humic acids admixed with mineral matter such as gypsum, silica and clay."[7] Lastly, the New Mexico deposits are the most common and widely talked about reserves but they also have a high ash content of 35.7%. This is a result of being associated with carbonaceous claystone, mudstone and shale.[8]

Other key parameters like cation exchange capacity (CEC) are very important when evaluating the quality or purity of organic materials. It has been reported that higher CEC is often associated with highly decomposed organic matter. CEC was measured in all deposits with the New Mexico deposits ranging from 55-70 mmol kg^{-1} and Alberta greater than 200. Deposits from North Dakota and Wyoming were somewhat lower and in the range 100-140 mmol kg^{-1}.

Sulfur content also was significantly different between the deposits. Humalite from Alberta had the lowest sulfur content of 0.5% or less with the Wyoming and New Mexico reserves much higher at 2-3%. Furthermore, the Wyoming deposit also showed a higher humic acid and heat content than the New Mexico deposits even though these deposits have many similar characteristics.

Wax content is often overlooked when comparing different deposits, but it can affect the overall purity of the raw material, especially when extracting the humic acid material from the ore. North Dakota based deposits have higher wax content compared to Alberta, in which waxes were non-detectable. Sodium is another factor that varied between deposits. This depends on whether the deposits originated in bodies of salt or fresh water. The highest amounts were reported in the North Dakota and Saskatchewan based deposits.

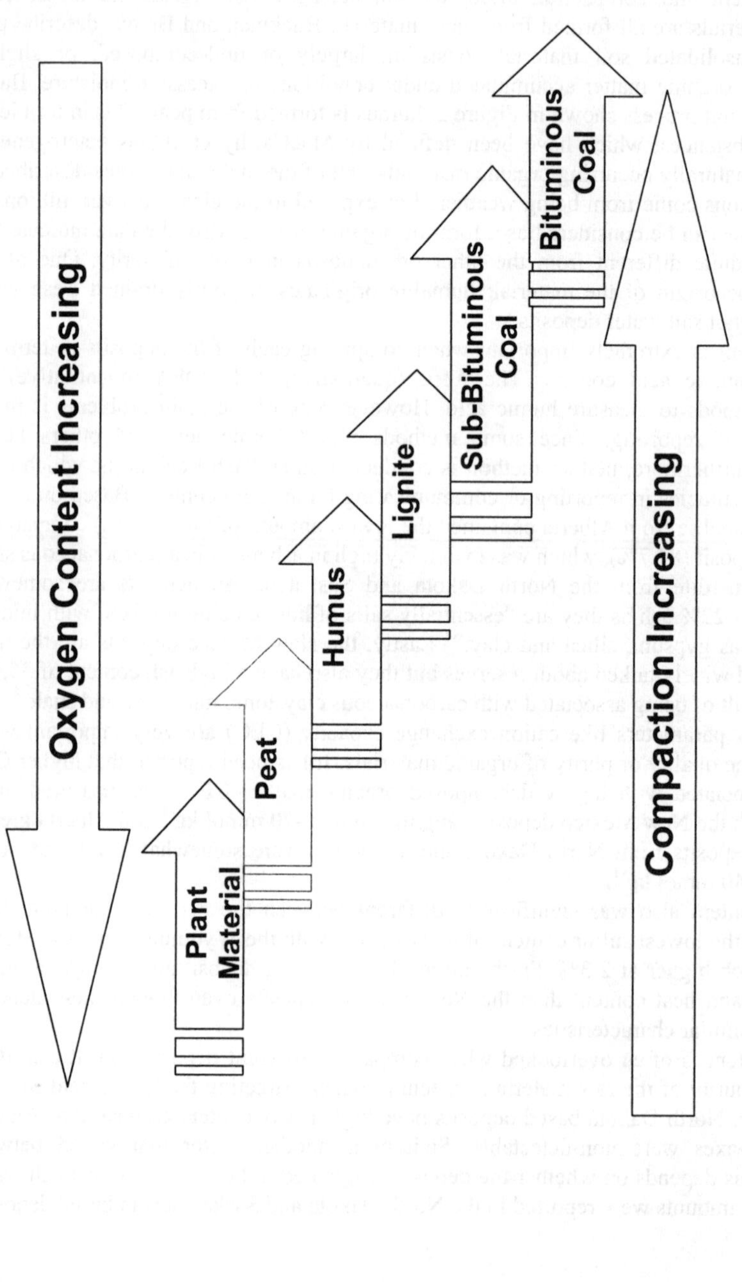

Figure 2 *The formation of organic matter and coal*

4 CONCLUSIONS

Based on the results and work over the last 25 years, the application and usage of humic substances have been evident but reserved. Opportunities exist today to further exploit and demonstrate the value of such products. In the discussion above, there are many similar characteristics among the deposits but some have distinct advantages over the others depending on their application in the agricultural, horticultural, environmental or industrial areas. Overall, humalite has shown to have the highest CEC and lowest ash content, thus indicating a higher purity than some of the other deposits. This becomes important when evaluating the deposits and their uses.

There are many issues today that make the use and understanding of humic substances a problem. One of these is quality. Quality products that are readily available on a commercial scale must be consistent to achieve success. Key parameters need to be identified and analysis methods for such deposits have to be more quantifiable. Education on the benefits of such substances and the interpretation of data are critical to the future. If industry wants to survive or progress further, we must have products that are beneficial to the end users, contain low levels of metals, improve plant growth and minimize environmental impact. Environmentally friendly and value-added products are the future.

References

1. M. Schnitzer, H. Dinel, T. Pare, H.-R. Schulten and D. Ozdoba, this volume, p. 315.
2. F.J. Stevenson, *Humus Chemistry; Genesis, Composition, Reactions*, Wiley, New York, 1982.
3. M.H.B. Hayes and F.L. Himes, in *Interactions of Soil Minerals with Natural Organics and Microbes*, eds. P.M. Huang and M. Schnitzer, Special Publication 17, Soil Science Society of America, Madison, WI, 1986, p. 104.
4. G.K. Hoffman, J.M. Barker and G.S. Austin, *New Mexico Bureau of Mines and Mineral Resources*, Campus Station, Socorro, New Mexico, 1995, p. 55.
5. H.O. Buckman and N.C. Brady, *The Nature and Properties of Soils*, The Macmillan Company, Toronto, Canada, 1969.
6. P. MacCarthy, R.L. Malcom, C.E. Clapp and P.R. Bloom, eds., *Humic Substances in Soil and Crop Science*, Soil Science Society of America, Madison, WI, 1990.
7. W.W. Fowkes and C.M. Frost, *Leonardite*, North Dakota Geological Survey, Bulletin 63, 1973, p. 72.
8. G.L. Hoffman, D.J. Nikols, S. Stuhec and R.A. Wilson, *Evaluation of Leonardite (Humalite) Resources of Alberta*, Energy, Mines and Resources Canada, Alberta Research Council, Canada, 1993.

SOME CHEMICAL AND SPECTROSCOPIC CHARACTERISTICS OF SIX ORGANIC ORES

M. Schnitzer,[1] H. Dinel,[1] T. Paré,[1] H.-R. Schulten[2] and D. Ozdoba[3]

[1] Eastern Cereals and Oilseed Research Centre, Agriculture and Agri-Food Canada, Ottawa, Ontario, Canada K1A 0C6
[2] University of Rostock, Rostock, Germany
[3] Luscar Ltd, Edmonton, Alberta, Canada T5J 3J1

1 INTRODUCTION

Six organic ore samples from different geographic locations in North America (North Dakota, Utah, Wyoming, New Mexico and Alberta) were submitted to us by Luscar Ltd. of Edmonton, Alberta. In order to provide chemical background information on these materials, each ore sample was analyzed by chemical methods, Fourier-transform infrared spectrophotometry (FTIR), CP-MAS ^{13}C NMR spectroscopy and by Curie Point-pyrolysis-gas chromatography-mass spectrometry (Cp-Py-GC-MS). It was hoped that a more comprehensive knowledge of the chemical composition and analytical characteristics of these ores would provide us with better insight as to how to use these materials either in agriculture or in other areas. Simultanuously, we were interested in comparing the analytical data on the six ores with those obtained previously on a representative soil humic acid (HA) and a fulvic acid (FA).[1]

2 MATERIALS AND METHODS

2.1 Origins of Samples

The organic ores samples were collected at the locations listed in Table 1. Geological characteristics of the ores are described in the preceding paper.[2] The HA was extracted from the Ah horizon of the Beaverhill soil, a Haploboroll in Northern Alberta, while the FA was separated from the Bh horizon of a Spodosol in Prince Edward Island, Canada. Analytical data for the soil HA and FA were taken from a previous publication.[1] Each sample was air-dried, finely ground and stored in a tightly stoppered glass bottle.

2.2 Chemical Analyses

Triplicate samples were oven-dried and then heated at 700°C for 3h to determine the ash content. All results were expressed on a moisture- and ash-free basis. Total C, H and N analyses are made by dry-combustion and total S was measured by oxygen-flask combustion. Coefficients of variations between replicate analyses were ± 0.1%. Carboxylic groups and total acidity were determined by the Ca-acetate and $Ba(OH)_2$ methods, respectively.[3] Coefficients of variations between replicate analyses were ± 2.5%. Phenolic OH groups were computed by subtracting (in mmol/g) COOH groups from the total acidity.

Table 1 *Geographic origins of ores and humic materials*

Ore No. or humic material	*Geographic origin*
L9	North Dakota
L25	Utah
L26	New Mexico
L27	Wyoming
L28	North Dakota
L31	Alberta
Soil HA	Udic Boroll, Alberta
Soil FA	Spodosol, Prince Edward Island

2.3 FTIR Analyses

Fourier-transform infrared spectra were recorded on a Perkin-Elmer Model 16F PC FTIR spectrophotometer. KBr pellets were prepared by pressing under vacuum a mixture of 1 mg of oven-dried ore and 400 mg of KBr at 10,000 kg cm^{-2} pressure for 30 min.

2.4 CP-MAS ^{13}C NMR Analyses

CP-MAS ^{13}C NMR spectra were recorded on 300 mg of each moisture-free ore with a Bruker CXP-180 spectrometer equipped with a Doty Scientific probe at a frequency of 45.28 MHz. Single-shot cross-polarization contacts of 2 ms were used with matching radio frequency field amplitudes of 75 MHz. Up to 120,000 500-word induction decays were co-added with a delay time of 1s. These were zero-filled to 4K before Fourier transformation. Magic angle spinning rates were about 4 KHz. For a more detailed analysis of the CP-MAS ^{13}C NMR data, the total area (0-190 ppm) under the spectrum was integrated and divided into the following regions:[4] 0 to 40 ppm (aliphatic C), 41-60 ppm (C in OCH_3 and amino acids), 61-105 ppm (C in carbohydrates and other aliphatic structures substituted by OH groups), 106-150 ppm (aromatic C), 151-170 (phenolic C) and 171-190 ppm (C in CO_2H groups, esters and amides). Aromaticities were computed by expressing aromatic C (106-170 ppm) as a percentage of aliphatic C+ aromatic C (0-170 ppm).

2.5 Curie-Point Pyrolysis-Gas Chromatography-Mass Spectrometry

A Fischer Curie point pyrolyzer (Type 0316) was used for this purpose. The ore sample, suspended in distilled water, was coated on the wire and allowed to dry. The wire + ore were heated to 500°C in 9.9 sec. The pyrolyzer was connected to a 45-m DB5 column in a Varian 3700 gas chromatograph programmed from 40 to 280°C at a rate of 3°C/min. Mass spectra of the chromatographically separated peaks were run on a FinniganMAT mass spectrometer in the electron impact (EI) mode. The mass spectra were identified with the aid of NBS, Wiley, and in-house mass spectral libraries, and if available by comparisons with mass spectra of known compounds.

3 RESULTS AND DISCUSSION

3.1 Chemical Characteristics of Ore Samples, HA and FA

Except for sample L25, which contained close to 85.0% ash, elemental analyses of the five remaining ore samples were similar (Table 2). On a moisture- and ash-free basis, elemental analyses for samples L9, L26, L27, L28 and L31 averaged 63.67 ± 1.23% C, 3.97 ± 0.09% H, 1.32 ± 0.13% N, 1.52 ± 1.23% S and 29.49 ± 1.58% O. The five ore samples contained considerably more C than the HA and especially the FA; the latter, however, was richer in O than the HA and the five ore samples. As to S, ore samples L9, L25 and L28 contained between 2.0 and 3.0% S, while samples L26, L27 and L31 contained only between 0.61 and 0.63% S. Sample L25 differed from the other ore samples in that it contained a very high ash content and also yielded very high values for H and O, which suggest that some of the ash components contributed to the elemental analyses.

Table 2 *Elemental analyses and ash contents of six ore samples, a soil HA and a soil FA expressed on moisture, ash-free basis*

Sample	*C*	*H*	*N*	*S*	*O*	*ash*
			% w/w			
L9	65.2	3.97	1.10	2.96	26.8	20.6
L25	36.7	11.0	1.83	2.03	48.5	84.7
L26	63.6	3.92	1.43	0.61	30.4	35.7
L27	63.0	4.12	1.45	0.63	30.7	17.0
L28	62.0	3.96	1.24	2.79	30.0	22.5
L31	64.4	3.90	1.46	0.63	29.6	11.1
Soil HA	56.4	5.50	4.10	1.10	32.9	2.00
Soil FA	50.9	3.30	0.70	0.30	44.8	1.00

As shown in Table 3, atomic H/C ratios for the ore samples, except L25, ranged from 0.72 to 0.78, atomic O/C ratios from 0.31 to 0.37, atomic N/C ratios were constant at 0.02, while S/C ratios varied from 0.01 to 0.05. Note that the atomic H/C ratio for FA is of the same order of magnitude as those of the five ores, while the same ratio for HA is higher than those of FA and the five ores.

Carboxyl groups in the ores and humic substances were determined chemically by ion exchange with calcium acetate.[3] If the data for L25 are omitted, CO_2H groups ranged from 2.0 to 3.6 mmol/g (Table 4). All of the values for COOH groups were lower than those for the same groups in HA and especially in FA. Concentrations of phenolic OH groups in the five ores were higher than that in HA but similar to that in FA. Total acidities of the five ore samples approached that for HA but were significantly lower than that for FA. Between 21.7 and 40.8 % of the total O in the ores was found in COOH groups (Table 5), but only between 14.9 and 19.5 % of the O occurred in phenolic OH groups. Thus, a smaller proportion of the total O in the ores occurred in COOH groups than in HA and FA, but the reverse was true for phenolic OH groups.

Table 3 *Atomic ratios of six ore samples, a soil HA and a soil FA*

Sample	*H/C*	*O/C*	*N/C*	*S/C*
L9	0.73	0.31	0.02	0.05
L25	3.59	0.99	0.05	0.05
L26	0.74	0.36	0.02	0.01
L27	0.78	0.37	0.02	0.01
L28	0.76	0.36	0.02	0.04
L31	0.72	0.34	0.02	0.01
Soil HA	1.18	0.45	0.07	0.02
Soil FA	0.79	0.67	0.01	0.01

Table 4 *Functional group analyses of the six ores, a soil HA and a soil FA (mmol/g)*

Sample	*CO_2H*	*ph-OH*	*Total Acidity*
L9	3.4	2.5	5.9
L25	2.7	2.5	5.2
L26	3.1	3.7	6.8
L27	3.6	3.3	6.9
L28	3.1	3.6	6.7
L31	2.0	2.9	4.9
Soil HA	4.9	2.1	7.0
Soil FA	7.8	3.1	10.9

Table 5 *Distribution of oxygen in functional groups in ores and in HA and FA*

Sample	*% of total O in CO_2H*	*% of total O in ph-OH*	*% of total O in CO_2H+ph-OH*
L9	40.6	14.9	55.5
L26	32.6	19.5	52.1
L27	37.5	17.2	54.7
L28	33.0	19.2	52.2
L31	21.7	17.9	39.6
Soil HA	47.7	10.2	57.9
Soil FA	55.7	11.1	66.8

3.2 FTIR Spectroscopy of Ores

FTIR spectra of ores L9, L25, L26, L27 and L31 are shown in Figure 1. All spectra exhibit broad bands due to extensive overlapping of individual absorbances. The major absorbances are broad bands near 3400 cm^{-1} (H-bonded OH), shoulders near 1700 cm^{-1} (C-O stretch of C = O and CO_2H), strong peaks near 1600 cm^{-1} (C-O stretch of COO^-), shoulders near 1380 cm^{-1} (CH_2- and CH_3- bending, C-OH deformation of CO_2H, COO^- symmetric stretch), and small broad absorbances near 1200 cm^{-1} (C-O asymmetric stretch and OH deformation of CO_2H, C-OH bending of phenols and of tertiary alcohols). In addition, the spectra of all ores show small but distinct bands near 2900 cm^{-1} (aliphatic CH_3 stretch) and near 2820 cm^{-1} (aliphatic C-H stretch). It is somewhat surprising that the latter two bands are not stronger in view of the high aliphaticity indicated for these materials by CP-MAS ^{13}C NMR and Cp-Py-GC/MC. The overall impression conveyed by the FTIR spectra is that of materials rich in OH, CO_2H and mainly COO^- groups, that is CO_2H groups complexed with metals. Unfortunately, little structural information on these materials is provided by FTIR. Strong absorbances in the FTIR spectra of ores L25 and L26, which contain 84.7% and 35.7% ash, respectively, in the 3600 cm^{-1} region and below 1200 cm^{-1} indicate the predominance of vermiculite and kaolinite, respectively, among the ash components.

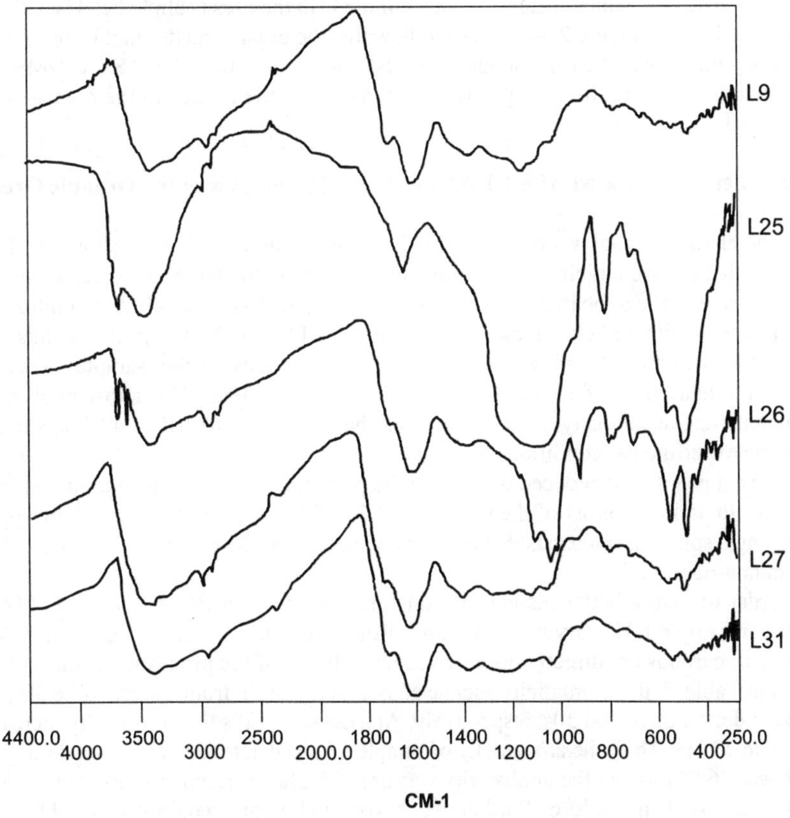

Figure 1 *FTIR spectra of ores L9, L25, L26, L27 and L31*

3.3 CP-MAS ^{13}C NMR Spectra of Ore Samples

The most prominent resonance in all spectra (Figure 2) is at 30 ppm, due to $(CH_2)_n$ in long aliphatic chains, which appear to be important chemical structures in these materials. Other shoulders or small peaks can be seen in all spectra near 40 ppm, which could include contributions from both n-alkyl C and amino acid C. The spectra of ores L27 and L28 show small resonances near 55 ppm that may indicate the presence of C in OCH_3 or amino acids. The spectrum of L9 shows a well developed signal at 72 ppm, that of L25 exhibits small signals at 60, 72, 80, 85, 94, 98 and 104 ppm and that of L28 shows a small but broad resonance near 72 ppm, while the spectrum of L31 shows small signals near 65, 75, 90 and 104 ppm. All of the signals in the 60-105 ppm region arise from O-alkyl (carbohydrate) C. The strong resonances at 130 ppm in all the spectra are due to aromatic C not substituted by N or O. It is likely that these resonances arise from alkylaromatics.[1] The small signals near 153 ppm are due to C in phenolic structures. Note that the spectrum of L25 shows several small peaks at 147, 153, 157 and 165 ppm that appear to be due to phenolic OH groups bonded to different ring carbons. The broad signals in the 170-180 ppm region are attributed to C in CO_2H groups. Amides and esters could also contribute to these peaks.

The CP-MAS ^{13}C NMR data are summarized in Table 6. Inspection of the data shows a similar C distribution in all ores except L25, which contains 85% ash. The aromaticity ranges from 27.3 to 42.8%, which means that most of the C in the ores is aliphatic. The CP-MAS ^{13}C NMR data for L25 (Figure 2 and Table 6) show that the organic matter in this material is richer in carbohydrates than the organic matter in the other ores. The CP-MAS ^{13}C NMR data also show that the concentrations of phenolic C and C in CO_2H groups in the ores are of similar magnitudes.

3.4 Problems Associated with CP-MAS ^{13}C NMR Analyses of the Organic Ores

From the elemental analysis data in Table 2 and the atomic H/C ratios shown in Table 3, it appeared that the aromaticities of the ores computed from the CP-MAS spectra were low and that, conversely, the aliphaticities were too high. In order to obtain additional information on these points we focused on sample L31 and ran CP-MAS ^{13}C NMR spectra on this sample at different delay times. As shown in Table 7, the aromaticity of this sample increased from 32.0% at a delay time of 1 s to 58.0% at a delay time of 300s. This substantial increase in aromaticity and accompanying decrease in aliphaticity (from 68.0% to 42.0%) suggests that under the experimental conditions employed in this investigation the CP (cross polarization) efficiency appears to be reduced because of the presence of unprotonated aromatic carbons in the ores.[5] In addition, most of the CP-MAS ^{13}C NMR spectra in Figure 2 show appreciable magic-angle spinning sidebands that reduce the intensity of the centerband and make CP-MAS spectra non-quantitative.[5]

In order to gain a better insight into this problem, we ran Bloch decay ^{13}C MAS NMR spectra on sample L31, varying delay times from 2 to 120s. In Bloch decay ^{13}C MAS NMR spectra, all carbons are directly observed independently of the proximity of the carbons.[6] As shown in Table 7, the aromaticity increased at a delay of 2s from 65.9% to 76.3%, and was 76.0% at delays of 10 and 30s, respectively. At a delay of 120s the aromaticity increased only slightly to 78.1%. Thus, the aromaticity of sample L31 as determined by Bloch decay ^{13}C MAS NMR was 76.8% so that the aliphaticity was only 23.2%, far removed from the CP-MAS ^{13}C NMR data listed in Table 6. Further ^{13}C NMR studies are required to establish whether quantitative analyses by CP-MAS ^{13}C NMR of materials like the organic ores studied here are possible and, if so, under what conditions.

Table 6 *Distribution of C(%) in six ore samples, a soil HA and a soil FA as determined by CP-MAS ^{13}C NMR*

Chemical shift range	*% C*							
(ppm)	*L9*	*L25*	*L26*	*L27*	*L28*	*L31*	*Soil HA*	*Soil FA*
0-40	41.5	22.2	33.9	34.4	32.1	44.3	24.0	15.6
41.60	11.8	14.6	10.2	8.9	12.3	13.5	12.5	12.8
61-105	14.9	23.2	9.7	12.2	13.9	7.8	13.5	19.3
106-150	21.0	23.2	33.3	31.2	28.9	26.6	35.0	30.3
151-170	4.6	8.1	7.0	6.3	7.0	4.2	4.5	3.7
171-190	6.2	8.7	5.9	6.9	5.9	3.6	10.5	18.3
Aliphatic (0-105 ppm)	68.2	60.0	53.8	55.5	58.3	65.6	50.0	47.7
Aromatic (106-150 ppm)	21.0	23.2	33.3	31.2	28.9	26.6	39.5	34.0
Phenolic (151-170 ppm)	4.6	8.1	7.0	6.3	7.0	4.2	4.5	3.7
Aromaticity[a]	27.3	34.3	42.8	40.3	38.2	32.0	44.1	41.6

[a] [(aromatic C + phenolic C) x 100]/[(aromatic C + phenolic C + aliphatic C)]

Table 7 *^{13}C NMR studies on sample L31*

Method	*Contact time (ms)*	*Delay (s)*	*% aromaticity*
CP-MAS	2	1	32.0
	2	300	58.0
Bloch decay-MAS		2	65.9
		10	76.3
		30	76.0
		120	78.1

3.5 Cp-Py-GC/MS of Ores

Two types of organic compounds were identified in the Cp-Py-GC/MS spectra of the ores: aliphatics and aromatics. As shown in Figure 3 and Table 8, n-alkanes ranged from C_6 to C_{29} and alkenes from $C_{5:1}$ to $C_{26:1}$. In addition, a C_{13} branched alkane and three branched alkenes ($C_{5:1}$, $C_{6:1}$ and $C_{12:1}$) (Figure 3 and Table 8) were also identified. It is noteworthy that no fatty acids were detected in the Cp-Py-GC/MS spectra of the ores although they were prominent, ranging from n-C_5 to n-C_{55}, in the Py-FIMS spectra (not shown here) of the same materials. Fatty acids apparently were decarboxylated to alkanes and alkenes on the Curie wire. Thus, the major aliphatic components of the ores are saturated and unsaturated fatty acids, alkanes and alkenes.

The major aromatic compounds identified in the spectra of the ores were benzene, methyl, dimethyl- and trimethylbenzenes, ethylmethylbenzene, styrene and benzofuran (see Figure 4 and Table 9). Other aromatics identified were methylbenzofuran, methylindene, dimethylbenzofuran and coumaranone. Additional aromatics identified (see Figure 5 and Table 9) were phenol, methyl and dimethylphenols, naphthalene and methyl-, dimethyl- and trimethylnaphthalenes. Thiophene was the only organic S-compound that we were able to identify.

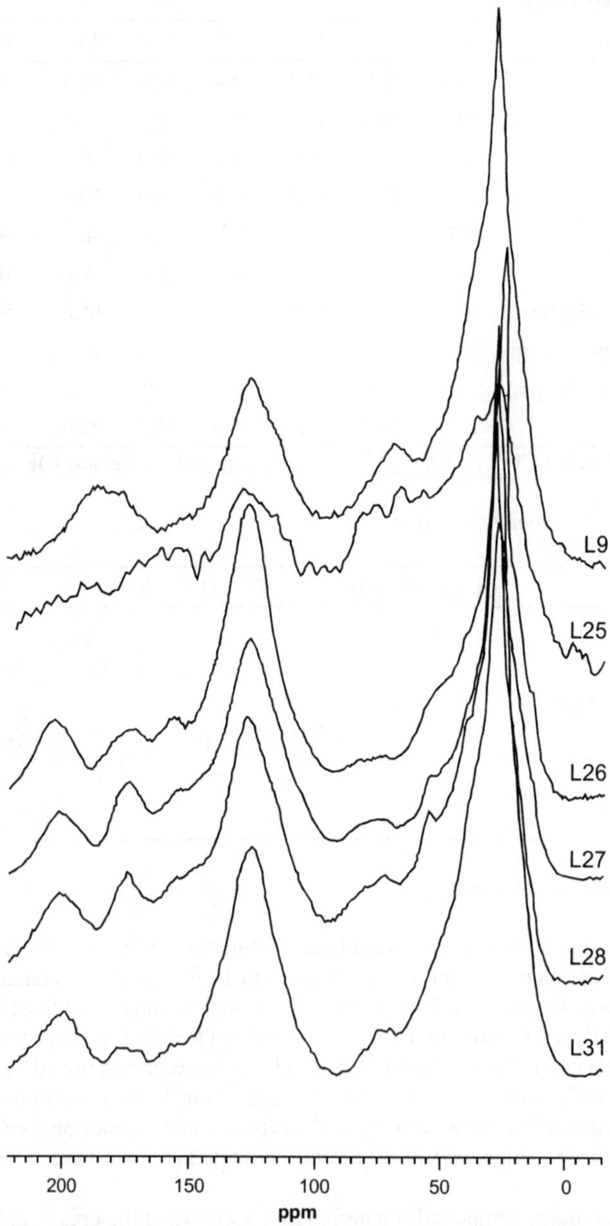

Figure 2 *CP-MAS ^{13}C NMR spectra of ores L9, L25, L26, L27, L28 and L31*

	Type of compound	*Relative Intensity*
	n-alkanes	XX
Ia	CH_3-$(CH_2)_4CH_3$	X
Ib	CH_3-$(CH_2)_5H_3$	X
Ic	CH_3-$(CH_2)_6H_3$	XXX
Id	CH_3-$(CH_2)_8H_3$	XX
Ie	CH_3-$(CH_2)_9H_3$	XX
If	CH_3-$(CH_2)_{10}CH_3$	XX
Ig	CH_3-$(CH_2)_{12}CH_3$	XX
Ih	CH_3-$(CH_2)_{13}H_3$	XX
Ii	CH_3-$(CH_2)_{14}H_3$	XX
Ij	CH_3-$(CH_2)_{15}H_3$	XX
Ik	CH_3-$(CH_2)_{16}H_3$	XX
Il	CH_3-$(CH_2)_{17}CH_3$	XX
Im	CH_3-$(CH_2)_{18}CH_3$	XX
In	CH_3-$(CH_2)_{19}CH_3$	X
Io	CH_3-$(CH_2)_{20}CH_3$	X
Ip	CH_3-$(CH_2)_{21}CH_3$	X
Iq	CH_3-$(CH_2)_{22}CH_3$	X
Ir	CH_3-$(CH_2)_{23}CH_3$	X
Is	CH_3-$(CH_2)_{24}CH_3$	X
It	CH_3-$(CH_2)_{25}CH_3$	X
Iu	CH_3-$(CH_2)_{26}CH_3$	X
Iv	CH_3-$(CH_2)_{27}CH_3$	X
	alkenes	
IIa	CH_2=CH-CH=CH-CH_3	XX
IIb	CH_3-CH=CH-CH=CH_2-CH_3	XX
IIp	CH_2=CH-$(CH_2)_{17}$-CH_3	XX
IIq	CH_2=CH-$(CH_2)_{18}$-CH_3	XX
IIr	CH_2=CH-$(CH_2)_{19}$-CH_3	X
IIs	CH_2=CH-$(CH_2)_{23}$-CH_3	X
	Branched alkanes	
IIIa	CH_3-CH_3-CH-$(CH_2)_9$-CH_3	X
	Branched alkenes	
IVa	CH_2=CH_3-C-CH=CH_2	X
IVb	CH_2=CH-CH_3-CH-CH_2-CH_3	X
IVc	CH_2=CH $(CH_2)_2$-CH_3-CH-$(CH_2)_5$-CH_3	X

Figure 3 *Chemical structures of aliphatics identified by Cp-Py-GC-MS*

Table 8 *Alkanes and alkenes identified in ores by Cp-Py-GC/MS*

Alkanes	*Alkenes*
Ia. Hexane	IIa. 1,3-Pentadiene
Ib. Heptane	IIb. 2,4-Hexadiene
Ic. Octane	IIc. 1-Hexene
Id. Decane	IId. 1-Heptene
Ie. Undecane	IIe. 1-Octene
If. Dodecane	IIf. 1-Nonene
Ig. Tetradecane	IIg. 1-Decene
Ih. Pentadecane	IIh. 1-Undecene
Ii. Hexadecane	IIi. 1-Dodecene
Ij. Heptadecane	IIj. 1-Tetradecene
Ik. Octadecane	IIk. 1-Pentadecene
Il. Nonadecane	Iil. 1-Hexadecene
Im. Eicosane	IIm. 1-Heptadecene
In. Heneicosane	IIn. 1-Octadecene
Io. Docosane	IIo. 1-Nonadecene
Ip. Tricosane	IIp. 1-Eicosene
Iq. Tetracosane	IIq. 1-Heneicosene
Ir. Pentacosane	IIr. 1-Docosene
Is. Hexacosane	IIs. 1-Hexacosene
It. Heptacosane	
Iu. Octacosane	
Iv. Nonacosane	
Branched Alkanes	
IIIa. 2-methyl-tridecane	
Branched Alkenes	
IVa. 2-methyl-1,3-butadiene	
IVb. 3-methyl-1-pentene	
IVc. 5-methyl-1-undecene	

Many of the aromatic rings in Figures 4 and 5 are substituted with alkyl groups. It is likely that the latter are remnants of longer alkyl chains that act as bridges between aromatic rings. This is in line with earlier observations that alkylaromatics are important "building blocks" of humified organic matter in soils.[1]

Because of the considerable degradation on the Curie wire it is not possible to ascribe the organic compounds identified to any specific precursors. It is likely that many of these compounds originate from highly aliphatic biopolymers,[7] lignins, polyphenols and polycyclic aromatics in the ores.

3.6 Composition of L31 Ash

The ash composition of L31 is compared with that of bituminous coal in Table 10.[8] The data show that the L31 ash contains fewer acidic (SiO_2, Al_2O_3, Fe_2O_3, etc) components but more basic (CaO, MgO, Na_2O and K_2O) ash components than does bituminous coal. This suggests that L31 is in the early stage of coalification.[8]

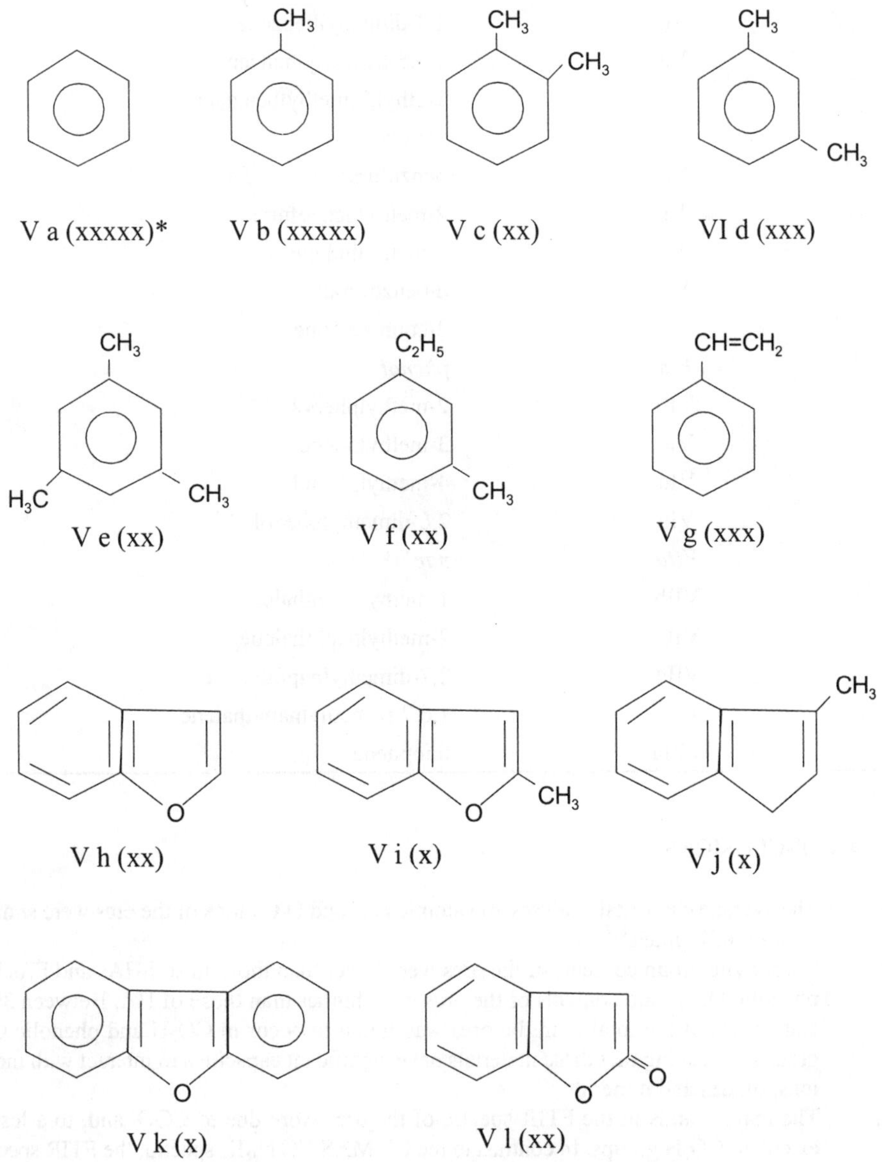

V a (xxxxx)* V b (xxxxx) V c (xx) VI d (xxx)

V e (xx) V f (xx) V g (xxx)

V h (xx) V i (x) V j (x)

V k (x) V l (xx)

Figure 4 *Chemical structures of aromatics identified by Cp-Py-GC-MS*

Table 9 *Aromatic compounds identified in ores by Cp-Py-GC/MS*

	Compound
Va	*benzene*
Vb	1-methylbenzene
Vc	1,2-dimethylbenzene
Vd	1,3-dimethylbenzene
Ve	1,3,5-trimethylbenzene
Vf	1-ethyl,3-methylbenzene
Vg	*styrene*
Vh	benzofuran
Vi	2-methylbenzofuran
Vj	1-methylindene
Vk	dibenzofuran
Vl	2-coumaranone
VIa	*phenol*
VIb	2-methylphenol
VIc	3-methylphenol
VId	4-methylphenol
VIe	2,6-dimethylphenol
VIIa	*naphthalene*
VIIb	1-methylnaphthalene
VIIc	2-methylnaphthalene
VIId	2,7-dimethylnaphthalene
VIIe	1,6,7-trimethylnaphthalene
VIIIa	thiophene

4 CONCLUSIONS

1. The average elemental analyses and atomic H/C and O/C ratios of the ores were similar to those of lignites.[8,9]
2. Carboxylic group contents of the ores were lower than those in soil HAs and FAs but phenolic OH group contents of the ores were higher than those of HA. Between 39.6 and 55.5% of the total O in the ores was found to occur in CO_2H and phenolic OH groups, indicating that these materials have significant capacities to interact with metal ions, oxides and minerals.
3. The major bands in the FTIR spectra of the ores were due to COO^- and, to a lesser extent, to CO_2H groups. In contrast to the CP-MAS ^{13}C NMR spectra, the FTIR spectra did not show the dominant occurrence of aliphatic CH_3 and CH_2 groups in the ores. The reasons for these divergencies are not well understood at this time.

4. CP-MAS ^{13}C NMR spectra of the ores showed that the most prominent components of the ores were long-chain aliphatic carbons such as those present in n-fatty acids, n-alkanes and n-alkyl esters. The second strongest resonance was due to aromatic carbons. The aromaticity of ores ranged from 27.3 to 42.8%. From comparisons with Bloch decay ^{13}C MAS NMR data presented in this paper, it does not appear that CP-MAS ^{13}C NMR as used in this study provided quantitative information on the C distribution in the organic ores that we investigated.
5. Cp-Py-GC-MS spectra of all ores showed the presence of two dominant components in the ores: aliphatics and aromatics. The major aliphatic components identified were C_6 to C_{29} n-alkanes, $C_{5:1}$ to $C_{26:1}$ alkenes, a C_{13} branched alkane and several branched alkenes. No fatty acids were identified, probably because they were decarboxylated to alkanes and methyl-, dimethyl-, trimethyl- and ethylmethylbenzenes, styrene, benzofuran, methyl- and dimethylbenzofurans, methyl indene, and coumaranone. Other aromatics identified were phenol, methyl- and dimethylphenols and naphthalene, methyl-, dimethyl- and trimethylnaphthalenes. The only organic S-compound identified was thiophene.

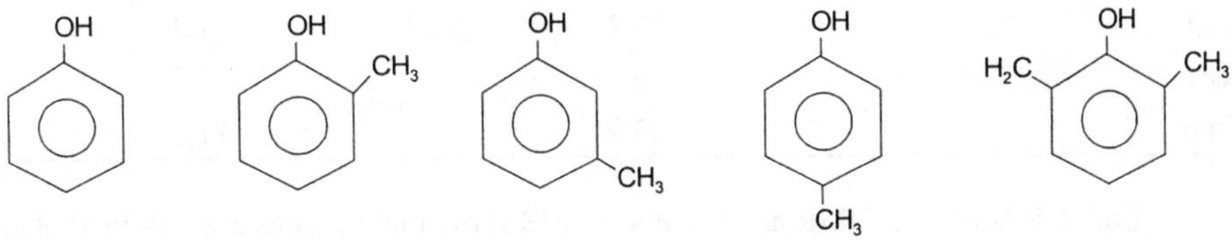

VI a (xxx) VI b (xx) VI c (xx) VI d (xx) VI e (x)

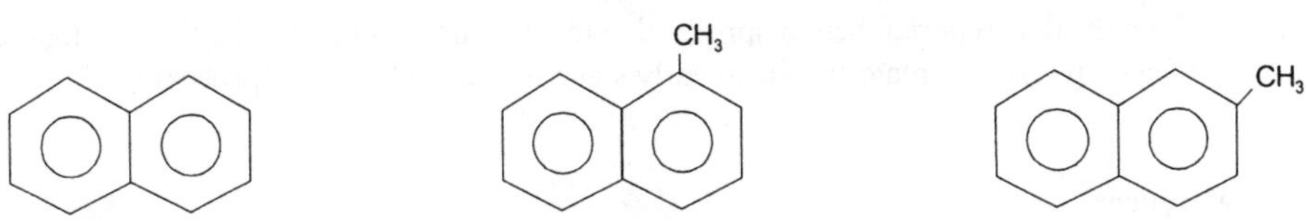

VII a (xx) VII b (xx) VII c (xx)

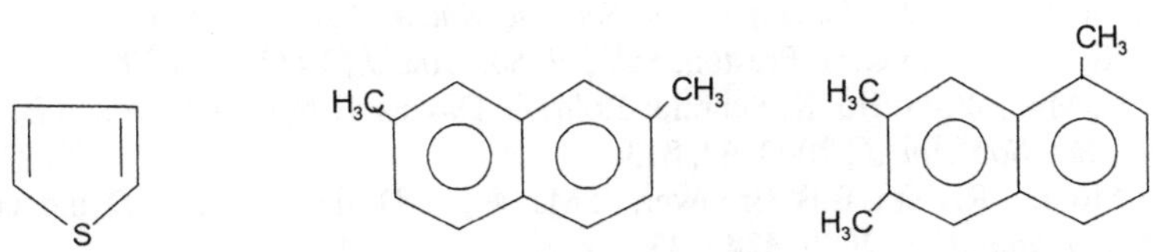

VIII a (x) VII d (xx) VIII d (xx)

Figure 5 *Chemical structures of phenolics, additional aromatics and one S-compounds identified by Cp-Py-GC-MS*

Table 10 *Ash composition of L31 and a bituminous coal*[8]

Type of components	*L31* % w/w	*Bituminous coal1*[8] % w/w
Acidic		
SiO_2	36.7	48.1
Al_2O_3	19.9	24.9
Fe_2O_3	8.4	14.9
TiO_2	0.5	1.1
P_2O_5	0.6	0.0
Total	66.1	89.0
Basic		
CaO	22.8	6.6
MgO	5.0	1.7
Na_2O	5.4	1.2
K_2O	0.7	1.5
Total	33.9	11.0

6. Both CP-MAS ^{13}C NMR and Cp-Py-GC/MS show that the ores are rich in aliphatic hydrocarbons and in aromatics. The wide occurrence of alkyl-bearing aromatics among the compounds identified points to the likely presence of alkylaromatics in the ores with long alkyl chains bridging aromatic rings.
7. From the data reported here it appears that the organic ores can be classified as lignites, which are organic materials in an early stage of the coalification process.[9]

References

1. M. Schnitzer, *Adv. Agron.*, 2000, **68**, 1.
2. D.M. Ozdoba, C. Blyth, R.F. Engler, H. Dinel and M. Schnitzer, this volume, p. 309.
3. M. Schnitzer and U.C. Gupta, *Soil Sci. Soc. Am. J.*, 1965, **29**, 274.
4. M. Schnitzer and C.M. Preston, *Soil Sci. Soc. Am. J.*, 1987, **51**, 639.
5. J.-D. Mao, W.-G. Hu, K. Schmidt-Rohr, G. Davies, E.A. Ghabbour and B. Xing, *Soil Sci. Soc. Am. J.*, 2000, **64**, 873.
6. F. Monteil-Riviera, E.B. Brouwer, S. Masset, Y. Deslandes and J. Dumonceau, *Anal. Chim. Acta*, 2000, **424**, 243.
7. M. Nip, E.W. Tegelaar, H. Brinkluis, J.W. De Leeuw, P.A. Schenck and P.J.H. Holloway, *Org. Geochem.*, 1985, **10**, 760.
8. W.R. Kobe, H.H. Schobert, S.A. Benson and F.R. Karner, in *The Chemistry of Low Rank Coals*, ed. H.H. Schobert, American Chemical Society, Washington, DC, 1984, p. 39.
9. V. Mallya and R.A. Zingaro, in *The Chemistry of Low Rank Coals*, ed. H.H. Schobert, American Chemical Society, Washington, DC, 1984, p. 132.

INTERPRETATION BY PRINCIPAL COMPONENT ANALYSIS OF PYROLYSIS-FIELD IONIZATION MASS SPECTRA OF LIGNITE ORES

H. Dinel,[1] M. Schnitzer,[1] T. Paré,[1] H.-R. Schulten,[2] D. Ozdoba[3] and T. Marche[4]

[1] Eastern Cereal and Oilseed Research Centre, Agriculture and Agri-Food Canada, Ottawa, Canada K1A 0C6
[2] University of Rostock, Rostock, Germany
[3] Specialty Products Division, Luscar Ltd, Edmonton, Alberta, Canada T5J 3J1
[4] Department of Environmental and Civil Engineering, Carleton University, Ottawa, Canada

1 INTRODUCTION

Pyrolysis-field ionization mass spectrometry (Py-FIMS) provides information on the building blocks of the organic components of lignite ores. In many Py-FIMS spectra over one thousand signals can be detected by mass spectrometry. One of the major problems facing users of these ores is to find rapid and reliable means to classify them in relation to their potential uses. Data provided by Py-FIMS are so complex that it is virtually impossible to differentiate between ores unless an unbiased tool is used to extract the most discriminatory features.

Principal-component analysis (PCA) is a multivariate technique for examining relationships among several quantitative descriptors or variables.[1] This approach is especially valuable in exploratory analyses to assess resemblances and differences among objects under study.[2] In order to successfully apply PCA to Py-FIMS data, we have developed strategies to handle the large numbers of peaks (descriptors) that exceed commonly available computer capacity.

The objective of this work was to apply PCA to isolate the most discriminating building blocks from the Py-FIMS information and to classify lignite ores on the basis of resemblances of their chemical compositions.

2 MATERIALS AND METHODS

2.1 Materials

Seven ores from North American lignite deposits were characterized by Py-FIMS spectrometry to obtain the chemical building block composition. The origins of the ores and the nature of the deposits are described in Ozdoba et al.[3] Some physical and chemical characteristics of the ores are presented in Table 1. Additional chemical and spectroscopic characteristics are reported by Schnitzer et al.[4]

2.2 Py-FIMS Analyses

About 100 μg of each dry ore was transferred to a quartz micro-oven and heated linearly from

373 to 823 K in the direct inlet system of the mass spectrometer at a rate of 10 $Kmin^{-1}$. A double-focusing Finnigan MAT 731 mass spectrometer was used for analyses. The ion source was kept at a pressure below 1mPa and at a temperature of 523 K. To avoid condensation of the volatilized products, the emitter was flashed-heated to 1773 K between magnetic scans. Between 35 to 40 spectra were recorded in the mass range m/z 18 to 1000. The FI signals of all spectra were integrated and plotted with the aid of a Finnigan SS200 data system to produce summed mass spectra.

Table 1 *Some physical and chemical characteristics of the lignite ores*

Ores	*Source*	*C*	*N*	*Ash*
L9	North Dakota	65.2	1.1	20.6
L23	"(chemically modified)"	42.5	1.0	37.3
L25	Utah	36.7	1.8	84.7
L26	New Mexico	63.6	1.4	35.7
L27	Wyoming	63.0	1.4	17.0
L28	North Dakota	62.0	1.2	22.5
L31	Alberta	64.4	1.5	11.1

2.3 Principal-Component Analysis

PCA was used to find the most significant peak signals of the Py-FIMS spectra of lignite ores. These analyses were performed with PRINCOMP (SAS Institute, Inc.) and the interpretation was in accordance with the conceptual approach of Gabriel[5] and Dinel et al.[2]

Briefly, each principal component is a linear combination of the original variables with coefficients equal to eigenvectors of the covariance matrix. The eigenvectors are orthogonal, so that the principal components represent jointly perpendicular directions through the space of the original variables.

For this study, we completed two successive PCA studies in order to select the most discriminating peaks or mass regions of the Py-FIMS spectra. This approach was necessary because the database for each spectrum exceeded the capacity of the computer system available. For the first PCA, the Py-FIMS spectra were divided into 45 mass regions of 20 peaks each. For the second PCA, 16 mass regions from the 45 that were significantly contributing to differentiate the ores were subdivided into 5 peak mass signals and statistically analyzed following the same approach as for the first analysis.

3 RESULTS AND DISCUSSION

3.1 Py-FIMS of the Ores

Identifications of the major signals in the Py-FIMS spectra of the lignite ores are based on the extensive researches by Schulten and Simmleit[6] and Schnitzer and Schulten.[7,8] The compounds identified in the whole spectra (m/z 0 to 900) are classified as saturated and unsaturated fatty acids, n-alkanes, alkenes, lignin dimers, alkyl aromatics, sterols and n-alkyl mono, di-, and tri-esters (Figures 1-3). For the purposes of this discussion, our attention will be focused on the mass region between m/z 250 and 350 since PCA defines this region as

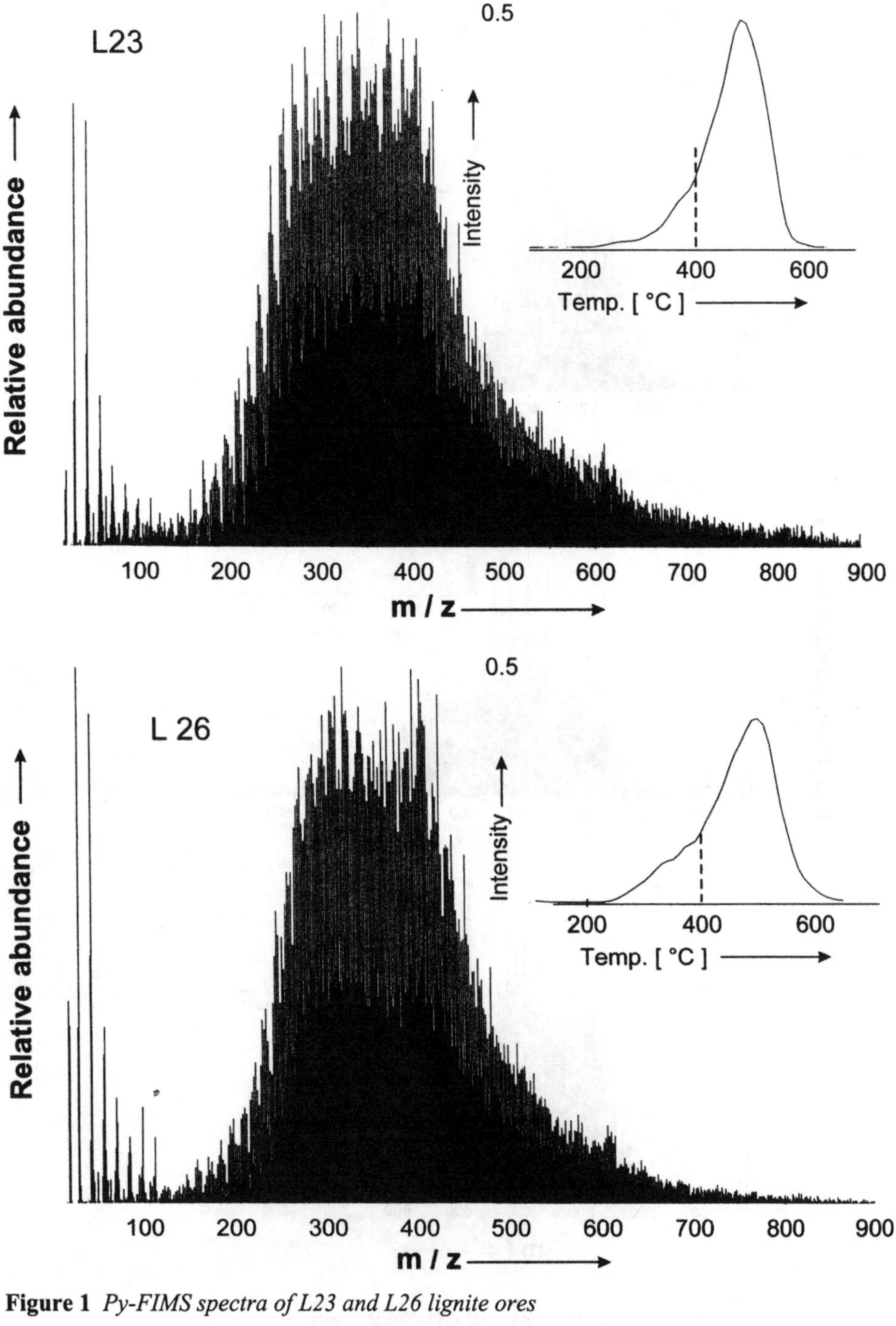

Figure 1 *Py-FIMS spectra of L23 and L26 lignite ores*

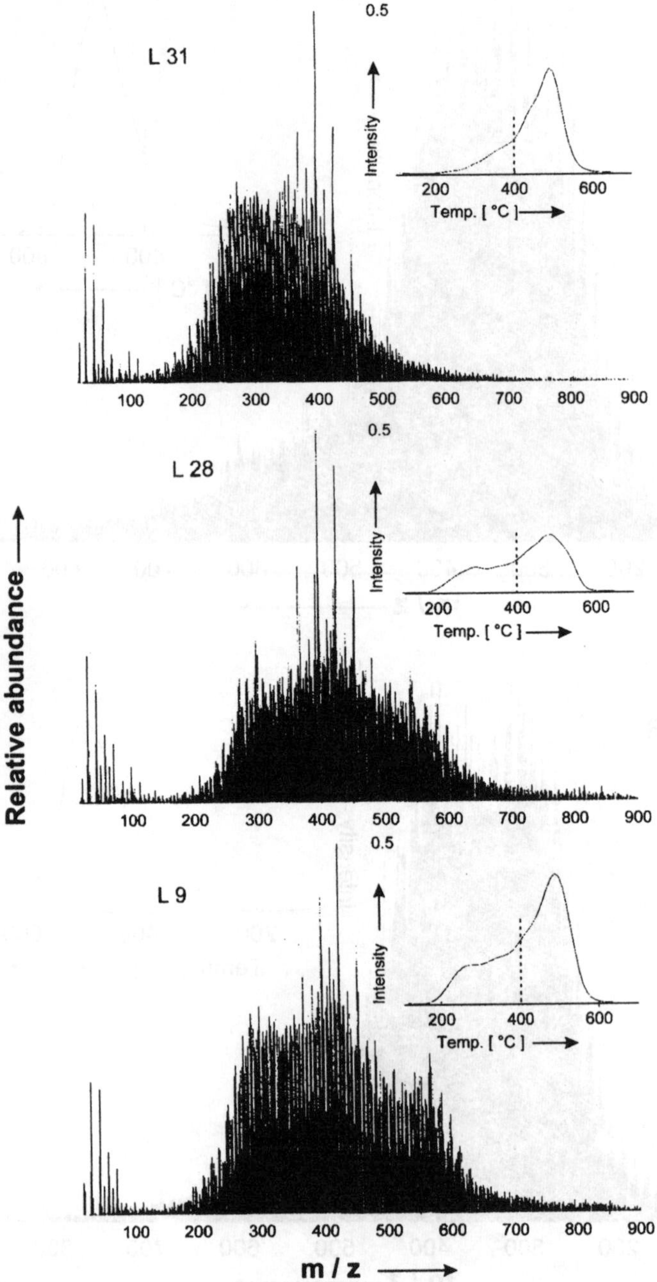

Figure 2 *Py-FIMS spectra of L9, L28 and L31 lignite ores*

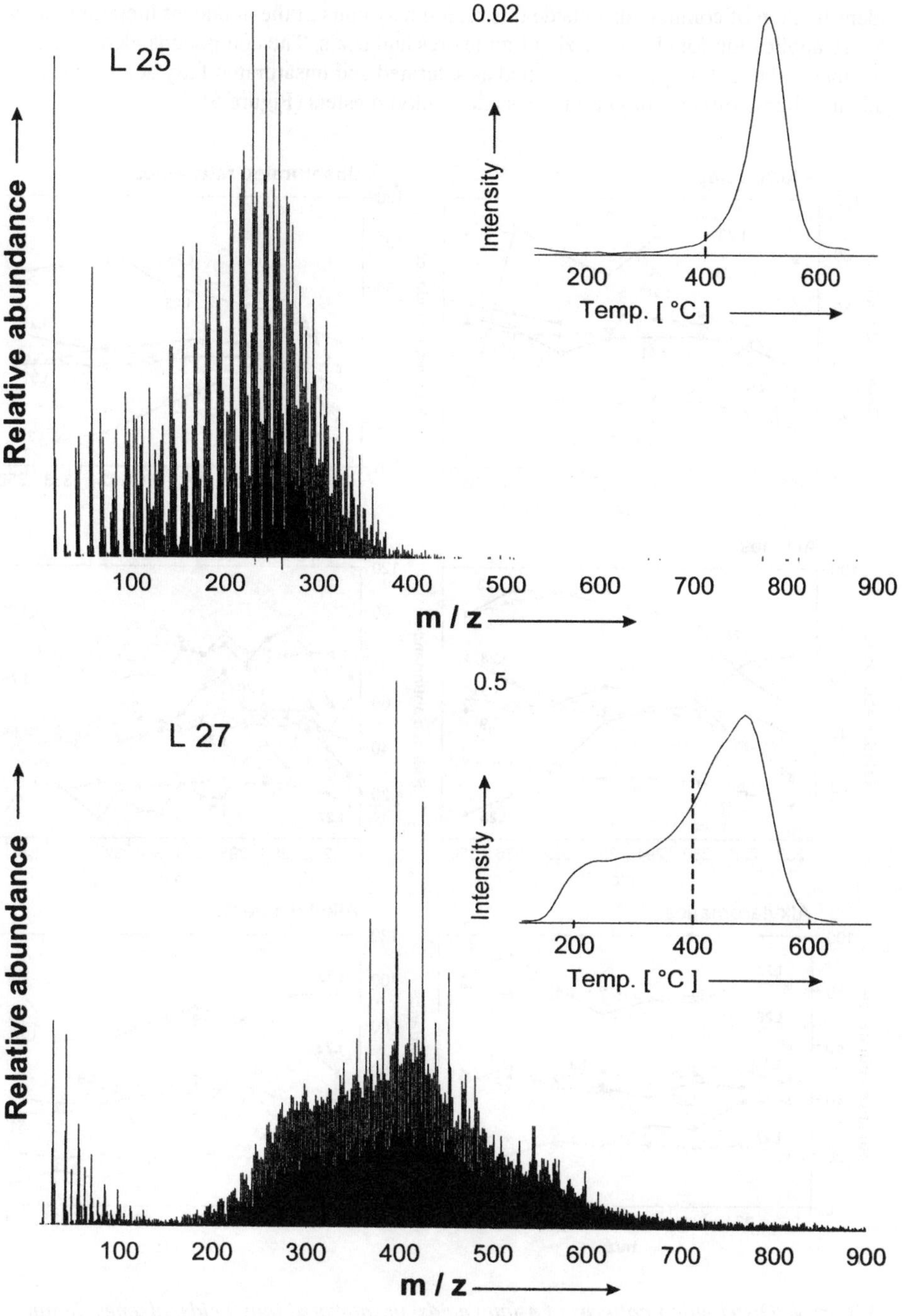

Figure 3 *Py-FIMS spectra of L25 and L27 lignite ores*

the most discriminating one for the seven lignite ores (Figure 4). Those interested in the identification of compounds outside this region may consult the abundant literature on Py-FIMS application for characterizing lignite ores and coals. The compounds identified in the m/z 250 to 350 region are classified as saturated and unsaturated fatty acids, n-alkanes, alkenes, lignin dimers, alkyl aromatics and n-alkyl diesters (Figure 5).

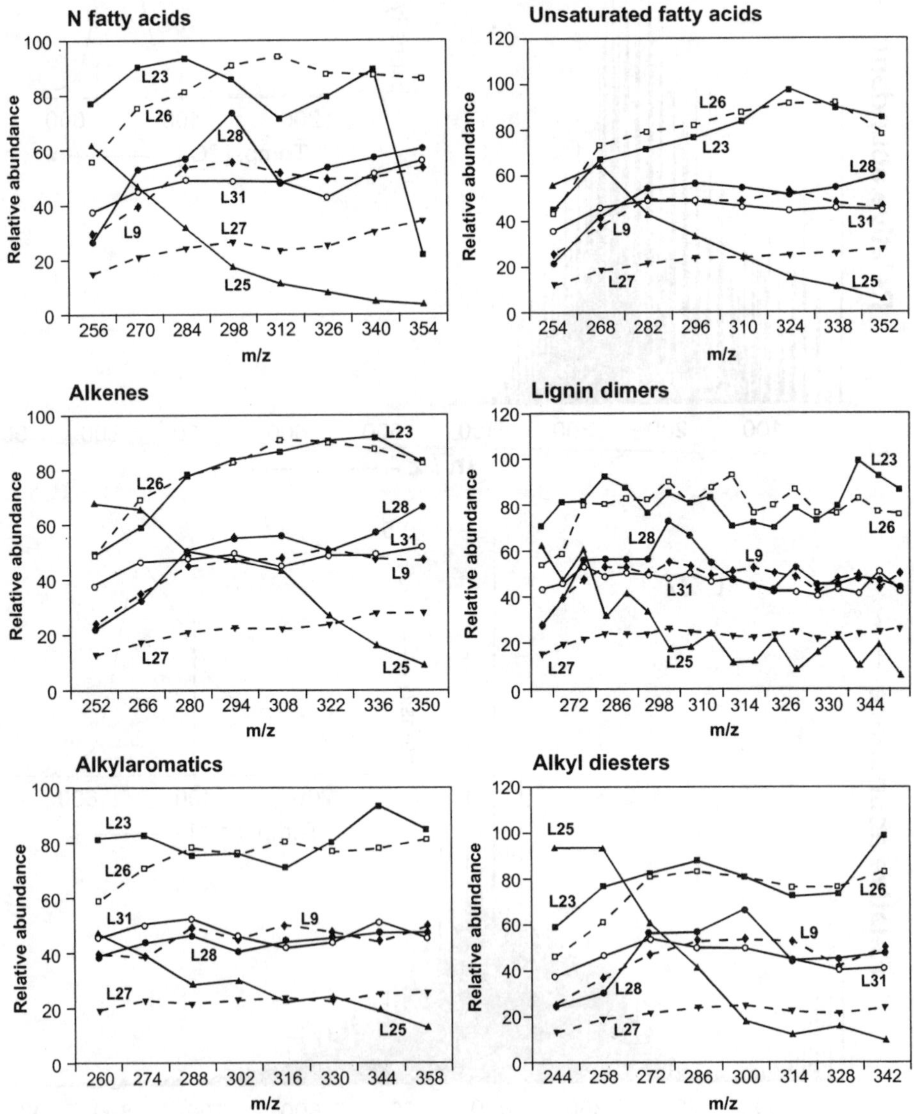

Figure 4 *Distribution patterns of n-fatty acids, unsaturated fatty acids, alkenes, lignin dimers, alkyl aromatics and alkyl diesters for the mass region from m/z 250 to 350*

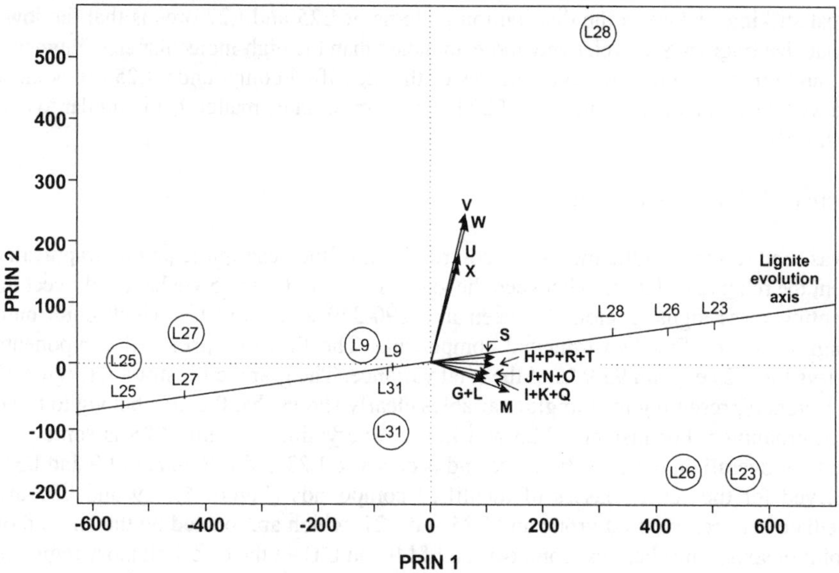

Figure 5 *Orthogonal distribution of the principal component (PRIN) analysis for the mass region m/z 250 to 310 and 600 to 619 of the seven lignite ores. PRIN1 vs PRIN2 explained 95.9% of the total variance. Letters G toT and U to X represent 5-peak mass region vectors from 290 to 359 and from 600 to 619, respectively. L9, L23, L25, L26, L27, L28 and L31 are lignite ores*

The signals corresponding to n-fatty acids in this region are at m/z 256 (C_{16}), 270 (C_{17}), 284 (C_{18}), 298 (C_{19}), 312 (C_{20}), 326 (C_{21}), 340 (C_{22}) and 354 (C_{23}). Thus, the fatty acids range from n-C_{16} to n-C_{23}, with n-C_{18} and n-C_{20} being the most abundant components. Signals typical of unsaturated fatty acids in this region are m/z 254 ($C_{16:1}$), 268 ($C_{17:1}$), 282 ($C_{18:1}$), 296 ($C_{19:1}$), 310 ($C_{20:1}$), 324 ($C_{21:1}$), 338 ($C_{22:1}$) and 352 ($C_{23:1}$). The most prominent unsaturated fatty acid was $C_{21:1}$.

These ores also contained n-alkanes and alkenes. The presence of n-alkanes is indicated by m/z 254 (C_{18}), 268 (C_{19}), 282 (C_{20}), 296 (C_{21}), 310 (C_{22}), 324 (C_{23}), 338 (C_{24}) and 352 (C_{25}). The following alkenes are present in these ores: m/z 252 ($C_{18:1}$), 266 ($C_{19:1}$), 280 ($C_{20:1}$), 294 ($C_{21:1}$), 308 ($C_{22:1}$), 322 ($C_{23:1}$), 336 ($C_{24:1}$) and 350 ($C_{25:1}$). The most abundant n-alkanes and alkenes in these ores are C_{23} and $C_{23:1}$, respectively.

The most prominent components of the ores are lignin dimers with 18 molecular fragments in our discriminatory region. These include m/z 246, 260, 272, 284, 286, 296, 298, 300, 310, 312, 314, 316, 326, 328, 330, 342, 344 and 356.

Other significant components of the discriminatory region are alkyl aromatics and n-alkyl diesters. Alkyl aromatics include m/z 260, 274, 288, 302, 316, 330, 344 and 358. The presence of n-alkyl diesters is indicated by m/z 244, 258, 272, 286, 300, 314, 328 and 342.

The ranking of these ores on the basis of the distribution patterns for the seven classes of identified compounds shows the same ranking order for all ores (Figure 4). Since the distribution of unsaturated fatty acids and n-alkanes were identical, distribution curves of n-alkanes are not shown. For instance, ores L26 and L23 are very similar and are located in the upper part of the graph, indicating that they contain high concentrations of the seven classes of identified compounds. A second group including L9, L28 and L31 is somewhat intermediate between the first group and L27 and L25. The latter belong to different types of ores as suggested by the distribution patterns of the seven classes of identified compounds.

The most striking feature in the distribution patterns of L25 and L27 ores is that the low-molecular size fragments are relatively more abundant than the high-molecular size fragments in L25, and this is true for the seven classes of the identified compounds. L25 ore is most likely a well developed lignite similar to L27 but contains organic matter that is similar to that found in soils.

3.2 Principal Component Analysis

PCA was used to separate the individual contributions of the descriptors and to graphically represent the relative differences between the various ores. In Figure 5 we have only vectors representing mass signal regions between m/z 290-359 and 600-619, which contribute meaningfully to the first two principal components. The first two principal components (PRIN1 vs PRIN2) explain 95.9 % of the total variance. The graphic illustration (Figure 5) of the vectors representing meaningful variables clearly shows that the ores belong to three different groupings. For instance, L23 and L26 are very similar, while L28 is somewhat different but overall on the evolution axis and is closer to L23 and L26 than to L9 and L31, as observed for the seven classes of identified compounds (Figure 4). L9 and L31 are intermediate between the first group and L25 and L27, which are located on the left end of the evolution axis. The chemical composition of L9 and L31 in the discriminatory region is similar to that of soil humic substances (data not shown). Based on the distribution pattern of the seven classes of identified compounds, L25 and L27 are considered to be different, probably due to the presence of organic matter similar to soil organic matter. The evolution of the ores based on the PCA can be ranked in the following relative order: L23>L26>L28>>>L9=L31>>>L27>L25. Except for L28, the distribution pattern of the seven classes of identified compounds confirms the ranking based on PCA.

4 CONCLUSIONS

The two-step PCA clearly shows that the analysis of a full Py-FIMS spectrum can be simplified by concentrating the analysis and interpretation on the most discriminatory regions located between m/z 250 and 310, and 600 and 619. In addition, the relative abundance of n-fatty acids, unsaturated fatty acids, alkanes, alkenes, lignin dimers, alkyl aromatics and n-alkyl diesters in the m/z 250-350 mass region shows a similar qualitative distribution pattern for all ores, confirming the validity of the PCA. The evolutionary ranking of lignite ores can be established by using the mass region between m/z 250 and 310.

References

1. A.M. Kshirsagar, *Multivariate Analysis*, Dekker, New York, 1972.
2. H. Dinel, G.R. Mehuys and M. Lévesque, *Soil Sci.*, 1991, **151**, 146.
3. D. Ozdoba, J. Blyth, R.F. Engler, H. Dinel and M. Schnitzer, this volume, p. 309.
4. M. Schnitzer, H. Dinel, T. Paré, H.-R. Schulten and D. Ozdoba, this volume, p. 315.
5. K.R. Gabriel, *Biometrika*, 1971, **58**, 453.
6. H.-R. Schulten and N. Simmleit, *Naturwissenschaften*, 1986, **73**, 618.
7. M. Schnitzer and H.-R. Schulten, *Sci. Total Environ.*, 1989, **81/82**, 19.
8. M. Schnitzer and H.-R. Schulten, *Soil Sci. Soc Am. J.*, 1992, **56**,1811.

A COMPARATIVE EVALUATION OF KNOWN LIQUID HUMIC ACID ANALYSIS METHODS

A.K. Fataftah, D.S. Walia, B. Gains and S.I. Kotob

Arctech, Inc., Chantilly, VA 20151, USA

1 INTRODUCTION

Humic substances (HSs) are brown or black multifunctional organic natural products with major agricultural and environmental roles. They are one of Earth's richest carbon reservoirs. HSs are complex aromatic macromolecules with various linkages between the aromatic groups. The different linkages include amino acids, amino sugars, peptides, aliphatic acids and other aliphatic compounds. The functional groups in humic substances include carboxylic (COOH), phenolic, aliphatic- and enolic-OH and carbonyl (C=O) structures of various types.[1]

Fulvic acids are the smallest of the humic substances and are soluble at all pH. Humin is insoluble at all pH. Humin is further along in the natural progression from live plants towards "dead" coals and carbon. Humic acids (HAs), the most important fraction of the HSs, are soluble at high pH and can be precipitated at pH <1. HAs are highly functionalized, carbon-rich biopolymers that stabilize soils as soil organic matter. Elemental analysis of humic acids shows that C, H, O, N, P and S generally account for 100% of their composition.[2] Humic acids are colloids and behave somewhat like clays, even though the nomenclature suggests they are acids that form true salts. When fully protonated HAs are acids and are named accordingly. The material is called a humate when the predominant cation on the exchange sites is other than hydrogen.[3]

The application of humic acids in agriculture is advantageous because they slowly release micronutrients to plants, enhance microbial growth, have high water-holding and buffering capacity, reduce soil erosion and stimulate plant growth. HAs increase the availability of phosphate to the plant by breaking the bonds between phosphate and iron or calcium. Phosphate is a stimulator of seed germination and root initiation in plants. In addition, HAs are very effective in converting iron to suitable forms to protect plants from chlorosis, even in the presence of high concentrations of phosphate. Humic acids also contribute to mineralization and immobilization of nitrogen in soil. The complexes formed between ammonia and humic acid are reported to release nitrogen slowly into the soil.[4]

Besides being a source of nutrients for plants, humic acids also affect the physico-chemical properties of soil. These properties are important in controlling the uptake of nutrients by the plant, their retention in the soil and counteracting soil acidity.[3]

Manufacturers worldwide are promoting the use of HA products in agriculture. These products are gaining acceptance by the agriculture community. Also, governments such as the United Arab Emirates and Turkey are including these products as part of their import specifications, which designate humic acid content. In the United States, all the states require registration of products specifying guaranteed analysis to comply with U.S. weights and measures laws. However, due to the lack of a standard method the HA content of the products is labeled based on arbitrary humic acid analysis methods.

HA optimum efficacy is known to depend critically on the concentration at the time of applications.[5] The optimum application rates of humic acid for plant growth have been reported to range from 50 to 350 ppm.[5] Testing these reported optimum application rates with three different humic acid products with HA contents of 3, 10 and 20 % resulted in a large range of the amount of water needed to dilute these products at the time of application. The number of gallons of water needed to dilute one gallon of 3 and 20% HA in these products ranges from 85 to 4000 gallons. This results in using a large amount of water to dilute products with high humic acid content. We noted in this study that the claimed HA content of a product depends on the method of analysis and does not necessarily indicate the actual value. Using the claimed value will lead to a false dilution factor, which will result in either no benefit of using the product if the dilution is too high or reaching a toxic level of HA if the dilution is too low at the time of application.

In this study three known methods of analysis were evaluated with HA products from different vendors. The methods evaluated were the acid precipitation method, the barium chloride precipitation method and the spectrophotometric method. The acid precipitation method is a gravimetric technique adapted from the Standard Methods for Soil Analysis, Part 2[6] and the California Department of Food and Agriculture (method # Ha4/JC, 2/86). The barium chloride method is another gravimetric method adapted from the American Colloid Company (Skokie, IL). The spectrophotometric method was adapted from procedures from the North Carolina Department of Agriculture.

2 MATERIALS AND METHODS

Ten samples of humic acid products from several manufacturers were used in this study. The source of HA in all products is coal and all the products are available in liquid form. Humic acids from Aldrich were used to prepare standard solutions. The acid precipitation, barium chloride precipitation and spectrophotometric methods were compared to evaluate HA contents of the ten different samples.

The acid precipitation method is a gravimetric technique adapted from Standard Methods for Soil Analysis, Part 2[6] and from the California Department of Food and Agriculture (method # Ha4/JC, 2/86). About 10 gm of each product was accurately weighed in a 50-mL centrifuge tube. Each sample was then treated with concentrated nitric acid solution until a pH of about 1.0 was reached. Acidified water was added to reach a total volume of 35 mL. Sample suspensions were vortexed for 1 min and centrifuged at 2000 rpm for 10 min to separate the precipitated humic acid. Supernatants were discarded without disturbing the HA precipitates, which were washed three times with equal volumes (35 mL) of acidified water. Tubes containing humic acid precipitates were dried in an oven at 100-105°C for 2 hours. Tubes were removed from the oven and weighed after cooling for 10 min. Dry weights of precipitates were calculated and humic acid percentage was calculated as shown in Figure 1A.

The barium chloride method is another gravimetric method adapted from the American Colloid Company. About 10 gm of each product was accurately weighed in a 250-mL centrifuge bottle. Each sample was diluted with 90 mL distilled water and the humic acid was precipitated by the addition of 10 mL of 12% barium chloride solution. Sample suspensions were centrifuged at 2000 rpm for 15 min to separate the precipitated humic acid, which was washed three times with distilled water. Bottles containing humic acid precipitates were dried in an oven at 100-105°C for 2 hours, removed from the oven and weighed after cooling for 10 min. Dry weights of precipitates were calculated and the humic acid percentage was calculated as shown in Figure 1B.

The spectrophotometric method was adapted from the North Carolina Department of Agriculture. A set of stock solutions was prepared for use in the standard curve preparation. All solutions were prepared in 1000 mL volumes using 1 L volumetric flasks. Stock solutions were prepared to contain 0, 3, 6, 9, 12 and 15 % humic acid on a weight to volume basis. All HAs were dissolved in 0.1 N KOH. Prior to analysis, a 40 mL aliquot of each stock solution was centrifuged at 2000 rpm for 10 min to remove any insoluble material. The supernatant of each stock solution was diluted 1:1000 and the absorbance was recorded at 450 nm. A calibration curve was created from the absorbance measurements. 40 mL aliquots of each of the ten humic acid samples were centrifuged at 2000 rpm for 10 min to remove any insoluble materials. Supernatants were diluted 1:1000 using water. The absorbance at 450 nm of diluted samples was recorded and used to calculate % humic acids from the calibration curve. A schematic diagram of the method is shown in Figure 1C.

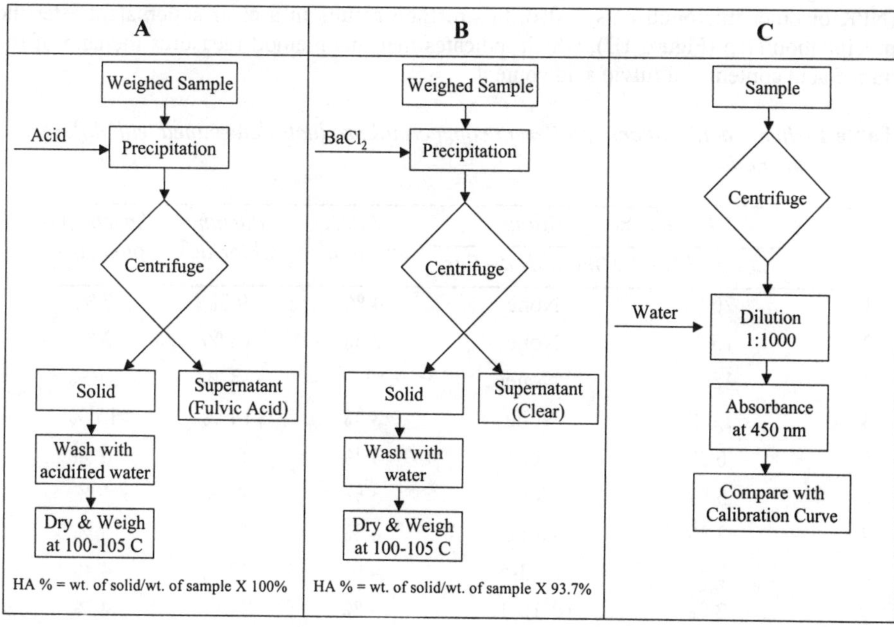

Figure 1 *Schematic diagram of A) Acid Precipitation Method, B) Barium Chloride Precipitation Method and C) Spectrophotometric Method*

3 RESULTS AND DISCUSSION

The results of the humic acid analysis of the ten commercial products using the three known methods are shown in Table 1.

3.1 Acid Precipitation Method

In this method, humic acid is precipitated as an acid. The method is based on the definitions of humic and fulvic acids. Humic acid precipitates at low pH but fulvic acid is soluble over the whole range of pH. This method precipitates humic acid, while fulvic acid remains in the supernatant. The acid precipitation method is part of the International Humic Substance Society (IHSS) standard method for the extraction of humic acid from solids and the extraction of the organic matter from soil.[6] The presence of mineral nutrients in the products does not affect the humic acid analysis results, which are consistent with the claimed value in most of the products.

3.2 Barium Chloride Method

In this method the humic acid is precipitated as barium humate, which results in higher apparent humic acid content as compared to other methods because of the high atomic weight of barium. The equation used for calculating the humic acid content in the barium chloride method is based on the assumption that the equivalent weight of the humic acid is 400 gm/eq. This assumption has no basis and the barium chloride method consistently gives high humic acid values, especially with products formulated with mineral nutrients (NPK or other micronutrients). Also, this method results in a clear supernatant after the precipitation step (Figure 1B), which indicates that this method measures the sum of the humic acid content and fulvic acid content.

Table 1 *Humic acid content of different commercial products determined with different methods*

Product	*Product compositions*		*Acid Precipitation*[a]	*Barium Chloride*[b]	*Spectrophotometric*[c]
	Claimed HA	*Mineral Nutrients*			
1	20 %	None	4 %	9 %	7 %
2	15 %	None	2 %	16 %	2 %
3	15 %	None	12 %	18 %	31 %
4	12 %	None	8 %	14 %	15 %
5	6 %	None	3 %	6 %	6 %
6	5 %	None	3 %	4 %	4 %
7	3 %	None	4 %	7 %	5 %
8	3 %	15-3-5	4 %	27 %	4 %
9	3 %	10-10-10	3 %	29 %	4 %
10	3 %	2 % Fe	3 %	14 %	4 %

[a] Acid precipitation method (California method); [b] Barium chloride method (American Colloid Co. method); [c] Spectrophotometric method (North Carolina method)

3.3 Spectrophotometric Method

In this method the humic acid content of liquid products is determined by measuring the absorbance using a spectrophotometer at a wavelength of 450 nm. This method sometimes gives higher humic acid content. The results are not consistent with the claimed humic acid content of some of the products. This method does not include any humic and fulvic acid separation and so it is measuring both humic and fulvic acid available in these products. This method requires standard curve preparation and includes more operational steps. The presence of mineral nutrients in commercial products also affects the results, although less so than in the barium chloride method.

3.4 Attempts at Fulvic Acid Determination

All the humic acid products obtained by alkali coal extraction contain different amounts of fulvic acid. Due to the lack of a standard method for fulvic acid analysis, different attempts were made to determine the fulvic acid content of some of the products evaluated in this study. These attempts were performed only on the acidic supernatant produced from the precipitation of the humic acid in the acid precipitation method. First, since the barium chloride method of humic acid analysis produced a clear supernatant, (indicating the precipitation of fulvic acid), barium chloride was added to the acidic supernatant. The result was no precipitation caused by the addition of barium chloride. Thereafter, the pH of the acidic mixture was increased to 8.0-8.5, which resulted in forming a precipitate and a clear supernatant. Only increasing the pH of the acidic supernatant, without the addition of barium chloride, formed the same precipitate. This indicates that the precipitate produced is due to increasing the pH of the mixture only and not to the addition of barium chloride. To characterize the precipitate produced, it was digested and analyzed for metals. The metal analysis showed that it contains mainly iron and potassium salts, which usually precipitate at high pH. This attempt did not precipitate a pure fulvic acid. The second attempt was to adsorb the fulvic acid on activated charcoal and determine the fulvic acid by the weight difference of the activated charcoal before and after adsorbing fulvic acid. This method was not accurate due to the small amount of fulvic acid absorbed on the activated charcoal. The result of fulvic acid determination using this approach on product number 4 (Table 1) was only 0.07 %, while the product was labeled with 2 % fulvic acid.

4 CONCLUSIONS

The results of this study indicate that the humic acid content claimed for a number of commercial products depends on the method of analysis. Humic acid analysis with barium chloride (and sometimes spectrophotometric) methods gives higher values for a given sample. The commercial humic acid products formulated with mineral nutrients give inaccurate analysis of the humic acid content with the barium chloride method. The acid precipitation method, which is also known as the California method, is based on the operational definition of humic acid. Also, this method is part of the IHSS standard method of extraction of humic acids from solids and is included in the Standard Methods for Soil Analysis.[6] Since the efficacy of HA on plant growth critically depends on the HA concentration of the solution at the time of application,[5] it is important for the end user to know the actual humic acid content of the product.

References

1. F.J. Stevenson, *J. Environ. Quality*, 1972, **1**, 333.
2. A.K. Fataftah, PhD Thesis, Northeastern University, Boston, 1997.
3. T.L. Senn and A.R. Kingman, *A Review of Humus and Humic Acid Research*,

Series No. 145, S.C. Agricultural Experiments Station, Clemson, SC, 1973.
4. G.H. Beames, in *Source Book on Humic Acids*, ed. D. S. Keen, American Colloid Company, Skokie, IL, 1986.
5. Y. Chen and T. Aviad, in *Humic Substances in Soil and Crop Sciences: Selected Readings*, eds. P. MacCarthy, C.E. Clapp, R.L. Malcolm and P.R. Bloom, Soil Science Society of America, Madison, WI, 1990, p. 161.
6. M. Schnitzer, in *Methods of Soil Analysis, Part 2, Chemical and Microbiological Properties*, 2nd Edn., eds. B.L. Page, R.H. Miller and R.D. Keeney, Agronomy Monograph No. 9, Soil Science Society of America, Madison, WI, 1982, p. 581.

Plant Growth Stimulation and Antimutagenesis

RESPONSE OF ALFALFA TO CALCIUM LIGNITE FERTILIZER

T. Paré,[1] M. Saharinen,[1] M.J. Tudoret,[1] H. Dinel,[1] M. Schnitzer[1] and D. Ozdoba[2]

[1] Eastern Cereals and Oilseeds Research Centre, Agriculture and Agri-Food Canada, Ottawa, ON, Canada K1A 0C6
[2] Luscar Ltd, Specialty Products Division, Edmonton, Alberta, Canada T5L 3C1

1 INTRODUCTION

Lignite is a low rank coal with low calorific value but it is potentially rich in humic substances (HSs), which are known to be non-polluting organic biostimulants particularly with regard to plant growth. HSs positively affect the nitrogen (N) and phosphorus (P) dynamics in soil,[1] stimulate nutrient uptake by plants[2-4] by affecting some enzyme activities[1] and so increase dry matter production.[2-4] Several authors assume that it is the functional groups (hydroxyl and carboxyl groups) contained in HSs that are mainly responsible for the responses obtained,[5-7] because HSs naturally chelate and transport cations and trace metals and make them more available to roots. These HSs properties are utilized in commercial organo-chemical fertilizers.

In the present study we compared the effects of a calcium fertilizer produced from lignite to those of EDTA-Ca and calcium chloride ($CaCl_2$) on the growth of alfalfa (*Medicago sativa* L.) and its nutrient uptake. We also compared the effects of the fertilizers when foliarly sprayed or applied on a growth substrate.

2 MATERIALS AND METHODS

2.1 Calcium Treatments

Three Ca treatments were used in this study: $CaCl_2$, EDTA-Ca and CaLF (a Ca fertilizer produced from lignite (Luscar Ltd)). Three levels of treatments were used: half of recommended rate of Ca for alfalfa (128.5 kg Ca ha^{-1}), full rate (257 kg Ca ha^{-1}) and 1.5 times the full rate (385.5 kg Ca ha^{-1}). These rates correspond to 78.3, 156.5 and 234.7 mg of Ca/pot, respectively.

2.2 Greenhouse Experiment

Alfalfa (*Medicago sativa* L., cv. Nitro) was used as the test crop because of its high Ca requirements. Plants were grown in 1.6 L pots filled with 1.2 kg (dry mass basis) of a substrate consisting of a mixture of vermiculite and silica sand No. 24 (1:1 v/v). Each pot was sown with 10 g of seeds. The plants were watered with tap water and once a week with modified Hoagland solution without Ca. In addition, alfalfa was fertilized 17 days after sowing with half the amount of its N requirement (35 N kg ha^{-1} applied as NH_4NO_3), with the full amount of its P_2O_5 requirement (135 kg ha^{-1}) and the full amount of its K_2O requirement (275 kg ha^{-1}). The rest of the N (35 kg N ha^{-1}) was applied as NH_4NO_3 35 days after sowing. Plants were first cut 35 days after sowing. Ca application was made in three stages, i.e., 35, 37 and 39 days after sowing. During each application, 50 mL of the appropriate Ca level was applied as a foliar spray or on the substrate. Each treatment and the control (no Ca application) was replicated 5 times in a randomized complete block design. Plants were harvested 67 days after sowing and separated as shoots and roots.

2.3 Laboratory Analyses

Shoots and roots were oven dried at 80°C for three days, weighed and ground to 60 mesh. They were digested with HNO_3 + $HClO_4$ + HF[8] prior to ICP analysis for Ca, Mg, P and K. Nitrogen concentrations of shoots were determined as Kjedhal N. Because of the low dry masses of roots in some treatments, N and P in roots were not analyzed. Nutrient uptakes were calculated from concentrations and dry masses.

3 RESULTS AND DISCUSSION

3.1 Dry Masses

3.1.1 Applied as Foliar Spray. CaLF and $CaCl_2$ produced similar amounts of shoot dry masses (Figure 1a). The highest shoot dry masses were obtained with the intermediate levels of CaLF (156.5 mg Ca pot^{-1}) and the highest rate of $CaCl_2$ (234.7 mg Ca pot^{-1}). By contrast, EDTA-Ca application led to a decrease of shoot mass production. Shoot masses obtained with EDTA-Ca were on average 75 and 66% lower than those obtained with the application of CaLF and $CaCl_2$, respectively (Figure 1a). All levels of CaLF and $CaCl_2$ produced similar amounts of root dry masses (Figure 2a). By contrast, EDTA application decreased root dry mass by 64 to 83% compared to CaLF, and by 68 to 87% compared to $CaCl_2$ (Figure 2a). Alfalfa root dry mass production decreased with increasing rates of Ca applied as EDTA-Ca (Figure 2a). When foliarly sprayed, CaLF and $CaCl_2$ produced similar whole plant dry masses, whereas EDTA-Ca led to a decrease of alfalfa whole plant dry masses (Figure 3a).

3.1.2 Applied on the Substrate. When applied on the substrate, CaLF and $CaCl_2$ produced similar amounts of alfalfa shoot dry masses (Figure 1b). On the other hand, increasing EDTA-Ca rates led to a linear decrease of shoot masses. On the average, shoot dry mass was 62 and 71% lower when plant were fertilized with EDTA-Ca than with CaLF and $CaCl_2$, respectively. CaLF and $CaCl_2$ produced similar root dry masses, but on average 67 % more root dry masses than did EDTA-Ca (Figure 2b). Root mass reduction was dramatic and linear with the rates of EDTA-Ca applied. On the average, alfalfa whole-plant dry masses were, respectively, 88 and 101% higher when fertilized with CaLF and $CaCl_2$ than with EDTA-Ca (Figure 3b).

3.1.3 Comparison of both Modes of Ca Application. CaLF and $CaCl_2$ produced more shoot masses when applied as foliar spray than when applied on the substrate, except at the lowest Ca rate, where the mode of application did not make a difference (Figures 1a and b).

Deleterious effects by EDTA-Ca on alfalfa shoot production were similar when this fertilizer was applied by spraying or on the substrate (Figures 1a and b). Foliar spray of

EDTA-Ca on average decreased root dry mass production by 39% compared to its application on substrate (Figures 2 a and b). Applied as spray or on the substrate, CaLF and $CaCl_2$ produced similar alfalfa root and whole-plant dry masses (Figures 2a and b; 3a and b)

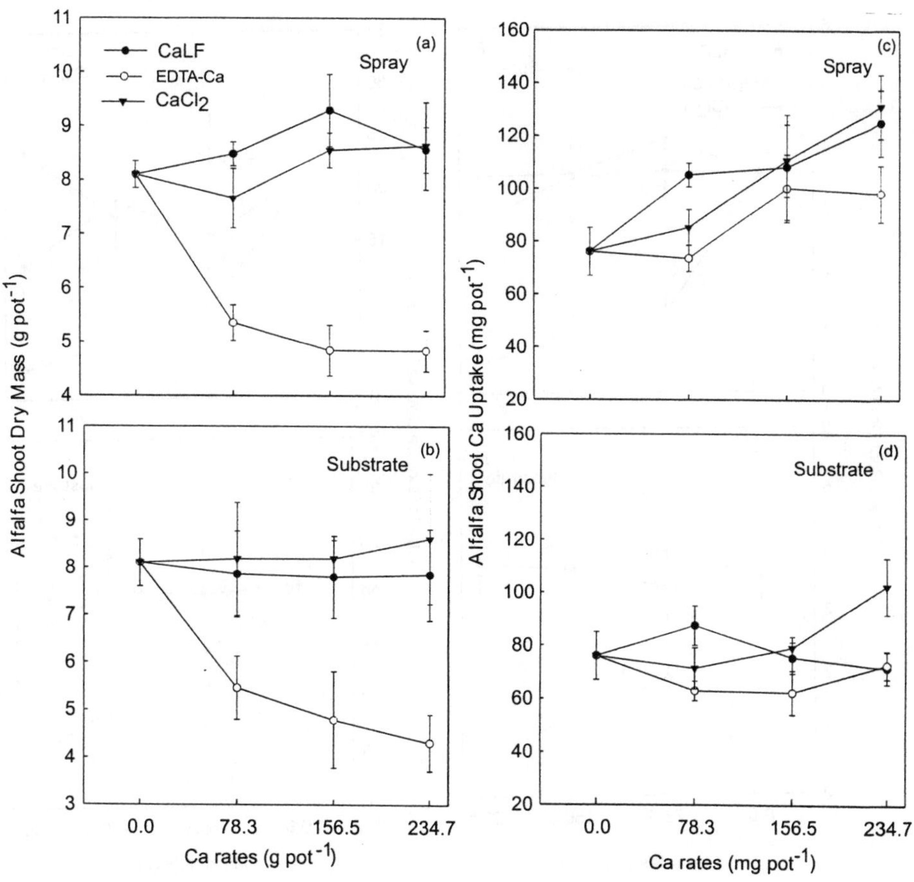

Figure 1 *Dry masses and Ca uptake of alfalfa shoots following fertilisation with CaLF, EDTA-Ca and $CaCl_2$ by foliar spray or on substrate (means of 5 replications ± SEM)*

3.2. Calcium Uptake

3.2.1 Applied by Spraying. For all fertilizers, alfalfa shoot Ca uptake increased with the rates of Ca applied (Figues 1c). Calcium uptake by shoots was on average similar (ca. 94 mg pot^{-1}) when plants were fertilized with CaLF and $CaCl_2$ and on average 21% higher than EDTA-Ca (Figure 1c). Calcium uptake by roots was similarly variable when CaLF and $CaCl_2$ were sprayed on alfalfa plants (Figure 2c) indicating that foliarly applied Ca was not translocated. By contrast, Ca uptake by roots decreased with increasing the rates of EDTA-Ca sprayed on plants (Figure 2c), which was due to weak root growth. Whole-plant Ca uptake increased with the rates of Ca applied with CaLF and $CaCl_2$, and there was no difference between the two fertilizers (Figure 3c). Except for a slight increase of uptake by whole plants following the foliar spray of 156.5 mg ca pot^{-1}, EDTA-Ca did not stimulate any Ca uptake.

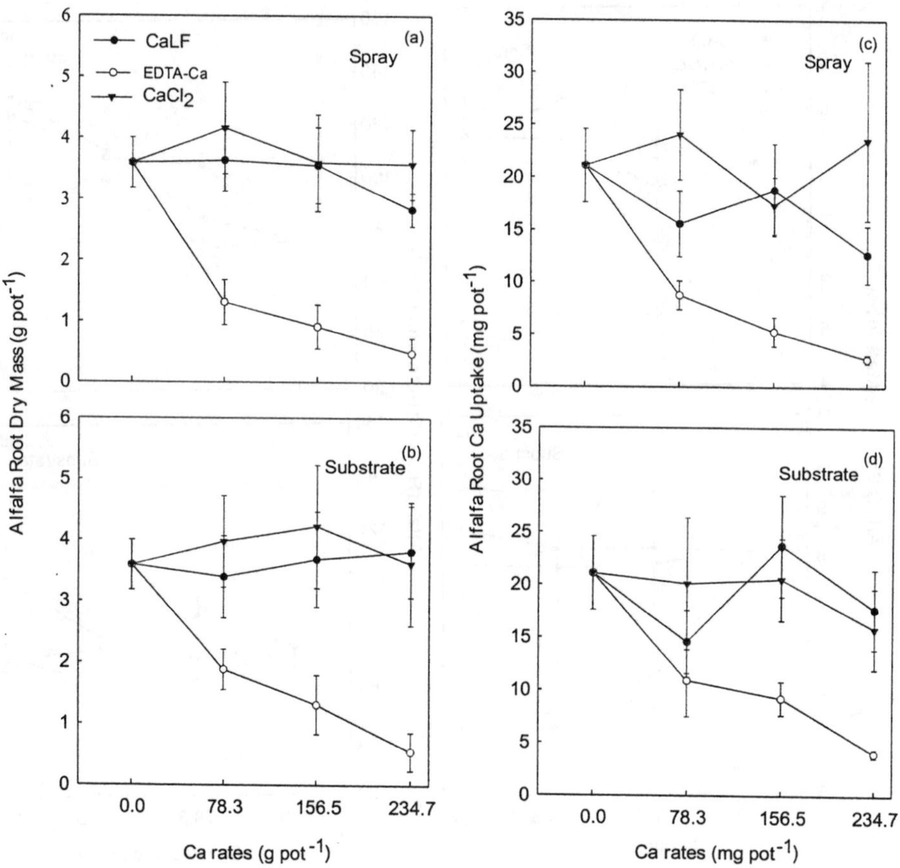

Figure 2 *Dry masses and Ca uptake of alfalfa roots following fertilisation with CaLF, EDTA-Ca and $CaCl_2$ by foliar spray or on substrate (means of 5 replications ± SEM)*

3.2.2 Applied on the Substrate. As shown in Figure 1d, Ca uptake by alfalfa shoots fertilized with CaLF decreased with increasing Ca rates. By contrast, Ca uptake by shoots increased with increasing $CaCl_2$ rates. Ca uptake by shoots following CaLF application was on average 23% lower than that of plants fertilized with $CaCl_2$. Shoots of alfalfa plants fertilized with CaLF on the average took up 17% more Ca than those fertilized with EDTA-Ca (Figure 1d). On the other hand, Ca uptake by shoots of plants fertilized with $CaCl_2$ was on average 45% higher than those fertilized with EDTA-Ca. Root Ca uptake was similar (ca.

20 mg pot^{-1}) when CaLF and $CaCl_2$ were applied on the substrate. Application of the highest rate of both fertilizers led to a slight decrease of Ca uptake by roots (Figure 2d). EDTA-Ca applied on the substrate drastically decreased Ca uptake by roots with increasing Ca rates. Calcium uptake by whole plants was similar (95 mg pot^{-1}) when CaLF and $CaCl_2$ were applied on the substrates (Figure 3d). EDTA-Ca application decreased Ca uptake by whole plants compared to CaLF and $CaCl_2$.

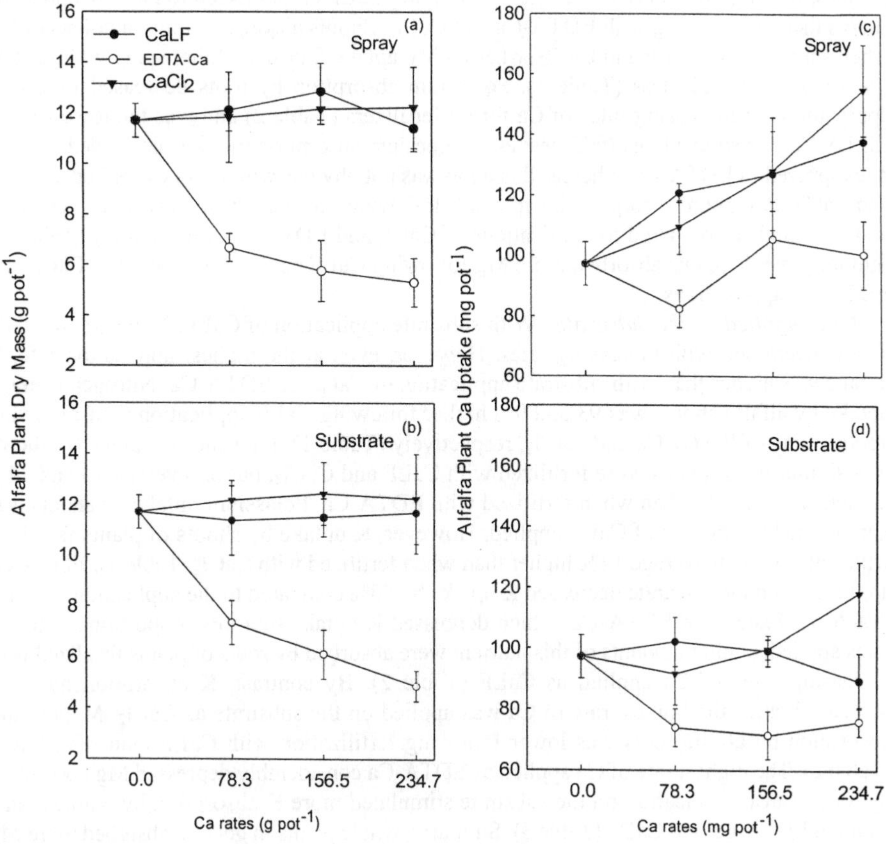

Figure 3 *Dry masses and Ca uptake of alfalfa whole plants fertilised with CaLF, EDTA-Ca and $CaCl_2$ by foliar spray or on substrate (means of 5 replications ± SEM)*

3.2.3 Comparison of both Modes of Calcium Application. Calcium uptake by shoots was higher when all Ca fertilizers were sprayed than when applied on substrates (Figures 1c and 1d). Indeed, it was on average 44, 38 and 15% higher for CaLF, EDTA-Ca and $CaCl_2$, respectively. Calcium uptake by shoots increased with the rates of Ca applied by spraying over those applied on substrates (Figures 1c and d). Calcium uptake by roots was not affected by the application mode, but its uptake by whole plants was on average 30 % higher when sprayed than applied on the substrate (Figures 2c-d and 3c-d).

3.3 Nitrogen, Phosphorus, Potassium and Magnesium Uptake

3.3.1 Applied by Spraying. Nitrogen uptake by shoots of alfalfa plants sprayed with CaLF did not vary with the Ca rates applied (Table 1). However, N uptake by alfalfa shoots was higher with CaLF than with EDTA-Ca and $CaCl_2$. Shoots of plants sprayed with CaLF and $CaCl_2$ absorbed similar amounts of P and on average 32 and 27% more, respectively, than plants sprayed with EDTA-Ca (Table 1). Spraying CaLF on plants led to more K uptake by shoots than did spraying with EDTA-Ca and $CaCl_2$. Shoots absorbed similar amounts of Mg when sprayed with CaLF and $CaCl_2$, whereas Mg uptake following foliar spray with EDTA-Ca was considerably less (Table 1). Potassium absorption by roots decreased following applications of increasing rates of Ca for all fertilizers (Table 2) and was the lowest when EDTA-Ca was sprayed on alfalfa leaves. Magnesium absorption by roots decreased with Ca rates sprayed as EDTA-Ca, whereas this trend was not obvious with Ca rates applied as CaLF and $CaCl_2$. Potassium absorption by all plants following CaLF application was on average 10 and 78% higher compared to applications of $CaCl_2$ and EDTA-Ca, respectively (Table 3). Similarly whole plants absorbed more Mg following CaLF application than after $CaCl_2$ and EDTA-Ca applications.

3.3.2 Applied on the Substrate. With substrate application of CaLF, N uptake in alfalfa shoots decreased with increasing rates. However, even at the highest application rate, N uptake was higher than with substrate application of $CaCl_2$ or EDTA-Ca. Nitrogen average uptake by alfalfa shoots was 93 and 9% higher following CaLF application than following applications of EDTA-Ca and $CaCl_2$, respectively (Table 1). Phosphorus uptake by shoots was similar when plants were fertilized with CaLF and $CaCl_2$, but on average 13 and 24% higher, respectively, than when fertilized with EDTA-Ca. Potassium uptake by shoots was not affected by the rate of CaLF applied. However, K uptake by shoots of plants fertilized with CaLF was on average 14% higher than when fertilized with $CaCl_2$ (Table 1). EDTA-Ca application on the substrate decreased K uptake by 50% compared to the application of CaLF (Table 1). Except for EDTA-Ca, which depressed K uptake by roots proportionally to the rates applied, similar amounts of this nutrient were absorbed by roots of plants fertilized with increasing rates of Ca applied as CaLF (Table 2). By contrast, K absorption by roots decreased when the highest rate of Ca was applied on the substrate as $CaCl_2$. Magnesium absorption by alfalfa roots was lower following fertilization with CaLF than with $CaCl_2$ (Table 2). The highest rate of Ca applied as EDTA-Ca considerably depressed Mg absorption by roots. CaLF application on the substrate stimulated more K absorption by whole plants than did EDTA-Ca and $CaCl_2$ (Table 3). Similarly, whole plants in general absorbed more Mg when CaLF was applied on the substrate compared to EDTA-Ca, but 12% less than when $CaCl_2$ was applied.

3.3.3 Comparison of both Modes of Ca Application. In general, foliar application of CaLF fertilizers stimulated more nutrient uptake in shoots compared to substrate application (Table 1). On the average, P and Mg uptake by shoots were 14 and 36% higher, respectively, when sprayed on leaves than when applied on the substrate (Table 1). Root K and Mg uptake were not affected by the mode of CaLF and $CaCl_2$ applications, whereas the uptake of these two nutrients by roots was higher when EDTA-Ca was applied on the substrate than when sprayed on leaves (Table 2). Whole-plant Mg absorption was 28% higher when CaLF was sprayed on leaves than when applied on the substrate, whereas K absorption was not affected by the two modes of fertilization (Table 3).

Table 1 *Nitrogen, phosphorus, potassium and magnesium uptake (mg pot^{-1}) by alfalfa (Medicago sativa L.) shoots fertilized by spray or on substrate with different rates of Ca fertilizers*[a]

Fertilizer treatments	*Rates mg pot^{-1}*	*Nitrogen spray*	*uptake substrate*	*Phosphorus spray*	*uptake substrate*	*Potassium spray*	*uptake substrate*	*Magnesium spray*	*uptake substrate*
Control	0.0	323±8	323±8	22±1	22±1	330±19	330±19	103±7	103±7
CaLF	78.3	333±9	323±16	24±2	21±1	347±14	333±23	108±6	98±7
	156.5	339±13	312±21	24±1	19±1	367±14	313±29	161±61	85±5
	234.7	309±26	267±1	23±2	21±1	323±26	303±17	85±7	78±6
EDTA-Ca	78.3	185±7	178±13	21±2	22±1	227±6	239±16	73±1	74±4
	156.5	164±18	145±20	18±1	18±1	208±25	216±28	73±9	78±11
	234.7	165±18	141±14	17±1	16±2	178±27	186±24	77±5	71±6
$CaCl_2$	78.3	258±18	279±22	18±1	24±2	303±26	270±37	83±9	95±13
	156.5	281±11	265±13	25±2	21±1	330±24	293±10	102±12	84±2
	234.7	277±29	278±19	23±2	23±2	283±27	267±18	91±13	104±12

[a] Means of 5 replications ± SEM.

Table 2 *Potassium and magnesium uptake (mg pot^{-1}) by alfalfa (Medicago sativa L.) roots fertilized by spray or on substrate with different rates of Ca*[a]

Fertilizer treatments	*Rates mg pot^{-1}*	*Potassium spray*	*uptake substrate*	*Magnesium spray*	*uptake substrate*
Control	0.0	53±6	53±6	49±20	49±20
CaLF	78.3	53±9	48±8	30±2	27±4
	156.5	44±3	44±6	38±8	33±8
	234.7	37±9	47±7	31±9	34±7
EDTA-Ca	78.3	18±2	31±5	16±2	23±7
	156.5	16±3	20±4	14±3	23±4
	234.7	10±2	9±2	8±2	8±2
$CaCl_2$	78.3	59±4	50±9	48±12	40±6
	156.5	48±9	59±7	33±9	31±5
	234.7	40±6	44±6	51±15	42±10

[a] Means of 5 replicates ± SEM

Table 3 *Potassium and magnesium uptake (mg pot^{-1}) by alfalfa (Medicago sativa L.) plants fertilized by spray or on substrate with different rates of Ca*[a]

Fertilizer treatments	*Rates mg pot^{-1}*	*Potassium spray*	*uptake substrate*	*Magnesium spray*	*uptake substrate*
Control	0.0	383±20	383±20	152±14	152±14
CaLF	78.3	401±15	382±22	139±7	124±6
	156.5	412±14	358±23	199±57	118±8
	234.7	360±28	350±24	116±14	112±9
EDTA-Ca	78.3	246±6	271±21	89±3	97±10
	156.5	224±28	236±31	87±9	101±11
	234.7	187±29	194±24	83±6	78±6
$CaCl_2$	78.3	363±26	320±33	131±15	136±7
	156.5	378±22	353±16	135±14	115±5
	234.7	323±29	311±21	142±23	146±20

[a] Means of 5 replicates ± SEM

It is obvious from the results reported in this study that Ca fertilizer produced with lignite (CaLF) stimulated plant growth and nutrient uptake compared to the chemical fertilizers commonly used. Thus, our observations confirm earlier findings that HSs can increase dry matter production and nutrients uptake.[2-4] A plausible explanation for the effects of CaLF is that its functional groups complex Ca and so increase its availability to alfalfa plants. CaLF applied on substrates also could have increased the permeability of alfalfa root membranes and so enhanced nutrients uptake.[2] According to Rauthan and Schnitzer,[2] HSs, particularly fulvic acid, contain structures that act like hormones that could facilitate the translocation of nutrients throughout the plants.

4 CONCLUSIONS

The data presented in this study show that

1. Calcium lignite fertilizer and $CaCl_2$ produced similar shoot, root and whole plant dry masses, which were much higher than when EDTA-Ca was applied.
2. Calcium uptake by shoots increased with Ca levels when foliarly applied. The levels and patterns of increase were similar when plants were sprayed with CaLF and $CaCl_2$, and on average 22% higher than when sprayed with EDTA-Ca.
3. Calcium lignite fertilizer did not decrease N uptake, as observed with EDTA-Ca and $CaCl_2$. N uptake by alfalfa shoots fertilized with CaLF was on average 91 and 20% higher than when fertilized with EDTA-Ca and $CaCl_2$, respectively.
4. When sprayed on alfalfa leaves, CaLF stimulated more Mg and P uptake than did EDTA-Ca.
5. In general, foliar sprays of Ca fertilizers were more effective than substrate applications.
6. Calcium lignite fertilizer has a significant potential for not only correcting Ca deficiency but also for stimulating the absorption of other nutrients.

References

1. F.A. Biondi, A. Figholia, R. Indiati and C. Issa, in *Humic Substances in the Global Environment and Implications on Human Health*, eds. N. Senesi and T.M. Miano, Elsevier, New York, 1994, p. 239
2. B.S. Rauthan and M. Schnitzer, *Plant and Soil*, 1981, **63**, 491.
3. K.H. Tan and D. Tantiwiramanond, *Soil Sci. Soc. Am. J.*, 1983, **47**, 1121.
4. M. Ayuso, T. Hernández, C. Garcia and J.A. Pascual, *Biores. Techn.*, 1996, **57**, 251.
5. M. Schnitzer and P.A. Poapst, *Nature*, 1967, **213**, 598.
6. A. Albuzio, G. Ferrari and S. Nardi, *Can. J. Soil Sci.*, 1986, **66**, 731.
7. A. Piccolo, S. Nardi and G. Concheri, *Soil Biol. Biochem.*, 1992, **24**, 373.
8. J.A. McKeague, in *The Manual of Soil Sampling and Methods of Analysis*, Canadian Society of Soil Science, Ottawa, 1978, p. 250.

EFFECTS OF HUMIC ACIDS AND NITROGEN MINERALIZATION ON CROP PRODUCTION IN FIELD TRIALS

Mir-M Seyedbagheri and James M. Torell

University of Idaho, Elmore County Extension Center, Mountain Home, ID 83647, USA

1 INTRODUCTION

Soil organic matter (SOM) is derived from plant and animal residues in various stages of decomposition. SOM is important to crop production because it provides nutrients and improves the physical properties of soils. The nutrient-providing role of SOM is especially important for organic growers and conventional growers striving to create more sustainable farming systems with reduced reliance on chemical fertilizers.[1]

When organic matter enters the soil, a community of soil organisms known as the soil foodweb degrades it. The soil foodweb is a diverse biological community including bacteria, fungi, protozoa, arthropods and earthworms. As these organisms use SOM as a substrate for cellular metabolism, the biomolecules of plants and animals are broken and rearranged repeatedly as organic matter is passed from one organism to another in the soil foodweb. Humic substances, a major constituent of SOM, have long been known to have beneficial effects on the physical properties of soils because of their ability to aid the formation of soil aggregates. These benefits also include increased water holding capacity, reduced crusting and better tilth. In addition to these effects, recent research has shown that humic substances provide direct stimulation of plant growth under laboratory conditions. The practical significance of these results in the field is unclear since it is difficult to separate these direct effects from indirect effects in field trials. The result of this process is that the nutrients in SOM are released and humic substances remain that are resistant to further degradation. This release of inorganic nutrients is termed mineralization.[2-9]

In recent years, several commercial humic products have been marketed as soil amendments. Research is being conducted to determine if supplementation of humic substances provided through natural biological cycles of the soil foodweb results in an economical crop response.

2 METHODS

Trials were established in 1990 and 1999 to evaluate the efficacy of humic acid treatments for potato production in Elmore County, Idaho. For each trial, the humic treatments were arranged in a randomized complete block design with four replications and were applied with a solo backpack sprayer. Plots were harvested by hand and potatoes were graded and weighed to determine yield. Data from both years were subjected to an analysis of variance. Data from 1990 were also subjected to a regression analysis.

The 1999 trial was established on an irrigated potato field and the treatments consisted of various combinations of dry and liquid humic acid products. These products are detailed in Table 1. The hills were opened to just above the seedpiece prior to application of pre-emergence treatments on April 22, 1999. Granular humic acid was weighed and spread by hand as evenly as possible to treated rows according to the randomization. Liquid humic acid was applied in 3.7 L of water with a solo sprayer to opened furrows. Furrows were closed immediately after application of treatments. On May 24, 1999, treatments 3 & 4 were applied. Liquid humic acid was applied with a solo sprayer followed by 2 passes of water to incorporate. The plot area was cultivated immediately after treatment application. On June 17, 1999, treatments 3 and 4 were applied by pulling the plants back and banding at the base of the plants. Treatments for the rest of the season were applied in this way. Applications for treatment 4 were made on June 30, July 15 and July 30, 1999. On September 22, 1999 the plots were harvested by hand. The potatoes were graded and 20 tuber samples of potatoes, which were at least 226.8 g in size were kept for determination of quality factors.

Nitrogen mineralization was determined using the buried bag method. At the beginning of the growing season a random soil sample was taken from each field and an initial sample was analyzed for NO_3^- and NH_4^+. Mineralization samples were buried in polyethylene bags and placed in the soil at their respective depths. Samples were retrieved at monthly intervals and analyzed for NO_3^- and NH_4^+. After converting NO_3^- and NH_4^+ values in part per million to kgN.ha^{-1}, nitrogen mineralization was determined by subtraction from the values for the initial sample.

Table 1 *Description of humic acid trials, 1999*

Treatment[a]	*Treatment description*	*Notes*
1: Control	Control	
2: Granular humate only	4.48 kg.ha^{-1} of 70% granular humic acid	Applied as a single preemergence treatment
3: Granular humate + 46.5 L.ha^{-1} of liquid humic acids	4.48 kg.ha^{-1} of 70% granular humic acid + 3 applications of Liquid humic acid, consisting of 2 applications at the rate of 18.6 L.ha^{-1} and 1 at the rate of 9.3 L.ha^{-1}	The liquid humic acid was applied postemergence at different times during the growing season (see text)
4: Granular humate + 93.0 L.ha^{-1} of liquid humic acids	4.48 kg.ha^{-1} of 70% granular humic acid + 93.0 L.ha^{-1} applications of liquid humic acid, consisting of 5 applications at the rate of 18.6 L.ha^{-1}	The liquid humic acid was applied postemergence at different times during the growing season (see text)

[a] Humic products supplied by Horizon-Ag, Inc.

3 RESULTS

Data from the humic acid trials showed that, at the rates recommended by commercial suppliers, yield responses were minimal and inconsistent (Tables 2 and 3). On the other hand, for a few high quality products yield increased with increased application rate up to a maximum yield increase of about 14 percent, corresponding to an application rate of about 84 kg.ha^{-1} (Figure 1). Yield decreased at the highest application rates. Observations of plants in the field showed stimulation of root growth that was consistent with results reported by Chen.[2] Nitrogen mineralization from organic matter varied from 40.3 to 184 kg/ha (Tables 4 and 5).

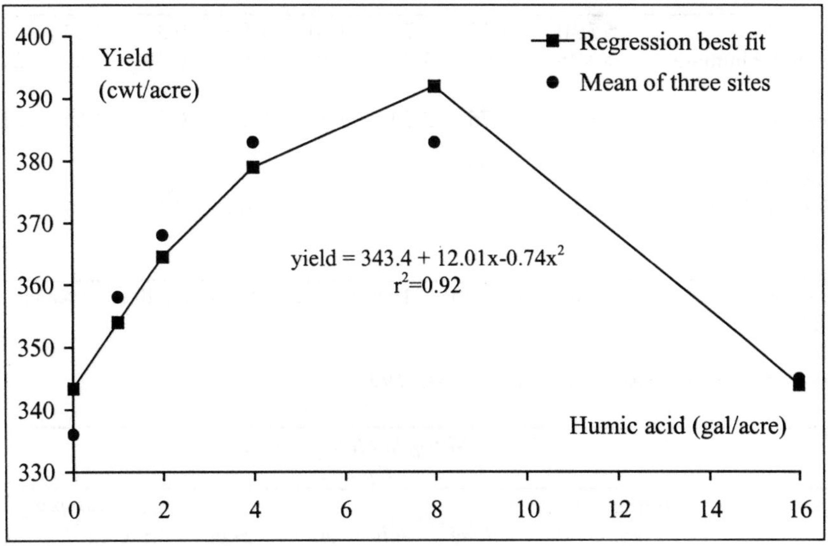

Figure 1 *Effect of humic acid on potato yield, 1990*

Table 2 *Effects of humic acid on potato yield, 1999*[a]

Mass, g / *Treatment*[b]	*Yield (MT.ha^{-1})*[a]					
	0-113.4	*113.4 – 220.8*	*220.8 – 340.2*	*>340.2*	*Culls*	*Total*
1: Control	10.2a[c]	17.9a	8.2a	5.0a	3.4a	44.6a
2: Granular Humate Only	10.7a	16.7a	8.4a	5.5a	2.8a	45.1a
3: Granular Humate + 46.5 L.ha^{-1} Liquid Humic Acid	11.5a	16.7a	7.0a	5.0a	4.5a	44.7a
4: Granular Humate + 93.0 L.ha^{-1} Liquid Humic Acid	11.0a	15.3a	6.4a	4.2a	4.6a	41.5a

[a] MT = metric ton; [b] humic products supplied by Horizon-Ag, Inc.; [c] Means followed by the same letter in the same column are not significantly different at the 0.05 level according to the Neumann-Keuls test.

Table 3 *Effect of humic acid on potato quality factors, 1999*[a,b]

Treatment	Fry Color Rating (% of Samples)			Count/Sample		Sugars ($mg.g^{-1}$)		
	FC 0	FC 1	FC 2	Dark end	Sugar end	Dextrose	Sucrose	% Solids
1: Control	77.50	18.75	3.75	0.50	1.00	0.31	1.64	21.9
2: Granular humate only	88.75	7.50	3.75	0.25	1.00	0.29	1.52	21.9
3: Granular humate + 46.5 $L.ha^{-1}$ liquid Humic acid	78.75	18.75	2.50	0.00	0.25	0.34	1.65	21.5
4: Granular humate + 93.0 $L.ha^{-1}$ liquid humic acid	82.50	13.75	3.75	0.75	2.00	0.39	1.63	22.5

[a] Humic products supplied by Horizon-Ag, Inc.; [b] Data were collected from a 20 tuber sample from each plot.

Table 4 *Nitrogen mineralization in potatoes, 1995*

Site	Depth (cm)	Initial N level ($kg.ha^{-1}$)[a]	Nitrogen Mineralization ($kg.ha^{-1}$) June	July	Aug.	Sept.	Soil N Supply ($kg.ha^{-1}$)[b]
1	0-30.5	67.0	9.0	22.4	53.8	44.8	112
	30.5-61	107.5	-13.4	4.5	80.6	- 4.5	103
	Total	174.5	- 4.5	26.9	134.4	40.3	215
2	0-30.5	147.8	-13.4	85.1	- 26.9	40.3	188
	30.5-61	183.7	-13.4	143.4	4.5	13.4	197
	Total	331.5	-26.9	228.5	- 22.4	53.8	385

[a] Initial soil nitrogen (N) = NH_4^+-N + NO_3^--N; [b] Total nitrogen (N) supply = Initial N + mineralization

Table 5 *Effects of compost on mineralized nitrogen: conventional farm (Location 2) and organic farm (Location 3), 2000*

	Treatment	Mineralized Nitrogen ($kg.ha^{-1}$) July 2000	September 2000
Location 2	Control	60.9	130
	11.2 $MT.ha^{-1}$ Compost	46.1	101
	56.0 $MT.ha^{-1}$ Compost	93.2	162
Location 3	Control	68.5	153
	11.2 $MT.ha^{-1}$ Compost	53.3	147
	56.0 $MT.ha^{-1}$ Compost	63.2	184

4 CONCLUSIONS

Growers have both realistic and unrealistic expectations with regard to humic acid based products. The profitability of humic acid applications of any rate is questionable at current costs. In many cases, growers may realize greater benefits from humic substances by improving their management of the natural cycles that produce humic substances in the agroecosystem. Yield results may vary with growers' cultural practices and environmental factors. Field observations showing stimulation of root growth by humic substances are consistent with the hypothesis that humic substances have hormonal activity on plants. The soil foodweb provides a useful framework for analyzing the complex biological and chemical processes that produce humic substances in the agroecosystem and that mediate plant responses to application of commercially prepared humic substances. More research needs to be done to better understand the complexities of the soil ecosystem, including the role of humic substances.

References

1. N.C. Brady, in *The Nature and Properties of Soils*, 8th Edn., MacMillan, New York, 1974, p.137.
2. Y. Chen and T. Aviad, in *Humic Substances in Soil and Crop Sciences: Selected Readings*, eds. P. MacCarthy, C.E. Clapp, R.L. Malcolm and P.R. Bloom, Soil Science Society of America, Madison, WI, 1990, p. 161.
3. P. Hendrix, R. Parmelee, D. Crossley, Jr., D. Coleman, E. Odem and P. Groffman, *Bioscience,* 1986, **36**, 374.
4. E. Ingham, J. Trofymow, R. Ames, H. Hunt, C. Morley, J. Moore and D. Coleman, *J. Applied Ecol.*, 1986, **23**, 615.
5. P. MacCarthy, R.L. Malcolm, C.E. Clapp and P.R. Bloom, in *Humic Substances in Soil and Crop Sciences: Selected Readings*, eds. P. MacCarthy, C.E. Clapp, R.L. Malcolm and P.R. Bloom, Soil Science Society of America, Madison, WI, 1990, p. 1.
6. P. MacCarthy, P.R. Bloom, C.E. Clapp and R.L. Malcom, in *Humic Substances in Soil and Crop Sciences: Selected Readings*, eds. P. MacCarthy, C.E. Clapp, R.L. Malcolm and P.R. Bloom, Soil Science Society of America, Madison, WI, 1990, p. 261.
7. T. Stieber, C. Shock, E. Feibert, M. Thornton, B. Brown, W. Cook, M. Seyedbagheri and D. Westermann, *Nitrogen Mineralization in Treasure Valley Soils, 1993 and 1994 Results*, Malheur County Crop Research Annual Research Annual Report, 1994.
8. D. Westermann and S.E. Crothers, *Agron. J.*, 1980, **72**, 1009.
9. F. Stevenson and X. He, in *Humic Substances in Soil and Crop Sciences: Selected Readings*, eds. P. MacCarthy, C.E. Clapp, R.L. Malcolm and P.R. Bloom, Soil Science Society of America, Madison, WI, 1990, p. 91.

ANTIMUTAGENIC AND ANTITOXIC ACTIONS OF HUMIC SUBSTANCES ON SEEDLINGS OF MONOCOTYLEDON AND DICOTYLEDON PLANTS

Giuseppe Ferrara, Elisabetta Loffredo and Nicola Senesi

Dipartimento di Biologia e Chimica Agroforestale ed Ambientale, Università di Bari, 70126 Bari, Italy

1 INTRODUCTION

Humic substances (HSs) can produce various morphological, physiological, biochemical and genetic effects on higher plants.[1-4] The capacity of humic acids (HAs) and fulvic acids (FAs) to enhance the growth and interfere with the development of whole plants or single plant organs has been demonstrated.[1] Further, results of some studies have revealed a synergistic or antagonistic effect of HSs on plant growth when they are used in combination with some xenobiotic molecules.[5-7] In particular, the presence of 200 mg L^{-1} of soil HAs in the nutrient medium was shown to attenuate growth depression and other toxic symptoms of tomato seedlings treated with the herbicides alachlor, rimsulfuron and imazethapyr.[7]

Relatively little information is available on the genetic effects that HSs can exert on plants, although a mutagenic or an antimutagenic action on whole living organisms or single cells has been observed to depend on the origin and nature of HSs and the organism examined. In particular, the mutagenic activity of aquatic HAs and FAs has been investigated on bacteria and animal cells,[8-10] and the antimutagenic behavior of HSs also has been studied.[2-4,10-14] Antimutagenic action can be defined as the suppression or reduction of mutagenic events, including breakage and translocation of chromosomes and spindle disturbances, and is generally investigated in root tip cells. For example, Gichner et al.[3] reported an antimutagenic action of potassium humate in plants of *Arabidopsis thaliana* treated with the pro-mutagen propoxur. A reduced frequency of genetic anomalies was observed in seedlings of *V. faba* treated with various herbicides when grown in an organic soil compared to a sandy soil.[4]

The objective of this paper is to discuss the antimutagenic and antitoxic effects exerted by HSs of different origin on several monocotyledon and dicotyledon herbaceous plant species treated with different mutagenic and phytotoxic compounds.

2 MATERIALS AND METHODS

The origin and nature of HSs samples used in this work with the corresponding codes, abbreviations and concentrations used are shown in Table 1. All samples were obtained from the Standard and Reference Collection of HAs and FAs of the International Humic

Table 1 *Origin and nature of humic substances used with corresponding codes, abbreviations and concentrations*

Origin and nature	*IHSS Code*	*Abbreviation*	*Concentrations used (mg L^{-1})*
Soil (mollisol) humic acid	1S102H	SHA	20, 50, 200, 500
Soil (mollisol) fulvic acid	1S102F	SFA	20, 50, 200, 500
Leonardite humic acid	1S104H	LHA	20, 50, 200, 500
Suwannee river fulvic acid	1R101F	SRFA	50, 500
Peat humic acid	1R103H	PHA	20, 50, 200, 500
Peat fulvic acid	1R103F	PFA	20, 50, 200, 500
Nordic aquatic humic acid	1R105H	NHA	50, 500
Nordic aquatic fulvic acid	1R105F	NFA	50, 500
Summit Hill soil humic acid	1R106H	SHHA	20, 200
Alluvial soil humic acid		ASHA	10, 100

Table 2 *Mutagenic compounds used with corresponding abbreviations and concentrations*

Mutagenic compound	*Producer*	*Abbreviation*	*Concentration (mg L^{-1})*
Maleic hydrazide	Sigma-Aldrich S.r.l., Milan, Italy	MH	5.3, 10, 53
Colchicine	Sigma-Aldrich S.r.l., Milan, Italy	COL	1, 10, 100
Alachlor	Lab. Dr. Ehrenstorfer, Augsburg, Germany	ALA	1, 10
2,4-D	Sigma-Aldrich S.r.l., Milan, Italy	2,4-D	0.01, 0.1, 1
Glyphosate	Sigma-Aldrich S.r.l., Milan, Italy	GLY	10, 100, 1000

Substances Society (IHSS), with the exception of the HA from an alluvial soil. The mutagenic compounds used in the experiments with the corresponding abbreviations and concentrations used are listed in Table 2. Eleven plant species were initially tested for their response to the Feulgen method (described below) in order to select those to be used in subsequent experiments with mutagenic compounds. The plant species tested and their response to the Feulgen technique, together with the kind of mutagenic compound used and the mutagenicity test adopted, are shown in Table 3.

In order to evaluate preliminarily the response of the various plant species to the Feulgen staining method, which is essential for the efficient microscopic observation of genetic anomalies of cells, a defined number of seeds of each plant were germinated in Petri dishes kept in a Phytotron growth chamber at 21 ± 1°C in the dark. Root tips (~ 2mm) were collected after 5 or 7 days of germination depending on the plant species, and subjected to the Feulgen staining procedure before the preparation of permanent slides for observation on an Olympus CX40 microscope. The Feulgen staining procedure essentially consists of fixation of root tips in Carnoy's solution I (ethanol and acetic acid, 3:1 v/v), staining with Schiff's reagent, and two successive immersions in 95% ethanol and histolemon Erba baths.[14] Only four plant species, *V. faba*, *A. cepa*, *P. sativum* and *T. turgidum* gave an adequate response for use in the subsequent experiments.

Table 3 *Plant species tested and corresponding responses to the Feulgen method, mutagenic compounds used and mutagenicity tests applied*

Common name	*Plant Species*	*Botanical group*	*Feulgen method response*	*Mutagenic compound*	*Mutagenicity test*[a]
Broad bean	*Vicia faba* L.	D[b]	positive	MH	MN, AAT
				COL	HC, PC
				2,4-D	MN, AAT
				GLY	MN, AAT
Onion	*Allium cepa* L.	M[c]	positive	MH	MN, AAT
				COL	HC, PC
				2,4-D	MN, AAT
				GLY	MN, AAT
Pea	*Pisum sativum* L.	D	positive	MH	MN, AAT
Durum wheat	*Triticum turgidum* L. convar *durum* (Desf.)	M	positive	MH	MN, AAT
				ALA	MN, AAT
Bean	*Phaseolus vulgaris* L.	D	negative		
Rape	*Brassica napus* L. var. *oleifera* D.C.	D	negative		
White mustard	*Sinapis alba* L.	D	negative		
Flax	*Linum usitatissimum* L.	D	negative		
Tomato	*Lycopersicon esculentum* Mill.	D	negative		
Melon	*Cucumis melo* L.	D	negative		
Sunflower	*Helianthus annuus* L.	D	negative		

[a] see text for abbreviations; [b] D: dicotyledon; [c]M: monocotyledon.

Seeds of plant species successful in the Feulgen technique were then germinated for 5 to 7 days in the presence of 12 mL (*V. faba*) or 8 mL (*A. cepa, P. sativum, T. turgidum*) of the following test solutions: (a) distilled H_2O (positive control); (b) mutagen only (negative control); (c) each HA or FA alone; and (d) each HA or FA in combination with the mutagenic compound. Before addition to the seeds, the HSs + mutagen combinations were interacted by mechanical shaking for 24 h at room temperature (20 ± 1 °C). The pH value of all solutions used ranged from 6 to 7. All experiments were triplicated. Root tips collected were subjected to the Feulgen staining procedure (described above), then prepared adequately as permanent slides and finally subjected to microscope observation. For each treatment, fifteen root tips (5 x 3 replicates) were prepared and 30,000 cells (2,000 cells per root tip) were examined in the case of MH, and 150 metaphases (5 x 3 replicates x 10 metaphases per root tip) were examined in the case of COL.

The mutagenicity level was estimated by two different assays depending on the mutagen tested: a) counting of frequencies of micronuclei (MN), aberrant anatelophases (AAT) and regular anatelophases (RAT) in the case of MH;[11,14] and b) counting of polyploid cells (PC) and hyperdiploid cells (HC) in the case of COL.[12] The MN consist of small portions of extranuclear DNA generated by altered replication or segregation of chromosomes, AAT are characterized by various anomalies in the process of cell division, and HC and PC are cells with a chromosome number between 2n and 3n, and >3n, respectively.

Some HSs samples used for the genetic study were also tested, either alone or in combination with the mutagenic and phytotoxic compound, for their possible antitoxic effect on plant seedlings. In these experiments, the sample ASHA at a concentration of 10 or 100 mg L^{-1}, alone or in combination with 1 or 10 mg L^{-1} of ALA, was tested on *T. turgidum*, and samples SHHA, PHA, PFA, LHA, SHA and SFA at concentrations of 20 or 200 mg L^{-1}, alone or in combination with 10 mg L^{-1} MH, were tested on *V. faba*. The antitoxic effect was evaluated by measuring some biometrical parameters such as length and dry weight of shoots and roots. In the case of *V. faba*, biometrical parameters were measured on 5-days old seedlings before cutting root tips for the antimutagenic observations. For *T. turgidum*, the measurements were made on germinated seedlings grown for 14 days in glass pots in the presence of the same test solutions used for seed germination.

Experimental data obtained were statistically analyzed by one-way analysis of variance (ANOVA) and the mean values were separated by using the least significant difference (LSD) test. For both antimutagenic and antitoxic evaluations, the mean values measured in HSs treatments were statistically compared to the value of the positive control (H_2O), whereas the mean values obtained in the combinations HSs + mutagen were compared to the value of the negative control (MH, COL or ALA).

3 RESULTS AND DISCUSSION

3.1 Antimutagenic Effects

Results of preliminary experiments showed that only root tip cells of *V. faba*, *A. cepa*, *P. sativum* and *T. turgidum* responded positively to the Feulgen technique, with an evident appearance of chromosomes and genetic anomalies at the microscopic level (Table 3 and Figure 1). The other plant species examined exhibited poor staining of the nuclear material, thus discouraging their use in succeeding experiments. Of the mutagenic compounds tested on each plant species (Table 3), those that produced evident mutagenic alterations were: MH on *V. faba*, *A. cepa* and *P. sativum* and COL on *V. faba* and *A. cepa*, whereas 2,4-D, GLY and ALA did not appear to induce statistically significant mutagenic damages in the plant species tested.

In agreement with previous findings,[4,13] both MN and AAT were observed in all treatments with MH, including the positive control (H_2O treatment). Treatments of any plant species with each HS sample alone did not result in MN and AAT frequencies statistically different from the positive control (data not shown), thus suggesting the absence of mutagenic action due to HS.[2,4,13] In contrast, with respect to the positive control, the treatment with MH produced a marked mutagenic effect in *V. faba*, *A. cepa* and *P. sativum*, with an increase of frequencies of MN, respectively up to 9.8, 10 and 14 times, and of AAT, respectively up to 7.6, 2.8 and 1.8 times.

All the combinations of HSs + MH substantially reduced the genetic anomalies caused by MH, indicating antimutagenic activity exerted by HSs in the three species. The effect of the combinations HS + MH on the relative frequencies (%) of MN and AAT in root tip cells of *V. faba*, *P. sativum* and *A. cepa*, referred to the frequency in the negative control

(MH) as 100%, are shown in Figures 2, 3 and 4, respectively. With respect to the MH treatment, the reduction of both anomalies in *V. faba* was highly significant for almost all the combinations. The greatest reduction of the MN frequency was measured for the combinations of NFA at 50 mg L^{-1} + MH (77.6%) and PHA and PFA at 20 mg L^{-1} + MH (74.5 and 68.5%, respectively), and of the AAT frequency for the combinations NHA at 500 mg L^{-1} + MH and NFA at 50 mg L^{-1} + MH (75% for both combinations).

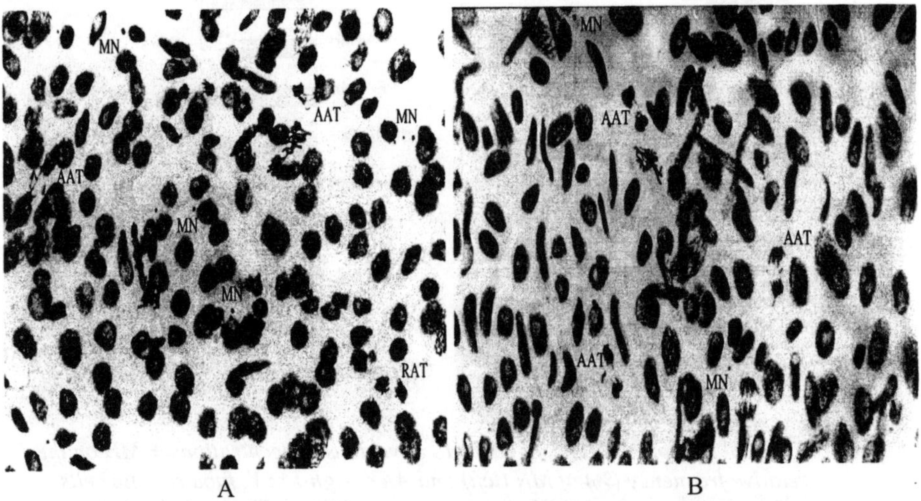

Figure 1 *Photomicrographs of root tip cells of V. faba (A) and A. cepa (B) treated with 10 mg L^{-1} of MH showing the presence of micronuclei (MN), aberrant anatelophases (AAT) and regular anatelophases (RAT). Magnification 400×*

In the experiments with *P. sativum,* the greatest reductions of MN, and especially AAT, were obtained at high HSs concentration, whereas at low HSs concentration the greatest reductions were obtained for the combinations PHA + MH and PFA + MH. In the case of *A. cepa,* the various HSs tested at high concentration (500 mg L^{-1}) behaved almost the same in reducing the frequency of either MN or AAT, and were more efficient in reducing AAT frequency than MN frequency.

Compared to the treatment with MH alone, any combination HSs + MH did not modify the mitotic activity of cells (RAT) in *V. faba*, whereas it caused a slight reduction of RAT in *P. sativum* and yielded contrasting results in *A. cepa* (data not shown).

The presence of PC was only detected in COL-treated cells (negative control) of *A. cepa* and *V. faba*, whereas PC were totally absent in cells grown in the positive control (H_2O). An example of the PC produced by COL treatment in root tip cells of *A. cepa* is shown in Figure 5 and the PC frequencies induced in cells of *A. cepa* treated with various HSs + COL combinations, relative to COL treatment alone, are shown in Figure 6.

Apparently, the combinations HS + COL produced less pronounced antimutagenic effects than those of HSs + MH combinations described above. In particular, only the combinations of LHA, PHA and PFA at lower concentration with COL produced a statistically significant reduction of PC (Figure 6).

With regard to HC behavior, their number in root tips of *A. cepa* and *V. faba* treated with COL (negative control) was much higher than that in the positive control. No significant reduction of HC frequencies was observed for any combination HSs + COL.

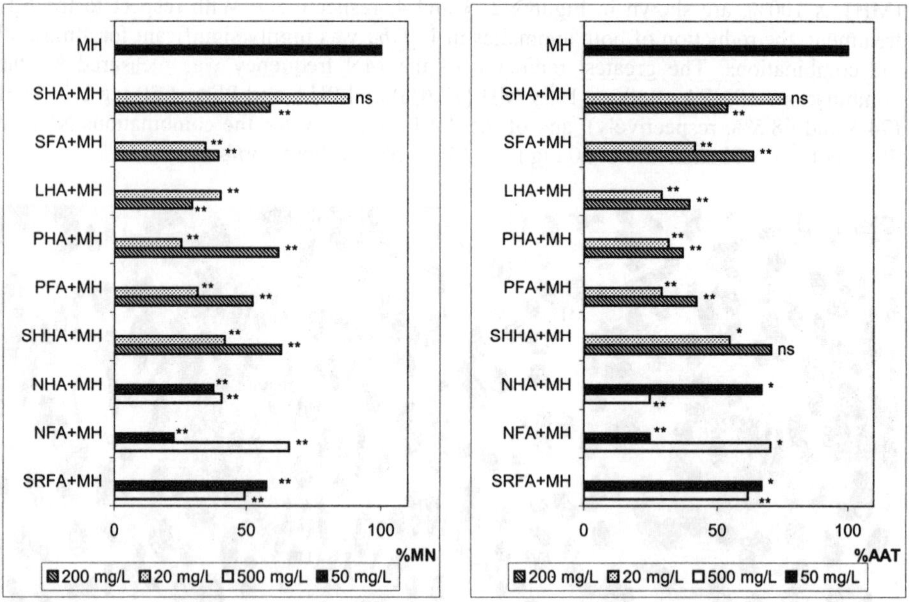

Figure 2 *Effect of the combinations of each HS at various concentrations + MH on the relative frequency (%) of MN (left) and AAT (right) in V. faba root tip cells referred to the control (MH alone, frequency 100%). The symbols **, *, and ns refer, respectively, to a difference significant at 0.01P, 0.05P, and no significant difference according to the LSD test*

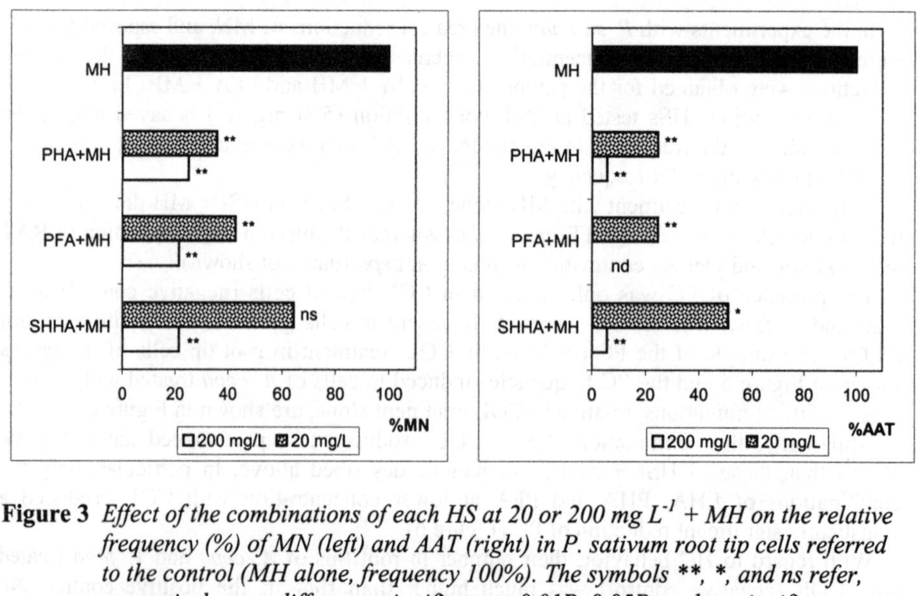

Figure 3 *Effect of the combinations of each HS at 20 or 200 mg L^{-1} + MH on the relative frequency (%) of MN (left) and AAT (right) in P. sativum root tip cells referred to the control (MH alone, frequency 100%). The symbols **, *, and ns refer, respectively, to a difference significant at 0.01P, 0.05P, and no significant difference according to the LSD test*

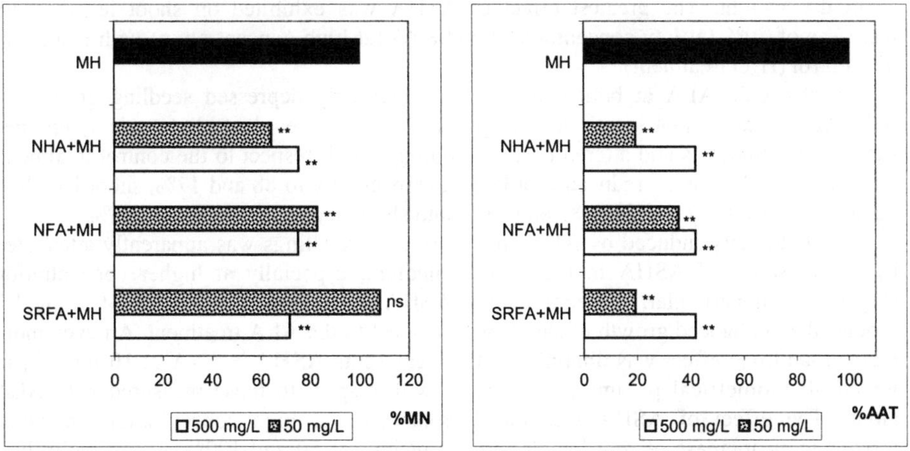

Figure 4 *Effect of the combinations of each HS at 50 or 500 mg L^{-1} + MH on the relative frequency (%) of MN (left) and AAT (right) in A. cepa root tip cells referred to the control (MH alone, frequency 100%). The symbols **, *, and ns refer, respectively, to a difference significant at 0.01P, 0.05P, and no significant difference according to the LSD test*

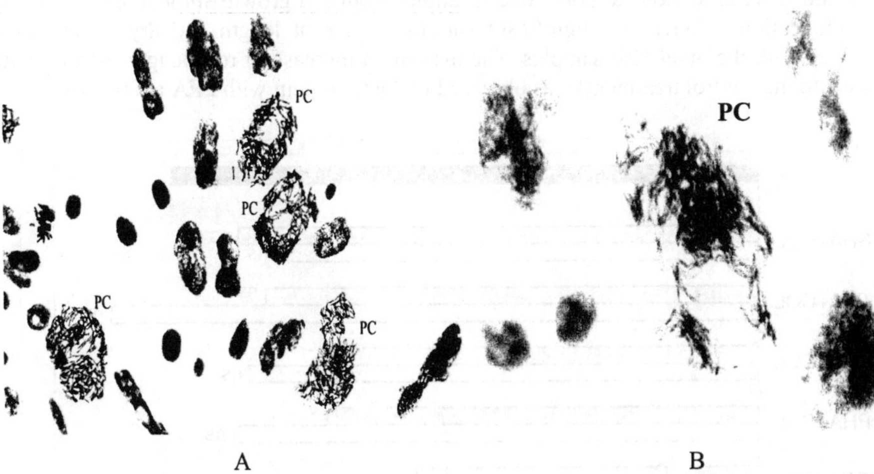

Figure 5 *Photomicrographs of root tip cells of A. cepa treated with 100 mg L^{-1} COL showing the presence of polyploid cells (PC) at magnifications of 400× (A) and 1000× (B)*

3.2 Antitoxic Effects

The effect of ASHA at concentrations of 10 or 100 mg L^{-1}, alone and in combination with ALA at concentrations of 1 and 10 mg L^{-1}, on some biometrical parameters of 14-days old seedlings of *T. turgidum* is shown in Figure 7. The presence in the growth medium of ASHA alone at both concentrations produced an increase of root and shoot lengths and shoot dry weight. The greatest effect of ASHA was exhibited on shoot length, with increases of 70% (at low concentration) and 80% (at high concentration) with respect to the control (H_2O treatment).

The herbicide ALA at both concentrations markedly depressed seedling growth by reducing root and shoot length and weight, and produced evident phytotoxic symptoms, such as leaf chlorosis and altered root morphology. With respect to the control treatment, ALA at 1 and 10 mg L^{-1} reduced root length, respectively to 88 and 17%, shoot length to 72 and 40%, root dry weight to 83 and 18%, and shoot dry weight to 86 and 46%.

The phytoxicity induced by ALA on *T. turgidum* seedlings was apparently attenuated by the presence of ASHA in the growth medium, especially at higher concentration (Figure 7). In particular, the combination ASHA at 100 mg L^{-1} + ALA at 1 mg L^{-1} produced an enhanced growth of shoots with respect to the ALA treatment. An even more marked antitoxic effect was obtained in the treatments ASHA + ALA at 10 mg L^{-1}, in which all biometrical parameters increased with respect to those measured with ALA alone. The effect of ASHA was more pronounced at high concentration on roots, producing an increase of root length and dry weight of 50 and 100%, respectively. In a previous study on tomato seedlings treated with ALA and other herbicides,[5] a marked attenuation of the toxic symptoms was found when soil HAs were added to the herbicide in the growth medium.

The effects of samples SHHA, PFA, PHA, LHA, SFA and SHA at concentrations of 20 and 200 mg L^{-1}, either alone or in combination with 10 mg L^{-1} of MH, on root length and dry weight of 5-days old seedlings of *V. faba* are shown in Figure 8. Only samples PFA, PHA and LHA used alone at both concentrations produced growth stimulation, especially root elongation, whereas no significant variations of root length and dry weight were observed with the other HSs samples. The maximum increase of root length (~150% with respect to the control treatment) was observed in the treatment with PFA at 20 mg L^{-1}.

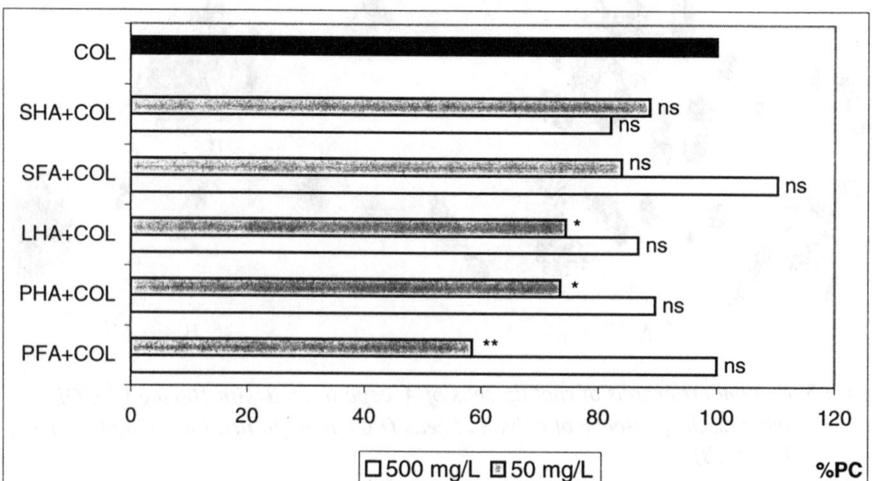

Figure 6 *Effect of the combinations of each HS at 50 or 500 mg L^{-1} + COL on the relative frequency (%) of PC in A. cepa root tip cells referred to the control (COL alone, frequency 100%). The symbols **, *, and ns refer, respectively, to a difference significant at 0.01P, 0.05P, and no significant difference according to the LSD test*

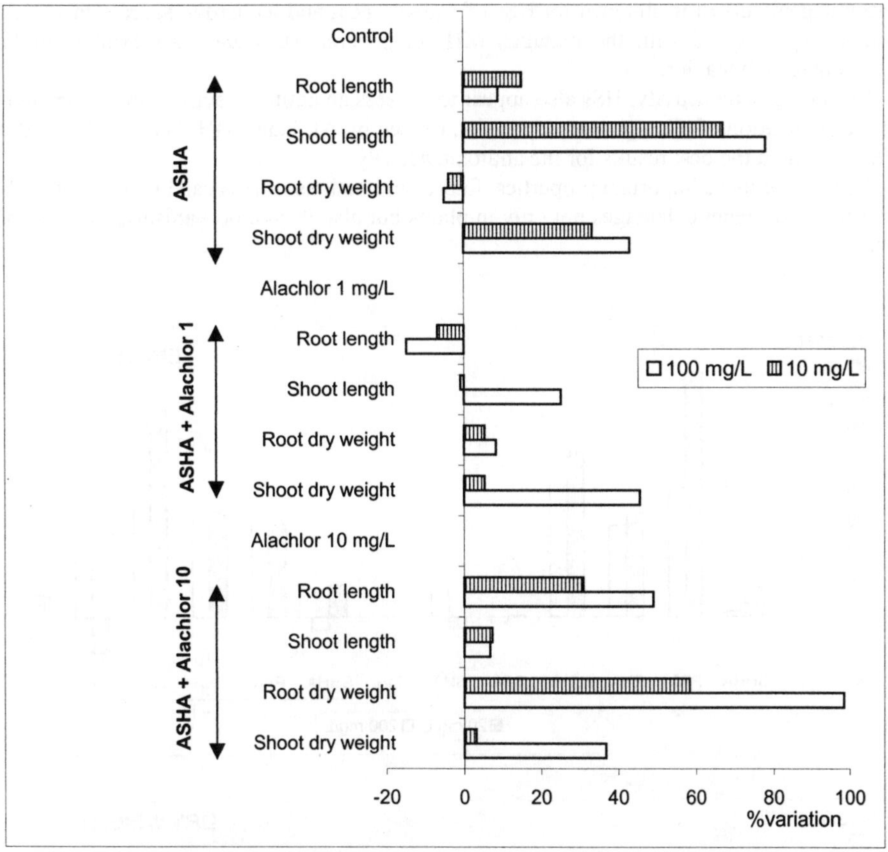

Figure 7 *Effects of ASHA at 10 or 100 mg L^{-1}, alone and in combination with 1 or 10 mg L^{-1} ALA, on biometrical parameters of T. turgidum seedlings. Data of ASHA and ASHA + ALA treatments are expressed as the percentage of the variation observed with respect to the control (H_2O treatment) and to the ALA treatments, respectively*

Besides genotoxic effects, MH also produced phytotoxic effects on *V. faba* seedlings by reducing root length and dry weight, respectively to 70 and 62%, with respect to the control treatment. Only the combinations with PFA, PHA and LHA at both concentrations + MH produced a reduction of the toxic effect of MH, which was highest in the case of LHA + MH, with about 140 and 100% increase of root length and dry weight, respectively, with respect to the MH treatment.

4 CONCLUSIONS

The HSs samples studied exert an apparent antimutagenic action in plant species. This action differs in intensity and type of the genetic damage that is reduced as a function of the source and dose of HS applied, the plant species and the mutagen used. The largest antimutagenic effect is obtained for HSs of aquatic, peat and leonardite sources in *V. faba* and *A. cepa* treated with the mutagen MH. In general, HAs and FAs exhibit similar antimutagenic behavior.

Besides genetic activity, HSs also appear to possess an antitoxic activity that is apparent in the attenuation of plant growth depression caused by ALA and MH. Peat and leonardite HSs also yield the best results for the antitoxic activity.

Because of these important properties, future applications of HSs can be expected in the prevention of genetic damages not only in plants but also in microorganisms, animals and even humans.

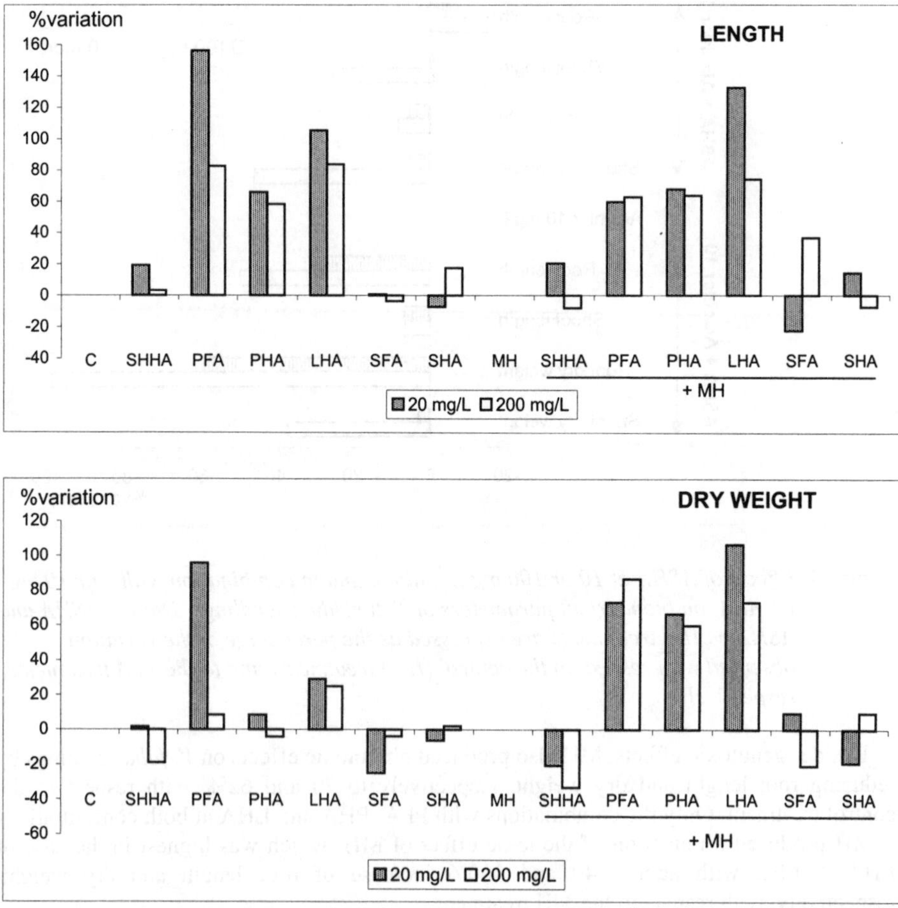

Figure 8 *Effects of different HS samples at 20 and 200 mg L^{-1}, used either alone or in combination with MH at 10 mg L^{-1}, on root length and dry weight of V. faba seedlings. Data of HS and HS + MH treatments are expressed as the percentage of the variation observed with respect to the H_2O treatment (C) and to the MH treatment, respectively*

References

1. Y. Chen and T. Aviad, in *Humic Substances in Soil and Crop Sciences: Selected Readings*, eds. P. MacCarthy, C.E. Clapp, R.L. Malcolm and P.R. Bloom, American Society of Agronomy, Madison, WI, 1990, p. 161.
2. T. Sato, Y. Ose and H. Nagase, *Mutat. Res.*, 1986, **162**, 173.
3. T. Gichner, F. Badaev, F. Pospisil and J. Veleminsky, *Mutat. Res.*, 1990, **229**, 37.
4. A. De Marco, C. De Simone, M. Raglione and P. Lorenzoni, *Mutat. Res.*, 1995, **344**, 5.
5. N. Senesi, E. Loffredo and G. Padovano, *Plant Soil*, 1990, **127**, 41.
6. N. Senesi and E. Loffredo, *J. Plant Nutr.*, 1994, **17**, 493.
7. E. Loffredo, N. Senesi and V. D'Orazio, *Z. Pflanz. Bodenkunde*, 1997, **160**, 455.
8. J.R. Meier, R.D. Lingg and R.J. Bull, *Mutat. Res.*, 1983, **118**, 25.
9. H. Matsuda, Y. Ose, H. Nagase, T. Sato, H. Kito and K. Sumida, *Sci. Total Environ.*, 1991, **103**, 129.
10. R. Cozzi, M. Nicolai, P. Perticone, R. De Salvia and F. Spuntarelli, *Mutat. Res.*, 1993, **299**, 37.
11. G. Ferrara, E. Loffredo, R. Simeone and N. Senesi, *Environ. Toxicol.*, 2000, **15**, 513.
12. I. Sbrana, A. Di Sibio, A. Lomi and V. Scarcelli, *Mutat. Res.*, 1993, **287**, 57.
13. A. De Marco, C. De Simone, C. D'Ambrosio and M. Owczarek, *Mutat. Res.*, 1999, **438**, 89.
14. G. Ferrara, E. Loffredo and N. Senesi, *J. Agric. Food Chem.*, 2001, **49**, 1652.

Subject Index

A&L method, 310
Aberrant anatelophases, 363, 365
Absorbance coefficients, 179
Absorption spectra, 124, 137, 255
Acetate anions, 212
Acetic acid, 41
Acid
    dissociation, 20
    precipitation method, 338, 340, 341
Acid-base, 158, 160
Acidic polyelectrolytes, 271
Acquisition time, 33, 74, 284
Activated charcoal, 341
Adsorption, 15, 136, 153, 158, 209, 214, 215, 241-244, 250, 255, 269
    isotherm, 255
Affinity distribution, 153, 159, 161
Aggregate, 3, 4, 40, 95, 99, 165, 242, 244, 247, 248, 250, 355
Aggregation, 76, 79, 133, 134, 173, 201, 242, 264, 267, 268
Agriculture, 3, 4, 9, 13, 73, 166, 168, 172, 173-175, 179, 278, 313, 315, 337
Agricultural soils, 166, 174, 175, 278
Agronomy, 3
Agronomy Research Farm, 272
Al, 254, 264
    oxide, 122-125, 255-259, 261-265, 268, 325, 328
Alachlor, 361, 362
Albumin, 53
Aldehydes, 135, 148, 274
Aldrich HA, 52, 53, 56, 58, 136, 137, 142, 145, 146, 148, 221, 222, 233-235, 297, 338
Alfalfa (*Medicago sativa L.*), 6, 10, 345-353
Aliphatic
    alcohol, 170, 224, 228, 258, 259, 276, 277
    carbon, 34, 165, 224, 275
    chains, 79, 102, 147, 320
    components, 22, 34, 78, 79, 81, 158, 175, 168, 321, 327
    -aromatic backbone, 133
Aliphaticity, 320
Alkaline solution, 143, 145, 148
Alkenes, 274, 321, 323, 324, 327, 330, 334-336
Alkoxyl, 32, 97
Alkyl groups, 41, 43-45, 324
Alkylbenzenes, 224, 291
Allium cepa L., 363
Al-O bond, 267
Aluminum, 211-215, 217
    -acetate, 212
    –citrate, 212
    -oxalate, 212
    toxicity, 211
American beech (*Fagus grandifolia*), 282
Amide, 22, 169, 170, 199, 275, 276, 289, 316
Amino acids, 22, 32, 34, 39, 221, 224, 229, 230, 288-290, 309, 316, 320, 337
Amino sugars, 337
Amorphous
    Builder, 226
    character, 23, 226
β-amylase, 53
Analysis methods, 313
Analysis of variance (ANOVA), 364
Analytical data, 86, 222, 223, 235, 300, 303, 315
Andersen pressure control method, 226

Animal
cells, 361
residues, 63, 355
Annelid compost piles, 166
Antagonistic effect, 361
Antarctic organic matter, 123
Antarctica, 121, 124, 130
Anthracene, 222
Antimutagenic action, 361, 364, 369
Antitoxic effect, 361, 364, 368
Apoferritin, 53
Apparent
molecular weight at peak maximum, 52, 53, 58, 59, 171, 172, 175
molecular weight, 156, 171
Aprotinin, 53
Aquifers, 153
Arabidopsis thaliana, 361
Arctangent function, 41
Aromatic
carbon, 276
moieties, 78, 79
ring, 44, 195, 231
stacking interactions, 194
Aromaticity, 51, 85, 87, 133, 137-139, 141, 145, 148, 153, 158, 276, 320, 321, 327
Arthropods, 355
As, 165, 174, 175
^{75}As, 169, 173
Ash components, 317, 319, 325
Ash content, 64, 85, 154, 157, 167, 168, 180, 255, 263, 284, 310, 311, 313, 315, 317
Aspen (*Populus tremuloides*), 32-37
Association, 156, 200, 212, 300
Atomic absorption spectroscopy, 255
Atomic force microscopy (AFM), 241-244, 255, 260, 263, 264
Atomic ratios, 223, 230, 288
Atomic surface tensions, 204
Automated parameterization procedure, 199
Autooxidation, 133
Average binding constant, 161

Background electrolyte, 124, 136, 138, 244, 247, 250
Background subtraction, 53, 58
Bacteria, 179, 355, 361
Balsam fir (*Abies balsamea*), 282
Barium chloride precipitation method, 338
Barium humate, 340
Baseline disturbance, 111
Bayerite, 254, 258, 264
Becke-Lee-Young-Parr method, 210
Beidelite, 214
Benzene, 22, 52, 204, 224, 321, 326
Benzenecarboxylic acid, 98
Berendsen thermal coupling method, 226
Bidentate triacetate complex, 212
Bilayer structures, 32
Binding sites, 153, 154, 161
Bioavailability, 153, 221
Biodegradation, 276, 292
Biogeochemical processes, 281
Biological
activity, 183
cycles, 355
Biomatter, 19, 20, 21, 24, 28
Biometrical parameters, 364, 367-369
Biopolymers, 20, 39, 201, 324, 337
Birnessite, 254, 257, 258
Black humic acids, 305
Bloch decay, 320, 321, 327
Bond
length, 194
topology, 223
Bovine thyroglobulin, 53
Branched
alkane, 321, 327
alkenes, 321, 327
Bridging groups, 58, 328
Broad bean, 363
Buffers, 4, 110, 111, 165, 337
Building blocks, 28, 193, 198, 199, 224, 230, 324, 329
Bulk density, 222, 226, 231, 235

Buried bag method, 356

^{13}C NMR spectra, 33-35, 40, 45, 95, 97, 98, 106, 157, 159, 161, 163, 271, 273, 274, 282, 290, 292, 304, 316, 320, 322, 326, 327
C content, 168, 261, 262, 287, 288, 299-301
C/N atomic ratio, 167, 168, 180, 182-184, 282, 288
Calcium
 acetate, 84, 317
 Ca^{2+}, 221, 281
 CaLF, 345-353
 chloride, 345-353
 CaO, 325, 328
 uptake, 347, 349, 353
Calculated energies, 198
Calibration, 51, 52, 56-60, 109, 117, 124, 138, 154, 156, 171, 243, 339
Capillary
 coating, 136
 electrophoresis, 110, 117, 165
 zone electrophoresis, 109-112, 115, 117, 121, 124, 130, 133, 134, 136-138, 145, 146
Capucine, 136
Carbohydrate, 34, 36, 37, 95, 97, 207, 258, 275, 276, 288-292, 304, 316, 320
Carbon
 distribution, 157
 normalized binding constants, 233, 235
 source, 24
Carbonaceous shales, 309, 310
Carbonic anhydrase, 53
Carbonyl carbons, 34
Carboxyl groups, 21, 41, 44, 58, 83, 87, 89, 90, 92, 96, 116, 134, 136, 158, 228, 267, 276, 289, 291, 345
Carboxylate
 carbons, 34
 groups, 255, 258
Carboxylic acids, 148, 199, 214, 224, 230, 326
Carotenoids, 36
Carr-Purcell pulse sequence, 64
Catechol, 40, 41, 253-268
Cation exchange capacity, 281, 282, 311, 313
Cd, 165, 253
^{114}Cd, 169, 173
CDFA (California) method, 310
Cell
 division, 364
 membrane, 32
 volume, 226
Cellulose, 5, 31, 53, 110, 167, 169, 276
Channel flow, 53, 168
Characterization, 4, 10, 15, 31, 95, 136, 154, 166, 175, 295, 310
Charge
 effects, 112
 equilibration, 226
 interactions, 215-217
Charge-to-mass ratio, 109, 111
Charred plant
 fragments, 300, 305
 residues, 297, 305
Chelsea HA, 222, 223, 226, 227, 229-235
Chemapex HA Standard, 123, 125-127, 130
Chemical
 catalysis, 24
 composition, 153, 154, 156, 161, 163, 315, 329, 336
 derivatization, 22
 methods, 207, 209, 315
 pretreatment, 271
 reactivity, 20, 21, 26
 shift anisotropy, 74
 speciation program, 156
 structures, 31, 304, 320
Chemisorption, 209
Chromophore systems, 186
Chromosomes, 361, 364
Chrysene, 222, 233, 234
Cis-chlordane, 221
Citric acid, 52, 211

Clay-humic complexes, 241, 250
Clays, 241, 337
Clear-cutting, 281, 282, 286, 289, 293, 294
Climax vegetation, 297
Coals, 21, 39, 63, 64, 66, 70, 125, 126, 130, 136, 137, 148, 309, 310-312, 325, 334, 337, 338, 341, 345
Coalification, 325, 328
Coastal oases, 121
Coated capillaries, 111
Colchicine, 362
Commercial
    HAs, 14, 235, 341
    humic products, 355
Compaction, 311
Competitive adsorption, 153, 154, 160- 163
Complex formation constants, 174
Complexation, 11, 21, 153, 165, 194, 211, 212, 258, 265, 281
Complexing cations, 241
Compositional differences, 12, 24, 271
Computational
    chemistry, 207, 209, 217, 223
    efficiency, 197
Computer assisted structure elucidation, 222, 235
Condensation, 24, 31, 134, 135, 148, 168, 172, 297, 330
Condensed
    aromatics, 102
    domains, 63
    phase strain energy, 226, 231
    Conformation, 197, 201, 204, 241, 250
Conformational
    flexibility, 193, 194, 201, 204, 207
    space, 197, 204
Constant-force scanning microscope, 243
Constrained NICA-Donnan model, 161
Contact-mode AFM, 242, 255
Contaminants, 4, 95, 204, 223, 241
Continuous variation, 22
Continuum mode, 97
Conventional rotation, 271
COOH groups, 41, 86, 88-92, 169, 170, 225, 229, 258, 259, 265, 266, 268, 274, 316, 317, 337
Corn (*Zea mays L.*), 5, 16
Correlation, 44, 64, 74, 76, 89, 116, 133, 204, 221
Correlative microscopy, 39
Coulomb radii, 204
Coumaranone, 322, 326, 327
Coupling reactions, 31
Covariance matrix, 330
Cover crops, 271, 272, 274-278
Cow manure, 166
CP-MAS ^{13}C NMR spectroscopy, 12, 13, 34, 40, 45, 85, 95, 97, 98, 106, 154, 157, 158, 271, 273, 274, 278, 284, 290, 315, 316, 319, 320-322, 326-328
CR–CR bonds, 229
CR–O bonds, 229
Cramer/Truhlar SM5.4/P model, 204
Creeping bentgrass (*Agrostis palustris Huds.*), 14
Critical interaction parameter, 228, 233
Crop production, 355
Cross-flow hollow fiber ultrafiltration, 154
Csp^3–CR bonds, 229
Csp^3–O bonds, 229
Cu, 153, 155, 165, 174, 175
    binding, 161-163
    ^{63}Cu, 169, 173, 174
    Cu(II), 153, 161
Curie point pyrolyzer, 316
Curie-Point Pyrolysis-Gas Chromatography-Mass Spectrometry, 315, 321, 323-328
Cutin, 275
α-cyano-4-hydroxycinnamic acid, 86
Cytochrome C, 53

Δlog K value, 299
2,4-D, 211, 362-364
d_6-DMSO, 74, 76-81
Decarboxylation, 136, 267
Deciduous trees, 31

Decomposition, 16, 20, 21, 31, 117, 133, 135, 139, 147, 153, 275, 276, 281, 286, 294, 355
Degradation, 22, 24, 28, 31, 32, 133, 147, 148, 173, 224, 292, 304, 305, 324, 355
Degree of decomposition of peats, 117
Delay time, 74, 300, 316, 320
Denaturation, 165, 166
Density, 55, 203, 212, 226, 228, 267, 268
Density function theory, 207, 209, 210, 212
Deschampsia antarctica, 122
Desorption, 39, 51, 63, 121, 209
Detoxification, 73
Dextran, 4, 52, 53, 55, 56
Diazomethane, 83-85, 90, 92
Dicarboxylic acids, 224
Dichromate oxidation method, 299
Dickite, 210, 211, 214, 215
Dicotyledon, 361, 363
Diffuse reflectance infrared Fourier transform (DRIFT) spectrometry, 167, 168, 170, 171, 175, 271-273, 276-278
Diffusion, 53, 76, 166, 171, 175, 209, 217, 265
    coefficient, 53, 166, 171, 175
Dimethylbenzofuran, 322
Dipolar interactions, 73, 74, 79, 81
Direct polarization, 33
Dispersion interactions, 228
Dissociation, 25, 99, 134, 212, 268
Dissolved organic carbon (DOC), 179-, 184, 188, 189, 286
Dissolved organic matter (DOM), 165, 181, 221, 222
Distribution
    coefficients, 209
    probabilities, 215
Domain
    analysis, 66
    distribution, 63, 66, 67, 69, 70
Donnan model, 153, 154, 160, 161, 162
DPPH, 86
Dry mass production, 346, 347
Dual-mode sorption model, 63
Durum wheat, 363
Dynamic equilibrium, 25

E_4/E_6 ratio, 85, 137, 138, 140, 141, 157, 167, 168
Early diagenesis, 19, 21
Earthworms, 167, 355
Ecological disturbances, 294
EDTA, 53, 124, 137, 138, 345-353
Eigenvectors, 330
Electron
    density, 194
    donor-acceptor, 194
    energy-loss spectroscopy (EELS), 41
    impact (EI) mode, 316
    microscopy, 242
    spin resonance, 83, 86-88
    -accepting capacity, 133, 139
Electronic
    delocalisation, 199
    distributions, 194
    donor-acceptor, 194
Electrophoresis, 4, 5, 21, 109-111, 115, 117-119, 126, 133, 136, 137, 141-145
Electrophoretic mobility, 4, 109, 111, 117
Electrospray ionization, 28, 51, 55, 95-101, 103, 105, 106
Electrostatic
    deficiencies, 196
    polarization, 204
    potential map, 201
Elemental
    analysis, 124, 125, 180, 299, 317, 326
    compositions, 24, 167, 284, 286, 300, 301
Emission, 180, 181
Empirical
    data, 19, 198
    functions, 194
Energy minimization, 193, 222, 226, 231, 235
Enolic structures, 148, 277
Entanglement threshold, 110, 112
Entropic interaction parameter, 228, 232

Environment, 14, 19, 28, 73, 121, 153, 166, 189, 209, 223, 241, 242, 250, 253, 271, 273, 313, 337, 359
Environmental
factors, 271, 359
quality, 73, 253
roles, 19, 337
Enzymatic control, 23
Epoxidation, 136
Equilibrium binding constant, 227
Equivalent weight, 340
Esterification, 83, 90-92
Esters, 22, 133, 136, 158, 159, 161, 224, 275, 291, 305, 316, 320, 327, 330
Ethers, 224
Evolution axis, 336
Ewald summation, 226
Excitation, 67, 73, 180, 181
Excitation-emission wavelength pair (EEWP), 186, 189
Exponential function, 65-67
Extended skeletal units, 23
Extraction, 4, 6, 7, 11, 23, 60, 64, 96, 110, 123, 167, 173-175, 282, 285, 299, 340, 341

Face-to-face interactions, 196
Fatty acids, 95, 98, 99, 102, 106, 274, 321, 327, 330, 334-336
Fe, 254, 264
^{57}Fe, 169, 173
Fe oxide, 124, 125, 255-258, 260, 262-268, 325, 328
Fe-O bond, 267
Ferrihydrite
aggregates, 40
gel, 40
Fertilizers, 275-277, 345, 347, 349, 350, 352, 353, 355
Feulgen staining method, 362-364
Field-flow fractionation, 53, 165-169, 171-173, 175
First Principle of himic substances, 19-24, 26
Flory-Huggins solution theory, 222, 227, 228, 235
Flow field-flow fractionation, 51, 53, 56, 58-60
Fluorene, 222
Fluorescence, 39, 154, 179, 181, 188, 189, 221
Fluorobenzoic acid, 201, 203
Folding, 79, 81
Foliar spraying, 345, 346
Follin-Ciocalteu method, 84
Force field, 194-197, 204, 205, 209, 214, 226
Forest soils, 31, 294
Formation, 19, 21, 23-27, 51, 52, 73, 110, 115, 133-135, 139, 146, 148, 211-, 213, 216, 217, 241, 253, 255, 258, 267, 268, 297, 305, 311, 312
Formic acid, 41
Fourier transform, 51, 75, 82, 85, 96, 99, 300, 315, 316
Fractional precipitation, 21
Fractionation, 21, 25, 26, 27, 52, 53, 56, 109, 110, 154, 159, 165, 166, 171, 173
Fractions, 12, 13, 21, 22, 25, 26, 40-44, 51, 109-117, 144, 154, 156, 158, 161, 165, 171, 175, 179-181, 184, 186, 189, 241, 272, 273, 275-278, 282, 284, 286, 288-291, 298, 300
Fragmentation pathways, 28
Free radicals, 21, 32, 83, 86-92, 267
Fresh water, 311
Frictional coefficient, 112
FTIR, 39, 85, 87, 89-91, 154, 158, 159, 161, 163, 167, 254, 255, 257-261, 263, 300, 315, 316, 319, 326
Full mass range, 86, 99
Fulvic acid, 14, 21-25, 28, 39, 40, 51-60, 78, 79, 96, 97, 99, 101, 153-159, 161, 163, 165, 182, 200, 221, 242, 244, 247, 250, 271, 282, 286, 288, 289, 292, 309, 315, 340, 341, 352, 361, 362
Functional
features, 281
group, 22, 24, 27, 39, 40, 74, 75, 77, 78, 79, 81, 83, 95, 134, 139, 143, 153, 158, 159, 161, 163, 165, 166, 168-171, 175, 193-195, 197, 204, 215, 217, 223-225, 265, 266, 268, 273, 276, 282, 284, 288, 289, 318, 337, 345, 352
Fungi, 355
'fuzzy' molecular shape, 199

G values, 86, 87
Gallic acid, 41
Gas phase
strain energy, 226, 231
structures, 196, 210
Gaussian
affinity distribution, 153
decay function, 64
functions, 41, 67
Gaussian98, 210, 218
Gel
matrices, 110
shrinking, 109
-filled capillaries, 109
Genetic
anomalies, 361, 362, 364
toxic effects, 361, 369
Geochemical modeling, 153
stability, 241
Geochemistry, 212
Geometric optimization, 194, 196, 199, 206
Gibbs free energies, 210, 213
Gibbsite, 214, 254, 258, 264
Gill-Meibohm phase cycling, 64
Glacier activity, 121
Glass, 67, 123, 155, 180, 242, 243, 255, 273, 298, 315, 364
Glassy (rigid) domain, 63
Glyphosate, 362
Graphite, 242, 254, 300, 304
Grassland, 7, 297, 299, 304
Groundwater, 51, 52, 56, 58, 153
Growth substrate, 345
Guaiacylpropanoid, 36
Guiding principles, 19

H/C atomic ratios, 124, 125, 288, 317, 318, 320, 326
H/O atomic ratios, 102
HA content, 338
HA standard, 125, 126, 130
Hakomori method, 83, 84, 87, 92
Half tennis ball, 201
Haploboroll, 315
Hardwood forest, 281, 282
Harmonic oscillator, 210
Hartree-Fock theory, 212
Hemicellulose, 276
Heterocyclic aromatic compounds, 224
Heterogeneity, 19, 20-24, 26-28, 32, 51, 67, 68, 70, 73, 92, 153, 161, 172, 173, 194, 271, 281, 311
Heterotrophic dynamics, 179
Hexaaquoaluminum complex, 212
Hierarchical approach to modeling, 222, 232, 235
High affinity binding sites, 63
High performance size exclusion chromatography (HPSEC), 51, 52, 55-60, 165
Hildebrand-Scatchard theory, 228
HNO_3 treatment, 299, 304
Hoagland solution, 346
Hollow-fibre ultrafiltration, 40
Homolytic cleavage, 133
Hooke's law, 194
Hormonal activity, 352, 359
Horse manure, 166, 168
Horticulture, 13, 313
Hubbard Brook Experimental Forest, 282, 283, 290, 294
Humalite, 310, 311, 313
Humans, 370
Humic
-fulvic mixture, 28
-humic interactions, 206
-mica, 242
-plant stimulation, 4
treatments, 355
-water, 206
Humic acid, 11-13, 22, 25, 26, 28, 40, 44, 52, 53, 58, 60, 63, 64, 67, 70, 74, 76, 78, 79, 83-92, 95, 97, 99, 100, 102-106, 121, 130, 133-137, 139, 141, 142, 144, 145, 148, 153-159, 161, 163, 165-168, 170, 172, 175, 182, 188, 194, 221, 224, 225, 229-235, 242, 244, 247, 250, 253, 271, 282, 288, 289, 297, 299, 300, 301, 303, 304, 311, 315, 337-341, 355-359, 361, 362
analysis methods, 338
colloids, 242, 244, 248, 250
fractions, 21, 154, 284, 290
macromolecules, 4, 22, 134, 265
materials, 19, 21, 23, 24, 133, 136-138, 194, 217, 242, 258, 260, 264-266, 316

models, 194, 207
molecules, 20, 22, 23, 110, 113, 115-117, 198, 241, 291
products, 356, 358
structures, 22, 96, 133, 134, 194, 197-199, 201, 204, 207
Humic substances, 4, 10, 12, 13, 15-17, 19-28, 31, 39, 51, 53, 56, 58-61, 63, 64, 67, 73, 79, 81, 83, 95-97, 99, 102, 106, 109, 110-118, 121, 123-126, 130, 133-, 143, 145-148, 153, 154, 156, 158, 165, 175, 193, 194, 199, 215-217, 221-224, 227, 231, 232, 241, 253, 268, 269, 271, 276, 279, 281, 282, 284-289, 291, 294, 295, 297, 304, 305, 309-311, 313, 317, 336, 337, 345, 352, 355, 359, 361-365, 368-370
association, 51
formation, 21, 23, 24, 253
surface interactions, 241
Humification, 12, 31, 32, 37, 83, 121, 130, 137, 172, 253, 255, 260, 263, 265-, 268, 275, 276
Humified organic matter, 309, 324
Humin, 25, 34, 165, 221, 271, 272, 309
Hydrazine, 134-140, 142-144
Hydrazones, 135
Hydrodynamic diameters, 51, 52, 56, 58-60, 109, 171, 172, 175
Hydrogen adducts, 96
Hydrogen bonding, 21, 25, 58, 77, 194, 196, 197, 199, 200, 202, 206
Hydrolysis constants, 156
Hydrophilic character, 21, 52, 110, 112, 133, 153, 207
Hydrophobic character, 21, 58, 133, 194, 207, 221, 233, 235
Hydrophobic organic contaminants, 63, 70, 221, 228, 232, 234, 235
Hydroquinones, 41, 133, 143, 225
Hydroxyl group, 22, 32, 83, 84, 87, 90, 92, 136, 197, 214, 224, 225, 228, 273, 345
Hyperdiploid cells, 363
Image resolution, 242
3-D images, 245-249
Imazethapyr, 361
Imines, 135
Immature compost HAs, 173
Import specifications, 338
Induction decays, 316
Inductively coupled plasma-mass spectrometry (ICP-MS), 165, 166, 168, 169, 173-, 175
Industrial uses of HSs, 166, 313
Infrared laser desorption mass spectrometry (LD-MS), 51, 55
Inner filter effect, 180, 186
Inorganic nutrients, 355
Inositols, 36
Input parameters, 224, 225
Instrumental data, 23
Interaction parameters, 193, 232
Intercalation, 214
Interfragment bonds, 223, 225, 228-230
Internal energy, 226
International Humic Substances Society, 4, 10, 11, 14, 15, 17, 23, 30, 52, 64, 83, 86, 123-127, 129, 130, 136, 137, 142, 148, 154, 242, 243, 260, 261, 340, 341, 362
Intramolecular
conversions, 186
interactions, 74
Inversion recovery pulse sequence, 74, 76
Ion
binding, 153, 154
cyclotron resonance mass spectrometry (ICR-MS), 51, 96
exchange, 20, 84, 165, 317
exclusion, 52, 109
selective electrodes, 155, 156, 161
Ionic strength, 52, 59, 110, 111, 115, 116, 133, 155, 156, 159, 180, 221, 222, 241, 242, 244, 250
Iron, 13, 17, 123, 124, 253, 273, 337, 341
Iron(III) oxides, 123, 124, 253, 254
Isoflavone structures, 134
Isolation of HSs, 4, 10, 26, 123, 165, 285, 287, 288, 295
Isopropanol, 97
Isotachophoretic mechanism, 109
Isotherm equation, 153, 161

K^+, 96, 281
K_2O, 325, 328, 346
Kaolinite, 210, 214, 215, 319
Kenezasa (*Pleioblastus pubescens Nakai*), 299
Ketones, 135, 274
$KMnO_4$, 299
Kubleka-Munk function, 167

Laser desorption ionization-Fourier-transform mass spectrometry (LDI-FTMS), 51
Leaf cells, 32
Leonardite, 14, 136, 309-311, 313, 362, 370
Lichens, 121, 122
Lignin, 24, 28, 31, 36, 133, 147, 224, 275, 288, 290, 297, 324, 334-336
    dimers, 330
Lignite, 309, 329-333, 335, 336, 345, 352, 353
    ores, 329-333, 335, 336
Lindane, 221
Line widths, 87
Linear positive mode, 124, 125, 127, 128
Lipophilic environments, 216
Long lived complexes, 216
Lowest energy conformational structure, 195

Macromolecular
    conformation, 171
    -solvent interactions, 171
    structures, 165
Magic-angle spinning, 32
Magnetic
    particles, 272
    susceptibility, 32
Matrix assisted laser desorption time of flight mass spectrometry (MALDI-TOF MS), 51, 86, 136, 138, 141, 143, 145
Maleic hydrazide, 362
Malonic acid, 41
Malus 'Manbeck Weeper', 32-37
Managed watershed, 282
Manganese(IV), 253
Manure materials, 173
Mass
    discrimination, 96, 98, 102, 106
    distribution, 51, 52, 55, 60
    integral, 54
    range, 99, 100-102, 105, 106, 330
    spectra, 53, 54, 96-100, 102, 103, 105, 126, 130, 144-147, 224, 316
    spectrometry, 39, 51, 53, 95, 96, 98, 106, 121, 136, 166, 254, 315, 316, 329
Matrix hydrolysis, 110
Mature HAs, 168, 175
Maximum signal intensity, 75
MD simulation codes, 226
Membranes, 52, 110, 211, 298, 352
Mesopores, 264
Metabolism, 73, 355
Metal complexation, 20, 175
Metal oxides, 156, 253-255, 257, 261, 262, 264-266, 268
Metal
    -HA associations, 4, 175
    -ligand interactions, 165
Metals, 51, 73, 165, 174, 253, 281, 313, 319, 341
Metaphases, 363
Methanol, 52, 84, 92, 97
Methylation, 53, 57-59, 83-85, 87, 89, 90, 92, 136
Methylbenzofuran, 322, 326
Methylindene, 322, 326
Methylsalicylate, 222
$Mg(ClO_4)_2$, 273
Mg^{2+}, 281
MgO, 86, 325, 328
Mica (muscovite, 241, 243, 244, 247
    surface, 242-250
Micelles, 51, 58
Microbial
    activity, 9, 165, 281
    decomposition, 24, 153, 305
    degradation, 20, 21, 24
    organic matter, 133
    organisms, 20, 21, 121, 165, 254, 370
Micronuclei, 363, 365

Micronutrients, 337, 340
Micropores, 63, 264, 265
Micro-spectroscopies, 39
Migration times, 111, 112, 116
Minerals, 21, 63, 64, 67, 153, 154, 169, 210, 211, 214, 241, 244, 247, 250, 271, 273, 276, 286, 288, 289, 292, 293, 300, 311, 340, 341
  peaks, 169
Mineralization, 5, 337, 355-359
Minimum energy, 198, 204
Mitotic activity of cells, 365
MM+, 195, 197
MM2, 196
Mn, 54-56, 58, 165, 174, 175, 227, 254, 264
^{55}Mn, 169, 173
Mn oxide, 253-258, 260-268
Mn (IV) oxide, 267
Mn-O bond, 267
Mobile domains, 66, 67
Mobilization, 175, 211
Models, 9, 23, 24, 63, 153, 199, 201, 204, 209, 210, 215, 217, 222-232, 235, 241
Modifiers, 116, 136
Modulation decoupling, 97
M-OH groups, 266-268
M-OH_2, 265-268
Molar
  mass, 230, 231
  volume, 226, 228, 231-233
Molecular
  conformation, 267, 268
  dynamics, 3, 4, 31, 73, 81, 214, 222, 226, 231, 235
  formula, 230, 231
  fractions, 165, 166, 173
  fragment, 23, 231
  fragments bond, 223-225, 228, 229, 230, 335
  mass, 51, 52, 54-58, 60, 110, 112, 114
  mechanics, 193, 194, 196-198, 204, 226
  modeling, 193, 207, 209
  motion, 31, 32, 34, 64, 74, 79
  nature, 19, 28
  shape, 73
  sieving, 265
  size, 51, 52, 55, 56, 58-60, 109-113, 115, 117, 137, 158, 159, 165, 166, 172, 179, 187-189, 336
  size fractions, 179
  structure control, 23
  structures, 23, 27, 63, 96, 106, 215
  weights, 15, 22, 26-28, 31, 32, 44, 63, 68, 73, 86, 87, 95, 99, 109, 110, 116, 117, 133, 137, 138, 140, 148, 153, 154, 156-, 159, 165-167, 171, 173, 175, 223, 227, 271, 277, 309
Møller-Plesset perturbation theory, 212
Monocotyledon, 361, 363
Monosaccharides, 32
Monounsaturated fatty acids, 106
Monte Carlo methods, 193, 198, 214
MP2, 212
Mule manure, 166, 167
Multiple peaks, 96
Muscovite, 241-243, 250
Mutagenic
  compounds, 361-364
  level, 363
Mutagenicity test, 362
M_w, 54-56, 58

^{15}N, 4, 5, 9, 12
N/C atomic ratios, 317
Na_2O, 325, 328
$Na_4P_2O_7$, 6, 7, 272
N-alkanes, 321, 323, 327, 330, 334, 335
Nano.NMR probes, 32
Nano3 electrolyte, 158, 160
Naphthalene, 222, 322, 326, 327
Natural organic matter (NOM), 31, 32, 39, 40, 44, 45, 153, 179, 181, 183-189, 221
Natural product, 28
N-decane, 222, 233, 234
N-dodecane, 222, 233, 234
Near-edge X-ray absorption fine structure (NEXAFS), 39-46
Nebulizer, 95, 168

Negative ionization mode, 53
Negative mass defect, 96
Net surface charge, 255, 265, 266
NH_4^+, 96, 356, 358
NH_4NO_3, 272, 346
N-heptane, 222
^{60}Ni, 169, 173, 174
Nitrogen
content, 276, 303
fertilizers, 276-278
mineralization, 356
uptake, 9, 350
N-nonane, 222, 233, 234
NO_3^-, 281, 356, 358
N-octane, 222, 233, 234
Nominal molecular weight, 179, 181
Nonpolar interactions, 21
Non-quantitative spectra, 320
N-tridecane, 222, 233, 234
Nuclear Overhauser Enhancement Spectroscopy (NOESY), 73
Number average molar mass, 223, 224, 230
Nunataks, 121
N-undecane, 222, 233, 234
Nutrients, 63, 73, 121, 253, 268, 281, 337, 340, 341, 350, 352, 353, 355
uptake, 5, 345, 350, 352

O/C atomic ratios, 288, 317, 318, 326
O/R ratio, 168-171, 277, 278
O=C-OH• radicals, 133
Onion, 363
Organelles, 32
Organic biostimulants, 345
Organic
carbon, 10, 110
carbon cycle, 281
growers, 355
matter, 355
Matter Trail, 3, 4, 14, 15
molecules, 11, 22, 74, 194, 195
ores, 130, 315, 320, 327, 328
Overhauser enhancement, 34
Oxidation, 4, 15, 98, 133, 135, 136, 143, 154, 255, 257, 267, 268
Oxygen content, 96, 137, 138, 140, 157, 288, 299, 311
Oxygen-flask combustion, 310, 315

Π* transition, 43
P,p'-DDT, 221
P_2O_5, 272, 328, 346
Pair correlation function, 226
Particle shape, 242
Partitioning, 63, 111, 217, 222, 233
coefficients, 232
^{208}Pb, 169, 173, 174
Pb, 153, 155, 161-163, 165, 168, 174, 175
Pb(II), 153, 161
PCBs, 221, 222
p-dichlorobenzene, 221
Pea, 363
Peak
elution times, 156
heights, 99, 273, 276, 278
position shifts, 189
Peat, 14, 25, 52, 60, 64, 67, 83, 86, 87- 90, 92, 110, 112, 115, 117, 130, 141, 144, 145, 147, 148, 311, 370
Pectinic acid, 27
Peptides, 34, 39, 337
Perilene, 222
Peroxyl radical, 32
Pesticides, 36, 51, 214, 216, 217
pH, 6, 7, 40, 52, 53, 56, 59, 74, 78, 81, 97, 109-117, 123, 124, 133, 134, 138, 145, 154, 155, 161, 163, 165, 167, 169, 180, 188, 211, 212, 221, 222, 241-250, 254, 265, 267, 268, 271, 272, 281, 282, 299, 337, 338, 340, 341, 363
Phenanthrene, 222
Phenolic OH, 21, 43, 44, 86, 115, 266, 268, 316, 317, 320, 326
Phenols, 135, 224, 319
Phosphate, 4, 111-117, 180, 253, 337
Phosphorus uptake, 350
Photo
decarboxylation, 133

degradation, 136
diode array, 53
Phthalic acid, 41
Phyllosilicate minerals, 241
Physical gels, 110, 112, 113, 115, 117
Physisorption, 209
Phytotoxic compounds, 361
Pisum sativum L., 363
pK values, 265, 266
Plant
degradation products, 31
growth stimulation, 13, 17, 63, 165, 313, 337, 338, 341, 345, 352, 361, 370
inputs, 276
tissue, 31, 32
PM3, 199-206
Point of zero salt effect, 253, 255, 265- 268
Polar interactions, 216
Polarized continuum model, 210, 212
Pollutants, 51, 166, 179, 253, 268
Polyacrylamide (PAA), 110, 111
Polyacrylic acids, 53, 57, 59, 60
Polycarboxylic acids, 52, 57, 59
Polycondensates, 267
Polycyclic aromatic compounds, 324
Polydispersity, 26, 27, 51, 110, 171, 173, 175
Polyethylene glycol (PEG), 110
3-D polymeric matrices, 222, 227
Polymerization, 172, 173, 253
Polyphenols, 324
Polyploid cells, 363, 367
Polysaccharide, 3, 4, 15, 32, 109, 154, 158, 169, 221, 276, 284, 287, 288
Polystyrene sulfonates (PSS), 52, 53, 55, 56, 58, 60, 167, 171
Polyunsaturated fatty acid, 32
Polyvinyl alcohol (PVA), 110
Pore water, 52, 56, 58
Porous matrix, 52, 110
Positive ion mode, 96, 97
Potassium uptake, 350
Potato production, 355
Potential energy
functions, 194
surfaces, 198
Precipitation, 242, 244, 250, 294, 338, 340, 341
Pre-optimization calculations, 195
Pre-saturation pulse, 76
Principal-component analysis (PCA), 329, 330, 335, 336
Propionic acid, 41
Propoxur, 361
Proteins, 32, 34, 53, 55, 56, 58-60, 76, 95, 109, 115, 119, 154, 156, 201, 209, 274, 288
Protozoa, 355
^{195}Pt, 173
Pulse delay, 33, 74
Purification procedure, 168, 299
Purines, 224
Purity, 20, 22, 26, 27, 52, 244, 311, 313
Pyrazole derivatives, 134
Pyrene, 222, 233, 234
Pyrethrins, 36
Pyrimidines, 224
Pyrolysis, 39, 136, 315
Pyrolysis-field ionization mass spectrometry (PyFIMS), 321, 329-334, 336

Quadrupole time-of-flight mass spectrometry, (Q-TOF MS), 28, 51, 55, 95-99, 102, 106
Quantitative descriptors, 329
Quantum
efficiency, 181
mechanical methods, 193, 194
mechanics, 194, 198, 209
Quinones, 133-135, 139, 274, 277

Radicals, 32, 133, 135, 139, 143, 145, 267
Ramp cross polarization, 95, 97
Random coil conformation, 134, 171
Reaction entropies, 212
Red spruce (*Picea rubens*), 282
Redox reactions, 20, 139, 267
Reduction, 9, 36, 133, 135, 139, 143, 148, 209, 346, 361, 365, 369

Reed canarygrass (*Phalaris arundinacea*), 5
Reference humic materials, 23
Refractory, 19-21, 23, 24, 28, 95
Regular mixture, 27
Relative fluorescence intensities (RFI), 179, 186, 188, 189
Release of water, 63, 213, 216, 217
Residence time effect, 253
Resistant aliphatic biopolymers, 34
Resorcinol, 224
RF value, 299
Rhizosphere, 211
Rigid domain, 65-67, 69
Rigid networks, 31
Rimsulfuron, 361
Roots, 8, 14, 16, 116, 211, 292, 345- 350, 352, 353, 357, 359, 361-370
Root initiation, 337
Rotating-frame Nuclear Overhauser Enhancement Spectroscopy (ROESY), 73, 76, 79-81
Roth HA, 222, 231, 233-235
Rubbery (mobile) domain, 63, 67
Rydberg/valence, 39, 45

Σ* transitions, 41, 43
Salicylic acid, 41, 229
Salt water, 311
Scanning transmission X-ray microscopy (STXM), 39, 40, 45
Sedimentation, 166, 242, 244, 250
Sediments, 51, 63, 73, 95, 166, 174, 221, 241, 250
Seedling growth, 368
Selectivity coefficient, 112
Self-assembly, 193, 194, 201, 203
Semi-empirical methods, 194, 196, 198-202, 204- 206
Semiquinone radical, 32, 83, 133, 134, 139, 267
Senescence, 32, 36, 37
Separation, 21, 27, 109-113, 115-117, 126, 136, 141, 142, 146, 166, 180, 297, 341
Sewage sludges, 3
SF^{5+} ion beam, 53
Shape, 12, 54, 55, 159, 204, 247, 260
Shoots, 14, 346-348, 353, 364, 367, 368
SIGNATURE, 222-232, 235
Significant peak signals, 330
Silica, 111, 124, 138, 311, 345
Silicon nitride, 40, 255
Single quantum coherence, 33, 36, 37
SiO_2, 124, 125, 325, 328
Size, 9, 12, 39-44, 51, 52, 56, 58, 59, 73, 117, 122, 123, 158, 179, 181, 186, 189, 209, 224, 228, 250, 336
  distributions, 51, 56, 122, 158, 179
  distribution maxima, 56, 58
  exclusion chromatography, 51- 53, 55-59, 109, 154, 156, 157, 161, 165, 173
  fraction data, 179
  fractionation, 53, 154
  fractions, 154-163, 179, 181-189
  to mass ratio, 58, 60
Sodium hydroxide, 74, 96, 123, 130, 134-137, 141, 143, 145
Sodium polytungstate, 298
Sodium pyrophosphate, 110, 167
Soft ionization, 106
Soil
  acidity, 73, 337
  aggregates, 73
  carbon, 9
  crumbs, 3, 15
  depth, 11, 282, 286, 292
  derived HAs, 166, 168
  erosion, 337
  fertility, 28, 281
  foodweb, 355, 359
  formation, 121
  horizon, 63, 110, 154, 166, 282, 284, 286-289, 291-293, 297, 299, 315
  humus, 31, 34
  managements, 271
  moisture, 73
  organic matter (SOM), 16, 34, 39, 63, 175, 221, 222, 271-273, 276, 277, 281, 282, 284, 286, 293, 295, 306, 309, 336, 337, 355

organisms, 355
Science, 3, 4, 12, 29, 30, 73, 207, 219, 251, 253, 268, 294, 313, 342, 353
structure, 3, 4, 15
surface, 281, 299
types, 130, 271
Solubility parameter, 222, 226-228, 231, 233, 235
Solute
activity coefficients, 227, 232
polarization, 204
Solute-HS enthalpic interaction parameter, 232
Solute-solvent interaction, 206
Solvation effects, 193, 204-206, 217
Solvation energies, 204, 206
Solvent effects, 193, 204, 207, 210, 212
Sorption, 20, 52, 63, 70, 136, 209, 214- 216, 222, 231, 234, 235, 241, 244, 247
Source material, 25
Spatial
segregation, 32
variability, 243
Specific
absorbance, 179, 184, 186, 188
gravity, 297, 299
sorption interactions, 209
Spectral integration, 284
Spectrophotometric method, 338, 339
Spectroscopic properties, 166, 297, 304
Sphagnum, 86, 110, 117, 118
Spheroidal shape, 260
Spin diffusion, 73
Spin-lattice relaxation time, 73-79, 81
Spinning sidebands, 273, 300, 320
Spin-spin relaxation, 63
Split-valence, 210
Spodosols, 281, 282
Stability constants, 156
Standard curve, 339, 341
Standard Sample Collection, 4
Standard Soil HA, 242, 243
Static mode, 64
Statistical mechanics, 226
Steric factors, 194
Sterols, 330
Stimulation of plant growth, 355
Stochastic approach, 223
Stochastic generator of chemical structure (SIGNATURE), 223
Strain energies, 222, 231, 235
Structural
changes, 272, 293
moieties, 22
nature, 26, 97
Sub-bituminous coal, 309, 310
Suberin, 275
Substituted aromatics, 230
Sugar, 358
Sugar maple (*Acer saccharum*), 282
Sugar-amine, 297
Sugars, 4, 5, 15, 95, 102, 309
Sulfides, 224
Sulfur content, 311
Summed spectra, 330
Supermixture, 26-28
Superoxide radical, 32
Superposition principle, 199
Supramolecules, 139
Surface
area, 204, 207, 253, 255, 264, 265, 268
chemistry, 253
coverage, 242
morphology, 244, 253
properties, 253, 265, 266, 268
structure, 265
Susuki (*Miscanthus sinensis Anderss*), 299
Sybyl®, 195-197, 204-206
Synergistic effect, 361

T. Turgidum, 362-364, 367-369
Tapping-mode AFM, 242

Temperature, 63-67, 70, 76, 84, 97, 110, 111, 123, 124, 138, 140, 143, 146, 180, 198, 212, 222, 226, 228, 235, 243, 255, 272, 299, 300, 330, 363
Tetramethylammonium hydroxide, 134- 140, 142-145, 147
Thermal
    degradative methods, 22
    expansion coefficient, 226
Thermochemolysis, 136
Thermodynamic
    data, 209, 212, 217, 222, 226, 228
    parameters, 222, 232, 235
Thimerosal, 254
Thiophene, 322, 326, 327
Three-dimensional images, 242
Tightly bound metals, 166
Time-of-flight secondary ion mass spectrometry (SIMS), 51, 53-56, 58
Titration, 155, 156, 158-161, 254, 255, 282, 284
TOSS, 271, 273, 274, 278, 300
Total acidity, 289, 316, 318
Total luminescence, 180, 189
Toxicity, 211, 221, 281
Trace metals, 153, 165, 166, 173, 175, 345
Treatments with HSs, 4, 6, 10, 23, 250, 272, 277, 299, 304, 305, 346, 363-365, 368-370
1,2,3-Trichlorobenzene, 221
1,2,4-Trichlorobenzene, 221
Trimethoxymethane, 83-85, 90, 92
Triticum turgidum L., 363
Turbomole, 210, 218
Turfgrass, 13, 17
Two size fractions, 162
Typic Haplorthods, 282

^{238}U, 173
Ultrafiltration, 4, 109, 110, 156, 157, 163, 179, 180
    membranes, 109
    technique, 154
Ultraviolet-visible spectroscopy, 121, 123, 124, 136-138, 179-181, 184, 185, 188, 303
Uniformity, 20, 24, 28
Unique class, 20, 28
United Atom Topological Model, 210
Universal Force Field, 226
Unresolved peaks, 99, 115
UV-fractograms, 172, 173

Value-added products, 313
Van der Waals interactions, 58, 193, 226
Vapor pressure osmometry, 55
Vegetable compost, 166
Vermiculite, 319, 345
Vetch/rye, 271, 272, 275-278
Vicia faba L., 361-366, 368-370
Volcanic ash soils, 297, 301, 304, 305

WALTZ decoupling, 33
Waste management, 3, 5, 15
Waste water, 3, 8, 9, 166
Water quality, 5, 9
Watersheds, 5, 179, 181, 183, 188, 189, 282, 283
Waxy coatings, 32
Weights and measures laws, 338
White birch (*Betula papyrifera*), 282
Wild fires, 184, 297, 305

XAD-8 resin, 154
Xenobiotic compounds, 194, 201, 361
X-ray diffraction, 63, 193, 300, 305
X-ray powder diffraction, 254, 258, 259

Yellow birch (*Betula alleghaniensis*), 282

Zero-crossing points, 76
Zinc, 165, 174, 175
^{64}Zn, 169, 173, 174

T